Advanced Mechanics of SOLIDS

Third Edition

About the Author

LS Srinath received his PhD from Illinois Institute of Technology, Chicago, and has served as Professor of Mechanics and Aerospace Engineering at University of Kansas, Lawrence. He is a distinguished alumnus of IIT, Chicago. He has also served as Professor of Mechanical Engineering at the Indian Institute of Technology, Kanpur, and also the Indian Institute of Science, Bangalore. Besides these, he was the Director of Indian Institute of Technology, Madras, Chennai.

Professor Srinath has authored several books and papers and has been on Advisory Boards of quite a few national bodies dealing with engineering education.

Advanced Mechanics of

SOLIDS

Third Edition

L S Srinath

Former Director
Indian Institute of Technology Madras
Chennai

Tata McGraw-Hill Publishing Company Limited
NEW DELHI

McGraw-Hill Offices

New Delhi New York St Louis San Francisco Auckland Bogotá Caracas Kuala Lumpur Lisbon London Madrid Mexico City Milan Montreal San Juan Santiago Singapore Sydney Tokyo Toronto

Tata McGraw-Hill

Published by the Tata McGraw-Hill Publishing Company Limited,
7 West Patel Nagar, New Delhi 110 008.

ISBN 13: 978-0-07-070260-8
ISBN 10: 0-07-070260-8

Managing Director: *Ajay Shukla*

General Manager: Publishing—SEM & Tech Ed: *Vibha Mahajan*
Sponsoring Editor: *Shukti Mukherjee*
Jr. Editorial Executive: *Surabhi Shukla*
Executive—Editorial Services: *Sohini Mukherjee*
Senior Production Manager: *P L Pandita*

General Manager: Marketing—Higher Education & School: *Michael J Cruz*
Product Manager: SEM & Tech Ed: *Biju Ganesan*

Controller—Production: *Rajender P Ghansela*
Asst. General Manager—Production: *B L Dogra*

Typeset at Astral Pre Media Pvt. Ltd., A-39, Sector-58, Noida (UP) and printed at Rashtriya Printers, M-135, Panchsheel Garden, Naveen Shahdara, Delhi 110 032

Cover: Rashtriya Printers

RCXDCRQXRYDXL

Contents

Preface

The present edition of the book is a completely revised version of the earlier two editions. The second edition provided an opportunity to correct several typographical errors and wrong answers to some problems. Also, in addition, based on many suggestions received, a chapter on composite materials was also added and this addition was well received. Since this is a second-level course addressed to senior level students, many suggestions were being received to add several specialized topics. While it was difficult to accommodate all suggestions in a book of this type, still, a few topics due to their importance needed to be included and a new edition became necessary. As in the earlier editions, the first five chapters deal with the general analysis of mechanics of deformable solids. The contents of these chapters provide a firm foundation to the mechanics of deformable solids which will enable the student to analyse and solve a variety of strength-related design problems encountered in practice. The second reason is to bring into focus the assumptions made in obtaining several basic equations. Instances are many where equations presented in handbooks are used to solve practical problems without examining whether the conditions under which those equations were obtained are satisfied or not.

The treatment starts with Analysis of stress, Analysis of strain, and Stress–Strain relations for isotropic solids. These chapters are quite exhaustive and include materials not usually found in standard books. Chapter 4 dealing with Theories of Failure or Yield Criteria is a general departure from older texts. This treatment is brought earlier because, in applying any design equation in strength related problems, an understanding of the possible factors for failure, depending on the material properties, is highly desirable. Mohr's theory of failure has been considerably enlarged because of its practical application. Chapter 5 deals with energy methods, which is one of the important topics and hence, is discussed in great detail. The discussions in this chapter are important because of their applicability to a wide variety of problems. The coverage is exhaustive and discusses the theorems of Virtual Work, Castigliano, Kirchhoff, Menabria, Engesser, and Maxwell–Mohr integrals. Several worked examples illustrate the applications of these theorems.

Bending of beams, Centre of flexure, Curved Beams, etc., are covered in Chapter 6. This chapter also discusses the validity of Euler–Bernoulli hypothesis in the derivations of beam equations. Torsion is covered in great detail in Chapter 7. Torsion of circular, elliptical, equilateral triangular bars, thin-walled multiple cell sections, etc., are discussed. Another notable inclusion in this chapter is the torsion of bars with multiply connected sections which, in spite of its importance, is not found in standard texts. Analysis of axisymmetric problems like composite tubes under internal and external pressures, rotating disks, shafts and cylinders can be found in Chapter 8.

Stresses and deformations caused in bodies due to thermal gradients need special attention because of their frequent occurrences. Usually, these problems are treated in books on Thermoelasticity. The analysis of thermal stress problems are not any more complicated than the traditional problems discussed in books on Advanced Mechanics of Solids. Chapter 9 in this book covers thermal stress problems.

Elastic instability problems are covered in Chapter 10. In addition to topics on Beam Columns, this chapter exposes the student to the instability problem as an eigenvalue problem. This is an important concept that a student has to appreciate. Energy methods as those of Rayleigh–Ritz, Timoshenko, use of trigonometric series, etc., to solve buckling problems find their place in this chapter.

Introduction to the mechanics of composites is found in Chapter 11. Modern–day engineering practices and manufacturing industries make use of a variety of composites. This chapter provides a good foundation to this topic. The subject material is a natural extension from isotropic solids to anisotropic solids. Orthotropic materials, off-axis loading, angle-ply and cross-ply laminates, failure criteria for composites, effects of Poisson's ratio, etc., are covered with adequate number of worked examples.

Stress concentration and fracture are important considerations in engineering design. Using the theory-of-elasticity approach, problems in these aspects are discussed in books solely devoted to these. However, a good introduction to these important topics can be provided in a book of the present type. Chapter 12 provides a fairly good coverage with a sufficient number of worked examples. Several practical problems can be solved with confidence based on the treatment provided.

While SI units are used in most of numerical examples and problems, a few can be found with kgf, meter and second units. This is done deliberately to make the student conversant with the use of both sets of units since in daily life, kgf is used for force and weight measurements. In those problems where kgf units are used, their equivalents in SI units are also given.

The web supplements can be accessed at http://www.mhhe.com/srinath/ams3e and it contains the following material:

For Instructors

- Solution Manual
- PowerPoint Lecture Slides

For Students

- MCQ's (interactive)
- Model Question Papers

I am thankful to all the reviewers who took out time to review this book and gave me their suggestions. Their names are given below.

K S R K Murthy	*Department of Mechanical Engineering, Indian Institute of Technology, Guwahati, Assam*
P K Sarkar	*Department of Mechanical Engineering, Indian School of Mines, Dhanbad, Bihar*
S R Pandey	*Department of Applied Mechanics, NIT Jamshedpur, Jharkhand*
Dr Amit Kumar	*Department of Mechanical Engineering, National Institute of Technology, Patna, Bihar*
M K Singha	*Department of Applied Mechanics, Indian Institute of Technology, New Delhi*
P Venkitanarayanan	*Department of Applied Mechanics, Indian Institute of Technology, Kanpur, Uttar Pradesh*
G Rajesh Kumar	*Department of Mechanical Engineering, Rizvi College of Engineering, Mumbai, Maharashtra*
C A Akhadkar	*Department of Mechanical Engineering, SSVPS's B S Deora College of Engineering, Dhule, Maharashtra*
D Prasanna Venkatesh	*Department of Mechanical Engineering, SRM Institute of Science and Technology, Chennai, Tamil Nadu*
E V M Sargunar	*Sree Sastha Institute of Engineering and Technology, Chennai, Tamil Nadu*

In addition to this, I am also thankful to the staff at McGraw-Hill Education India, especially Ms Vibha Mahajan, Ms Shukti Mukherjee, Ms Surabhi Shukla, Ms Sohini Mukherjee and Mr P L Pandita, for their cooperation during the different stages of this project.

Lastly, I wish to thank my family members for their patience, support and love given to me during the preparation of this manuscript.

Feedback and suggestions are always welcome at srinath_ls@sify.com.

L S Srinath

List of Symbols
(In the order they appear in the text)

σ	normal stress
F	force
$\overset{n}{\boldsymbol{T}}$	force vector on a plane with normal n
$\overset{n}{\boldsymbol{T}}_{x,y,z}$	components of force vector in x, y, z directions
A	area of section
A	normal to the section
τ	shear stress
$\sigma_{x,y,z}$	normal stress on x-plane, y-plane, z-plane
$\tau_{xy,yz,zx}$	shear stress on x-plane in y-direction, shear stress on y-plane in z-direction, shear stress on z-plane in x-direction
n_x, n_y, n_z	direction cosines of n in x, y, z directions
$\sigma_1, \sigma_2, \sigma_3$	principal stresses at a point
I_1, I_2, I_3	first, second, third invariants of stress
σ_{oct}	normal stress on octahedral plane
τ_{oct}	shear stress on octahedral plane
$\sigma_r, \sigma_\theta, \sigma_z$	normal stresses in radial, circumferential, axial (polar) direction
γ, θ, φ	spherical coordinates
$\tau_{\gamma\theta}, \tau_{\gamma z}, \tau_{\theta z}$	shear stresses in polar coordinates
u_x, u_y, u_z	displacements in x, y, z directions
E_{xx}, E_{yy}, E_{zz}	linear strains in x-direction, y-direction, z-direction (with non-linear terms)
$\varepsilon_{xx}, \varepsilon_{yy}, \varepsilon_{zz}$	linear strains (with linear terms only)
E_{xy}, E_{yz}, E_{zx}	shear strain components (with non-linear terms)
$\gamma_{xy}, \gamma_{yz}, \gamma_{zx}$	shear strain components (with linear terms only)
$\omega_x, \omega_y, \omega_z$	rigid body rotations about x, y, z axes
$\Delta = \varepsilon_{xx} + \varepsilon_{yy} + \varepsilon_{zz}$	cubical dilatation
$\varepsilon_1, \varepsilon_2, \varepsilon_3$	principal strains at a point
J_1, J_2, J_3	first, second, third invariants of strain

$\varepsilon_{\gamma}, \varepsilon_{\theta}, \varepsilon_z$	strains in radial, circumferential, axial directions
λ, μ	Lame's constants
$G = \mu$	rigidity modulus
μ	engineering Poisson's ratio
E	modulus of elasticity
K	bulk modulus; stress intensity factor
P	pressure
υ	Poisson's ratio
σ_y	yield point stress
U	elastic energy
U^*	distortion energy; complementary energy
σ_{ut}	ultimate stress in uniaxial tension
σ_{ct}	ultimate stress in uniaxial compression
a_{ij}	influence coefficient; material constant
b_{ij}	compliance component
M_x, M_y, M_z	moments about x, y, z axes
δ	linear deflection; generalized deflection
I_x, I_y, I_z	moments of inertia about x, y, z axes
I_ρ	polar moment of inertia
I_{xy}, I_{yz}	products of inertia about xy and yz coordinates
T	torque; temperature
Ψ	warping function
α	coefficient of thermal expansion
Q	lateral load
P	axial load
V	elastic potential
ν_{ij}	Poisson's ratio in i-direction due to stress in j-direction
b, w	width
t	thickness
K_t	theoretical stress concentration factor
N	normal force
ϕ	stream function
ρ	fillet radius
D, d	radii
q	notch sensitivity
K_c, K_{Ic}	fracture toughness in mode I
S_y	offset yield stress
ω	angular velocity
R	fracture resistance
σ_{fr}	fracture stress
Γ	boundary
J	J-integral

SI Units
(Systeme International d'Unit'es)

(a) Base Units

Quantity	*Unit (Symbol)*
length	meter (m)
mass	kilogram (kg)
time	second (s)
force	newton (N)
pressure	pascal (Pa)

force is a derived unit: kgm/s^2
pressure is force per unit area: N/m^2: kg/ms^2
kilo-watt is work done per second: kNm/s

(b) Multiples

giga (G)	1 000 000 000
mega (M)	1 000 000
kilo (k)	1 000
milli (m)	0.001
micro (μ)	0.000 001
nano (n)	0.000 000 001

(c) Conversion Factors

To Convert	*to*	*Multiply by*
kgf	newton	9.8066
kgf/cm^2	Pa	9.8066×10^4
kgf/cm^2	kPa	98.066
newton	kgf	0.10197
Pa	N/m^2	1
kPa	kgf/cm^2	0.010197
HP	kW	0.746
HP	kNm/s	0.746
kW	kNm/s	1

Typical Physical Constants
(As an Aid to Solving Problems)

Material	*Ultimate Strength (MPa)*			*Yield Strength (MPa)*		*Elastic Modulus (GPa)*		*Poisson's Ratio*	*Coeff. Therm Expans. per °C*
	Tens.	*Comp*	*Shear*	*Tens or Comp*	*Shear*	*Tens*	*Shear*		$\times 10^{-6}$
Aluminium alloy	414	414	221	300	170	73	28	0.334	23.2
Cast iron, gray	210	825	—	—	—	90	41	0.211	10.4
Carbon steel	690	690	552	415	250	200	83	0.292	11.7
Stainless steel	568	568	—	276	—	207	90	0.291	17.0

For more accurate values refer to hand-books on material properties

CHAPTER 1

Analysis of Stress

1.1 INTRODUCTION

In this book we shall deal with the mechanics of deformable solids. The starting point for discussion can be either the analysis of stress or the analysis of strain. In books on the theory of elasticity, one usually starts with the analysis of strain, which deals with the geometry of deformation without considering the forces that cause the deformation. However, one is more familiar with forces, though the measurement of force is usually done through the measurement of deformations caused by the force. Books on Strength of Materials, begin with the analysis of stress. The concept of stress has already been introduced in the elementry strength of materials. When a bar of uniform cross-section, say a circular rod of diameter *d*, is subjected to a tensile force *F* along the axis of the bar, the average stress induced across any transverse section perpendicular to the axis of the bar and away from the region of loading is given by

$$\sigma = \frac{F}{\text{Area}} = \frac{4F}{\pi d^2}$$

It is assumed that the reader is familiar with the elementary flexural stress and torsional stress concepts. In general, a structural member or a machine element will not possess uniform geometry of shape or size, and the loads acting on it will also be complex. For example, an automobile crankshaft or a piston inside an engine cylinder or an aircraft wing are subject to loadings that are both complex as well as dynamic in nature. In such cases, one will have to introduce the concept of the state of stress at a point and its analysis, which will be the subject of discussion in this chapter. However, we shall not deal with forces that vary with time.

It will be assumed that the matter of the body that is being considered is continuously distributed over its volume, so that if we consider a small volume element of the matter surrounding a point and shrink this volume, in the limit we shall not come across a void. In reality, however, all materials are composed of many discrete particles, which are often microscopic, and when an arbitrarily selected volume element is shrunk, in the limit one may end up in a void. But in our analysis, we assume that the matter is continuously distributed. Such a body is called a continuous medium and the mechanics of such a body or bodies is called continuum mechanics.

1.2 BODY FORCE, SURFACE FORCE AND STRESS VECTOR

Consider a body B occupying a region of space referred to a rectangular coordinate system $Oxyz$, as shown in Fig. 1.1. In general, the body will be subjected to two types of forces—body forces and surface forces. The body forces act on each volume element of the body. Examples of this kind of force are the gravitational force, the inertia force and the magnetic force. The surface forces act on the surface or area elements of the body. When the area considered lies on the actual boundary of the body, the surface force distribution is often termed surface traction. In Fig. 1.1, the surface forces F_1, F_2, F_3 . . . F_r, are concentrated forces, while p is a distributed force. The support reactions R_1, R_2 and R_3 are also surface forces. It is explicitly assumed that under the action of both body forces and surface forces, the body is in equilibrium.

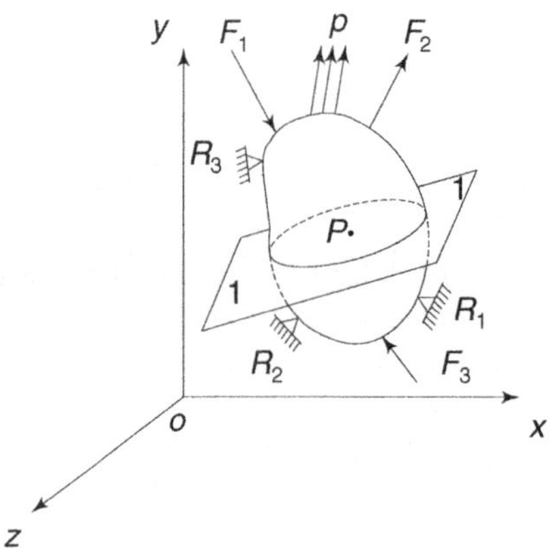

Fig. 1.1 *Body subjected to forces*

Let P be a point inside the body with coordinates (x, y, z). Let the body be cut into two parts C and D by a plane 1-1 passing through point P, as shown in Fig. 1.2. If we consider the free-body diagrams of C and D, then each part is in equilibrium under the action of the externally applied forces and the internally distributed forces across the interface. In part D, let ΔA be a small area surrounding the point P. In part C, the corresponding area at P' is $\Delta A'$. These two areas are distinguished by their outward drawn normals $\overset{1}{\boldsymbol{n}}$ and $\overset{1}{\boldsymbol{n}}'$. The action of part C on ΔA at point P can be represented by the force vector $\Delta\overset{1}{\boldsymbol{T}}$ and the action of part D on $\Delta A'$ at P' can be represented by the force vector $\Delta\overset{1}{\boldsymbol{T}}'$. We assume that as ΔA tends to zero, the ratio $\dfrac{\Delta\overset{1}{\boldsymbol{T}}}{\Delta A}$ tends to a definite limit, and

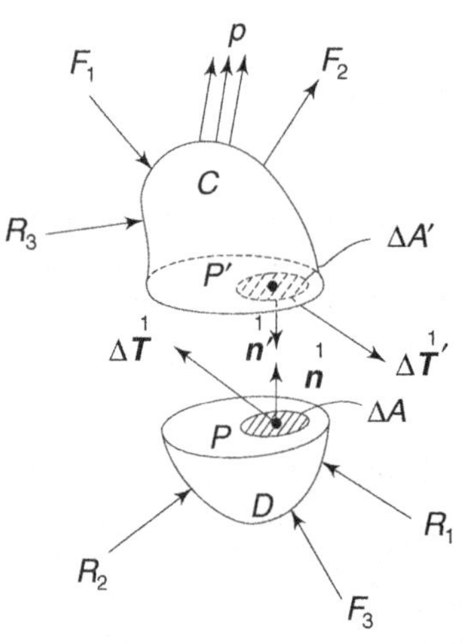

Fig. 1.2 *Free-body diagram of a body cut into two parts*

further, the moment of the forces acting on area ΔA about any point within the area vanishes in the limit. The limiting vector is written as

$$\lim_{\Delta A \to 0} \frac{\Delta \overset{1}{\boldsymbol{T}}}{\Delta A} = \frac{d\overset{1}{\boldsymbol{T}}}{dA} = \overset{1}{\boldsymbol{T}} \tag{1.1}$$

Similarly, at point P', the action of part D on C as $\Delta A'$ tends to zero, can be represented by a vector

$$\lim_{\Delta A' \to 0} \frac{\Delta \overset{1}{\boldsymbol{T}}{}'}{\Delta A'} = \frac{d\overset{1}{\boldsymbol{T}}{}'}{dA'} = \overset{1}{\boldsymbol{T}}{}' \tag{1.2}$$

Vectors $\overset{1}{\boldsymbol{T}}$ and $\overset{1}{\boldsymbol{T}}{}'$ are called the stress vectors and they represent forces per unit area acting respectively at P and P' on planes with outward drawn normals $\overset{1}{\boldsymbol{n}}$ and $\overset{1}{\boldsymbol{n}}{}'$.

We further assume that stress vector $\overset{1}{\boldsymbol{T}}$ representing the action of C on D at P is equal in magnitude and opposite in direction to stress vector $\overset{1}{\boldsymbol{T}}{}'$ representing the action of D on C at corresponding point P'. This assumption is similar to Newton's third law, which is applicable to particles. We thus have

$$\overset{1}{\boldsymbol{T}} = -\overset{1}{\boldsymbol{T}}{}' \tag{1.3}$$

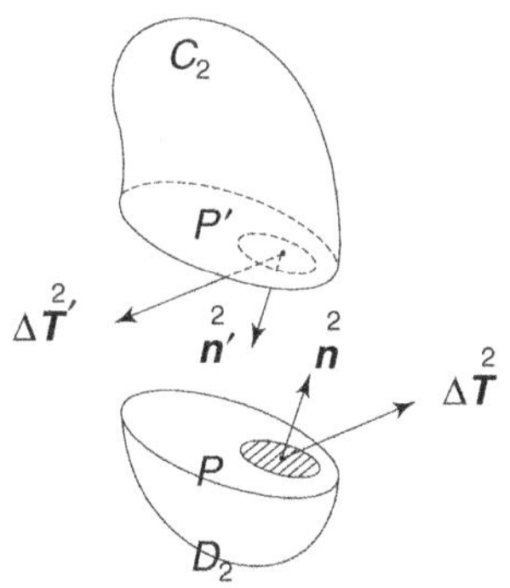

Fig. 1.3 *Body cut by another plane*

If the body in Fig. 1.1 is cut by a different plane 2-2 with outward drawn normals $\overset{2}{\boldsymbol{n}}$ and $\overset{2}{\boldsymbol{n}}{}'$ passing through the same point P, then the stress vector representing the action of C_2 on D_2 will be represented by $\overset{2}{\boldsymbol{T}}$ (Fig. (1.3)), i.e.

$$\overset{2}{\boldsymbol{T}} = \lim_{\Delta A \to 0} \frac{\Delta \overset{2}{\boldsymbol{T}}{}'}{\Delta A}$$

In general, stress vector $\overset{1}{\boldsymbol{T}}$ acting at point P on a plane with outward drawn normal $\overset{1}{\boldsymbol{n}}$ will be different from stress vector $\overset{2}{\boldsymbol{T}}$ acting at the same point P, but on a plane with outward drawn normal $\overset{2}{\boldsymbol{n}}$. Hence the stress at a point depends not only on the location of the point (identified by coordinates x, y, z) but also on the plane passing through the point (identified by direction cosines n_x, n_y, n_z of the outward drawn normal).

1.3 THE STATE OF STRESS AT A POINT

Since an infinite number of planes can be drawn through a point, we get an infinite number of stress vectors acting at a given point, each stress vector characterised by the corresponding plane on which it is acting. The totality of all stress vectors acting on every possible plane passing through the point is defined to be the state of stress at the point. It is the knowledge of this state of stress that is of importance to a designer in determining the critical planes and the respective critical stresses. It will be shown in Sec. 1.6 that if the stress vectors acting on three mutually perpendicular planes passing through the point are known, we can determine the stress vector acting on any other arbitrary plane at that point.

1.4 NORMAL AND SHEAR STRESS COMPONENTS

Let $\overset{n}{\boldsymbol{T}}$ be the resultant stress vector at point P acting on a plane whose outward drawn normal is $\boldsymbol{n}$ (Fig.1.4). This can be resolved into two components, one along the normal $\boldsymbol{n}$ and the other perpendicular to $\boldsymbol{n}$. The component parallel to $\boldsymbol{n}$ is called the normal stress and is generally denoted by $\overset{n}{\sigma}$. The component perpendicular to $\boldsymbol{n}$ is known as the tangential stress or shear stress component and is denoted by $\overset{n}{\tau}$. We have, therefore, the relation:

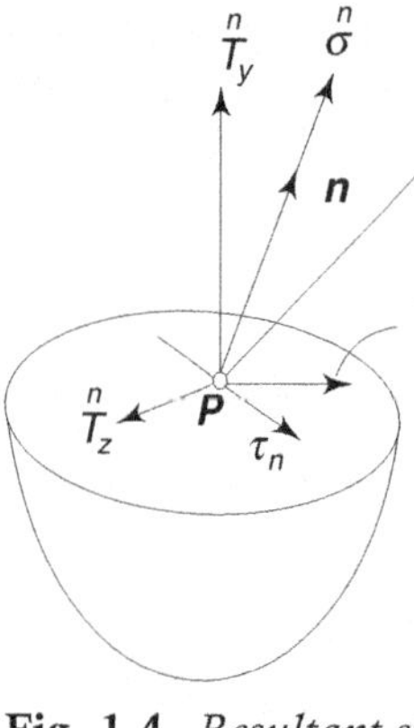

Fig. 1.4 *Resultant stress vector, normal and shear stress components*

$$\left|\overset{n}{\boldsymbol{T}}\right|^2 = \overset{n}{\sigma}{}^2 + \overset{n}{\tau}{}^2 \tag{1.4}$$

where $\left|\overset{n}{\boldsymbol{T}}\right|$ is the magnitude of the resultant stress. Stress vector $\overset{n}{\boldsymbol{T}}$ can also be resolved into three components parallel to the x, y, z axes. If these components are denoted by $\overset{n}{T}_x, \overset{n}{T}_y, \overset{n}{T}_z$, we have

$$\left|\overset{n}{\boldsymbol{T}}\right|^2 = \overset{n}{T}{}_x^2 + \overset{n}{T}{}_y^2 + \overset{n}{T}{}_z^2 \tag{1.5}$$

1.5 RECTANGULAR STRESS COMPONENTS

Let the body B, shown in Fig. 1.1, be cut by a plane parallel to the yz plane. The normal to this plane is parallel to the x axis and hence, the plane is called the x plane. The resultant stress vector at P acting on this will be $\overset{x}{\boldsymbol{T}}$. This vector can be resolved into three components parallel to the x, y, z axes. The component parallel to the x axis, being normal to the plane, will be denoted by σ_x (instead of by $\overset{x}{\sigma}$). The components parallel to the y and z axes are shear stress components and are denoted by τ_{xy} and τ_{xz} respectively (Fig. 1.5).

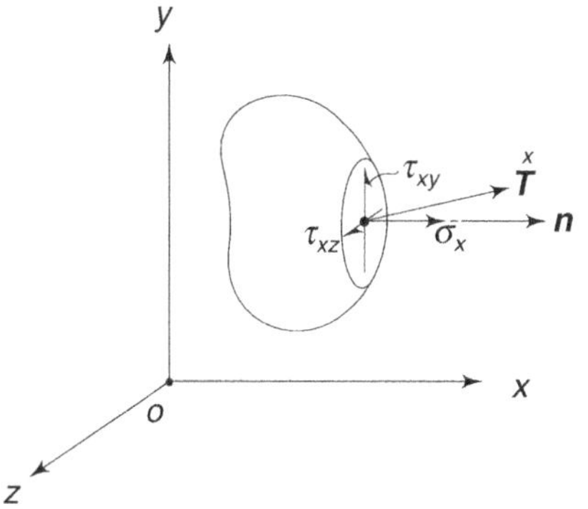

Fig. 1.5 *Stress components on x plane*

In the above designation, the first subscript x indicates the plane on which the stresses are acting and the second subscript (y or z) indicates the direction of the component. For example, τ_{xy} is the stress component on the x plane in y direction. Similarly, τ_{xz} is the stress component on the x plane in z direction. To maintain consistency, one should have denoted the normal stress component as τ_{xx}. This would be the stress component on the x plane in the x direction. However, to distinguish between a normal stress and a shear stress, the normal stress is denoted by σ and the shear stress by τ.

At any point P, one can draw three mutually perpendicular planes, the x plane, the y plane and the z plane. Following the notation mentioned above, the normal and shear stress components on these planes are

$$\sigma_x,\ \tau_{xy},\ \tau_{xz} \text{ on } x \text{ plane}$$

$$\sigma_y,\ \tau_{yx},\ \tau_{yz} \text{ on } y \text{ plane}$$

$$\sigma_z,\ \tau_{zx},\ \tau_{zy} \text{ on } z \text{ plane}$$

These components are shown acting on a small rectangular element surrounding the point P in Fig. 1.6.

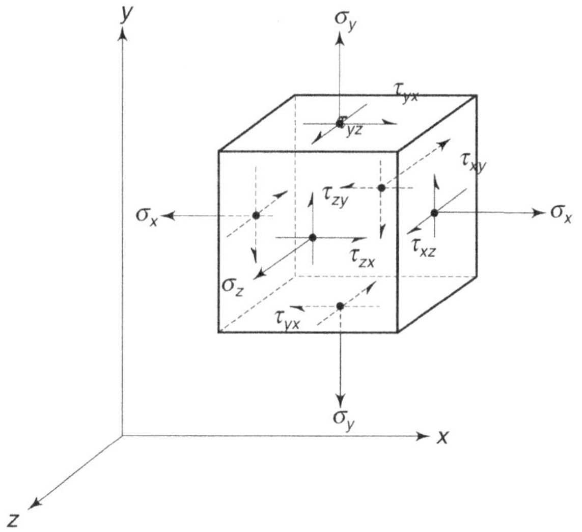

Fig. 1.6 *Rectangular stress components*

One should observe that the three visible faces of the rectangular element have their outward drawn normals along the positive x, y and z axes respectively. Consequently, the positive stress components on these faces will also be directed along the positive axes. The three hidden faces have their outward drawn normals in the negative x, y and z axes. The positive stress components on these faces will, therefore, be directed along the negative axes. For example, the bottom face has its outward drawn normal along the negative y axis. Hence, the positive stress components on this face, i.e., σ_y, τ_{yx} and τ_{yz} are directed respectively along the negative y, x and z axes.

1.6 STRESS COMPONENTS ON AN ARBITRARY PLANE

It was stated in Section 1.3 that a knowledge of stress components acting on three mutually perpendicular planes passing through a point will enable one to determine the stress components acting on any plane passing through that point. Let the three mutually perpendicular planes be the x, y and z planes and let the arbitrary plane be identified by its outward drawn normal $\boldsymbol{n}$ whose direction cosines are n_x, n_y and n_z. Consider a small tetrahedron at P with three of its faces normal to the coordinate axes, and the inclined face having its normal parallel to $\boldsymbol{n}$. Let h be the perpendicular distance from P to the inclined face. If the tetrahedron is isolated from the body and a free-body diagram is drawn, then it will be in equilibrium under the action of the surface forces and the body forces. The free-body diagram is shown in Fig. 1.7.

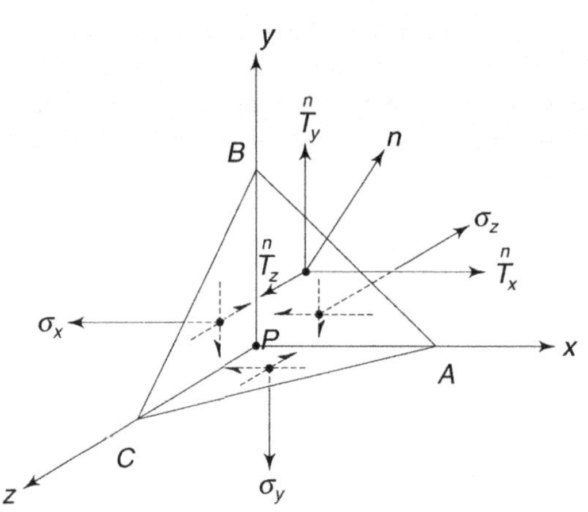

Fig. 1.7 *Tetrahedron at point P*

Since the size of the tetrahedron considered is very small and in the limit as we are going to make h tend to zero, we shall speak in terms of the average stresses over the faces. Let $\overset{n}{\boldsymbol{T}}$ be the resultant stress vector on face ABC. This can be resolved into components $\overset{n}{T}_x, \overset{n}{T}_y, \overset{n}{T}_z$, parallel to the three axes x, y and z. On the three faces, the rectangular stress components are σ_x, τ_{xy}, τ_{xz}, σ_y, τ_{yz}, τ_{yx}, σ_z, τ_{zx} and τ_{zy}. If A is the area of the inclined face then

$$\begin{aligned}
\text{Area of } BPC &= \text{projection of area } ABC \text{ on the } yz \text{ plane} \\
&= An_x \\
\text{Area of } CPA &= \text{projection of area } ABC \text{ on the } xz \text{ plane} \\
&= An_y \\
\text{Area of } APB &= \text{projection of area } ABC \text{ on the } xy \text{ plane} \\
&= An_z
\end{aligned}$$

Let the body force components in x, y and z directions be γ_x, γ_y and γ_z respectively, per unit volume. The volume of the tetrahedron is equal to $\frac{1}{3}\,Ah$, where h is the perpendicular distance from P to the inclined face. For equilibrium of the tetrahedron, the sum of the forces in x, y and z directions must individually vanish. Thus, for equilibrium in x direction

$$\overset{n}{T}_x A - \sigma_x A n_x - \tau_{yx} A n_y - \tau_{zx} A n_z + \frac{1}{3} A h \gamma_x = 0$$

Cancelling A,

$$\overset{n}{T}_x = \sigma_x n_x + \tau_{yx} n_y + \tau_{zx} n_z - \frac{1}{3} h \gamma_x \tag{1.6}$$

Similarly, for equilibrium in y and z directions,

$$\overset{n}{T}_y = \tau_{xy} n_x + \sigma_y n_y + \tau_{zy} n_z - \frac{1}{3} h \gamma_y \tag{1.7}$$

and

$$\overset{n}{T}_z = \tau_{xz} n_x + \tau_{yz} n_y + \sigma_z n_z - \frac{1}{3} h \gamma_z \tag{1.8}$$

In the limit as h tends to zero, the oblique plane ABC will pass through point P, and the average stress components acting on the faces will tend to their respective values at point P acting on their corresponding planes. Consequently, one gets from equations (1.6)–(1.8)

$$\begin{aligned}\overset{n}{T}_x &= n_x \sigma_x + n_y \tau_{yx} + n_z \tau_{zx} \\ \overset{n}{T}_y &= n_x \tau_{xy} + n_y \sigma_y + n_z \tau_{zy} \\ \overset{n}{T}_z &= n_x \tau_{xz} + n_y \tau_{yz} + n_z \sigma_z\end{aligned} \tag{1.9}$$

Equation (1.9) is known as Cauchy's stress formula. This equation shows that the nine rectangular stress components at P will enable one to determine the stress components on any arbitrary plane passing through point P. It will be shown in Sec. 1.8 that among these nine rectangular stress components only six are independent. This is because $\tau_{xy} = \tau_{yx}$, $\tau_{zy} = \tau_{yz}$ and $\tau_{zx} = \tau_{xz}$. This is known as the equality of cross shears. In anticipation of this result, one can write Eq. (1.9) as

$$\overset{n}{T}_i = n_x \tau_{ix} + n_y \tau_{iy} + n_z \tau_{iz} = \sum_j n_j \tau_{ij} \tag{1.10}$$

where i and j can stand for x or y or z, and $\sigma_x = \tau_{xx}$, $\sigma_y = \tau_{yy}$ and $\sigma_z = \tau_{zz}$.

If $\overset{n}{\boldsymbol{T}}$ is the resultant stress vector on plane ABC, we have

$$\left|\overset{n}{\boldsymbol{T}}\right|^2 = \overset{n}{T}_x{}^2 + \overset{n}{T}_y{}^2 + \overset{n}{T}_z{}^2 \tag{1.11a}$$

If σ_n and τ_n are the normal and shear stress components, we have

$$\left|\overset{n}{\boldsymbol{T}}\right|^2 = \sigma_n^2 + \tau_n^2 \tag{1.11b}$$

Since the normal stress is equal to the projection of $\overset{n}{\boldsymbol{T}}$ along the normal, it is also equal to the sum of the projections of its components $\overset{n}{T}_x$, $\overset{n}{T}_y$ and $\overset{n}{T}_z$ along $\boldsymbol{n}$. Hence,

$$\sigma_n = n_x \overset{n}{T}_x + n_y \overset{n}{T}_y + n_z \overset{n}{T}_z \tag{1.12a}$$

Substituting for $\overset{n}{T}_x$, $\overset{n}{T}_y$ and $\overset{n}{T}_z$ from Eq. (1.9)

$$\sigma_n = n_x^2 \sigma_x + n_y^2 \sigma_y + n_z^2 \sigma_z + 2n_x n_y \tau_{xy} + 2n_y n_z \tau_{yz} + 2n_z n_x \tau_{zx} \qquad (1.12b)$$

Equation (1.11) can then be used to obtain the value of τ_n

Example 1.1 *A rectangular steel bar having a cross-section 2 cm × 3 cm is subjected to a tensile force of 6000 N (612.2 kgf). If the axes are chosen as shown in Fig. 1.8, determine the normal and shear stresses on a plane whose normal has the following direction cosines:*

(i) $n_x = n_y = \dfrac{1}{\sqrt{2}}, n_z = 0$

(ii) $n_x = 0, n_y = n_x = \dfrac{1}{\sqrt{2}}$

(iii) $n_x = n_y = n_x = \dfrac{1}{\sqrt{3}}$

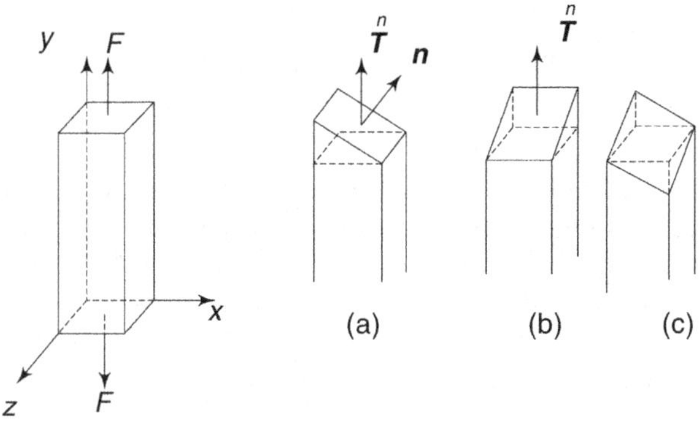

Fig. 1.8 *Example 1.1*

Solution Area of section = 2 × 3 = 6 cm². The average stress on this plane is 6000/6 = 1000 N/cm². This is the normal stress σ_y. The other stress components are zero.

(i) Using Eqs (1.9), (1.11b) and (1.12a)

$$\overset{n}{T}_x = 0, \qquad \overset{n}{T}_y = \frac{1000}{\sqrt{2}}, \qquad \overset{n}{T}_z = 0$$

$$\sigma_n = \frac{1000}{2} = 500 \text{ N/cm}^2$$

$$\tau_n^2 = \left|\overset{n}{\boldsymbol{T}}\right|^2 - \sigma_n^2 = 250{,}000 \text{ N}^2/\text{cm}^4$$

$$\tau_n = 500 \text{ N/cm}^2 \ (51 \text{ kgf/cm}^2)$$

(ii) $\overset{n}{T}_x = 0, \quad \overset{n}{T}_y = \frac{1000}{\sqrt{2}}, \quad \overset{n}{T}_z = 0$

$$\sigma_n = 500 \text{ N/cm}^2, \text{ and } \tau_n = 500 \text{ N/cm}^2 \text{ (51 kgf/cm}^2\text{)}$$

(iii) $\overset{n}{T}_x = 0, \quad \overset{n}{T}_y = \frac{1000}{\sqrt{3}}, \quad \overset{n}{T}_z = 0$

$$\sigma_n = \frac{1000}{3} \text{ N/cm}^2$$

$$\tau_n = 817 \text{ N/cm}^2 \text{ (83.4 kgf/cm}^2\text{)}$$

Example 1.2 *At a point P in a body, σ_x = 10,000 N/cm² (1020 kgf/cm²), σ_y = –5,000 N/cm² (–510 kgf/cm²), σ_z = –5,000 N/cm², $\tau_{xy} = \tau_{yz} = \tau_{zx}$ = 10,000 N/cm². Determine the normal and shearing stresses on a plane that is equally inclined to all the three axes.*

Solution A plane that is equally inclined to all the three axes will have

$$n_x = n_y = n_z = \frac{1}{\sqrt{3}} \quad \text{since } n_x^2 + n_y^2 + n_z^2 = 1$$

From Eq. (1.12)

$$\sigma_n = \frac{1}{3}\,[10000 - 5000 - 5000 + 20000 + 20000 + 20000]$$
$$= 20000 \text{ N/cm}^2$$

From Eqs (1.6)–(1.8)

$$\overset{n}{T}_x = \frac{1}{\sqrt{3}}\,(10000 + 10000 + 10000) = 10000\sqrt{3} \text{ N/cm}^2$$

$$\overset{n}{T}_y = \frac{1}{\sqrt{3}}\,(10000 - 5000 + 10000) = -5000\sqrt{3} \text{ N/cm}^2$$

$$\overset{n}{T}_z = \frac{1}{\sqrt{3}}\,(10000 - 10000 - 5000) = -5000\sqrt{3} \text{ N/cm}^2$$

$\therefore$ $$\left|\overset{n}{\boldsymbol{T}}\right|^2 = 3\,[(10^8) + (25 \times 10^6) + (25 \times 10^6)] \text{ N}^2/\text{cm}^4$$
$$= 450 \times 10^6 \text{ N}^2/\text{cm}^4$$

$\therefore$ $$\tau_n^2 = 450 \times 10^6 - 400 \times 10^6 = 50 \times 10^6 \text{ N}^2/\text{cm}^4$$

or $$\tau_n = 7000 \text{ N/cm}^2 \text{ (approximately)}$$

Example 1.3 *Figure 1.9 shows a cantilever beam in the form of a trapezium of uniform thickness loaded by a force P at the end. If it is assumed that the bending stress on any vertical section of the beam is distributed according to the elementary* *flexure formula, show that the normal stress σ on a section perpendicular to the top edge of the beam at point A is $\frac{\sigma_1}{\cos^2\theta}$, where σ_1 is the flexural stress $\frac{Mc}{I}$, as shown in Fig. 1.9(b).*

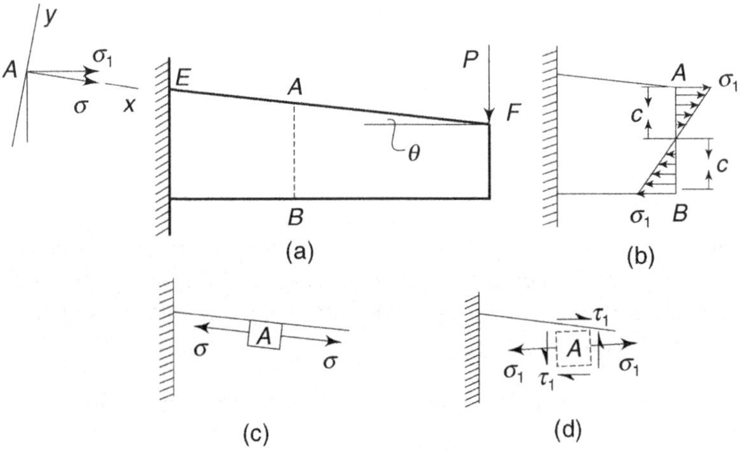

Fig. 1.9 *Example 1.3*

Solution At point A, let axes x and y be chosen along and perpendicular to the edge. On the x plane, i.e. the plane perpendicular to edge EF, the resultant stress is along the normal (i.e., x axis). There is no shear stress on this plane since the top edge is a free surface (see Sec. 1.9). But on plane AB at point A there can exist a shear stress. These are shown in Fig. 1.9(c) and (d). The normal to plane AB makes an angle θ with the x axis. Let the normal and shearing stresses on this plane be σ_1 and τ_1.

We have

$$\sigma_x = \sigma, \qquad \sigma_y = \sigma_z = 0, \qquad \tau_{xy} = \tau_{yz} = \tau_{zx} = 0$$

The direction cosines of the normal to plane AB are

$$n_x = \cos\theta, \qquad n_y = \sin\theta, \qquad n_z = 0$$

The components of the stress vector acting on plane AB are

$$\overset{n}{T}_x = \sigma_1 = n_x\,\sigma_x + n_y\,\tau_{yx} + n_z\,\tau_{zy} = \sigma\cos\theta$$

$$\overset{n}{T}_y = n_x\,\tau_{xy} + n_y\,\sigma_y + n_z\,\tau_{zy} = 0$$

$$\overset{n}{T}_z = n_x\,\tau_{xz} + n_y\,\tau_{yz} + n_z\,\sigma_z = 0$$

Therefore, the normal stress on plane $AB = \sigma_n = n_x \overset{n}{\boldsymbol{T}}_x + n_y \overset{n}{\boldsymbol{T}}_y + n_z \overset{n}{\boldsymbol{T}}_z = \sigma\cos^2\theta$.

Since $\sigma_n = \sigma_1$

$$\sigma = \frac{\sigma_1}{\cos^2\theta} = \frac{Mc}{I\cos^2\theta}$$

Further, the resultant stress on plane *AB* is

$$\left|\overset{n}{\boldsymbol{T}}\right|^2 = \overset{n}{T}_x^2 + \overset{n}{T}_y^2 + \overset{n}{T}_z^2 = \sigma^2\cos^2\theta$$

Hence
$$\tau^2 = \sigma^2\cos^2\theta - \sigma_n^2$$
$$= \sigma^2\cos^2\theta - \sigma^2\cos^4\theta$$

or
$$\tau = \frac{1}{2}\sigma\sin 2\theta$$

1.7 DIGRESSION ON IDEAL FLUID

By definition, an ideal fluid cannot sustain any shearing forces and the normal force on any surface is compressive in nature. This can be represented by

$$\overset{n}{\boldsymbol{T}} = -p\boldsymbol{n}, \quad p \geq 0$$

The rectangular components of $\overset{n}{\boldsymbol{T}}$ are obtained by taking the projections of $\overset{n}{\boldsymbol{T}}$ along the *x*, *y* and *z* axes. If n_x, n_y and n_z are the direction cosines of $\boldsymbol{n}$, then

$$\overset{n}{T}_x = -pn_x, \qquad \overset{n}{T}_y = -pn_y, \qquad \overset{n}{T}_z = -pn_z \tag{1.13}$$

Since all shear stress components are zero, one has from Eqs. (1.9),

$$\overset{n}{T}_x = n_x\,\sigma_x, \qquad \overset{n}{T}_y = n_y\,\sigma_y, \qquad \overset{n}{T}_z = n_z\sigma_z \tag{1.14}$$

Comparing Eqs (1.13) and (1.14)

$$\sigma_x = \sigma_y = \sigma_z = -p$$

Since plane $\boldsymbol{n}$ was chosen arbitrarily, one concludes that the resultant stress vector on any plane is normal and is equal to $-p$. This is the type of stress that a small sphere would experience when immersed in a liquid. Hence, the state of stress at a point where the resultant stress vector on any plane is normal to the plane and has the same magnitude is known as a hydrostatic or an isotropic state of stress. The word isotropy means 'independent of orientation' or 'same in all directions'. This aspect will be discussed again in Sec. 1.14.

1.8 EQUALITY OF CROSS SHEARS

We shall now show that of the nine rectangular stress components σ_x, τ_{xy}, τ_{xz}, σ_y, τ_{yx}, τ_{yz}, σ_z, τ_{zx} and τ_{zy}, only six are independent. This is because $\tau_{xy} = \tau_{yx}$, $\tau_{yz} = \tau_{zy}$ and $\tau_{zx} = \tau_{xz}$. These are known as cross-shears. Consider an infinitesimal rectangular parallelpiped surrounding point *P*. Let the dimensions of the sides be Δx, Δy and Δz (Fig. 1.10).

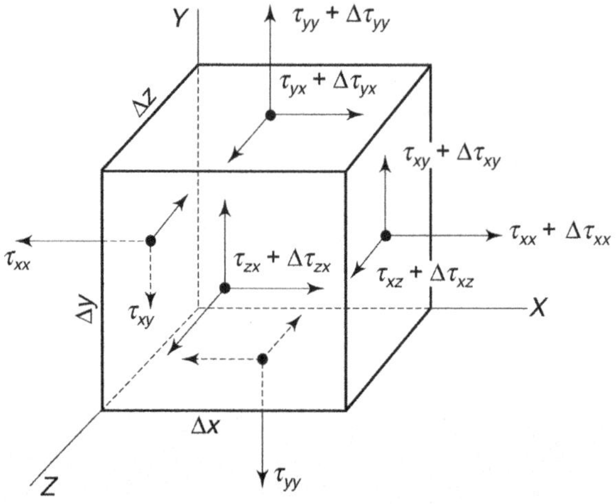

Fig. 1.10 *Stress components on a rectangular element*

Since the element considered is small, we shall speak in terms of average stresses over the faces. The stress vectors acting on the faces are shown in the figure. On the left x plane, the stress vectors are $\tau_{xx,}$ τ_{xy} and τ_{xz}. On the right face, the stresses are $\tau_{xx} + \Delta\tau_{xx}$, $\tau_{xy} + \Delta\tau_{xy}$ and $\tau_{xz} + \Delta\tau_{xz}$. These changes are because the right face is at a distance Δx from the left face. To the first order of approximation we have

$$\Delta\tau_{xx} = \frac{\partial\tau_{xx}}{\partial x}\Delta x, \qquad \Delta\tau_{xy} = \frac{\partial\tau_{xy}}{\partial x}\Delta x, \qquad \Delta\tau_{xz} = \frac{\partial\tau_{xz}}{\partial x}\Delta x$$

Similarly, the stress vectors on the top face are $\tau_{yy} + \Delta\tau_{yy}$, $\tau_{yx} + \Delta\tau_{yx}$ and $\tau_{yz} + \Delta\tau_{yz}$, where

$$\Delta\tau_{yy} = \frac{\partial\tau_{yy}}{\partial y}\Delta y, \qquad \Delta\tau_{yx} = \frac{\partial\tau_{yx}}{\partial y}\Delta y, \qquad \Delta\tau_{yz} = \frac{\partial\tau_{yz}}{\partial y}\Delta y$$

On the rear and front faces, the components of stress vectors are respectively

$$\tau_{zz}, \tau_{zx}, \tau_{zy}$$
$$\tau_{zz} + \Delta\tau_{zz}, \tau_{zx} + \Delta\tau_{zx}, \tau_{zy} + \Delta\tau_{zy}$$

where

$$\Delta\tau_{zz} = \frac{\partial\tau_{zz}}{\partial z}\Delta z, \qquad \Delta\tau_{zx} = \frac{\partial\tau_{zx}}{\partial z}\Delta z, \qquad \Delta\tau_{zy} = \frac{\partial\tau_{zy}}{\partial z}\Delta z$$

For equilibrium, the moments of the forces about the x, y and z axes must vanish individually. Taking moments about the z axis, one gets

$$\tau_{xx}\,\Delta y\,\Delta z\,\frac{\Delta y}{2} - (\tau_{xx} + \Delta\tau_{xx})\,\Delta y\,\Delta z\,\frac{\Delta y}{2} +$$

$$(\tau_{xy} + \Delta\tau_{xy})\,\Delta y\,\Delta z\,\Delta x - \tau_{yy}\,\Delta x\,\Delta z\,\frac{\Delta x}{2} +$$

$$(\tau_{yy} + \Delta\tau_{yy})\,\Delta x\,\Delta z\,\frac{\Delta x}{2} - (\tau_{yx} + \Delta\tau_{yx})\,\Delta x\,\Delta z\,\Delta y +$$

$$\tau_{zy}\,\Delta x\,\Delta y\,\frac{\Delta x}{2} - \tau_{zx}\,\Delta x\,\Delta y\,\frac{\Delta y}{2} - (\tau_{zy} + \Delta\tau_{zy})\,\Delta x\,\Delta y\,\frac{\Delta x}{2} +$$

$$(\tau_{zx} + \Delta\tau_{zx})\,\Delta x\,\Delta y\,\frac{\Delta y}{2} = 0$$

Substituting for $\Delta\tau_{xx}$, $\Delta\tau_{xy}$ etc., and dividing by $\Delta x\,\Delta y\,\Delta z$

$$-\frac{\partial\tau_{xx}}{\partial x}\frac{\Delta y}{2} + \tau_{xy} + \frac{\partial\tau_{xy}}{\partial x}\Delta x + \frac{\partial\tau_{yy}}{\partial y}\frac{\Delta y}{2} -$$

$$\tau_{yx} - \frac{\partial\tau_{yx}}{\partial y}\Delta y - \frac{\partial\tau_{zy}}{\partial z}\frac{\Delta x}{2} + \frac{\partial\tau_{zx}}{\partial z}\frac{\Delta y}{2} = 0$$

In the limit as Δx, Δy and Δz tend to zero, the above equation gives $\tau_{xy} = \tau_{yx}$. Similarly, taking moments about the other two axes, we get $\tau_{yz} = \tau_{zy}$ and $\tau_{zx} = \tau_{xz}$. Thus, the cross shears are equal, and of the nine rectangular components, only six are independent. The six independent rectangular stress components are $\tau_{xx} = \sigma_x$, $\tau_{yy} = \sigma_y$, $\tau_{zz} = \sigma_z$, τ_{xy}, τ_{yz} and τ_{zx}.

1.9 A MORE GENERAL THEOREM

The fact that cross shears are equal can be used to prove a more general theorem which states that if $\boldsymbol{n}$ and $\boldsymbol{n}'$ define two planes (not necessarily orthogonal but in the limit passing through the same point) with corresponding stress vectors $\overset{n}{\boldsymbol{T}}$ and $\overset{n'}{\boldsymbol{T}}$, then the projection of $\overset{n}{\boldsymbol{T}}$ along $\boldsymbol{n}'$ is equal to the projection of $\overset{n'}{\boldsymbol{T}}$ along $\boldsymbol{n}$, i.e. $\overset{n}{\boldsymbol{T}} \cdot \boldsymbol{n}' = \overset{n'}{\boldsymbol{T}} \cdot \boldsymbol{n}$ (see Fig. 1.11).

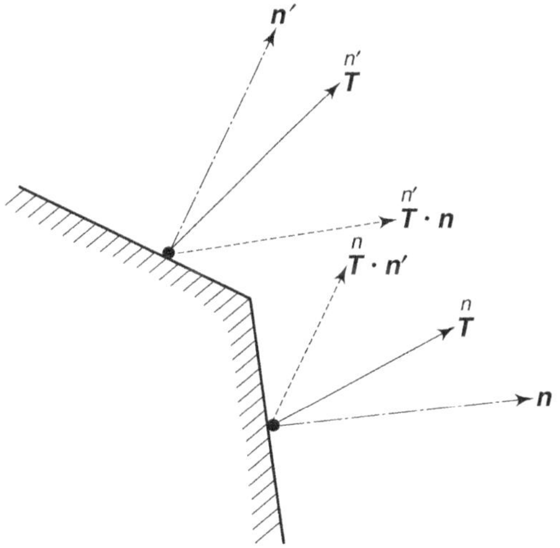

Fig. 1.11 *Planes with normals **n** and **n′***

The proof is straightforward. If n'_x, n'_y and n'_z are the direction cosines of $\boldsymbol{n}'$, then

$$\overset{n}{\boldsymbol{T}} \cdot \boldsymbol{n}' = \overset{n}{T}_x\, n'_x + \overset{n}{T}_y\, n'_y + \overset{n}{T}_z\, n'_z$$

From Eq. (1.9), substituting for $\overset{n}{T}_x$, $\overset{n}{T}_y$ and $\overset{n}{T}_z$ and regrouping normal and shear stresses

$$\overset{n}{\boldsymbol{T}} \cdot \boldsymbol{n}' = \sigma_x\, n_x\, n'_x + \sigma_y\, n_y\, n'_y + \sigma_z\, n_z\, n'_z + \tau_{xy}\, n_x\, n'_y + \tau_{yx}\, n_y\, n'_x +$$
$$\tau_{yz}\, n_y\, n'_z + \tau_{zy}\, n_z\, n'_y + \tau_{zx}\, n_z\, n'_x + \tau_{xz}\, n_x\, n'_z$$

Using the result $\tau_{xy} = \tau_{yx}$, $\tau_{yz} = \tau_{zy}$ and $\tau_{zx} = \tau_{zx}$

$$\overset{n}{\boldsymbol{T}} \cdot \boldsymbol{n}' = \sigma_x n_x n'_x + \sigma_y n_y n'_y + \sigma_z n_z n'_z + \tau_{xy} (n_x n'_y + n_y n'_x) + \tau_{yz} (n_y n'_z + n_z n'_y) + \tau_{zx} (n_z n'_x + n_x n'_z)$$

Similarly,

$$\overset{n'}{\boldsymbol{T}} \cdot \boldsymbol{n} = \sigma_x n_x n'_x + \sigma_y n_y n'_y + \sigma_z n_z n'_z + \tau_{xy} (n_x n'_y + n_y n'_x) + \tau_{yz} (n_y n'_z + n_z n'_y) + \tau_{zx} (n_z n'_x + n_x n'_z)$$

Comparing the above two expressions, we observe

$$\overset{n}{\boldsymbol{T}} \cdot \boldsymbol{n}' = \overset{n'}{\boldsymbol{T}} \cdot \boldsymbol{n} \tag{1.15}$$

Note: An important fact is that cross shears are equal. This can be used to prove that a shear cannot cross a free boundary. For example, consider a beam of rectangular cross-section as shown in Fig. 1.12.

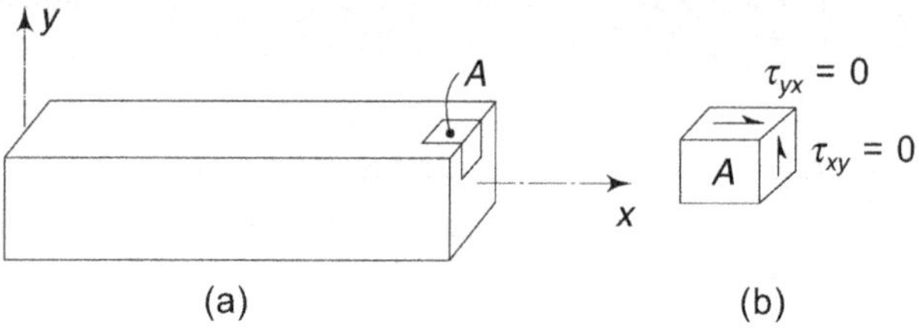

Fig. 1.12 *(a) Element with free surface; (b) Cross shears being zero*

If the top surface is a free boundary, then at point A, the vertical shear stres component $\tau_{xy} = 0$ because if τ_{xy} were not zero, it would call for a complementary shear τ_{yx} on the top surface. But as the top surface is an unloaded or a free surface, τ_{yx} is zero and hence, τ_{xy} is also zero (refer Example 1.3).

1.10 PRINCIPAL STRESSES

We have seen that the normal and shear stress components can be determined on any plane with normal $\boldsymbol{n}$, using Cauchy's formula given by Eqs (1.9). From the strength or failure considerations of materials, answers to the following questions are important:

(i) Are there any planes passing through the given point on which the resultant stresses are wholly normal (in other words, the resultant stress vector is along the normal)?

(ii) What is the plane on which the normal stress is a maximum and what is its magnitude?

(iii) What is the plane on which the tangential or shear stress is a maximum and what it is its magnitude?

Answers to these questions are very important in the analysis of stress, and the next few sections will deal with these. Let us assume that there is a plane $\boldsymbol{n}$ with direction cosines n_x, n_y and n_z on which the stress is wholly normal. Let σ be the magnitude of this stress vector. Then we have

$$\overset{n}{\boldsymbol{T}} = \sigma \boldsymbol{n} \tag{1.16}$$

The components of this along the x, y and z axes are

$$\overset{n}{T}_x = \sigma n_x, \qquad \overset{n}{T}_y = \sigma n_y, \qquad \overset{n}{T}_z = \sigma n_z \tag{1.17}$$

Also, from Cauchy's formula, i.e. Eqs (1.9),

$$\overset{n}{T}_x = \sigma_x n_x + \tau_{xy} n_y + \tau_{xz} n_z$$

$$\overset{n}{T}_y = \tau_{xy} n_x + \sigma_y n_y + \tau_{yz} n_z$$

$$\overset{n}{T}_z = \tau_{xz} n_x + \tau_{yz} n_y + \sigma_z n_z$$

Subtracting Eq. (1.17) from the above set of equations we get

$$\begin{aligned} (\sigma_x - \sigma) n_x + \tau_{xy} n_y + \tau_{xz} n_z &= 0 \\ \tau_{xy} n_x + (\sigma_y - \sigma) n_y + \tau_{yz} n_z &= 0 \\ \tau_{xz} n_x + \tau_{yz} n_y + (\sigma_z - \sigma) n_z &= 0 \end{aligned} \tag{1.18}$$

We can view the above set of equations as three simultaneous equations involving the unknowns n_x, n_y and n_z. These direction cosines define the plane on which the resultant stress is wholly normal. Equation (1.18) is a set of homogeneous equations. The trivial solution is $n_x = n_y = n_z = 0$. For the existence of a non-trivial solution, the determinant of the coefficients of n_x, n_y and n_z must be equal to zero, i.e.

$$\begin{vmatrix} (\sigma_x - \sigma) & \tau_{xy} & \tau_{xz} \\ \tau_{xy} & (\sigma_y - \sigma) & \tau_{yz} \\ \tau_{xz} & \tau_{yz} & (\sigma_z - \sigma) \end{vmatrix} = 0 \tag{1.19}$$

Expanding the above determinant, one gets a cubic equation in σ as

$$\sigma^3 - (\sigma_x + \sigma_y + \sigma_z)\sigma^2 + (\sigma_x \sigma_y + \sigma_y \sigma_z + \sigma_z \sigma_x - \tau_{xy}^2 - \tau_{yz}^2 - \tau_{zx}^2)\,\sigma -$$

$$(\sigma_x \sigma_y \sigma_z + 2\tau_{xy} \tau_{yz} \tau_{zx} - \sigma_x \tau_{yz}^2 - \sigma_y \tau_{xz}^2 - \sigma_z \tau_{xy}^2) = 0 \tag{1.20}$$

The three roots of the cubic equation can be designated as σ_1, σ_2 and σ_3. It will be shown subsequently that all these three roots are real. We shall later give a method (Example 4) to solve the above cubic equation. Substituting any one of these three solutions in Eqs (1.18), we can solve for the corresponding n_x, n_y and n_z. In order to avoid the trivial solution, the condition.

$$n_x^2 + n_y^2 + n_z^2 = 1 \tag{1.21}$$

is used along with any two equations from the set of Eqs (1.18). Hence, with each σ there will be an associated plane. These planes on each of which the stress vector is wholly normal are called the principal planes, and the corresponding stresses, the principal stresses. Since the resultant stress is along the normal, the tangential stress component on a principal plane is zero, and consequently, the principal plane is also known as the shearless plane. The normal to a principal plane is called the principal stress axis.

1.11 STRESS INVARIANTS

The coefficients of σ^2, σ and the last term in the cubic Eq. (1.20) can be written as follows:

$$l_1 = \sigma_x + \sigma_y + \sigma_z \tag{1.22}$$

$$l_2 = \sigma_x\sigma_y + \sigma_y\sigma_z + \sigma_z\sigma_x - \tau_{xy}^2 - \tau_{yz}^2 - \tau_{zx}^2$$

$$= \begin{vmatrix} \sigma_x & \tau_{xy} \\ \tau_{xy} & \sigma_y \end{vmatrix} + \begin{vmatrix} \sigma_y & \tau_{yz} \\ \tau_{yz} & \sigma_z \end{vmatrix} + \begin{vmatrix} \sigma_x & \tau_{xz} \\ \tau_{xz} & \sigma_z \end{vmatrix} \tag{1.23}$$

$$l_3 = \sigma_x\,\sigma_y\,\sigma_z + 2\tau_{xy}\,\tau_{yz}\,\tau_{zx} - \sigma_x\,\tau_{yz}^2 - \sigma_y\,\tau_{zx}^2 - \sigma_z\,\tau_{xy}^2$$

$$= \begin{vmatrix} \sigma_x & \tau_{xy} & \tau_{zx} \\ \tau_{xy} & \sigma_y & \tau_{yz} \\ \tau_{zx} & \tau_{yz} & \sigma_z \end{vmatrix} \tag{1.24}$$

Equation (1.20) can then be written as

$$\sigma^3 - l_1\sigma^2 + l_2\sigma - l_3 = 0$$

The quantities l_1, l_2 and l_3 are known as the first, second and third invariants of stress respectively. An invariant is one whose value does not change when the frame of reference is changed. In other words if x', y', z', is another frame of reference at the same point and with respect to this frame of reference, the rectangular stress competence are $\sigma_{x'}, \sigma_{y'}, \sigma_{z'}, \tau_{x'y'}, \tau_{y'z'}$ and $\tau_{z'x'}$, then the values of l_1, l_2 and l_3, calculated as in Eqs (1.22) – (1.24), will show that

$$\sigma_x + \sigma_y + \sigma_z = \sigma_x' + \sigma_y' + \sigma_z'$$

i.e. $$l_1 = l_1'$$

and similarly, $$l_2 = l_2' \quad \text{and} \quad l_3 = l_3'$$

The reason for this can be explained as follows. The principal stresses at a point depend only on the state of stress at that point and not on the frame of reference describing the rectangular stress components. Hence, if xyz and $x'y'z'$ are two orthogonal frames of reference at the point, then the following cubic equations

$$\sigma^3 - l_1\,\sigma^2 + l_2\,\sigma - l_3 = 0$$

and $$\sigma^3 - l'_1\,\sigma^2 + l'_2\,\sigma - l'_3 = 0$$

must give the same solutions for σ. Since the two systems of axes were arbitrary, the coefficients of σ^2, and σ and the constant terms in the two equations must be equal, i.e.

$$l_1 = l_1', \qquad l_2 = l_2' \qquad \text{and} \qquad l_3 = l_3'$$

In terms of the principal stresses, the invariants are

$$l_1 = \sigma_1 + \sigma_2 + \sigma_3$$

$$l_2 = \sigma_1\sigma_2 + \sigma_2\sigma_3 + \sigma_3\sigma_1$$

$$l_3 = \sigma_1\sigma_2\sigma_3$$

1.12 PRINCIPAL PLANES ARE ORTHOGONAL

The principal planes corresponding to a given state of stress at a point can be shown to be mutually orthogonal. To prove this, we make use of the general theorem in Sec. 1.9. Let $\boldsymbol{n}$ and $\boldsymbol{n}'$ be the two principal planes and σ_1 and σ_2, the corresponding principal stresses. Then the projection of σ_1 in direction $\boldsymbol{n}'$ is equal to the projection of σ_2 in direction $\boldsymbol{n}$, i.e.

$$\sigma_1 \boldsymbol{n}' \cdot \boldsymbol{n} = \sigma_2 \boldsymbol{n} \cdot \boldsymbol{n}' \tag{1.25}$$

If n_x, n_y and n_z are the direction cosines of $\boldsymbol{n}$, and n'_x, n'_y and n'_z those of $\boldsymbol{n}'$, then expanding Eq. (1.25)

$$\sigma_1 \left(n_x n_x' + n_y n_y' + n_z n_z'\right) = \sigma_2 \left(n_x n_x' + n_y n_y' + n_z n_z'\right)$$

Since in general, σ_1 and σ_2 are not equal, the only way the above equation can hold is

$$n_x n_x' + n_y n_y' + n_z n_z' = 0$$

i.e. $\boldsymbol{n}$ and $\boldsymbol{n}'$ are perpendicular to each other. Similarly, considering two other planes $\boldsymbol{n}'$ and $\boldsymbol{n}''$ on which the principal stresses σ_2 and σ_3 are acting, and following the same argument as above, one finds that $\boldsymbol{n}'$ and $\boldsymbol{n}''$ are perpendicular to each other. Similarly, $\boldsymbol{n}$ and $\boldsymbol{n}''$ are perpendicular to each other. Consequently, the principal planes are mutually perpendicular.

1.13 CUBIC EQUATION HAS THREE REAL ROOTS

In Sec. 1.10, it was stated that Eq. (1.20) has three real roots. The proof is as follows. Dividing Eq. (1.20) by σ^2,

$$\sigma - l_1 + \frac{l_2}{\sigma} - \frac{l_3}{\sigma^2} = 0$$

For appropriate values of σ, the quantity on the left-hand side will be equal to zero. For other values, the quantity will not be equal to zero and one can write the above function as

$$\sigma - l_1 + \frac{l_2}{\sigma} - \frac{l_3}{\sigma^2} = f(\sigma) \tag{1.26}$$

Since l_1, l_2 and l_3 are finite, $f(\sigma)$ can be made positive for large positive values of σ. Similarly, $f(\sigma)$ can be made negative for large negative values of σ. Hence, if one

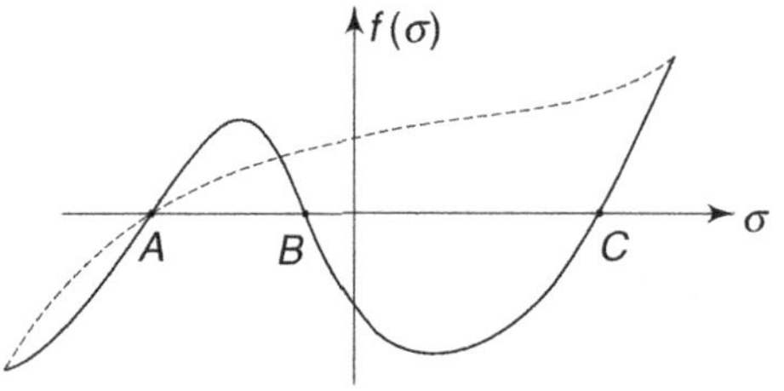

Fig. 1.13 *Plot of $f(\sigma)$ versus σ*

plots $f(\sigma)$ for different values of σ as shown in Fig. 1.13, the curve must cut the σ axis at least once as shown by the dotted curve and for this value of $\sigma, f(\sigma)$ will be equal to zero. Therefore, there is at least one real root.

Let σ_3 be this root and ***n*** the associated plane. Since the state of stress at the point can be characterised by the six rectangular components referred to any orthogonal frame of reference, let us choose a particular one, $x'y'z'$, where the z' axis is along ***n*** and the other two axes, x' and y', are arbitrary. With reference to this system, the stress matrix has the form.

$$\begin{bmatrix} \sigma_{x'} & \tau_{x'y'} & 0 \\ \tau_{x'y'} & \sigma_{y'} & 0 \\ 0 & 0 & \sigma_3 \end{bmatrix} \tag{1.27}$$

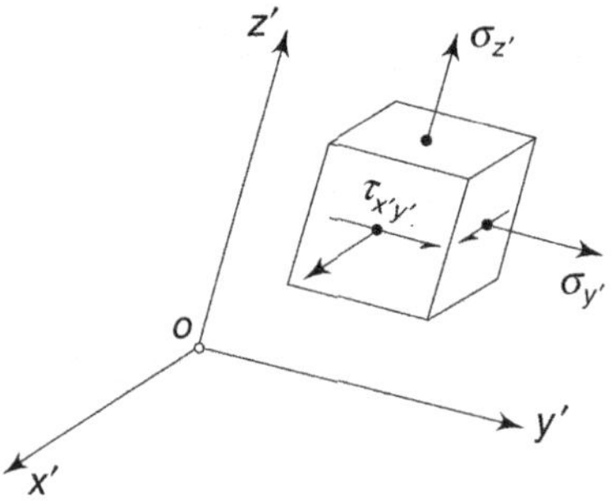

Fig. 1.14 *Rectangular element with faces normal to x', y', z' axes*

Figure 1.14 shows these stress vectors on a rectangular element. The shear stress components $\tau_{x'z'}$ and $\tau_{y'z'}$ are zero since the z' plane is chosen to be the principal plane. With reference to this system, Eq. (1.19) becomes

$$\begin{vmatrix} (\sigma_{x'} - \sigma) & \tau_{x'y'} & 0 \\ \tau_{x'y'} & (\sigma_{y'} - \sigma) & 0 \\ 0 & 0 & (\sigma_3 - \sigma) \end{vmatrix} = 0 \tag{1.28}$$

Expanding $\quad (\sigma_3 - \sigma)\,[\sigma^2 - (\sigma_{x'} + \sigma_{y'})\,\sigma + \sigma_{x'}\sigma_{y'} - \tau^2_{x'y'}] = 0$

This is a cubic in σ. One of the solutions is $\sigma = \sigma_3$. The two other solutions are obtained by solving the quadratic inside the brakets. The two solutions are

$$\sigma_{1,2} = \frac{\sigma_{x'} + \sigma_{y'}}{2} \pm \left[\left(\frac{\sigma_{x'} - \sigma_{y'}}{2} \right)^2 + \tau^2_{x'y'} \right]^{\frac{1}{2}} \tag{1.29}$$

The quantity under the square root $\left(\text{power } \frac{1}{2}\right)$ is never negative and hence, σ_1 and σ_2 are also real. This means that the curve for $f(\sigma)$ in Fig. 1.13 will cut the σ axis at three points A, B and C in general. In the next section we shall study a few particular cases.

1.14 PARTICULAR CASES

(i) If σ_1, σ_2 and σ_3 are distinct, i.e. σ_1, σ_2 and σ_3 have different values, then the three associated principal axes $\boldsymbol{n}_1$, $\boldsymbol{n}_2$ and $\boldsymbol{n}_3$ are unique and mutually perpendicular. This follows from Eq. (1.25) of Sec. 1.12. Since σ_1, σ_2 and σ_3 are distinct, we get three distinct axes $\boldsymbol{n}_1$, $\boldsymbol{n}_2$ and $\boldsymbol{n}_3$ from Eqs (1.18), and being mutually perpendicular they are unique.

(ii) If $\sigma_1 = \sigma_2$ and σ_3 is distinct, the axis of $\boldsymbol{n}_3$ is unique and every direction perpendicular to $\boldsymbol{n}_3$ is a principal direction associated with $\sigma_1 = \sigma_2$. This is shown in Fig. 1.15.

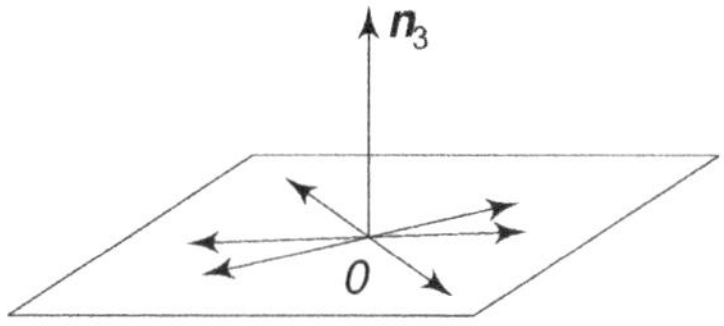

Fig. 1.15 *Case with $\sigma_1 = \sigma_2$ and σ_3 distinct*

To prove this, let us choose a frame of reference $Ox'y'z'$ such that the z' axis is along $\boldsymbol{n}_3$ and the x' and y' axes are arbitrary. From Eq. (1.29), if $\sigma_1 = \sigma_2$, then the quantity under the radical must be zero. Since this is the sum of two squared quantities, this can happen only if

$$\sigma_{x'} = \sigma_{y'} \quad \text{and} \quad \tau_{x'y'} = 0$$

But we have chosen x' and y' axes arbitrarily, and consequently the above condition must be true for any frame of reference with the z' axis along $\boldsymbol{n}_3$. Hence, the x' and y' planes are shearless planes, i.e. principal planes. Therefore, every direction perpendicular to $\boldsymbol{n}_3$ is a principal direction associated with $\sigma_1 = \sigma_2$.

(iii) If $\sigma_1 = \sigma_2 = \sigma_3$, then every direction is a principal direction. This is the hydrostatic or the isotropic state of stress and was discussed in Sec. 1.7. For proof, we can repeat the argument given in (ii). Choose a coordinate system $Ox'y'z'$ with the z' axis along $\boldsymbol{n}_3$ corresponding to σ_3. Since $\sigma_1 = \sigma_2$ every direction perpendicular to $\boldsymbol{n}_3$ is a principal direction. Next, choose the z' axis parallel to $\boldsymbol{n}_2$ corresponding to σ_2. Then every direction perpendicular to $\boldsymbol{n}_2$ is a principal direction since $\sigma_1 = \sigma_3$. Similarly, if we choose the z' axis parallel to $\boldsymbol{n}_1$ corresponding to σ_1, every direction perpendicular to $\boldsymbol{n}_1$ is also a principal direction. Consequently, every direction is a principal direction.

Another proof could be in the manner described in Sec. 1.7. Choosing $Oxyz$ coinciding with $\boldsymbol{n}_1$, $\boldsymbol{n}_2$ and $\boldsymbol{n}_3$, the stress vector on any arbitrary plane $\boldsymbol{n}$ has value σ, the direction of σ coinciding with $\boldsymbol{n}$. Hence, every plane is a principal plane. Such a state of stress is equivalent to a hydrostatic state of stress or an isotropic state of stress.

1.15 RECAPITULATION

The material discussed in the last few sections is very important and it is worthwhile to put it in the form of definitions and theorems.

Definition

For a given state of stress at point P, if the resultant stress vector $\overset{n}{T}$ on any plane n is along n having a magnitude σ, then σ is a principal stress at P, n is the principal direction associated with σ, the axis of σ is a principal axis, and the plane is a principal plane at P.

Theorem

In every state of stress there exist at least three mutually perpendicular principal axes and at most three distinct principal stresses. The principal stresses σ_1, σ_2 and σ_3 are the roots of the cubic equation

$$\sigma^3 - I_1\sigma^2 + I_2\sigma - I_3 = 0$$

where I_1, I_2 and I_3 are the first, second and third invariants of stress. The principal directions associated with σ_1, σ_2 and σ_3 are obtained by substituting σ_i $(i = 1, 2, 3)$ in the following equations and solving for n_x, n_y and n_z:

$$(\sigma_x - \sigma_i)\, n_x + \tau_{xy}\, n_y + \tau_{xz}\, n_z = 0$$

$$\tau_{xy}\, n_x + (\sigma_y - \sigma_i)\, n_y + \tau_{yz}\, n_z = 0$$

$$n_x^2 + n_y^2 + n_z^2 = 1$$

If σ_1, σ_2 and σ_3 are distinct, then the axes of n_1, n_2 and n_3 are unique and mutually perpendicular. If, say $\sigma_1 = \sigma_2 \neq \sigma_3$, then the axis of n_3 is unique and every direction perpendicular to n_3 is a principal direction associated with $\sigma_1 = \sigma_2$. If $\sigma_1 = \sigma_2 = \sigma_3$, then every direction is a principal direction.

Standard Method of Solution

Consider the cubic equation $y^3 + py^2 + qy + r = 0$, where p, q and r are constants.

Substitute $$y = x - \frac{1}{3}p$$

This gives $$x^3 + ax + b = 0$$

where $$a = \frac{1}{3}\left(3q - p^2\right), \qquad b = \frac{1}{27}\left(2p^3 - 9pq + 27r\right)$$

Put $$\cos\phi = -\frac{b}{2\left(-\dfrac{a^3}{27}\right)^{1/2}}$$

Determine ϕ, and putting $g = 2\sqrt{-a/3}$, the solutions are

$$y_1 = g\cos\frac{\phi}{3} - \frac{p}{3}$$

$$y_2 = g\cos\left(\frac{\phi}{3} + 120°\right) - \frac{p}{3}$$

$$y_3 = g\cos\left(\frac{\phi}{3} + 240°\right) - \frac{p}{3}$$

Example 1.4 *At a point P, the rectangular stress components are*

$$\sigma_x = 1, \sigma_y = -2, \quad \sigma_z = 4, \quad \tau_{xy} = 2, \quad \tau_{yz} = -3, \quad \text{and} \quad \tau_{xz} = 1$$

all in units of kPa. Find the principal stresses and check for invariance.

Solution The given stress matrix is

$$[\tau_{ij}] = \begin{bmatrix} 1 & 2 & 1 \\ 2 & -2 & -3 \\ 1 & -3 & 4 \end{bmatrix}$$

From Eqs (1.22)–(1.24),

$$I_1 = 1 - 2 + 4 = 3$$
$$I_2 = (-2 - 4) + (-8 - 9) + (4 - 1) = -20$$
$$I_3 = 1(-8 - 9) - 2(8 + 3) + 1(-6 + 2) = -43$$

$$\therefore \quad f(\sigma) = \sigma^3 - 3\sigma^2 - 20\sigma + 43 = 0$$

For this cubic, following the standard method,

$$y = \sigma, \quad p = -3, \quad q = -20, \quad r = 43$$

$$a = \frac{1}{3}(-60 - 9) = -23$$

$$b = \frac{1}{27}(-54 - 540 + 1161) = 21$$

$$\cos\phi = -\frac{\left(\frac{21}{2}\right)}{\left(\frac{12167}{27}\right)^{1/2}}$$

$$\therefore \quad \phi = -119° \, 40$$

The solutions are

$$\sigma_1 = y_1 = 4.25 + 1 = 5.25 \text{ kPa}$$
$$\sigma_2 = y_2 = -5.2 + 1 = -4.2 \text{ kPa}$$
$$\sigma_3 = y_3 = 0.95 + 1 = 1.95 \text{ kPa}$$

Renaming such that $\sigma_1 \geq \sigma_2 \geq \sigma_3$ we have,

$$\sigma_1 = 5.25 \text{ kPa}, \quad \sigma_2 = 1.95 \text{ kPa}, \quad \sigma_3 = -4.2 \text{ kPa}$$

The stress invariants are

$$I_1 = 5.25 + 1.95 - 4.2 = 3.0$$
$$I_2 = (5.25 \times 1.95) - (1.95 \times 4.2) - (4.2 \times 5.25) = -20$$
$$I_3 = -(5.25 \times 1.95 \times 4.2) = -43$$

These agree with their earlier values.

Example 1.5 *With respect to the frame of reference* Oxyz, *the following state of stress exists. Determine the principal stresses and their associated directions. Also, check on the invariances of* I_1, I_2, I_3.

$$[\tau_{ij}] = \begin{bmatrix} 1 & 2 & 1 \\ 2 & 1 & 1 \\ 1 & 1 & 1 \end{bmatrix}$$

Solution For this state

$$I_1 = 1 + 1 + 1 = 3$$
$$I_2 = (1 - 4) + (1 - 1) + (1 - 1) = -3$$
$$I_3 = 1(1 - 1) - 2(2 - 1) + 1(2 - 1) = -1$$
$$f(\sigma) = \sigma^3 - I_1\,\sigma^2 + I_2\,\sigma - I_3 = 0$$

i.e., $\sigma^3 - 3\sigma^2 - 3\sigma + 1 = 0$

or $(\sigma^3 + 1) - 3\sigma(\sigma + 1) = 0$

i.e., $(\sigma + 1)(\sigma^2 - \sigma + 1) - 3\sigma(\sigma + 1) = 0$

or $(\sigma + 1)(\sigma^2 - 4\sigma + 1) = 0$

Hence, one solution is $\sigma = -1$. The other two solutions are obtained from the solution of the quadratic equation, which are $\sigma = 2 \pm \sqrt{3}$.

$\therefore$ $\sigma_1 = -1, \quad \sigma_2 = 2 + \sqrt{3}, \quad \sigma_3 = 2 - \sqrt{3}$

Check on the invariance:
With the set of axes chosen along the principal axes, the stress matrix will have the form

$$[\tau_{ij}] = \begin{bmatrix} -1 & 0 & 0 \\ 0 & 2+\sqrt{3} & 0 \\ 0 & 0 & 2-\sqrt{3} \end{bmatrix}$$

Hence, $I_1 = -1 + 2 + \sqrt{3} + 2 - \sqrt{3} = 3$

$$I_2 = (-2 - \sqrt{3}) + (4 - 3) + (-2 + \sqrt{3}) = -3$$
$$I_3 = -1\,(4 - 3) = -1$$

Directions of principal axes:

(i) For $\sigma_1 = -1$, from Eqs (1.18) and (1.21)

$$(1 + 1)n_x + 2n_y + n_z = 0$$
$$2n_x + (1 + 1)n_y + n_z = 0$$
$$n_x + n_y + (1 + 1)n_z = 0$$

together with

$$n_x^2 + n_y^2 + n_z^2 = 1$$

From the second and third equations above, $n_z = 0$. Using this in the third and fourth equations and solving, $n_x = \pm\left(1/\sqrt{2}\right)$, $n_y = \pm\left(1/\sqrt{2}\right)$.

Hence, $\sigma_1 = -1$ is in the direction $\left(+1/\sqrt{2}, -1/\sqrt{2}, 0\right)$.

It should be noted that the plus and minus signs associated with n_x, n_y and n_z represent the same line.

(ii) For $\sigma_2 = 2 + \sqrt{3}$

$$(-1-\sqrt{3})n_x + 2n_y + n_z = 0$$
$$2n_x + (-1-\sqrt{3})n_y + n_z = 0$$
$$n_x + n_y(-1-\sqrt{3})n_z = 0$$

together with

$$n_x^2 + n_y^2 + n_z^2 = 1$$

Solving, we get

$$n_x = n_y = \left(1 + \frac{1}{\sqrt{3}}\right)^{1/2} \qquad n_z = \frac{1}{\left(3+\sqrt{3}\right)^{1/2}}$$

(iii) For $\sigma_3 = 2 - \sqrt{3}$

We can solve for n_x, n_y and n_z in a manner similar to the preceding one or get the solution from the condition that $\boldsymbol{n}_1$, $\boldsymbol{n}_2$ and $\boldsymbol{n}_3$ form a right-angled triad, i.e. $\boldsymbol{n}_3 = \boldsymbol{n}_1 \times \boldsymbol{n}_2$.

The solution is

$$n_x = n_y = -\frac{1}{2}\left(1 - \frac{1}{\sqrt{3}}\right)^{1/2}, \quad n_z = \frac{1}{\sqrt{2}}\left(1 + \frac{1}{\sqrt{3}}\right)^{1/2}$$

Example 1.6 *For the given state of stress, determine the principal stresses and their directions.*

$$[\tau_{ij}] = \begin{bmatrix} 0 & 1 & 1 \\ 1 & 0 & 1 \\ 1 & 1 & 0 \end{bmatrix}$$

Solution

$$I_1 = 0, I_2 = -3, I_3 = 2$$
$$f(\sigma) = -\sigma^3 + 3\sigma + 2 = 0$$
$$= (-\sigma^3 - 1) + (3\sigma + 3)$$
$$= -(\sigma + 1)(\sigma^2 - \sigma + 1) + 3(\sigma + 1)$$
$$= (\sigma + 1)(\sigma - 2)(\sigma + 1) = 0$$

$\therefore$ $\qquad \sigma_1 = \sigma_2 = -1 \quad \text{and} \quad \sigma_3 = 2$

Since two of the three principal stresses are equal, and σ_3 is different, the axis of σ_3 is unique and every direction perpendicular to σ_3 is a principal direction associated with $\sigma_1 = \sigma_2$. For $\sigma_3 = 2$

$$-2n_x + n_y + n_z = 0$$
$$n_x - 2n_y + n_z = 0$$

$$n_x + n_y - 2n_z = 0$$
$$n_x^2 + n_y^2 + n_z^2 = 1$$

These give $n_x = n_y = n_z = \dfrac{1}{\sqrt{3}}$

Example 1.7 *The state of stress at a point is such that*

$$\sigma_x = \sigma_y = \sigma_z = \tau_{xy} = \tau_{yz} = \tau_{zx} = \rho$$

Determine the principal stresses and their directions

Solution For the given state,

$$I_1 = 3\rho, \qquad I_2 = 0, \qquad I_3 = 0$$

Therefore the cubic is $\sigma^3 - 3\rho\sigma^2 = 0$; the solutions are $\sigma_1 = 3\rho$, $\sigma_2 = \sigma_3 = 0$. For $\sigma_1 = 3\rho$

$$(\rho - 3\rho)n_x + \rho n_y + \rho n_z = 0$$
$$\rho n_x + (\rho - 3\rho)\, n_y + \rho n_z = 0$$
$$\rho n_x + \rho n_y + (\rho - 3\rho)\, n_z = 0$$

or

$$-2n_x + n_y + n_z = 0$$
$$n_x - 2n_y + n_z = 0$$
$$n_x + n_y - 2n_z = 0$$

The above equations give

$$n_x = n_y = n_z$$

With $n_x^2 + n_y^2 + n_z^2 = 1$, one gets $n_x = n_y = n_z = 1/\sqrt{3}$.

Thus, on a plane that is equally inclined to *xyz* axes, there is a tensile stress of magnitue 3ρ. This is the case of a uniaxial tension, the axis of loading making equal angles with the given *xyz* axes. If one denotes this loading axis by z', the other two axes, x' and y', can be chosen arbitrarily, and the planes normal to these, i.e. x' plane and y' plane, are stress free.

1.16 THE STATE OF STRESS REFERRED TO PRINCIPAL AXES

In expressing the state of stress at a point by the six rectangular stress components, we can choose the principal axes as the coordinate axes and refer the rectangular stress components accordingly. We then have for the stress matrix

$$[\tau_{ij}] = \begin{bmatrix} \sigma_1 & 0 & 0 \\ 0 & \sigma_2 & 0 \\ 0 & 0 & \sigma_3 \end{bmatrix} \tag{1.30}$$

On any plane with normal $\boldsymbol{n}$, the components of the stress vector are, from Eq. (1.9),

$$\overset{n}{T}_x = \sigma_1 n_x, \qquad \overset{n}{T}_y = \sigma_2 n_y, \qquad \overset{n}{T}_z = \sigma_3 n_z \tag{1.31}$$

The resultant stress has a magnitutde

$$\left|\overset{n}{\boldsymbol{T}}\right|^2 = \sigma_1^2 n_x^2 + \sigma_2^2 n_y^2 + \sigma_3^2 n_z^2 \tag{1.32}$$

If σ is the normal and τ the shearing stress on this plane, then

$$\sigma = \sigma_1 n_x^2 + \sigma_2 n_y^2 + \sigma_3 n_z^2 \tag{1.33}$$

and $$\tau^2 = \left|\overset{n}{\boldsymbol{T}}\right|^2 - \sigma^2 \tag{1.34}$$

$$= n_x^2 n_y^2 (\sigma_1 - \sigma_2)^2 + n_y^2 n_z^2 (\sigma_2 - \sigma_3)^2 + n_z^2 n_x^2 (\sigma_3 - \sigma_1)^2$$

The stress invariants assume the form

$$\begin{aligned} I_1 &= \sigma_1 + \sigma_2 + \sigma_3 \\ I_2 &= \sigma_1\sigma_2 + \sigma_2\sigma_3 + \sigma_3\sigma_1 \\ I_3 &= \sigma_1\sigma_2\sigma_3 \end{aligned} \tag{1.35}$$

1.17 MOHR'S CIRCLES FOR THE THREE-DIMENSIONAL STATE OF STRESS

We shall now describe a geometrical construction that brings out some important results. At a given point P, let the frame of reference $Pxyz$ be chosen along the principal stress axes. Consider a plane with normal $\boldsymbol{n}$ at point P. Let σ be the normal stress and τ the shearing stress on this plane. Take another set of axes σ and τ. In this plane we can mark a point Q with co-ordinates (σ, τ) representing the values of the normal and shearing stress on the plane $\boldsymbol{n}$, Fig. 1.16. For different planes passing through point P, we get different values of σ and τ. Corresponding to each plane $\boldsymbol{n}$, a point Q can be located with coordinates (σ, τ). The plane with the σ axis and the τ axis is called the stress plane π. (No numerical value is associated with this symbol). The problem now is to determine the bounds for Q (σ, τ) for all possible directions $\boldsymbol{n}$.

Arrange the principal stresses such that algebraically

$$\sigma_1 \geq \sigma_2 \geq \sigma_3$$

Mark off σ_1, σ_2 and σ_3 along the σ axis and construct three circles with diameters $(\sigma_1 - \sigma_2)$, $(\sigma_2 - \sigma_3)$ and $(\sigma_1 - \sigma_3)$ as shown in Fig. 1.16.

It will be shown in Sec. 1.18 that the point $Q(\sigma, \tau)$ for all possible $\boldsymbol{n}$ will lie within the shaded area. This region is called Mohr's stress plane π and the three circles are known as Mohr's circles. From Fig. 1.16, the following points can be observed:

(i) Points A, B and C represent the three principal stresses and the associated shear stresses are zero.

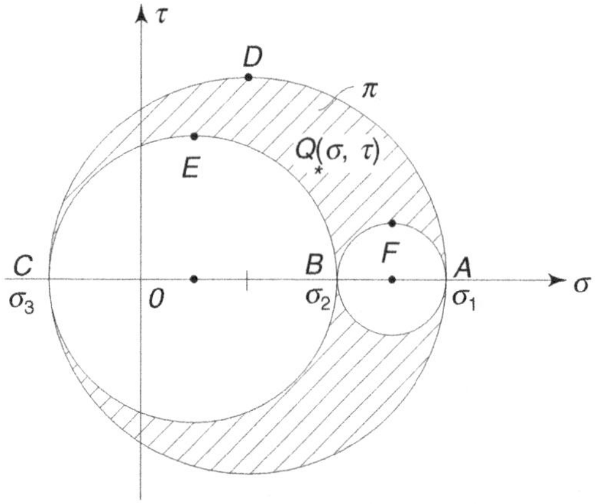

Fig. 1.16 *Mohr's stress plane*

(ii) The maximum shear stress is equal to $\frac{1}{2}(\sigma_1 - \sigma_3)$ and the associated normal stress is $\frac{1}{2}(\sigma_1 + \sigma_3)$. This is indicated by point D on the outer circle.

(iii) Just as there are three extremum values σ_1, σ_2 and σ_3 for the normal stresses, there are three extremum values for the shear stresses, these being $\frac{\sigma_1 - \sigma_3}{2}$, $\frac{\sigma_2 - \sigma_3}{2}$ and $\frac{\sigma_1 - \sigma_2}{2}$. The planes on which these shear stresses act are called the principal shear planes. While the planes on which the principal normal stresses act are free of shear stresses, the principal shear planes are not free from normal stresses. The normal stresses associated with the principal shears are respectively $\frac{\sigma_1 + \sigma_3}{2}$, $\frac{\sigma_2 + \sigma_3}{2}$ and $\frac{\sigma_1 + \sigma_2}{2}$. These are indicated by points D, E and F in Fig. 1.16. It will be shown in Sec. 1.19 that the principal shear planes are at 45° to the principal normal planes. The principal shears are denoted by τ_1, τ_2 and τ_3 where

$$2\tau_3 = (\sigma_1 - \sigma_2), \qquad 2\tau_2 = (\sigma_1 - \sigma_3), \qquad 2\tau_1 = (\sigma_2 - \sigma_3) \tag{1.36}$$

(iv) When $\sigma_1 = \sigma_2 \neq \sigma_3$ or $\sigma_1 \neq \sigma_2 = \sigma_3$, the three circles reduce to only one circle and the shear stress on any plane will not exceed $\frac{1}{2}(\sigma_1 - \sigma_3)$ or $\frac{1}{2}(\sigma_1 - \sigma_2)$ according as $\sigma_1 = \sigma_2$ or $\sigma_2 = \sigma_3$.

(v) When $\sigma_1 = \sigma_2 = \sigma_3$, the three circles collapse to a single point on the σ axis and every plane is a shearless plane.

1.18 MOHR'S STRESS PLANE

It was stated in the previous section that when points with coordinates (σ, τ) for all possible planes passing through a point are marked on the $\sigma - \tau$ plane, as in Fig. 1.16, the points are bounded by the three Mohr's circles. In this, section we shall prove this.

Choose the coordinate frame of reference *Pxyz* such that the axes are along the principal axes. On any plane with normal $\boldsymbol{n}$, the resultant stress vector $\overset{n}{\boldsymbol{T}}$ and the normal stress σ are such that from Eqs (1.32) and (1.33)

$$\left|\overset{n}{\boldsymbol{T}}\right|^2 = \sigma^2 + \tau^2 = \sigma_1^2 n_x^2 + \sigma_2^2 n_y^2 + \sigma_3^2 n_z^2 \tag{1.37}$$

$$\sigma = \sigma_1 n_x^2 + \sigma_2 n_y^2 + \sigma_3 n_z^2 \tag{1.38}$$

and also

$$1 = n_x^2 + n_y^2 + n_z^2 \tag{1.39}$$

The above three equations can be used to solve for n_x^2, n_y^2 and n_z^2 yielding

$$n_x^2 = \frac{(\sigma - \sigma_2)(\sigma - \sigma_3) + \tau^2}{(\sigma_1 - \sigma_2)(\sigma_1 - \sigma_3)} \tag{1.40}$$

$$n_y^2 = \frac{(\sigma - \sigma_3)(\sigma - \sigma_1) + \tau^2}{(\sigma_2 - \sigma_3)(\sigma_2 - \sigma_1)} \tag{1.41}$$

$$n_z^2 = \frac{(\sigma - \sigma_1)(\sigma - \sigma_2) + \tau^2}{(\sigma_3 - \sigma_1)(\sigma_3 - \sigma_2)} \tag{1.42}$$

Since n_x^2, n_y^2 and n_z^2 are all positive, the right-hand side expressions in the above equations must all be positive. Recall that we have arranged the principal stresses such that $\sigma_1 \geq \sigma_2 \geq \sigma_3$. There are three cases one can consider.

Case (i) $\sigma_1 > \sigma_2 > \sigma_3$

Case (ii) $\sigma_1 = \sigma_2 > \sigma_3$

Case (iii) $\sigma_1 = \sigma_2 = \sigma_3$

We shall consider these cases individually.

Case (i) $\sigma_1 > \sigma_2 > \sigma_3$

For this case, the denominator in Eq. (1.40) is positive and hence, the numerator must also be positive. In Eq. (1.41), the denominator being negative, the numerator must also be negative. Similarly, the numerator in Eq. (1.42) must be positive. Therefore.

$$(\sigma - \sigma_2)(\sigma - \sigma_3) + \tau^2 \geq 0$$
$$(\sigma - \sigma_3)(\sigma - \sigma_1) + \tau^2 \leq 0$$
$$(\sigma - \sigma_1)(\sigma - \sigma_2) + \tau^2 \geq 0$$

The above three inequalities can be rewritten as

$$\tau^2 + \left(\sigma - \frac{\sigma_2 + \sigma_3}{2}\right)^2 \geq \left(\frac{\sigma_2 - \sigma_3}{2}\right)^2$$

$$\tau^2 + \left(\sigma - \frac{\sigma_3 + \sigma_1}{2}\right)^2 \leq \left(\frac{\sigma_3 - \sigma_1}{2}\right)^2$$

$$\tau^2 + \left(\sigma - \frac{\sigma_1 + \sigma_2}{2}\right)^2 \geq \left(\frac{\sigma_1 - \sigma_2}{2}\right)^2$$

According to the first of the above equations, the point (σ, τ) must lie on or outside a circle of radius $\frac{1}{2}(\sigma_2 - \sigma_3)$ with its centre at $\frac{1}{2}(\sigma_2 + \sigma_3)$ along the σ axis (Fig. 1.16). This is the circle with *BC* as diameter. The second equation indicates that the point (σ, τ) must lie inside or on the circle *ADC* with radius $\frac{1}{2}(\sigma_1 - \sigma_3)$ and centre at $\frac{1}{2}(\sigma_1 + \sigma_3)$ on the σ axis. Similarly, the last equation indicates that the point (σ, τ) must lie on or outside the circle *AFB* with radius equal to $\frac{1}{2}(\sigma_1 - \sigma_2)$ and centre at $\frac{1}{2}(\sigma_1 + \sigma_2)$.

Hence, for this case, the point $Q(\sigma, \tau)$ should lie inside the shaded area of Fig. 1.16.

Case (ii) $\sigma_1 = \sigma_2 > \sigma_3$

Following arguments similar to the ones given above, one has for this case from Eqs (1.40)–(1.42)

$$\tau^2 + \left(\sigma - \frac{\sigma_2 + \sigma_3}{2}\right)^2 = \left(\frac{\sigma_2 - \sigma_3}{2}\right)^2$$

$$\tau^2 + \left(\sigma - \frac{\sigma_3 + \sigma_1}{2}\right)^2 = \left(\frac{\sigma_3 - \sigma_1}{2}\right)^2$$

$$\tau^2 + \left(\sigma - \frac{\sigma_1 + \sigma_2}{2}\right)^2 \geq \left(\frac{\sigma_1 - \sigma_2}{2}\right)^2$$

From the first two of these equations, since $\sigma_1 = \sigma_2$, point (σ, τ) must lie on the circle with radius $\frac{1}{2}(\sigma_1 - \sigma_3)$ with its centre at $\frac{1}{2}(\sigma_1 + \sigma_3)$. The last equation indicates that the point must lie outside a circle of zero radius (since $\sigma_1 = \sigma_2$). Hence, in this case, the Mohr's circles will reduce to a circle *BC* and a point circle *B*. The point *Q* lies on the circle *BEC*.

Case (iii) $\sigma_1 = \sigma_2 = \sigma_3$

This is a trivial case since this is the isotropic or the hydrostatic state of stress. Mohr's circles collapse to a single point on the σ axis.

See Appendix 1 for the graphical determination of the normal and shear stresses on an arbitrary plane, using Mohr's circles.

1.19 PLANES OF MAXIMUM SHEAR

From Sec. 1.17 and also from Fig. 1.16 for the case $\sigma_1 > \sigma_2 > \sigma_3$, the maximum shear stress is $\frac{1}{2}(\sigma_1 - \sigma_3) = \tau_2$ and the associated normal stress is $\frac{1}{2}(\sigma_1 + \sigma_3)$.

Substituting these values in Eqs.(1.37)–(1.39) in Sec. 1.18, one gets $n_x = \pm\sqrt{1/2}$, $n_y = 0$ and $n_z = \pm 1\sqrt{2}$. This means that the planes (there are two of them) on which the shear stress takes on an extremum value, make angles of 45° and 135° with the σ_1 and σ_2 planes as shown in Fig. 1.17.

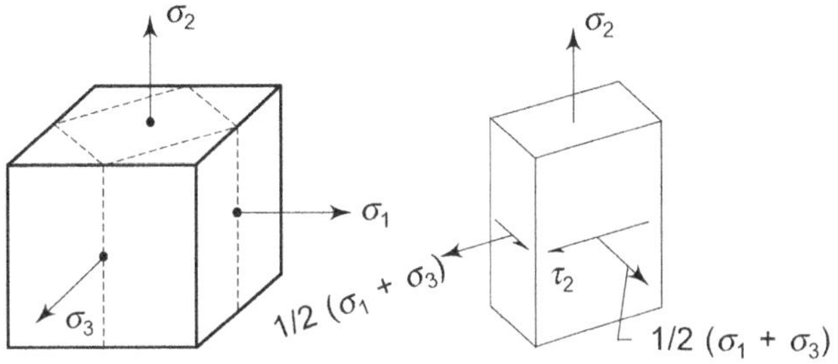

(a) (b)

Fig. 1.17 *(a) Principal planes (b) Planes of maximum shear*

If $\sigma_1 = \sigma_2 > \sigma_3$, then the three Mohr's circles reduce to one circle BC (Fig.1.16) and the maximum shear stress will be $\frac{1}{2}(\sigma_2 - \sigma_3) = \tau_1$, with the associated normal stress $\frac{1}{2}(\sigma_2 + \sigma_3)$. Substituting these values in Eqs (1.37)–(1.39), we get $n_x = 0/0$, $n_y = 0/0$ and $n_z = \pm 1\sqrt{2}$ i.e. n_x and n_y are indeterminate. This means that the planes on which τ_1 is acting makes angles of 45° and 135° with the σ_3 axis but remains indeterminate with respect to σ_1 and σ_2 axes. This is so because, since $\sigma_1 = \sigma_2 \neq \sigma_3$, the axis of σ_3 is unique, whereas, every direction perpendicular to σ_3 is a principal direction associated with $\sigma_1 = \sigma_2$ (Sec. 1.14). The principal shear plane will, therefore, make a fixed angle with σ_3 axis (45° or 135°) but will have different values depending upon the selection of σ_1 and σ_3 axes.

1.20 OCTAHEDRAL STRESSES

Let the frame of reference be again chosen along σ_1, σ_2 and σ_3 axes. A plane that is equally inclined to these three axes is called an octahedral plane. Such a plane will have $n_x = n_y = n_z$. Since $n_x^2 + n_y^2 + n_z^2 = 1$, an octahedral plane will be defined by $n_x = n_y = n_z = \pm 1/\sqrt{3}$. There are eight such planes, as shown in Fig.1.18.

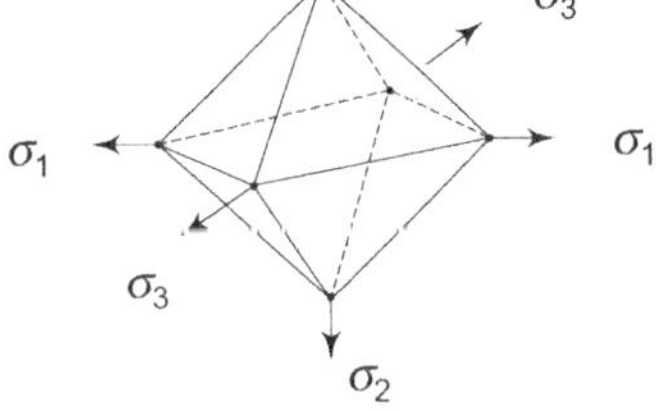

Fig.1.18 *Octahedral planes*

The normal and shearing stresses on these planes are called the octahedral normal stress and octahedral shearing stress respectively. Substituting $n_x = n_y = n_z = \pm 1/\sqrt{3}$ in Eqs (1.33) and (1.34),

$$\sigma_{\text{oct}} = \frac{1}{3}(\sigma_1 + \sigma_2 + \sigma_3) = \frac{1}{3}\, I_1 \tag{1.43}$$

and
$$\tau_{\text{oct}}^2 = \frac{1}{9}[(\sigma_1 - \sigma_2)^2 + (\sigma_2 - \sigma_3)^2 + (\sigma_3 - \sigma_1)^2] \tag{1.44a}$$

or
$$9\tau_{\text{oct}}^2 = 2(\sigma_1 + \sigma_2 + \sigma_3)^2 - 6(\sigma_1\sigma_2 + \sigma_2\sigma_3 + \sigma_3\sigma_1) \tag{1.44b}$$

or
$$\tau_{\text{oct}} = \frac{\sqrt{2}}{3}(I_1^2 - 3I_2)^{1/2} \tag{1.44c}$$

It is important to remember that the octahedral planes are defined with respect to the principal axes and not with reference to an arbitrary frame of reference. Since σ_{oct} and τ_{oct} have been expressed in terms of the stress invariants, one can express these in terms of σ_x, σ_y, σ_z, τ_{xy}, τ_{yz} and τ_{zx} also. Using Eqs (1.22) and (1.23),

$$\sigma_{\text{oct}} = \frac{1}{3}(\sigma_x + \sigma_y + \sigma_z) \tag{1.45}$$

$$9\tau_{\text{oct}}^2 = (\sigma_x - \sigma_y)^2 + (\sigma_y - \sigma_z)^2 + (\sigma_z - \sigma_x)^2 + 6\left(\tau_{xy}^2 + \tau_{yz}^2 + \tau_{zx}^2\right) \tag{1.46}$$

The octahedral normal stress being equal to 1/3 I_1, it may be interpreted as the mean normal stress at a given point in a body. If in a state of stress, the first invariant $(\sigma_1 + \sigma_2 + \sigma_3)$ is zero, then the normal stresses on the octahedral planes will be zero and only the shear stresses will act. This is important from the point of view of the strength and failure of some materials (see Chapter 4).

Example 1.8 *The state of stress at a point is characterised by the components*

$$\sigma_x = 100 \text{ MPa}, \sigma_y = -40 \text{ MPa}, \sigma_z = 80 \text{ MPa},$$
$$\tau_{xy} = \tau_{yz} = \tau_{zx} = 0$$

Determine the extremum values of the shear stresses, their associated normal stresses, the octahedral shear stress and its associated normal stress.

Solution The given stress components are the principal stresses, since the shears are zero. Arranging the terms such that $\sigma_1 \geq \sigma_2 \geq \sigma_3$,

$$\sigma_1 = 100 \text{ MPa}, \sigma_2 = 80 \text{ MPa}, \sigma_3 = -40 \text{ MPa}$$

Hence from Eq. (1.36),

$$\tau_1 = \frac{\sigma_2 - \sigma_3}{2} = \frac{80 + 40}{2} = 60 \text{ MPa}$$

$$\tau_2 = \frac{\sigma_3 - \sigma_1}{2} = \frac{-40 - 100}{2} = -70 \text{ MPa}$$

$$\tau_3 = \frac{\sigma_1 - \sigma_2}{2} = \frac{100 - 80}{2} = 10 \text{ MPa}$$

The associated normal stresses are

$$\sigma_1^* = \frac{\sigma_2 + \sigma_3}{2} = \frac{80 - 40}{2} = 20 \text{ MPa}$$

$$\sigma_2^* = \frac{\sigma_3 + \sigma_1}{2} = \frac{-40+100}{2} = 30 \text{ MPa}$$

$$\sigma_3^* = \frac{\sigma_1 + \sigma_2}{2} = \frac{100+80}{2} = 90 \text{ MPa}$$

$$\tau_{oct} = \frac{1}{3}\left[(\sigma_1 - \sigma_2)^2 + (\sigma_2 - \sigma_3)^2 + (\sigma_3 - \sigma_1)^2\right]^{1/2} = 61.8 \text{ MPa}$$

$$\sigma_{oct} = \frac{1}{3}(\sigma_1 + \sigma_2 + \sigma_3) = \frac{140}{3} = 46.7 \text{ MPa}$$

1.21 THE STATE OF PURE SHEAR

The state of stress at a point can be characterised by the six rectangular stress components referred to a coordinate frame of reference. The magnitudes of these components depend on the choice of the coordinate system. If, for at least one particular choice of the frame of reference, we find that $\sigma_x = \sigma_y = \sigma_z = 0$, then a state of pure shear is said to exist at point *P*. For such a state, with that particular choice of coordinate system, the stress matrix will be

$$[\tau_{ij}] = \begin{bmatrix} 0 & \tau_{xy} & \tau_{xz} \\ \tau_{xy} & 0 & \tau_{yz} \\ \tau_{xz} & \tau_{yz} & 0 \end{bmatrix}$$

For this coordinate system, $l_1 = \sigma_x + \sigma_y + \sigma_z = 0$. Since l_1 is an invariant, this must be true for any choice of coordinate system selected at *P*. Hence, the necessary condition for a state of pure shear to exist is that $l_1 = 0$, It can be shown (Appendix 2) that this is also a sufficient condition.

It was remarked in the previous section that when $l_1 = 0$, an octahedral plane is subjected to pure shear with no normal stress. Hence, for a pure shear stress state, the octahedral plane (remember that this plane is defined with respect to the principal axes and not with respect to an arbitrary set of axes) is free from normal stress.

1.22 DECOMPOSITION INTO HYDROSTATIC AND PURE SHEAR STATES

It will be shown in the present section that an arbitrary state of stress can be resolved into a hydrostatic state and a state of pure shear. Let the given state referred to a coordinate system be

$$[\tau_{ij}] = \begin{bmatrix} \sigma_x & \tau_{xy} & \tau_{xz} \\ \tau_{xy} & \sigma_y & \tau_{yz} \\ \tau_{xz} & \tau_{yz} & \sigma_z \end{bmatrix}$$

Let $$p = 1/3(\sigma_x + \sigma_y + \sigma_z) = 1/3 l_1 \tag{1.47}$$

The given state can be resolved into two different states, as shown:

$$\begin{bmatrix} \sigma_x & \tau_{xy} & \tau_{xz} \\ \tau_{xy} & \sigma_y & \tau_{yz} \\ \tau_{xz} & \tau_{yz} & \sigma_z \end{bmatrix} = \begin{bmatrix} p & 0 & 0 \\ 0 & p & 0 \\ 0 & 0 & p \end{bmatrix} + \begin{bmatrix} \sigma_x - p & \tau_{xy} & \tau_{xz} \\ \tau_{xy} & \sigma_y - p & \tau_{yz} \\ \tau_{xz} & \tau_{yz} & \sigma_z - p \end{bmatrix} \tag{1.48}$$

The first state on the right-hand side of the above equation is a hydrostatic state. [Refer Sec. 1.14(iii).]

The second state is a state of pure shear since the first invariant for this state is

$$\begin{aligned} l'_1 &= (\sigma_x - p) + (\sigma_y - p) + (\sigma_x - p) \\ &= \sigma_x + \sigma_y + \sigma_z - 3p \\ &= 0 \text{ from Eq. (1.47)} \end{aligned}$$

If the given state is referred to the principal axes, the decomposition into a hydrostatic state and a pure shear state can once again be done as above, i.e.

$$\begin{bmatrix} \sigma_1 & 0 & 0 \\ 0 & \sigma_2 & 0 \\ 0 & 0 & \sigma_3 \end{bmatrix} = \begin{bmatrix} p & 0 & 0 \\ 0 & p & 0 \\ 0 & 0 & p \end{bmatrix} + \begin{bmatrix} \sigma_1 - p & 0 & 0 \\ 0 & \sigma_2 - p & 0 \\ 0 & 0 & \sigma_3 - p \end{bmatrix} \tag{1.49}$$

where, as before, $p = 1/3(\sigma_1 + \sigma_2 + \sigma_3) = 1/3 l_1$.

The pure shear state of stress is also known as the deviatoric state of stress or simply as stress deviator.

Example 1.9 *The state of stress characterised by τ_{ij} is given below. Resolve the given state into a hydrostatic state and a pure shear state. Determine the normal and shearing stresses on an octahedral plane. Compare these with the σ_{oct} and τ_{oct} calculated for the hydrostatic and the pure shear states. Are the octahedral planes for the given state, the hydrostatic state and the pure shear state the same or are they different? Explain why.*

$$[\tau_{ij}] = \begin{bmatrix} 10 & 4 & 6 \\ 4 & 2 & 8 \\ 6 & 8 & 6 \end{bmatrix}$$

Solution $\quad l_1 = 10 + 2 + 6 = 18, \quad \frac{1}{3} l_1 = 6$

Resolving into hydrostatic and pure shear states, Eq. (1.48),

$$[\tau_{ij}] = \begin{bmatrix} 6 & 0 & 0 \\ 0 & 6 & 0 \\ 0 & 0 & 6 \end{bmatrix} + \begin{bmatrix} 4 & 4 & 6 \\ 4 & -4 & 8 \\ 6 & 8 & 0 \end{bmatrix}$$

For the given state, the octahedral normal and shear stresses are:

$$\sigma_{oct} = \frac{1}{3}\, I_1 = 6$$

From Eq. (1.44)

$$\tau_{oct} = \frac{\sqrt{2}}{3}\left(I_1^2 - 3I_2^2\right)^{1/2}$$

$$= \frac{\sqrt{2}}{3}\left[18^2 - 3\left(20 - 16 + 12 - 64 + 60 - 36\right)\right]^{1/2}$$

$$= \frac{\sqrt{2}}{3}\,(396)^{1/2} = 2\sqrt{22}$$

For the hydrostatic state, $\sigma_{oct} = 6$, since every plane is a principal plane with $\sigma = 6$ and consequently, $\tau_{oct} = 0$.

For the pure shear state, $\sigma_{oct} = 0$ since the first invariant of stress for the pure shear state is zero. The value of the second invariant of stress for the pure shear state is

$$I'_2 = (-16 - 16 + 0 - 64 + 0 - 36\,) = -132$$

Hence, the value of τ_{oct} for the pure shear state is

$$\tau_{oct} = \frac{\sqrt{2}}{3}\,(396)^{1/2} = 2\sqrt{22}$$

Hence, the value of σ_{oct} for the given state is equal to the value of σ_{oct} for the hydrostatic state, and τ_{oct} for the given state is equal to τ_{oct} for the pure shear state.

The octahedral planes for the given state (which are identified after determining the principal stress directions), the hydrostatic state and the pure shear state are all identical. For the hydrostatic state, every direction is a principal direction, and hence, the principal stress directions for the given state and the pure shear state are identical. Therefore, the octahedral planes corresponding to the given state and the pure shear state are identical.

Example 1.10 *A cylindrical boiler, 180 cm in diameter, is made of plates 1.8 cm thick, and is subjected to an internal pressure 1400 kPa. Determine the maximum shearing stress in the plate at point P and the plane on which it acts.*

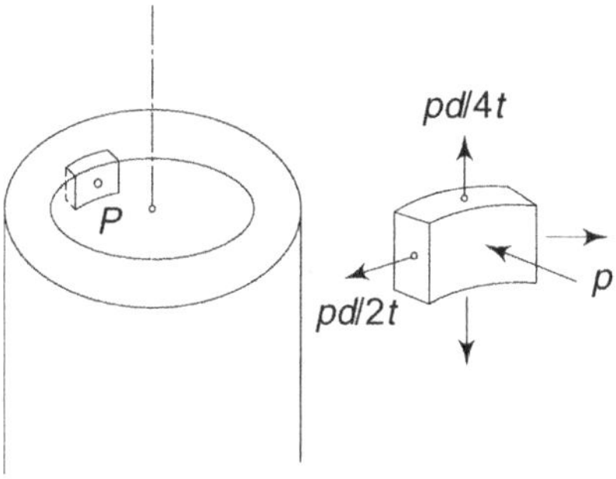

Fig. 1.19 *Example 1.10*

Solution From elementary strength of materials, the axial stress in the plate is $\frac{pd}{4t}$ where p is the internal pressure, d the diameter and t the thickness. The circumferential or the hoop stress is $\frac{pd}{2t}$. The state of stress acting on an element is as shown in Fig. 1.19.

The principal stresses when arranged such that $\sigma_1 \geq \sigma_2 \geq \sigma_3$ are

$$\sigma_1 = \frac{pd}{2t}; \quad \sigma_2 = \frac{pd}{4t}; \quad \sigma_3 = -p$$

The maximum shear stress is therefore,

$$\tau_{max} = \frac{1}{2}(\sigma_1 - \sigma_3) = \frac{1}{2}p\left(\frac{d}{2t}+1\right)$$

Substituting the values

$$\tau_{max} = \frac{1400}{2}\left(\frac{1.8\times 100}{2\times 1.8}+1\right) = 35{,}700 \text{ kPa}$$

1.23 CAUCHY'S STRESS QUADRIC

We shall now describe a geometrical description of the state of stress at a point P. Choose a frame of reference whose axes are along the principal axes. Let σ_1, σ_2 and σ_3 be the principal stresses. Consider a plane with normal $\boldsymbol{n}$. The normal stress on this plane is from Eq. (1.33),

$$\sigma = \sigma_1 n_x^2 + \sigma_2 n_y^2 + \sigma_3 n_z^2$$

Along the normal $\boldsymbol{n}$ to the plane, choose a point Q such that

$$PQ = R = 1/\sqrt{\sigma} \tag{1.50}$$

As different planes $\boldsymbol{n}$ are chosen at P, we get different values for the normal stress σ and correspondingly different PQs. If such Qs are marked for every plane passing through P, then we get a surface S. This surface determines the normal component of stress on every plane passing through P. This surface is known as the stress surface of Cauchy. This has a very interesting property. Let Q be a point on the surface, Fig. 1.20(a). By the previous definition, the length $PQ = R$ is such that the normal stress on the plane whose normal is along PQ is given by

$$\sigma = \frac{1}{R^2} \tag{1.51}$$

If $\boldsymbol{m}$ is a normal to the tangent plane to the surface S at point Q, then this normal $\boldsymbol{m}$ is parallel to the resultant stress vector $\overset{n}{\boldsymbol{T}}$ at P.

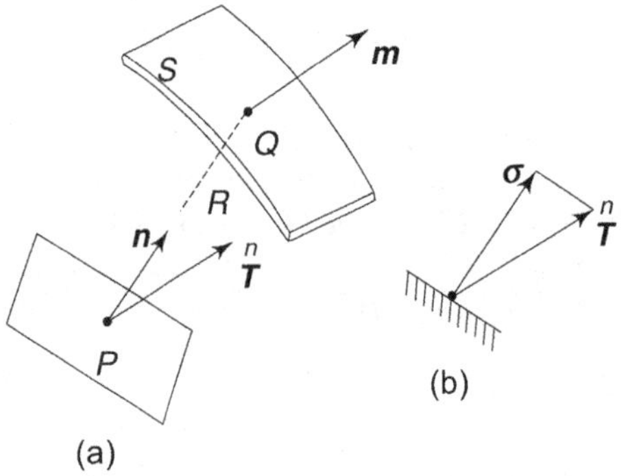

Fig. 1.20 *(a) Cauchy's stress quadric (b) Resultant stress vector and normal stress component*

Since the direction of the resultant vector $\overset{n}{\boldsymbol{T}}$ is known, and its component σ along the normal is known, the resultant stress vector $\overset{n}{\boldsymbol{T}}$ can be easily determined, as shown in Fig. 1.20(b).

We shall now show that the normal $\boldsymbol{m}$ to the surface S is parallel to $\overset{n}{\boldsymbol{T}}$, the resultant stress vector. Let $Pxyz$ be the principal axes at P (Fig. 1.21). $\boldsymbol{n}$ is the normal to a particular plane at P. The normal stress on this plane, as before, is

$$\sigma = \sigma_1 n_x^2 + \sigma_2 n_y^2 + \sigma_3 n_z^2$$

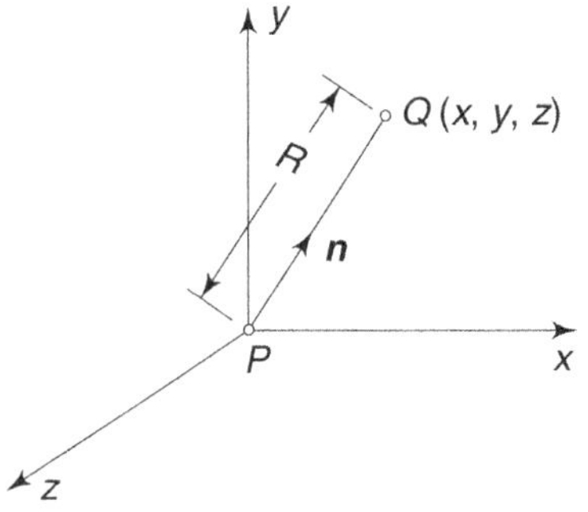

Fig. 1.21 *Principal axes at P and* ***n*** *to a plane*

If the coordinates of the point Q are (x, y, z) and the length $PQ = R$, then

$$n_x = \frac{x}{R}, \qquad n_y = \frac{y}{R}, \qquad n_z = \frac{z}{R} \tag{1.52}$$

Substituting these in the above equation for σ

$$\sigma R^2 = \sigma_1 x^2 + \sigma_2 y^2 + \sigma_3 z^2$$

From Eq. (1.51), we have $\sigma R^2 = \pm 1$. The plus sign is used when σ is tensile and the minus sign is used when σ is compressive. Hence, the surface S has the equations (a surface of second degree)

when σ is tensile

$$\sigma_1 x^2 + \sigma_2 y^2 + \sigma_3 z^2 = +1 \tag{1.53a}$$

when σ is compressive

$$\sigma_1 x^2 + \sigma_2 y^2 + \sigma_3 z^2 = -1 \tag{1.53b}$$

We know from calculus that for a surface with equation $F(x, y, z) = 0$, the normal to the tangent plane at a point Q on the surface has direction cosines proportional to $\frac{\partial F}{\partial x}, \frac{\partial F}{\partial y}$ and $\frac{\partial F}{\partial z}$. From Fig. (1.20), $\boldsymbol{m}$ is the normal perpendicular to the tangent plane to S at Q. Hence, if m_x, m_y, and m_z are the direction cosines of $\boldsymbol{m}$, then

$$m_x = \alpha \frac{\partial F}{\partial x}, \qquad m_y = \alpha \frac{\partial F}{\partial y}, \qquad m_z = \alpha \frac{\partial F}{\partial z}$$

From Eq. (1.53a) or Eq. (1.53b)

$$m_x = 2\alpha\sigma_1 x, \qquad m_y = 2\alpha\sigma_2 y, \qquad m_z = 2\alpha\sigma_3 z \tag{1.54}$$

where α is a constant of proportionality.

$\overset{n}{\boldsymbol{T}}$ is the resultant stress vector on plane $\boldsymbol{n}$ and its components $\overset{n}{T}_x, \overset{n}{T}_y$, and $\overset{n}{T}_z$ according to Eq. (1.31), are

$$\overset{n}{T}_x = \sigma_1 n_x, \qquad \overset{n}{T}_y = \sigma_1 n_y, \qquad \overset{n}{T}_z = \sigma_3 n_z$$

Substituting for n_x, n_y and n_z from Eq. (1.52)

$$\overset{n}{T}_x = \frac{1}{R}\sigma_1 x, \qquad \overset{n}{T}_y = \frac{1}{R}\sigma_2 y, \qquad \overset{n}{T}_z = \frac{1}{R}\sigma_3 z$$

or $$\sigma_1 x = R\overset{n}{T}_x, \qquad \sigma_2 y = R\overset{n}{T}_y, \qquad \sigma_3 z = R\overset{n}{T}_z$$

Substituting these in Eq. (1.54)

$$m_x = 2\alpha\, R\overset{n}{\boldsymbol{T}}_x, \qquad m_y = 2\alpha\, R\overset{n}{\boldsymbol{T}}_y, \qquad m_z = 2\alpha\, R\overset{n}{\boldsymbol{T}}_z$$

i.e. m_x, m_y and m_z are proportional to $\overset{n}{\boldsymbol{T}}_x$, $\overset{n}{\boldsymbol{T}}_y$ and $\overset{n}{\boldsymbol{T}}_z$.

Hence, $\boldsymbol{m}$ and $\overset{n}{\boldsymbol{T}}$ are parallel.

The stress surface of Cauchy, therefore, has the following properties:

(i) If Q is a point on the stress surface, then $PQ = 1/\sqrt{\sigma}$ where σ is the normal stress on a plane whose normal is PQ.

(ii) The normal to the surface at Q is parallel to the resultant stress vector $\overset{n}{\boldsymbol{T}}$ on the plane with normal PQ.

Therefore, the stress surface of Cauchy completely defines the state of stress at P. It would be of interest to know the shape of the stress surface for different states of stress. This aspect will be discussed in Appendix 3.

1.24 LAME'S ELLIPSOID

Let $Pxyz$ be a coordinate frame of reference at point P, parallel to the principal axes at P. On a plane passing through P with normal $\boldsymbol{n}$, the resultant stress vector is $\overset{n}{\boldsymbol{T}}$ and its components, according to Eq. (1.31), are

$$\overset{n}{T}_x = \sigma_1 n_x, \qquad \overset{n}{T}_y = \sigma_2 n_y, \qquad \overset{n}{T}_z = \sigma_3 n_z$$

Let PQ be along the resultant stress vector and its length be equal to its magnitude, i.e. $PQ = |\overset{n}{\boldsymbol{T}}|$. The coordinates (x, y, z) of the point Q are then

$$x = \overset{n}{T}_x, \qquad y = \overset{n}{T}_y, \qquad z = \overset{n}{T}_z$$

Since $n_x^2 + n_y^2 + n_z^2 = 1$, we get from the above two equations.

$$\frac{x^2}{\sigma_1^2} + \frac{y^2}{\sigma_2^2} + \frac{z^2}{\sigma_3^2} = 1 \tag{1.55}$$

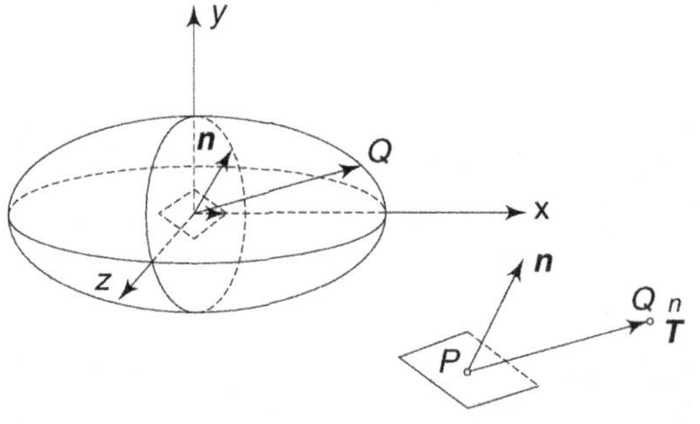

Fig. 1.22 *Lame's ellipsoid*

This is the equation of an ellipsoid referred to the principal axes. This ellipsoid is called the stress ellipsoid or Lame's ellipsoid. One of its three semi-axes is the longest, the other the shortest, and the third in-between (Fig.1.22). These are the extermum values.

If two of the principal stresses are equal, for instance

$\sigma_1 = \sigma_2$, Lame's ellipsoid is an ellipsoid of revolution and the state of stress at a given point is symmetrical with respect to the third principal axis Pz. If all the principal stresses are equal, $\sigma_1 = \sigma_2 = \sigma_3$, Lame's ellipsoid becomes a sphere.

Each radius vector PQ of the stress ellipsoid represents to a certain scale, the resultant stress on one of the planes through the centre of the ellipsoid. It can be shown (Example 1.11) that the stress represented by a radius vector of the stress ellipsoid acts on the plane parallel to tangent plane to the surface called the stress-director surface, defined by

$$\frac{x^2}{\sigma_1} + \frac{y^2}{\sigma_2} + \frac{z^2}{\sigma_3} = 1 \tag{1.56}$$

The tangent plane to the stress-director surface is drawn at the point of intersection of the surface with the radius vector. Consequently, Lame's ellipsoid and the stress-director surface together completely define the state of stress at point P.

Example 1.11 *Show that Lame's ellipsoid and the stress-director surface together completely define the state of stress at a point.*

Solution If σ_1, σ_2 and σ_3 are the principal stresses at a point P, the equation of the ellipsoid referred to principal axes is given by

$$\frac{x^2}{\sigma_1^2} + \frac{y^2}{\sigma_2^2} + \frac{z^2}{\sigma_3^2} = 1$$

The stress-director surface has the equation

$$\frac{x^2}{\sigma_1} + \frac{y^2}{\sigma_2} + \frac{z^2}{\sigma_3} = 1$$

It is known from analytical geometry that for a surface defined by $F(x, y, z) = 0$, the normal to the tangent plane at a point (x_0, y_0, z_0) has direction cosines proportional to $\frac{\partial F}{\partial x}, \frac{\partial F}{\partial y}, \frac{\partial F}{\partial z}$, evaluated at (x_0, y_0, z_0). Hence, at a point (x_0, y_0, z_0) on the stress ellipsoid, if $\boldsymbol{m}$ is the normal to the tangent plane (Fig.1.23), then

$$m_x = \alpha\frac{x_0}{\sigma_1}, \qquad m_y = \alpha\frac{y_0}{\sigma_2}, \qquad m_z = \alpha\frac{z_0}{\sigma_3}$$

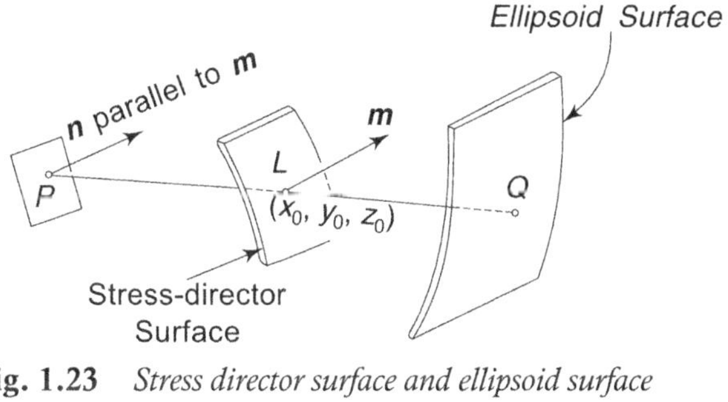

Fig. 1.23 *Stress director surface and ellipsoid surface*

Consider a plane through P with normal parallel to $\boldsymbol{m}$. On this plane, the resultant stress vector will be $\overset{n}{\boldsymbol{T}}$ with components given by

$$\overset{m}{T}_x = \sigma_1 m_x; \qquad \overset{m}{T}_y = \sigma_2 m_y; \qquad \overset{m}{T}_z = \sigma_3 m_z$$

Substituting for m_x, m_y and m_z

$$\overset{m}{T}_x = \alpha x_0, \qquad \overset{m}{T}_y = \alpha y_0, \qquad \overset{m}{T}_z = \alpha z_0$$

i.e. the components of stress on the plane with normal $\boldsymbol{m}$ are proportional to the coordinates (x_0, y_0, z_0). Hence the stress-director surface has the following property. Let $L(x_0, y_0, z_0)$ be a point on the stress-director surface. Let $\boldsymbol{m}$ be the normal to the tangent plane at L. On a plane through P with normal $\boldsymbol{m}$, the resultant stress vector is $\overset{m}{\boldsymbol{T}}$ with components proportional to x_0, y_0 and z_0. This means that the components of PL are proportional to $\overset{m}{T}_x$, $\overset{m}{T}_y$ and $\overset{m}{T}_z$.

PQ being an extension of PL and equal to $\overset{n}{\boldsymbol{T}}$ in magnitude, the plane having this resultant stress will have $\boldsymbol{m}$ as its normal.

1.25 THE PLANE STATE OF STRESS

If in a given state of stress, there exists a coordinate system $Oxyz$ such that for this system

$$\sigma_z = 0, \qquad \tau_{xz} = 0, \qquad \tau_{yz} = 0 \tag{1.57}$$

then the state is said to have a 'plane state of stress' parallel to the xy plane. This state is also generally known as a two-dimensional state of stress. All the foregoing discussions can be applied and the equations reduce to simpler forms as a result of Eq. (1.57). The state of stress is shown in Fig. 1.24.

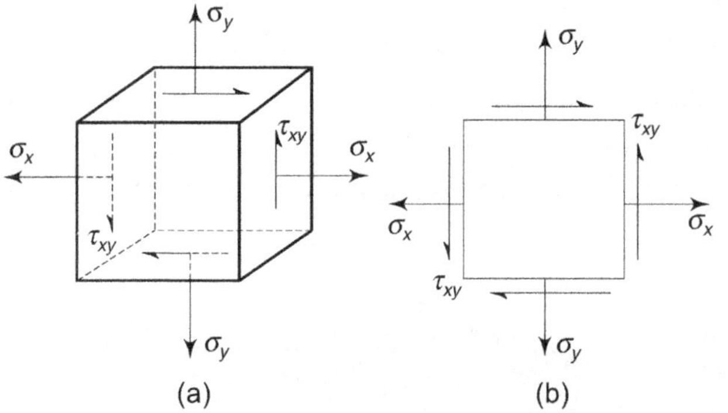

Fig. 1.24 *(a) Plane state of stress (b) Conventional representation*

Consider a plane with the normal lying in the xy plane. If n_x, n_y and n_z are the direction cosines of the normal, we have $n_x = \cos\theta$, $n_y = \sin\theta$ and $n_z = 0$ (Fig. 1.25). From Eq. (1.9)

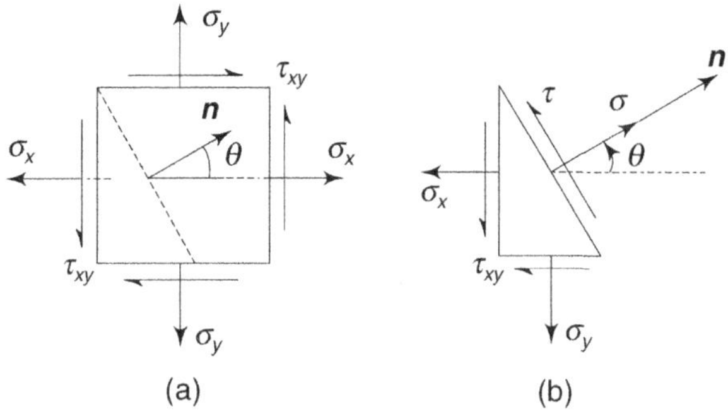

Fig. 1.25 *Normal and shear stress components on an oblique plane*

$$\overset{n}{T}_x = \sigma_x \cos\theta + \tau_{xy} \sin\theta$$

$$\overset{n}{T}_y = \sigma_y \sin\theta + \tau_{xy} \cos\theta \qquad (1.58)$$

$$\overset{n}{T}_z = 0$$

The normal and shear stress components on this plane are from Eqs (1.11a) and (1.11b)

$$\sigma = \sigma_x \cos^2\theta + \sigma_y \sin^2\theta + 2\tau_{xy} \sin\theta \cos\theta$$

$$= \frac{\sigma_x + \sigma_y}{2} + \frac{\sigma_x - \sigma_y}{2} \cos 2\theta + \tau_{xy} \sin 2\theta \qquad (1.59)$$

and $$\tau^2 = \overset{n}{T}{}_x^2 + \overset{n}{T}{}_y^2 - \sigma^2$$

or $$\tau = \frac{\sigma_x - \sigma_y}{2} \sin 2\theta + \tau_{xy} \cos 2\theta \qquad (1.60)$$

The principal stresses are given by Eq. (1.29) as

$$\sigma_1, \sigma_2 = \frac{\sigma_x + \sigma_y}{2} \pm \left[\left(\frac{\sigma_x - \sigma_y}{2}\right)^2 + \tau_{xy}^2\right]^{1/2} \qquad (1.61)$$

$$\sigma_3 = 0$$

The principal planes are given by

(i) the z plane on which $\sigma_3 = \sigma_z = 0$ and

(ii) two planes with normals in the xy plane such that

$$\tan 2\phi = \frac{2\tau_{xy}}{\sigma_x - \sigma_y} \qquad (1.62)$$

The above equation gives two planes at right angles to each other.

If the principal stresses σ_1, σ_2 and σ_3 are arranged such that $\sigma_1 \geq \sigma_2 \geq \sigma_3$, the maximum shear stress at the point will be

$$\tau_{max} = \frac{\sigma_1 - \sigma_3}{2} \qquad (1.63a)$$

In the xy plane, the maximum shear stress will be

$$\tau_{\max} = \frac{1}{2}(\sigma_1 - \sigma_2) \tag{1.63b}$$

and from Eq. (1.61)

$$\tau_{\max} = \left[\left(\frac{\sigma_x - \sigma_y}{2}\right)^2 + \tau_{xy}^2\right]^{1/2} \tag{1.64}$$

1.26 DIFFERENTIAL EQUATIONS OF EQUILIBRIUM

So far, attention has been focussed on the state of stress at a point. In general, the state of stress in a body varies from point to point. One of the fundamental problems in a book of this kind is the determination of the state of stress at every point or at any desired point in a body. One of the important sets of equations used in the analyses of such problems deals with the conditions to be satisfied by the stress components when they vary from point to point. These conditions will be established when the body (and, therefore, every part of it) is in equillibrium. We isolate a small element of the body and derive the equations of equilibrium from its free-body diagram (Fig. 1.26). A similar procedure was adopted in Sec. 1.8 for establishing the equality of cross shears.

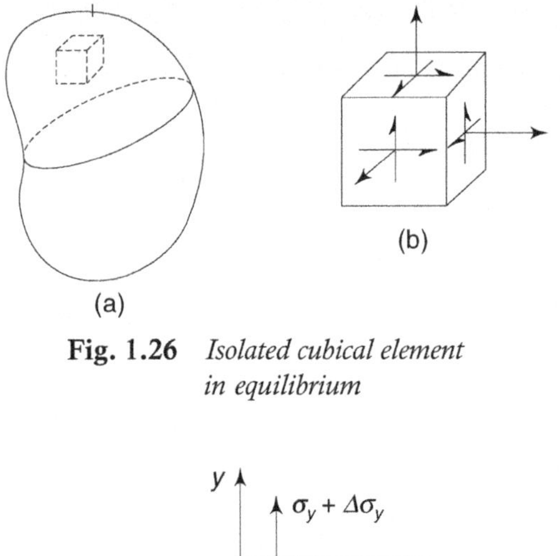

Fig. 1.26 *Isolated cubical element in equilibrium*

Consider a small rectangular element with sides Δx, Δy and Δz isolated from its parent body. Since in the limit, we are going to make Δx, y and Δz tend to zero, we shall deal with average values of the stress components on each face. These stress components are shown in Fig. 1.27.

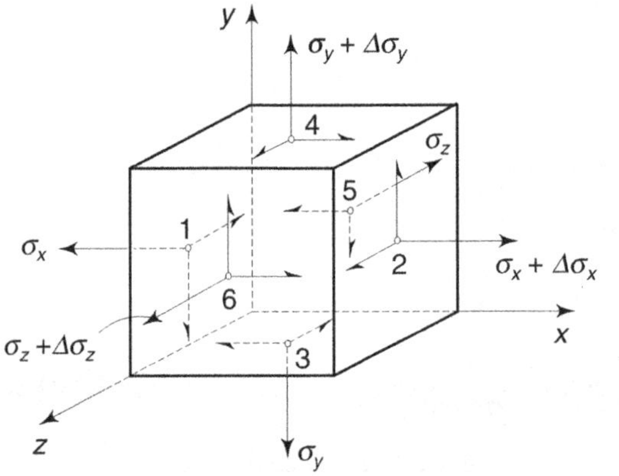

Fig. 1.27 *Variation of stresses*

The faces are marked as 1, 2, 3 etc. On the left hand face, i.e. face No. 1, the average stress components are σ_x, τ_{xy} and τ_{xz}. On the right hand face, i.e. face No. 2, the average stress components are

$$\sigma_x + \frac{\partial \sigma_x}{\partial x}\Delta x, \qquad \tau_{xy} + \frac{\partial \tau_{xy}}{\partial x}\Delta x, \qquad \tau_{xz} + \frac{\partial \tau_{xz}}{\partial x}\Delta x$$

This is because the right hand face is Δx distance away from the left hand face. Following a similar procedure, the stress components on the six faces of the element are as follows:

Face 1 $\quad \sigma_x, \quad \tau_{xy}, \quad \tau_{xz}$

Face 2 $\quad \sigma_x + \dfrac{\partial \sigma_x}{\partial x}\Delta x, \quad \tau_{xy} + \dfrac{\partial \tau_{xy}}{\partial x}\Delta x, \quad \tau_{xz} + \dfrac{\partial \tau_{xz}}{\partial x}\Delta x$

Face 3 $\quad \sigma_y, \quad \tau_{yx}, \quad \tau_{yz}$

Face 4

$$\sigma_y + \frac{\partial \sigma_y}{\partial y}\Delta y, \quad \tau_{yx} + \frac{\partial \tau_{yx}}{\partial y}\Delta y, \quad \tau_{yz} + \frac{\partial \tau_{yz}}{\partial y}\Delta y$$

Face 5 $\quad \sigma_z, \quad \tau_{zx}, \quad \tau_{zy}$

Face 6 $\quad \sigma_z + \dfrac{\partial \sigma_z}{\partial z}\Delta z, \quad \tau_{zx} + \dfrac{\partial \tau_{zx}}{\partial z}\Delta z, \quad \tau_{zy} + \dfrac{\partial \tau_{zy}}{\partial z}\Delta z$

Let the body force components per unit volume in the x, y and z directions be γ_x, γ_y, and γ_z. For equilibrium in x direction

$$\left(\sigma_x + \frac{\partial \sigma_x}{\partial x}\Delta x\right)\Delta y\,\Delta z - \sigma_x\,\Delta y\,\Delta z + \left(\tau_{yx} + \frac{\partial \tau_{yx}}{\partial y}\Delta y\right)\Delta z\,\Delta x - \tau_{yx}\,\Delta z\,\Delta x +$$

$$\left(\tau_{zx} + \frac{\partial \tau_{zx}}{\partial z}\Delta z\right)\Delta x\,\Delta y - \tau_{zx}\,\Delta x\,\Delta y + \gamma_x\,\Delta x\,\Delta y\,\Delta z = 0$$

Cancelling terms, dividing by Δx, Δy, Δz and going to the limit, we get

$$\frac{\partial \sigma_x}{\partial x} + \frac{\partial \tau_{yx}}{\partial y} + \frac{\partial \tau_{zx}}{\partial z} + \gamma_x = 0$$

Similarly, equating forces in the y and z directions respectively to zero, we get two more equations. On the basis of the fact that the cross shears are equal, i.e. $\tau_{xy} = \tau_{yx}$, $\tau_{yz} = \tau_{zy}$, $\tau_{zx} = \tau_{xz}$, we obtain the three differential equations of equilibrium as

$$\begin{aligned}
&\frac{\partial \sigma_x}{\partial x} + \frac{\partial \tau_{xy}}{\partial y} + \frac{\partial \tau_{zx}}{\partial z} + \gamma_x = 0\\
&\frac{\partial \sigma_y}{\partial y} + \frac{\partial \tau_{xy}}{\partial x} + \frac{\partial \tau_{yz}}{\partial z} + \gamma_y = 0\\
&\frac{\partial \sigma_z}{\partial z} + \frac{\partial \tau_{zx}}{\partial x} + \frac{\partial \tau_{yz}}{\partial y} + \gamma_z = 0
\end{aligned} \tag{1.65}$$

Equations (1.65) must be satisfied at all points throughout the volume of the body. It must be recalled that the moment equilibrium conditions established the equality of cross shears in Sec.1.8.

1.27 EQUILIBRIUM EQUATIONS FOR PLANE STRESS STATE

The plane stress has already been defined. If there exists a plane stress state in the xy plane, then $\sigma_z = \tau_{zx} = \tau_{yz} = \gamma_z = 0$ and only σ_x, σ_y, τ_{xy}, γ_x and γ_y exist. The differnetial equations of equilibrium become

$$\frac{\partial \sigma_x}{\partial x} + \frac{\partial \tau_{xy}}{\partial y} + \gamma_x = 0$$

$$\frac{\partial \sigma_y}{\partial y} + \frac{\partial \tau_{xy}}{\partial x} + \gamma_y = 0 \tag{1.66}$$

Example 1.12 *The cross-section of the wall of a dam is shown in Fig. 1.28. The pressure of water on face OB is also shown. With the axes Ox and Oy, as shown in Fig. 1.28, the stresses at any point (x, y) are given by (γ = specific weight of water and ρ = specific weight of dam material)*

$$\sigma_x = -\gamma y$$

$$\sigma_y = \left(\frac{\rho}{tan\,\beta} - \frac{2\gamma}{tan^3\,\beta}\right) x + \left(\frac{\gamma}{tan^2\,\beta} - \rho\right) y$$

$$\tau_{xy} = \tau_{yx} = -\frac{\gamma}{tan^2\,\beta}\, x$$

$$\tau_{yz} = 0, \quad \tau_{zx} = 0, \quad \sigma_z = 0$$

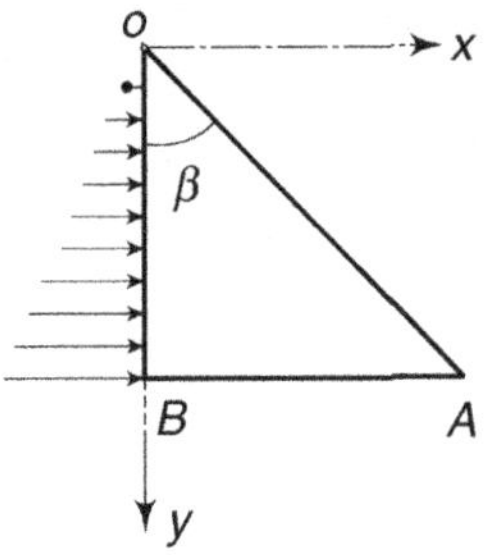

Fig. 1.28 *Example 1.12*

Check if these stress components satisfy the differential equations of equilibrium. Also, verify if the boundary conditions are satisfied on face OB.

Solution The equations of equilibrium are

$$\frac{\partial \sigma_x}{\partial x} + \frac{\partial \tau_{xy}}{\partial y} + \gamma_x = 0$$

and

$$\frac{\partial \sigma_y}{\partial y} + \frac{\partial \tau_{xy}}{\partial x} + \gamma_y = 0$$

Substituting and noting that $\gamma_x = 0$ and $\gamma_y = \rho$, the first equation is satisifed. For the second equation also

$$\frac{\gamma}{\tan^2\beta} - \rho - \frac{\gamma}{\tan^2\beta} + \rho = 0$$

On face OB, at any y, the stress components are $\sigma_x = -\gamma y$ and $\tau_{xy} = 0$. Hence the boundary conditions are also satisfied.

Example 1.13 *Consider a function $\phi(x, y)$, which is called the stress function. If the values of σ_x, σ_y and τ_{xy} are as given below, show that these satisfy the differential equations of equilibrium in the absence of body forces.*

$$\sigma_x = \frac{\partial^2 \phi}{\partial y^2}, \qquad \sigma_y = \frac{\partial^2 \phi}{\partial x^2}, \qquad \tau_{xy} = -\frac{\partial^2 \phi}{\partial x \, \partial y}$$

Solution Substituting in the differential equations of equilibrium

$$\frac{\partial^3 \phi}{\partial y^2 \, \partial x} - \frac{\partial^3 \phi}{\partial y^2 \, \partial x} = 0$$

$$\frac{\partial^3 \phi}{\partial x^2 \, \partial y} - \frac{\partial^3 \phi}{\partial x^2 \, \partial y} = 0$$

Example 1.14 *Consider the rectangular beam shown in Fig. 1.29. According to the elementary theory of bending, the 'fibre stress' in the elastic range due to bending is given by*

$$\sigma_x = -\frac{My}{I} = -\frac{12\,My}{bh^3}$$

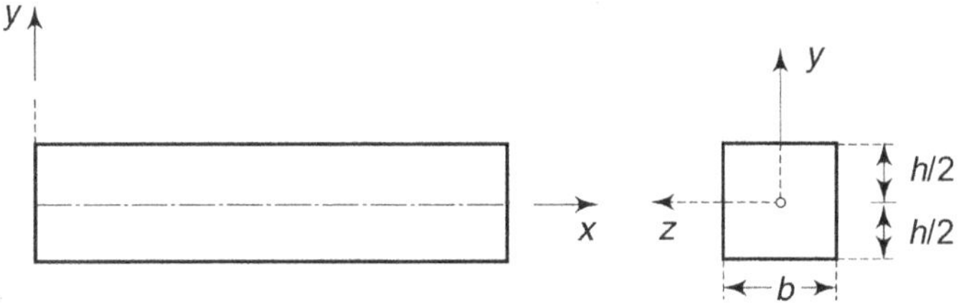

Fig. 1.29 *Exmaple 1.14*

where M is the bending moment which is a function of x. Assume that $\sigma_z = \tau_{zx} = \tau_{zy} = 0$ and also that $\tau_{xy} = 0$ at the top and bottom, and further, that $\sigma_y = 0$ at the bottom. Using the differential equations of equilibrium, determine τ_{xy} and σ_y. Compare these with the values given in the elementary strength of materials.

Solution From Eq. (1.65)

$$\frac{\partial \sigma_x}{\partial x} + \frac{\partial \tau_{xy}}{\partial y} + \frac{\partial \tau_{xz}}{\partial z} = 0$$

Since $\tau_{xz} = 0$ and M is a function of x

$$-\frac{12y}{bh^3}\frac{\partial M}{\partial x} + \frac{\partial \tau_{xy}}{\partial y} = 0$$

or
$$\frac{\partial \tau_{xy}}{\partial y} = \frac{12}{bh^3}\frac{\partial M}{\partial x}y$$

Integrating
$$\tau_{xy} = \frac{6}{bh^3}\frac{\partial M}{\partial x}y^2 + c_1 f(x) + c_2$$

where $f(x)$ is a function of x alone and c_1, c_2 are constants. It is given that $\tau_{xy} = 0$ at $y = \pm\frac{h}{2}$

$$\therefore \qquad \frac{6}{bh^3}\frac{h^2}{4}\frac{\partial M}{\partial x} = -c_1 f(x) - c_2$$

$$\text{or} \qquad c_1 f(x) + c_2 = -\frac{3}{2bh}\frac{\partial M}{\partial x}$$

$$\therefore \qquad \tau_{xy} = \frac{3}{2bh}\frac{\partial M}{\partial x}\left(\frac{4y^2}{h^2} - 1\right)$$

From elementary strength of materials, we have

$$\tau_{xy} = \frac{V}{Ib}\int_y^{h/2} y'\, dA$$

where $V = -\frac{\partial M}{\partial x}$ is the shear force. Simplifying the above expression

$$\tau_{xy} = -\frac{\partial M}{\partial x}\frac{12}{b^2\, h^3}\left(\frac{h^2}{4} - y^2\right)\frac{b}{2}$$

$$\text{or} \qquad \tau_{xy} = \frac{3}{2bh}\frac{\partial M}{\partial x}\left(\frac{4y^2}{h^3} - 1\right)$$

i.e. the same as the expression obtained above.

From the next equilibrium equation, i.e. from

$$\frac{\partial \sigma_y}{\partial y} + \frac{\partial \tau_{xy}}{\partial x} + \frac{\partial \tau_{yz}}{\partial z} = 0$$

$$\text{we get} \qquad \frac{\partial \sigma_y}{\partial y} = -\frac{3}{2bh}\left(\frac{4y^2}{h^2} - 1\right)\frac{\partial^2 M}{\partial x^2}$$

$$\therefore \qquad \sigma_y = -\frac{3}{2bh}\frac{\partial^2 M}{\partial x^2}\left(\frac{4y^3}{3h^2} - y\right) + c_3\, F(x) + c_4$$

where $F(x)$ is a function of x alone. It is given that $\sigma_y = 0$ at $y = -\frac{h}{2}$.

$$\text{Hence,} \qquad c_3 F(x) + c_4 = \frac{3}{2bh}\frac{\partial^2 M}{\partial x^2}\frac{h}{3}$$

$$= \frac{1}{2b}\frac{\partial^2 M}{\partial x^2}$$

Substituting

$$\sigma_y = -\frac{3}{2bh}\frac{\partial^2 M}{\partial x^2}\left(\frac{4y^3}{3h^2} - y - \frac{h}{3}\right)$$

At $y = +h/2$, the value of σ_y is

$$\sigma_y = \frac{1}{b}\frac{\partial^2 M}{\partial x^2} = \frac{w}{b}$$

where w is the intensity of loading. Since b is the width of the beam, the stress will be w/b as obtained above.

1.28 BOUNDARY CONDITIONS

Equation (1.66) must be satisfied throughout the volume of the body. When the stresses vary over the plate (i.e. the body having the plane stress state), the stress components σ_x, σ_y and τ_{xy} must be consistent with the externally applied forces at a boundary point.

Consider the two-dimensional body shown in Fig.1.30. At a boundary point P, the outward drawn normal is $\boldsymbol{n}$. Let F_x and F_y be the components of the surface forces per unit area at this point.

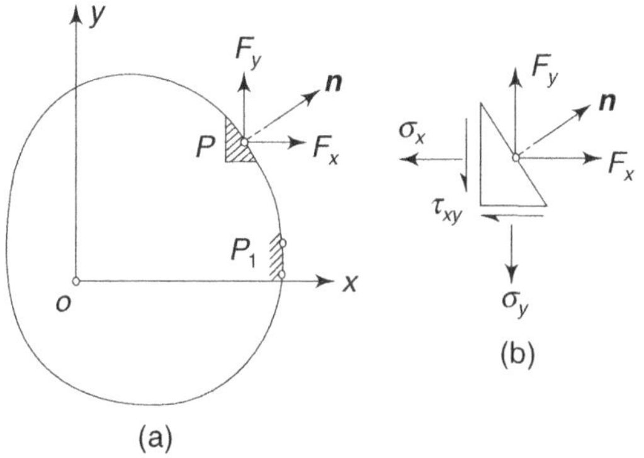

Fig. 1.30 *(a) Element near a boundary point (b) Free body diagram*

F_x and F_y must be continuations of the stresses σ_x, σ_y and τ_{xy} at the boundary. Hence, using Cauchy's equations

$$\overset{n}{T}_x = F_x = \sigma_x n_x + \tau_{xy} n_y$$

$$\overset{n}{T}_y = F_y = \sigma_y n_y + \tau_{xy} n_x$$

If the boundary of the plate happens to be parallel to y axis, as at point P_1, the boundary conditions become

$$F_x = \sigma_x \quad \text{and} \quad F_y = \tau_{xy}$$

1.29 EQUATIONS OF EQUILIBRIUM IN CYLINDRICAL COORDINATES

Till this section, we have been using a rectangular or the Cartesian frame of reference for analyses. Such a frame of reference is useful if the body under analysis happens to possess rectangular or straight boundaries. Numerous problems exist where the bodies under discussion possess radial symmetry; for example, a thick cylinder subjected to internal or external pressure. For the analysis of such problems, it is more convenient to use polar or cylindrical coordinates. In this section, we shall develop some equations in cylindrical coordinates.

Consider an axisymmetric body as shown in Fig. 1.31(a). The axis of the body is usually taken as the z axis. The two other coordinates are r and θ, where θ is measured counter-clockwise. The rectangular stress components at a point $P(r, \theta, z)$ are

$$\sigma_r,\ \sigma_\theta,\ \sigma_z,\ \tau_{\theta r},\ \tau_{\theta z} \text{ and } \tau_{zr}$$

These are shown acting on the faces of a radial element at point P in Fig.1.31(b). σ_r, σ_θ and σ_z are called the radial, circumferential and axial stresses respectively. If the stresses vary from point to point, one can derive the appropriate differential equations of equilibrium, as in Sec. 1.26. For this purpose, consider a cylindrical element having a radial length Δr with an included angle $\Delta\theta$ and a height Δz, isolated from the body. The free-body diagram of the element is shown in Fig.1.32(b). Since the element is very small, we work with the average stresses acting on each face.

The area of the face $aa'd'd$ is $r\,\Delta\theta\,\Delta z$ and the area of face $bb'c'c$ is $(r + \Delta r)\,\Delta\theta\,\Delta z$. The areas of faces $dcc'd'$ and $abb'e'$ are each equal to $\Delta r\,\Delta z$.

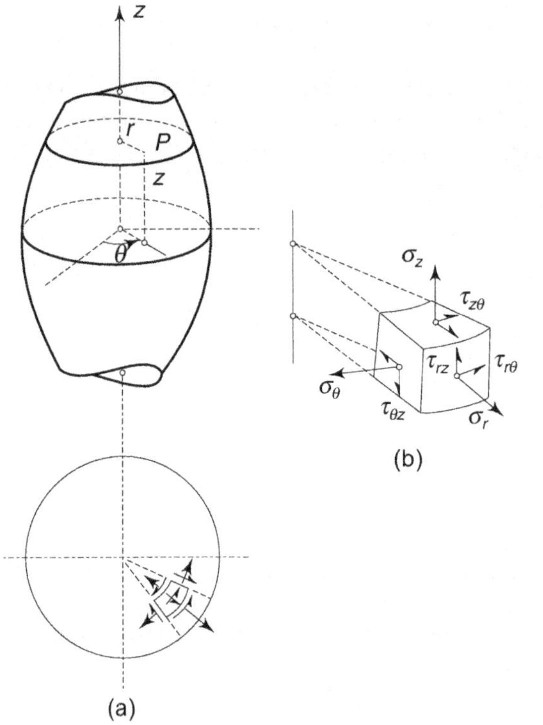

Fig. 1.31 *(a) Cylindrical coordinates of a point (b) Stresses on an element*

The faces $abcd$ and $a'b'c'd'$ have each an area $\left(r + \dfrac{\Delta r}{2}\right)\Delta\theta\,\Delta r$. The average stresses on these faces (which are assumed to be acting at the mid point of eace face) are

On face $aa'd'd$

normal stress σ_r

tangential stresses τ_{rz} and $\tau_{r\theta}$

On face $bb'c'c$

normal stress $\sigma_r + \dfrac{\partial \sigma_r}{\partial r}\Delta r$

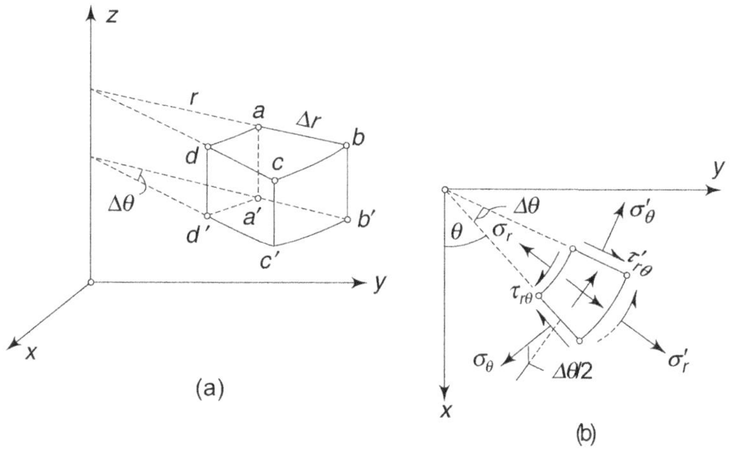

Fig. 1.32 *(a) Geometry of cylindrical element (b) Variation of stresses across faces*

tangential stresses $\tau_{rz} + \dfrac{\partial \tau_{rz}}{\partial r}\Delta r$ and $\tau_{r\theta} + \dfrac{\partial \tau_{\theta r}}{\partial r}\Delta r$

The changes are because the face $bb'c'c$ is Δr distance away from the face $aa'd'd$.

On face $dcc'd'$

normal stress σ_θ

tangential stresses $\tau_{r\theta}$ and $\tau_{\theta z}$

On face $abb'a$

normal stress $\sigma_\theta + \dfrac{\partial \sigma_\theta}{\partial \theta}\Delta\theta$

tangential stresses $\tau_{r\theta} + \dfrac{\partial \tau_{r\theta}}{\partial \theta}\Delta\theta$ and $\tau_{\theta z} + \dfrac{\partial \tau_{\theta z}}{\partial \theta}\Delta\theta$

The changes in the above components are because the face $abb'a$ is separated by an angle $\Delta\theta$ from the face $dcc'd'$.

On face $a'b'c'd'$

normal stress σ_z

tangential stresses τ_{rz} and $\tau_{\theta z}$

On face $abcd$

normal stress $\sigma_z + \dfrac{\partial \sigma_z}{\partial z}\Delta z$

tangential stresses $\tau_{rz} + \dfrac{\partial \tau_{rz}}{\partial z}\Delta z$ and $\tau_{\theta z} + \dfrac{\partial \tau_{\theta z}}{\partial z}\Delta z$

Let γ_r, γ_θ and γ_z be the body force components per unit volume. If the element is in equilibrium, the sum of forces in r, θ and z directions must vanish individually, Equating the forces in r direction to zero,

$$\left(\sigma_r + \frac{\partial \sigma_r}{\partial r}\Delta r\right)(r + \Delta r)\,\Delta\theta\Delta z + \left(\tau_{rz} + \frac{\partial \tau_{rz}}{\partial z}\Delta z\right)\left(r + \frac{\Delta r}{2}\right)\Delta\theta\Delta r$$

$$-\sigma_r\, r\,\Delta\theta\,\Delta z - \tau_{rz}\left(r + \frac{\Delta r}{2}\right)\Delta\theta\,\Delta r - \sigma_\theta \sin\frac{\Delta\theta}{2}\,\Delta r\,\Delta z$$

$$-\tau_{r\theta}\cos\frac{\Delta\theta}{2}\,\Delta r\,\Delta z - \left(\sigma_\theta + \frac{\partial\sigma_\theta}{\partial\theta}\Delta\theta\right)\sin\frac{\Delta\theta}{2}\,\Delta r\,\Delta z$$

$$+\left(\tau_{r\theta} + \frac{\partial\tau_{r\theta}}{\partial\theta}\Delta\theta\right)\cos\frac{\Delta\theta}{2}\,\Delta r\,\Delta z + \gamma_r\left(r + \frac{\Delta r}{2}\right)\Delta\theta\,\Delta r\,\Delta z = 0$$

Cancelling terms, dividing by $\Delta\theta\,\Delta r\,\Delta z$ and going to the limit with $\Delta\theta$, Δr and Δz, all tending to zero

$$r\frac{\partial\sigma_r}{\partial r} + r\frac{\partial\tau_{rz}}{\partial z} + \frac{\partial\tau_{r\theta}}{\partial\theta} + \sigma_r - \sigma_\theta + r\gamma_r = 0$$

or

$$\frac{\partial\sigma_r}{\partial r} + \frac{\partial\tau_{rz}}{\partial z} + \frac{1}{r}\frac{\partial\tau_{r\theta}}{\partial\theta} + \frac{\sigma_r - \sigma_\theta}{r} + \gamma_r = 0 \tag{1.67}$$

Similarly, for equilibrium in z and θ directions, we get

$$\frac{\partial\tau_{rz}}{\partial r} + \frac{\partial\sigma_z}{\partial z} + \frac{1}{r}\frac{\partial\tau_{\theta z}}{\partial\theta} + \frac{\tau_{rz}}{r} + \gamma_z = 0 \tag{1.68}$$

and

$$\frac{\partial\tau_{r\theta}}{\partial r} + \frac{\partial\tau_{\theta z}}{\partial z} + \frac{1}{r}\frac{\partial\sigma_\theta}{\partial\theta} + \frac{2\tau_{r\theta}}{r} + \gamma_\theta = 0 \tag{1.69}$$

Equations (1.67)–(1.69) are the differential equations of equilibrium expressed in polar coordinates.

1.30 AXISYMMETRIC CASE AND PLANE STRESS CASE

If an axisymmetric body is loaded symmetrically, the stress components do not depend on θ. Since the deformations are symmetric, $\tau_{r\theta}$ and $\tau_{\theta z}$ do not exist and consequently the above set of equations in the absence of body forces are reduced to

$$\frac{\partial\sigma_r}{\partial r} + \frac{\partial\tau_{rz}}{\partial z} + \frac{\sigma_r - \sigma_\theta}{r} = 0$$

$$\frac{\partial\tau_{rz}}{\partial r} + \frac{\partial\sigma_z}{\partial z} + \frac{\tau_{rz}}{r} = 0$$

A sphere under diametral compression or a cone under a load at the apex are examples to which the above set of equations can be applied.

If the state of stress is two-dimensional in nature, i.e. plane stress state, then only σ_r, σ_θ, $\tau_{r\theta}$, γ_r, and γ_θ exist. The other stress components vanish.These non-vanishing stress components depend only on θ and r and are independent of z in the absence of body forces. The equations of equilibrium reduce to

$$\begin{aligned} &\frac{\partial\sigma_r}{\partial r} + \frac{1}{r}\frac{\partial\tau_{r\theta}}{\partial\theta} + \frac{\sigma_r - \sigma_\theta}{r} = 0 \\ &\frac{\partial\tau_{r\theta}}{\partial r} + \frac{1}{r}\frac{\partial\sigma_\theta}{\partial\theta} + \frac{2\tau_{r\theta}}{r} = 0 \end{aligned} \tag{1.70}$$

Example 1.15 *Consider a function $\phi(r, \theta)$, which is called the stress function. If the values of σ_r, σ_θ, and $\tau_{r\theta}$ are as given below, show that in the absence of body forces, these satisfy the differential equations of equilibrium.*

$$\sigma_r = \frac{1}{r}\frac{\partial \phi}{\partial r} + \frac{1}{r^2}\frac{\partial^2 \phi}{\partial \theta^2}$$

$$\sigma_\theta = \frac{\partial^2 \phi}{\partial r^2}$$

$$\tau_{r\theta} = -\frac{1}{r}\frac{\partial^2 \phi}{\partial r \partial \theta} + \frac{1}{r^2}\frac{\partial \phi}{\partial \theta}$$

Solution The equations of equilibrium are

$$\frac{\partial \sigma_r}{\partial r} + \frac{1}{r}\frac{\partial \tau_{r\theta}}{\partial \theta} + \frac{\sigma_r - \sigma_\theta}{r} = 0$$

$$\frac{\partial \tau_{r\theta}}{\partial r} + \frac{1}{r}\frac{\partial \sigma_\theta}{\partial \theta} + \frac{2\tau_{r\theta}}{r} = 0$$

Substituting the stress function in the first equation of equilibrium,

$$-\frac{1}{r^2}\frac{\partial \phi}{\partial r} + \frac{1}{r}\frac{\partial^2 \phi}{\partial r^2} - \frac{2}{r^3}\frac{\partial^2 \phi}{\partial \theta^2} + \frac{1}{r^2}\frac{\partial^3 \phi}{\partial \theta^2 \, \partial r} + \frac{1}{r}\left(-\frac{1}{r}\frac{\partial^3 \phi}{\partial \theta^2 \, \partial r} + \frac{1}{r^2}\frac{\partial^2 \phi}{\partial \theta^2}\right)$$

$$+\frac{1}{r^2}\frac{\partial \phi}{\partial r} + \frac{1}{r^3}\frac{\partial^2 \phi}{\partial \theta^2} - \frac{1}{r}\frac{\partial^2 \phi}{\partial r^2} = 0$$

Hence, the first equation is satisfied. Similarly, it can easily be verified that the second condition also holds good.

Problems

1.1 It was assumed in Sec.1.2 that across any infinitesimal surface element in a solid, the action of the exterior material upon the interior is equipollent (i.e. equal in strength or effect) to only a force. It is also possible to assume that in addition to a force, there is also a couple, i.e. at any point across any plane $\boldsymbol{n}$, there is a stress vector $\overset{n}{\boldsymbol{T}}$ and a couple-stress vector $\overset{n}{\boldsymbol{M}}$, as shown in Fig. 1.33.

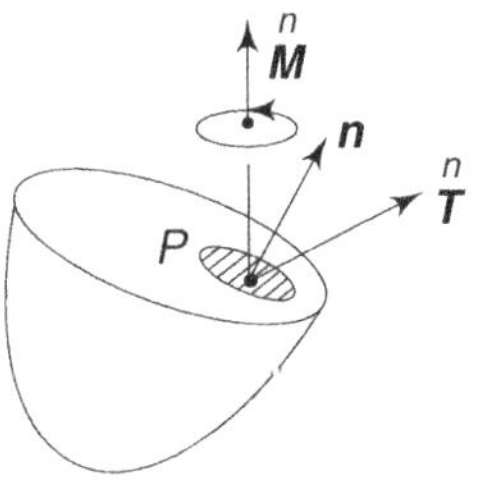

Fig. 1.33 *Problem 1.1*

Show that a set of equations similar to Cauchy's equations can be derived, i.e. if we know the couple-stress vectors on three mutually perpendicular planes passing through the point P, then we can determine the couple-stress vector on any plane $\boldsymbol{n}$

passing through the point. The equations are

$$\overset{n}{M}_x = M_{xx}\, n_x + M_{yx}\, n_y + M_{zx}\, n_z$$

$$\overset{n}{M}_y = M_{xy}\, n_x + M_{yy}\, n_y + M_{zy}\, n_z$$

$$\overset{n}{M}_z = M_{xz}\, n_x + M_{yz}\, n_y + M_{zz}\, n_z$$

$\overset{n}{\boldsymbol{M}}_x$, $\overset{n}{\boldsymbol{M}}_y$, $\overset{n}{\boldsymbol{M}}_z$ are the x, y and z components of the vector $\overset{n}{\boldsymbol{M}}$ acting on plane $\boldsymbol{n}$.

1.2 A rectangular beam is subjected to a pure bending moment M. The cross-section of the beam is shown in Fig. 1.34. Using the elementary flexure formula, determine the normal and shearing stresses at a point (x, y) on the plane AB shown.

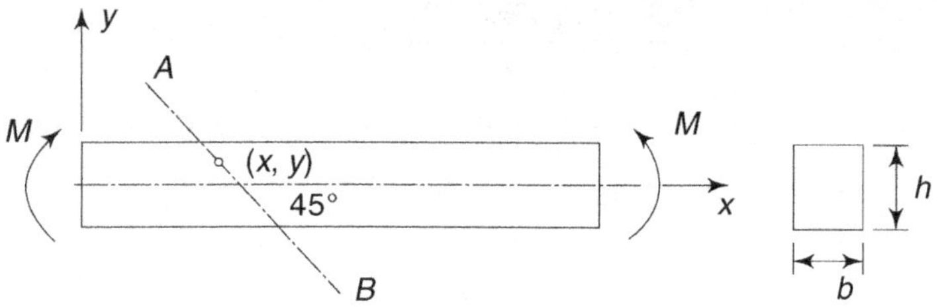

Fig. 1.34 *Problem 1.2*

$$\left[Ans.\quad \sigma_n = \tau_n = \frac{6My}{bh^3} \right]$$

1.3 Consider a sphere of radius R subjected to diametral compression (Fig. 1.35). Let σ_r, σ_θ and σ_ϕ be the normal stresses and $\tau_{r\theta}$, $\tau_{\theta\phi}$ and $\tau_{\phi r}$ the shear stresses at a point. At point $P(o, y, z)$ on the surface and lying in the yz plane, determine the rectangular normal stress components σ_x, σ_y and σ_z in terms of the spherical stress components.

[*Ans.* $\sigma_x = \sigma_\theta$; $\sigma_y = \sigma_\phi \cos^2\phi$; $\sigma_z = \sigma_\phi \sin^2\phi$]

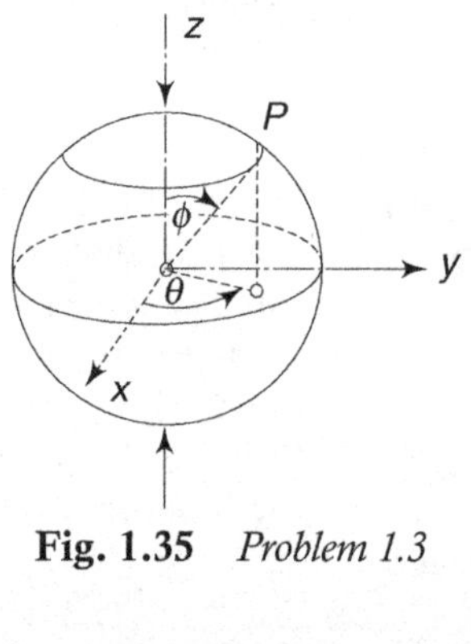

Fig. 1.35 *Problem 1.3*

1.4 The state of stress at a point is characterised by the matrix shown. Determine T_{11} such that there is at least one plane passing through the point in such a way that the resultant stress on that plane is zero. Determine the direction cosines of the normal to that plane.

$$\left[\tau_{ij}\right]=\begin{bmatrix} T_{11} & 2 & 1 \\ 2 & 0 & 2 \\ 1 & 2 & 0 \end{bmatrix}$$

$$\left[Ans.\quad T_{11}=2;\, n_x=\pm\frac{2}{3};\, n_y=\pm\frac{1}{3};\, n_z=\pm\frac{2}{3}\right]$$

1.5 If the rectangular components of stress at a point are as in the matrix below, determine the unit normal of a plane parallel to the z axis, i.e. $n_z = 0$, on which the resultant stress vector is tangential to the plane

$$\left[\tau_{ij}\right]=\begin{bmatrix} a & 0 & d \\ 0 & b & e \\ d & e & c \end{bmatrix}$$

$$\left[Ans.\quad n_x=\pm\left(\frac{b}{b-a}\right)^{1/2};\, n_y=\pm\left(\frac{a}{a-b}\right)^{1/2};\, n_z=0\right]$$

1.6 A cross-section of the wall of a dam is shown in Fig.1.36. The pressure of water on face OB is also shown. The stresses at any point (x, y) are given by the following expressions

$$\sigma_x = -\gamma y$$

$$\sigma_y = \left(\frac{\rho}{\tan\beta} - \frac{2\gamma}{\tan^3\beta}\right)x + \left(\frac{\gamma}{\tan^2\beta} - \rho\right)y$$

$$\tau_{xy} = \tau_{yx} = -\frac{\gamma x}{\tan^2\beta}$$

$$\tau_{yz} = \tau_{zx} = \sigma_z = 0$$

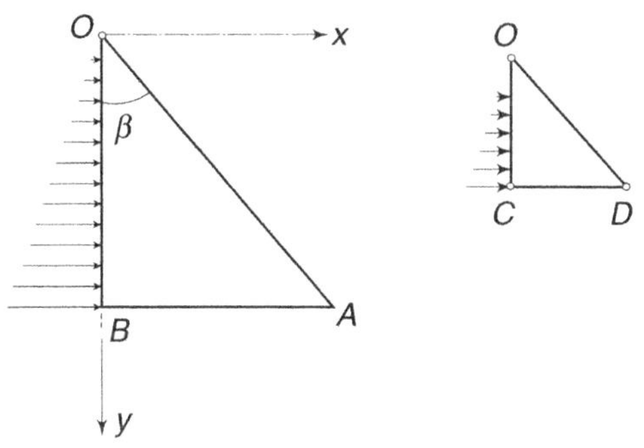

Fig. 1.36 *Problem 1.6*

where γ is the specific weight of water and ρ the specific weight of the dam material.

Consider an element OCD and show that this element is in equilibrium under the action of the external forces (water pressure and gravity force) and the internally distributed forces across the section CD.

1.7 Determine the principal stresses and their axes for the states of stress characterised by the following stress matrices (units are 1000 kPa).

(i) $\left[\tau_{ij}\right]=\begin{bmatrix} 18 & 0 & 24 \\ 0 & -50 & 0 \\ 24 & 0 & 32 \end{bmatrix}$

$$\begin{bmatrix} Ans.\ \sigma_1 = 50,\ n_x = 0.6,\ n_y = 0,\ n_z = 0.8 \\ \sigma_2 = 0,\ n_x = 0.8,\ n_y = 0,\ n_z = 0.6 \\ \sigma_3 = -50,\ n_x = n_z = 0,\ n_y = 1 \end{bmatrix}$$

(ii) $\left[\tau_{ij}\right]=\begin{bmatrix} 3 & -10 & 0 \\ -10 & 0 & 30 \\ 0 & 30 & -27 \end{bmatrix}$

$$\begin{bmatrix} Ans.\ \sigma_1 = 23,\ n_x = 0.394,\ n_y = 0.788,\ n_z = 0.473 \\ \sigma_2 = 0,\ n_x = 0.912,\ n_y = 0.274,\ n_z = 0.304 \\ \sigma_3 = -47,\ n_x = 0.941,\ n_y = 0.188,\ n_z = 0.288 \end{bmatrix}$$

1.8 The state of stress at a point is characterised by the components

$$\sigma_x = 12.31, \quad \sigma_y = 8.96, \quad \sigma_z = 4.34$$

$$\tau_{xy} = 4.20, \quad \tau_{yx} = 5.27, \quad \sigma_z = 0.84$$

Find the values of the principal stresses and their directions

$$\begin{bmatrix} Ans.\ \sigma_1 = 16.41,\ n_x = 0.709,\ n_y = 0.627,\ n_z = 0.322 \\ \sigma_2 = 8.55,\ n_x = 0.616,\ n_y = 0.643,\ n_z = 0.455 \\ \sigma_3 = 0.65,\ n_x = 0.153,\ n_y = 0.583,\ n_z = 0.798 \end{bmatrix}$$

1.9 For Problem 1.8, determine the principal shears and the associated normal stresses.

$$\begin{bmatrix} Ans.\ \tau_3 = 3.94,\ \sigma_n = 12.48 \\ \tau_2 = 7.88,\ \sigma_n = 8.53 \\ \tau_1 = 3.95,\ \sigma_n = 4.52 \end{bmatrix}$$

1.10 For the state of stress at a point characterised by the components (in 1000 kPa)

$$\sigma_x = 12, \quad \sigma_y = 4, \quad \sigma_z = 10, \quad \tau_{xy} = 3, \quad \tau_{yz} = \tau_{zx} = 0$$

determine the principal stresses and their directions.

$$\begin{bmatrix} Ans.\ \text{In 1000 kPa}\ \sigma_1 = 13;\ 18^\circ \text{ with } x \text{ axis};\ n_z = 0 \\ \sigma_2 = 10;\ n_x = 0;\ n_y = 0;\ n_z = 1 \\ \sigma_3 = 3;\ -72^\circ \text{ with } x \text{ axis};\ n_z = 0 \end{bmatrix}$$

1.11 Let $\sigma_x = -5c$, $\sigma_y = c$, $\sigma_z = c$, $\tau_{xy} = -c$, $\tau_{yz} = \tau_{zx} = 0$, where $c = 1000$ kPa. Determine the principal stresses, stress deviators, principal axes, greatest shearing stress and octahedral stresses.

$$\begin{bmatrix} Ans.\ \sigma_1 = (-2 + \sqrt{10})c;\ n_z = 0 \text{ and } \theta = 9.2^\circ \text{ with } y \text{ axis} \\ \sigma_2 = c,\ n_x = n_y = 0;\ n_z = 1 \\ \sigma_3 = (-2 - \sqrt{10})c;\ n_z = 0 \text{ and } \theta = 9.2^\circ \text{ with } x \text{ axis} \\ \tau_{max} = \sqrt{10}\,c;\ \sigma'_x = -4c;\ \sigma'_y = 2c;\ \sigma'_z = 2c \\ \sigma_{oct} = -c;\ \tau_{oct} = \dfrac{\sqrt{78}}{3}c \end{bmatrix}$$

1.12 A solid shaft of diameter $d = \sqrt{10}$ cm (Fig. 1.37) is subjected to a tensile force P = 10,000 N and a torque T = 5000 N cm. At point A on the surface, determine the principal stresses, the octahedral shearing stress and the maximum shearing stress.

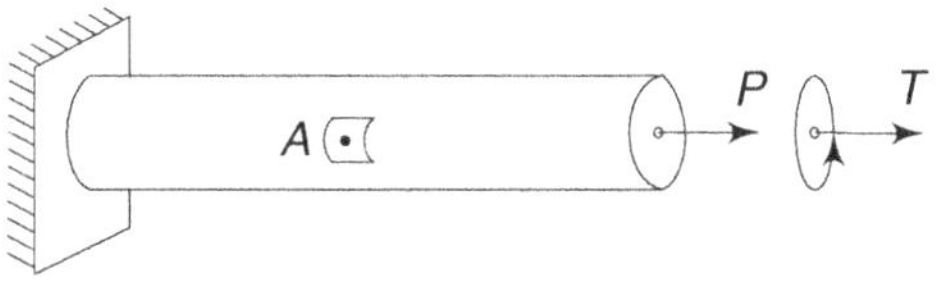

Fig. 1.37 *Problem 1.12*

$$\left[\begin{array}{l} \textit{Ans.} \quad \sigma_{1,2} = \dfrac{2000}{\pi}\left(1 \pm \sqrt{13/5}\right) Pa \\ \qquad\quad \tau_{max} = \dfrac{2000}{\pi}\sqrt{\dfrac{13}{5}}\ Pa \\ \qquad\quad \tau_{oct} = \dfrac{4000}{3\pi}\sqrt{\dfrac{22}{5}}\ Pa \end{array}\right]$$

1.13 A cylindrical rod (Fig. 1.38) is subjected to a torque T. At any point P of the cross-section LN, the following stresses occur

$$\sigma_x = \sigma_y = \sigma_z = \tau_{xy} = \tau_{yx} = 0,\ \tau_{xz} = \tau_{zx} = -G\theta y,\ \tau_{yz} = \tau_{zy} = G\theta x$$

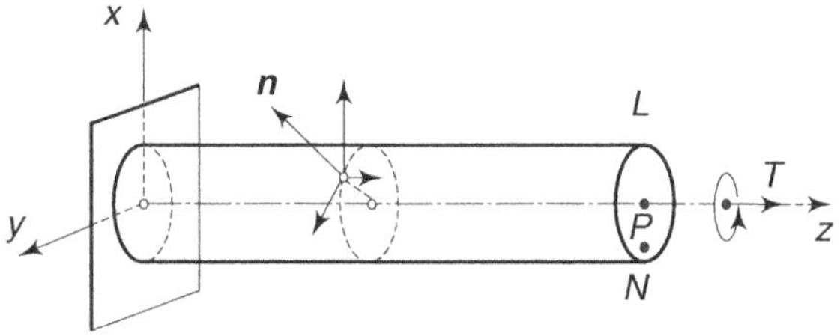

Fig. 1.38 *Problem 1.13*

Check whether these satisfy the equations of equilibrium. Also show that the lateral surface is free of load, i e. show that

$$\overset{n}{T}_x = \overset{n}{T}_y = \overset{n}{T}_z = 0$$

1.14 For the state of stress given in Problem 1.13, determine the principal shears, octahedral shear stress and its associated normal stress.

$$\left[\begin{array}{l} \textit{Ans.} \quad \tau_1 = \tau_3 = \dfrac{1}{2}\ G\theta\ \sqrt{x^2 + y^2}\ ;\ \tau_2 = -G\theta\ \sqrt{x^2 + y^2} \\ \qquad\quad \tau_{oct} = \dfrac{\sqrt{6}}{3}\ G\theta\left(\sqrt{x^2 + y^2}\right);\ \sigma_{oct} = 0 \end{array}\right]$$

Appendix 1

Mohr's Circles

It was stated in Sec. 1.17 that when points with coordinates (σ, τ) for all possible planes passing through a point are marked on the σ–τ plane, as in Fig. 1.16, the points are bounded by the three Mohr's circles. The same equations can be used to determine graphically the normal and shearing stresses on any plane with normal $\boldsymbol{n}$. Equations (1.40)–(1.42) of Sec.1.18 are

$$n_x^2 = \frac{(\sigma - \sigma_2)(\sigma - \sigma_3) + \tau^2}{(\sigma_1 - \sigma_2)(\sigma_1 - \sigma_3)} \tag{A1.1}$$

$$n_y^2 = \frac{(\sigma - \sigma_3)(\sigma - \sigma_1) + \tau^2}{(\sigma_2 - \sigma_3)(\sigma_2 - \sigma_1)} \tag{A1.2}$$

$$n_z^2 = \frac{(\sigma - \sigma_1)(\sigma - \sigma_2) + \tau^2}{(\sigma_3 - \sigma_1)(\sigma_3 - \sigma_2)} \tag{A1.3}$$

For the above equations, the principal axes coincide with the coordinate axes x, y and z. Construct a sphere of unit radius with P as the centre. P_1, P_2 and P_3 are the poles of this sphere (Fig.A1.1). Consider a point N on the surface of the sphere. The radius vector PN makes angles α, β and γ, respectively with the x, y and z axes. A plane through P with PN as normal will be parallel to a tangent plane at N to the unit sphere. If n_x, n_y and n_z are the direction cosines of the normal $\boldsymbol{n}$ to such a plane through P, then $n_x = \cos\alpha$, $n_y = \cos\beta$, $n_z = \cos\gamma$.

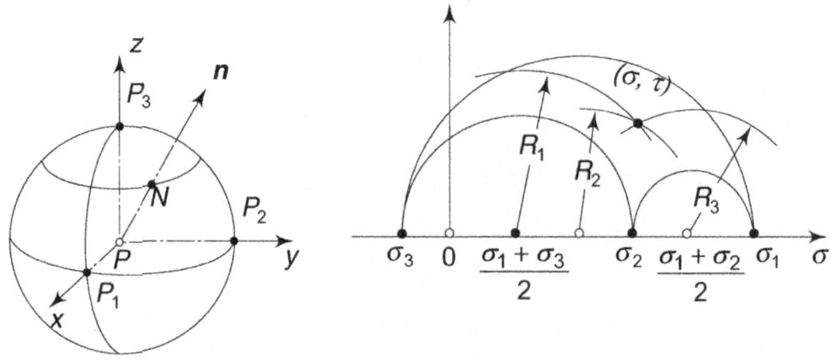

Fig. A1.1 *Mohr's circles for three-dimensional state of stress*

Let point N move in such a manner that γ remains constant. This gives a circle parallel to the equatorial circle P_1P_2.

From Eq. (A1.3)

$$(\sigma - \sigma_1)(\sigma - \sigma_2) + \tau^2 = n_z^2\ (\sigma_3 - \sigma_1)(\sigma_3 - \sigma_2)$$

or
$$\left(\sigma - \frac{\sigma_1 + \sigma_2}{2}\right)^2 + \tau^2 = n_z^2\ (\sigma_3 - \sigma_1)(\sigma_3 - \sigma_2) + \frac{(\sigma_1 - \sigma_2)^2}{4} = R_3^2$$

Since $n_z = \cos\gamma$ is a constant, the above equation describes a circle in the $\sigma - \tau$ plane with the centre at $\frac{\sigma_1 + \sigma_2}{2}$ on the σ axis and radius equal to R_3. This circle gives the values of σ and τ as N moves with γ constant. For different values of n_z, one gets a family of circles, all with centres at $\frac{\sigma_1 + \sigma_2}{2}$. If $n_z = 0$ we get a Mohr's circle.

Similarly, if $n_y = \cos\beta$ is kept constant, the point N on the unit sphere moves on a circle parallel to the circle P_1P_3. The values of σ and τ for different positions of N moving along this circle can be obtained again from (Eq. A1.2) as

$$(\sigma - \sigma_3)(\sigma - \sigma_1) + \tau^2 = n_y^2\ (\sigma_2 - \sigma_3)\ (\sigma_2 - \sigma_1)$$

or
$$\left(\sigma - \frac{\sigma_1 + \sigma_3}{2}\right)^2 + \tau^2 = n_y^2\ (\sigma_2 - \sigma_3)(\sigma_2 - \sigma_1) + \frac{(\sigma_1 - \sigma_3)^2}{4} = R_2^2$$

This describes a circle in the $\sigma - \tau$ plane with the centre at $\frac{(\sigma_1 + \sigma_3)}{2}$ and radius equal to R_2. For different values of n_y, we get a family of circles, all with centres at $\frac{(\sigma_1 + \sigma_3)}{2}$. With $n_y = 0$, we get the outermost circle. Similarly, with $n_x = \cos\alpha$ kept constant, we get another circle with centre at $\frac{(\sigma_2 + \sigma_3)}{2}$ and radius R_1. In order to determine the normal stress σ and shear stress τ on a plane with normal $\boldsymbol{n} = (n_x, n_y, n_z)$, we describe two circles with centres and radii as

centre at $\frac{\sigma_1 + \sigma_3}{2}$ and radius equal to R_2

centre at $\frac{\sigma_1 + \sigma_2}{2}$ and radius equal to R_3

where R_2 and R_3 are as given in the above equation. The intersection point of these two circles locates (σ, τ). The third circle with centre at $\frac{\sigma_2 + \sigma_3}{2}$ and radius R_1 is not an independent circle since among the three direction cosines n_x, n_y and n_z, only two are independent.

Appendix 2

The State of Pure Shear

Theorem: A necessary and sufficient condition for $\overset{n}{T}$ to be a state of pure shear is that the first invariant should be equal to zero, i.e. $I_1 = 0$.

Proof: By definition, $\overset{n}{T}$ is a state of pure shear at P, if there exists at least one frame of reference $Pxyz$, such that with respect to that frame,

$$[\tau_{ij}] = \begin{bmatrix} 0 & \tau_{xy} & \tau_{xz} \\ \tau_{xy} & 0 & \tau_{yz} \\ \tau_{xz} & \tau_{yz} & 0 \end{bmatrix}$$

Therefore, if the state of stress $\overset{n}{T}$ is a pure shear state, then I_1, an invariant, must be equal to zero. This is therefore a necessary condition. To prove that $I_1 = 0$ is also a sufficient condition, we proceed as follows:

Given $I_1 = \sigma_x + \sigma_y + \sigma_z = 0$. Let $Px'y'z'$ be the principal axes at P. If σ_1, σ_2 and σ_3, are the principal stresses then

$$I_1 = \sigma_1 + \sigma_2 + \sigma_3 = 0 \tag{A2.1}$$

From Cauchy's formula, the normal stress σ_n on a plane $\boldsymbol{n}$ with direction cosines $n_{x'}, n_{y'}, n_{z'}$ is

$$\sigma_n = \sigma_1 n_{x'}^2 + \sigma_2 n_{y'}^2 + \sigma_3 n_{z'}^2 \tag{A2.2}$$

We have to show that there exist at least three mutually perpendicular planes on which the normal stresses are zero. Let $\boldsymbol{n}$ be the normal to one such plane. Let $Q(x', y', z')$ be a point on this normal (Fig. A2.1).

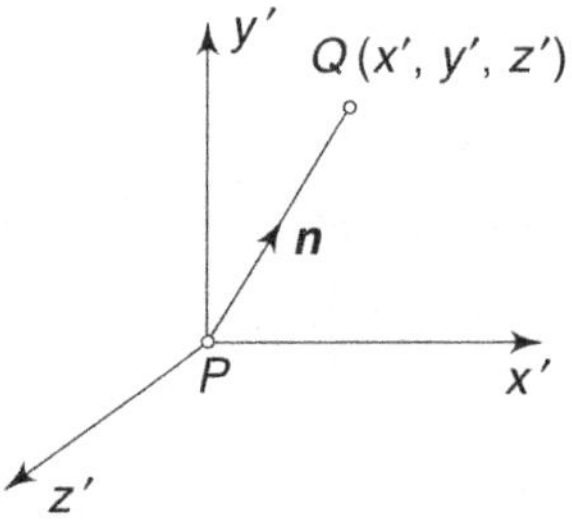

Fig. A2.1 *Normal* ***n*** *to a plane through P*

If $PQ = R$, then,

$$n_{x'} = \frac{x'}{R}, \qquad n_{y'} = \frac{y'}{R}, \qquad n_{z'} = \frac{z'}{R}$$

Since PQ is a pure shear normal, from Eq. (A2.2)

$$\sigma_1 x'^2 + \sigma_2 y'^2 + \sigma_3 z'^2 = R^2 \sigma_n = 0 \tag{A2.3}$$

The problem is to find the locus of Q. Since $I_1 = 0$, two cases are possible.

Case (i) If two of the principal stresses (say σ_1 and σ_2) are positive, the third principal stress σ_3 is negative, i.e.

$$\sigma_1 > 0, \quad \sigma_2 > 0, \quad \sigma_3 = -(\sigma_1 + \sigma_2) < 0$$

The case that σ_1 and σ_2 are negative and σ_3 is positive is similar to the above case, as the result will show.

Case (ii) One of the principal stresses (say σ_3) is zero, so that one of the remaining principal stress σ_1 is positive, and the other is negative, i.e.

$$\sigma_1 > 0, \qquad \sigma_2 = -\sigma_1 < 0, \qquad \sigma_3 = 0$$

The above two cases cover all posibilities. Let us consider case (ii) first since it is the easier one.

Case (ii) From Eq. (A2.3)

$$\sigma_1 x'^2 - \sigma_1 y'^2 = 0$$

or

$$x'^2 - y'^2 = 0$$

The solutions are

(i) $x' = 0$ and $y' = 0$. This represents the z' axis, i.e. the point Q, lies on the z' axis.

(ii) $x' = +y'$ or $x' = -y'$. These represent two mutually perpendicular planes, as shown in Fig. A2.2(a), i.e. the point Q can lie in either of these two planes.

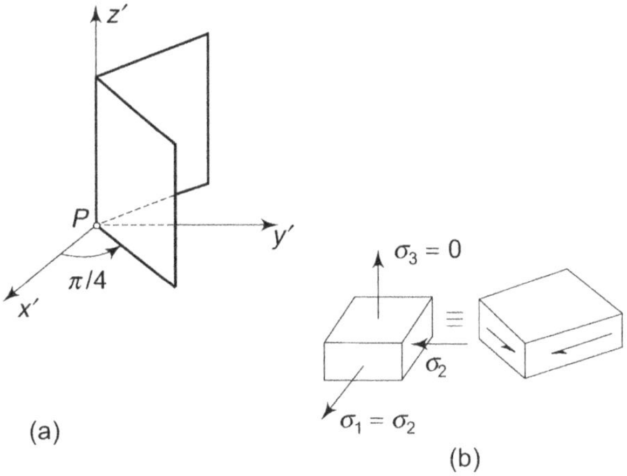

Fig. A2.2 *(a) Planes at 45° (b) Principal stress on an element under plane state of stress*

The above solutions show that for case (ii), i.e when $\sigma_3 = 0$ and $\sigma_1 = -\sigma_2$, there are three pure shear normals. These are the z' axis, an axis lying in the plane $x' = y'$ and another lying the plane $x' = -y'$. This is the elementary case usually discussed in a plane state of stress, as shown in Fig. A2.2(b).

Case (i) Since $\quad \sigma_3 = -(\sigma_1 + \sigma_2)$, Eq. (A2.3) gives

$$\sigma_1 x'^2 + \sigma_2 y'^2 - (\sigma_1 + \sigma_3) z'^2 = 0 \tag{A2.4}$$

This is the equation of an elliptic cone with vertex at P and axis along PZ' (Fig. A2.3). The point $Q(x', y', z')$ can be anywhere on the surface of the cone.

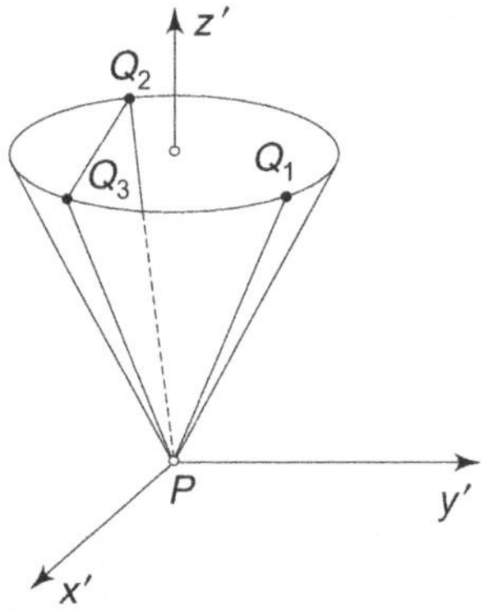

Fig. A2.3 *Cone with vertex at P and axis along PZ'*

Now one has to show that there are at least three mutually perpendicular generators of the above cone. Let $Q_1\,(x_1', y_1', 1)$ be a point on the cone and let S be a plane passing through P and perpendicular to PQ_1. We have to show that the plane S intersects the cone along PQ_2 and PQ_3 and that these two are perpendicular to each other.

Let $Q(x', y', 1)$ be a point in S. Then, S being perpendicular to PQ_1, PQ is perpendicular to PQ_1, i.e.

$$x_1'x' + y_1'y' + 1 = 0 \tag{A2.5}$$

If Q lies on the elliptic cone also, it must satisfy Eq. (A2.4), i.e.

$$\sigma_1 x'^2 + \sigma_2 y'^2 - (\sigma_1 + \sigma_2) = 0 \tag{A 2.6}$$

Multiply Eq. (A2.6) by $2y_1'^2$ and substitute for $y'y_1'$ from Eq. (A2.5). This gives

$$\sigma_1 x'^2\, y_1'^2 + \sigma_2 (x_1'\, x' + 1)^2 - (\sigma_1 + \sigma_2)\, y_1'^2 = 0$$

or

$$(\sigma_1\, y_1'^2 + \sigma_2\, x_1'^2)\, x'^2 + 2\sigma_2 x_1'\, x' + \left[\sigma_2 - (\sigma_1 + \sigma_2)\, y_1'^2\right] = 0 \tag{A 2.7}$$

Similarly, multiplying Eq. (A2.6) by $x_1'^2$ and substituting for $x'x'_1$ from Eq. (A2.5), we get

$$(\sigma_2\, x_1'^2 + \sigma_1\, y_1'^2) y'^2 + 2\sigma_1\, y'y_1' + \left[\sigma_1 - (\sigma_2 + \sigma_1)\, x_1'^2\right] = 0 \tag{A2.8}$$

If $Q(x', y_1', 1)$ is a point lying in S as well as on the cone, then it must satisfy Eqs (A2.5) and (A2.6) or equivalently Eqs (A2.7) and (A2.8). One can solve Eq. (A2.7) for x' and Eq. (A2.8) for y'. Since these are quadratic, we get two solutions for each. Let (x_2', y_2') and (x_3', y_3') be the solutions. Clearly

$$x_2'\, x_3' = \frac{\left[\sigma_2 - (\sigma_1 + \sigma_2)\, y_1'^2\right]}{\left[\sigma_2 x_1'^2 + \sigma_1 y_1'^2\right]} \tag{A 2.9}$$

$$y_2'\, y_3' = \frac{\left[\sigma_1 - (\sigma_2 + \sigma_1)\, x_1'^2\right]}{\left[\sigma_2 x_1'^2 + \sigma_1 y_1'^2\right]} \tag{A 2.10}$$

Adding the above two equations

$$x_2'\, x_3' + y_2'\, y_3' = \frac{\sigma_1 + \sigma_2 - \sigma_1 y_1'^2 - \sigma_2 y_1'^2 - \sigma_2 x_1'^2 - \sigma_1 x_1'^2}{\sigma_2 x_1'^2 + \sigma_1 y_1'^2}$$

Since $Q_1(x_1', y_1', 1)$ is on the cone and recalling that $\sigma_1 + \sigma_2 = -\sigma_3$, the right-hand side is equal to -1, i.e.

$$x_2' \, x_3' + y_2' \, y_3' + 1 = 0$$

Consequently, PQ_2 and PQ_3 are perpendicular to each other if $Q_2 = (x_2', \, y_2', \, 1)$ and $Q_3 \, (x_3', y_3', 1)$ are real. If x_2', x_3' and y_2', y_3', the solutions of Eqs (A2.7) and (A2.8), are to be real, then the descriminants must be greater than zero. For this, let $Q_1 \, (x_1', y_1', 1)$ be specifically $Q_1(1, \, 1, \, 1)$ i.e. choose $x_1' = y_1' = 1$. Both the descriminants of Eqs (A2.7) and (A2.8) then are

$$4(\sigma_1^2 + \sigma_1\sigma_2 + \sigma_2^2)$$

The above quantity is greater than zero, since $\sigma_1 > 0$ and $\sigma_2 > 0$. Therefore, x_2', x_3' and y_2', y_3' are real.

Appendix 3

Stress Quadric of Cauchy

Let $\overset{n}{\boldsymbol{T}}$ be the resultant stress vector at point P (see Fig. A3.1) on a plane with unit normal $\boldsymbol{n}$. The stress surface S associated with a given state of stress $\overset{n}{\boldsymbol{T}}$ is defined as the locus of all points Q, such that

$$\boldsymbol{PQ} = R\boldsymbol{n}$$

where

$$R = |\boldsymbol{PQ}| = \frac{1}{\left(|\sigma(\boldsymbol{n})|\right)^{1/2}}$$

and $\sigma(\boldsymbol{n})$ is the normal stress component on the plane $\boldsymbol{n}$. This means that a point Q is chosen along $\boldsymbol{n}$ such that $R = 1/\sqrt{\sigma}$. If such Qs are marked for every plane passing through P, then we get a surface S and this surface determines the normal component of stress on any plane through P. The surface consists of S_t and S_c—the tensile and the compressive branches of the surface.

The normal to the surface S at $Q(n)$ is parallel to $\overset{n}{\boldsymbol{T}}$. Thus, S completely determines the state of stress at P. The following cases are possible.

Case (i) $\sigma_1 \neq 0$, $\sigma_2 \neq 0$, $\sigma_3 \neq 0$; S_t and S_c are each a central quadric surface about $\boldsymbol{P}$ with axes along $\boldsymbol{n}_x$, $\boldsymbol{n}_y$ and $\boldsymbol{n}_z$.

(i) If σ_1, σ_2 and σ_3 all have the same sign, say $\sigma_1 > 0$, $\sigma_2 > 0$, $\sigma_3 > 0$ then $S = S_t$ is an ellipsoid with axes along $\boldsymbol{n}_x$, $\boldsymbol{n}_y$ and $\boldsymbol{n}_z$ at $\boldsymbol{P}$. There are two cases
   (a) If $\sigma_1 = \sigma_2 \neq \sigma_3$, then $S = S_t$ is a spheroid with polar axis along $\boldsymbol{n}_z$
   (b) If $\sigma_1 = \sigma_2 = \sigma_3$, then $S = S_t$ is a sphere.

(ii) If σ_1, σ_2 and σ_3 are not all of the same sign, say $\sigma_1 > 0$, $\sigma_2 > 0$ and $\sigma_3 < 0$, then S_t is a hyperboloid with one sheet and S_c is a double sheeted hyperboloid, the vertices of which are along the n_z axis. In particular, if $\sigma_1 = \sigma_2$, then S_t and S_c are hyperboloids of revolution with a polar axis along n_z.

Case (ii) Let $\sigma_1 \neq 0$, $\sigma_2 \neq 0$ and $\sigma_3 = 0$ (i.e. plane state). The S_t and S are right second-order cylinders whose generators are parallel to n_z and whose cross-sections have axes along $\boldsymbol{n}_x$ and $\boldsymbol{n}_y$. In this case, two possibilities can be considered.

(i) If $\sigma_1 > 0$, $\sigma_2 > 0$, then $S = S_t$ is an elliptic cylinder. In particular, If $\sigma_1 = \sigma_2$ then $S = S_t$ is a circular cylinder.

(ii) If $\sigma_1 > 0$ and $\sigma_2 < 0$, then S_t is a hyperbolic cylinder whose cross-section has vertices on the $\boldsymbol{n}_x$ axis and S_c is a hyperbolic cylinder.

Case (iii) If $\sigma_1 \neq 0$ and $\sigma_2 = \sigma_3 = 0$ (uniaxial state) and say $\sigma_1 > 0$ then $S = S_t$ consists of two parallel planes, each perpendicular to $\boldsymbol{n}_x$ and equidistant from P.

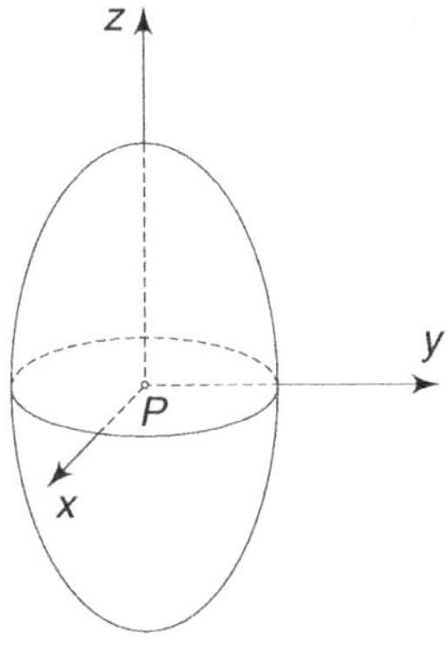

Fig. A3.1 *Ellipsoidal surface*

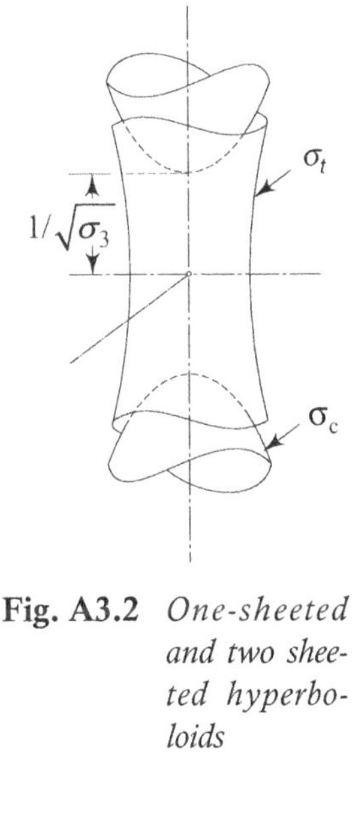

Fig. A3.2 *One-sheeted and two sheeted hyperboloids*

One can prove the above statements directly from Eqs (1.53) of Sec. 1.23. These equations are

$$S_t : \sigma_1 x^2 + \sigma_2 y^2 + \sigma_3 z^2 = 1$$
$$S_c : \sigma_1 x^2 + \sigma_2 y^2 + \sigma_3 z^2 = -1$$

These can be rewritten as

$$S_t : \frac{x^2}{\left(1/\sqrt{\sigma_1}\right)^2} + \frac{y^2}{\left(1/\sqrt{\sigma_2}\right)^2} + \frac{z^2}{\left(1/\sqrt{\sigma_3}\right)^2} = 1$$

$$S_c : \frac{x^2}{\left(1/\sqrt{\sigma_1}\right)^2} + \frac{y^2}{\left(1/\sqrt{\sigma_2}\right)^2} + \frac{z^2}{\left(1/\sqrt{\sigma_3}\right)^2} = -1 \quad \text{(A3.1)}$$

Case (i) $\sigma_1 \neq 0$, $\sigma_2 \neq 0$, $\sigma_3 \neq 0$

(i) $\sigma_1 > 0$, $\sigma_2 > 0$, $\sigma_3 > 0$

Equation (A3.1) shows that S_c is an imaginary surface and hence, $S = S_t$. This equation represents an ellipsoid.

(a) If $\sigma_1 = \sigma_2 \neq \sigma_3$ the central section is a circle

(b) If $\sigma_1 = \sigma_2 = \sigma_3$ the surface is a sphere

(ii) If $\sigma_1 > 0$, $\sigma_2 > 0$, $\sigma_3 < 0$

$$S_t : \frac{x^2}{\left(1/\sqrt{\sigma_1}\right)^2} + \frac{y^2}{\left(1/\sqrt{\sigma_2}\right)^2} - \frac{z^2}{\left(1/\sqrt{\sigma_3}\right)^2} = 1$$

$$S_c : -\frac{x^2}{\left(1/\sqrt{\sigma_1}\right)^2} - \frac{y^2}{\left(1/\sqrt{\sigma_2}\right)^2} + \frac{z^2}{\left(1/\sqrt{\sigma_3}\right)^2} = 1 \quad \text{(A3.2)}$$

Hence, S_t is a one-sheeted hyperboloid and S_c is a two-sheeted hyperbloid. This is shown in Fig. A3.2.

Case (ii) Let $\sigma_1 \neq 0$, $\sigma_2 \neq 0$ and $\sigma_3 = 0$. Then Eq. (1.53) reduces to

$$\sigma_1 x^2 + \sigma_2 y^2 = \pm 1 \quad \text{(A3.3)}$$

This is obviously a second-order cylinder, the surface of which is made of straight lines parallel to the z-axis, passing through every point of the curve in the xy plane, of which an equation in that plane is expressed by Eq. (A3.3).

(i) If $\sigma_1 > 0$ and $\sigma_2 > 0$, the above equation becomes

$$\sigma_1 x^2 + \sigma_2 y^2 + 1$$

or

$$\frac{x^2}{(1/\sigma_1)^2} + \frac{y^2}{(1/\sigma_2)^2} = 1$$

This is the equation of an ellipse in xy plane. Hence, $S = S_t$ is an elliptic cylinder.

In particular, if $\sigma_1 = \sigma_2$, the elliptic cylinder becomes a circular cylinder.

(ii) If $\sigma_1 > 0$ and $\sigma_2 < 0$, then the equation becomes

$$\sigma_1 x^2 - |\sigma_2|\, y^2 = \pm 1$$

or

$$x^2/(1/\sigma_1)^2 - y^2\,(1/\sigma_2^2) = \pm 1$$

This describes conjugate hyperbolas in the xy plane. S_t is given by a hyperbolic cylinder, the cross-sectional vertices of which lie on the $\boldsymbol{n}_x$ axis and S_c is given by a hyperbolic cylinder with its cross-sectional vertices lying on the $\boldsymbol{n}$ axis.

Case (iii) If $\sigma_1 \neq 0$, $\sigma_2 = \sigma_3 = 0$, Eq. (1.53) reduces to

$$\sigma_1\, x^2 = \pm 1$$

When $\sigma_1 > 0$, this becomes

$$x^2 = 1/\sigma_1$$

or

$$x = \pm 1/\sqrt{\sigma_1}$$

This represents two straight lines parallel to the y axis and equidistant from it. Hence, $S = S_t$ is given by two parallel planes, each perpendicular to $\boldsymbol{n}_x$ and equidistant from P.

CHAPTER 2

Analysis of Strain

2.1 INTRODUCTION

In this chapter the state of strain at a point will be analysed. In elementary strength of materials two types of strains were introduced: (i) the extensional strain (in x or y direction) and (ii) the shear strain in the xy plane. Figure 2.1 illustrates these two simple cases of strain. In each case, the initial or undeformed position of the element is indicated by full lines and the changed position by dotted lines. These are two-dimensional strains.

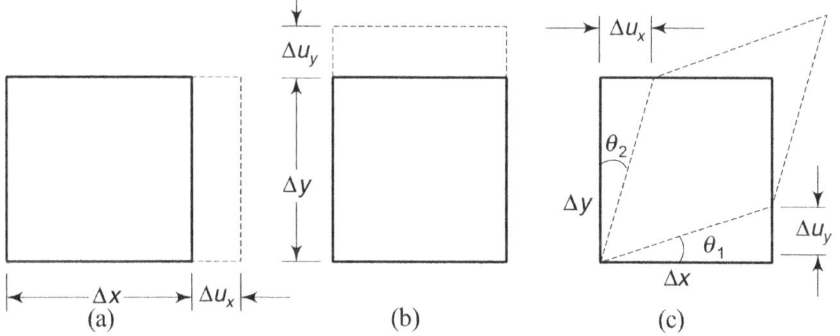

Fig. 2.1 *(a) Linear strain in x direction (b) linear strain in y direction (c) shear strain in xy plane*

In Fig. 2.1(a), the element undergoes an extension Δu_x in x direction. The extensional or linear strain is defined as the change in length per unit initial length. If ε_x denotes the linear strain in x direction, then

$$\varepsilon_x = \frac{\Delta u_x}{\Delta x} \tag{2.1}$$

Similarly, the linear strain in y direction [Fig. 2.1(b)] is

$$\varepsilon_y = \frac{\Delta u_y}{\Delta y} \tag{2.2}$$

Figure 2.1(c) shows the shear strain γ_{xy} in the xy plane. Shear strain γ_{xy} is defined as the change in the initial right angle between two line elements originally parallel to the x and y axes. In the figure, the total change in the angle is $\theta_1 + \theta_2$. If θ_1 and θ_2 are very small, then one can put

$$\theta_1 \text{ (in radians)} + \theta_2 \text{ (in radians)} = \tan\theta_1 + \tan\theta_2$$

From Fig. 2.1(c)

$$\tan\theta_1 = \frac{\Delta u_y}{\Delta x}, \qquad \tan\theta_2 = \frac{\Delta u_x}{\Delta y} \tag{2.3}$$

Therefore, the shear strain γ_{xy} is

$$\gamma_{xy} = \theta_1 + \theta_2 = \frac{\Delta u_y}{\Delta x} + \frac{\Delta u_x}{\Delta y} \tag{2.4}$$

Reduction in the initial right angle is considered to be a positive shear strain, since positive shear stress components τ_{xy} and τ_{yx} cause a decrease in the right angle.

In addition to these two types of strains, a third type of strain, called the volumetric strain, was also introduced in elementary strength of materials. This is change in volume per unit original volume. In this chapter, we will study strains in three dimensions and we will begin with the study of deformations.

2.2 DEFORMATIONS

In order to study deformation or change in the shape of a body, we compare the positions of material points before and after deformation. Let a point P belonging to the body and having coordinates (x, y, z) be displaced after deformations to P' with coordinates (x', y', z') (Fig. 2.2). Since P is displaced to P', the vector segment $\boldsymbol{PP'}$ is called the displacement vector and is denoted by $\boldsymbol{u}$.

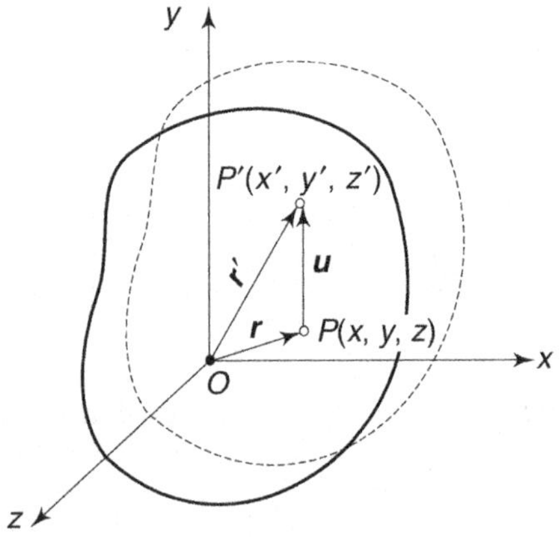

Fig. 2.2 *Displacement of point P to P′*

The displacement vector $\boldsymbol{u}$ has components u_x, u_y and u_z along the x, y and z axes respectively, and one can write

$$\boldsymbol{u} = \boldsymbol{i}u_x + \boldsymbol{j}u_y + \boldsymbol{k}u_z \tag{2.5}$$

The displacement undergone by any point is a function of its initial coordinates. We assume that the displacement is defined throughout the volume of the body, i.e. the displacement vector $\boldsymbol{u}$ (both in magnitude and direction) of any point P belonging to the body is known once its coordinates are known. Then we can say that a displacement vector field has been defined throughout the volume of the body. If $\boldsymbol{r}$ is the position vector of point P, and $\boldsymbol{r}'$ that of point P', then

$$\begin{aligned} \boldsymbol{r}' &= \boldsymbol{r} + \boldsymbol{u} \\ \boldsymbol{u} &= \boldsymbol{r}' - \boldsymbol{r} \end{aligned} \tag{2.6}$$

Example 2.1 *The displacement field for a body is given by*

$$\boldsymbol{u} = (x^2 + y)\boldsymbol{i} + (3 + z)\boldsymbol{j} + (x^2 + 2y)\boldsymbol{k}$$

What is the deformed position of a point originally at (3, 1, –2)?

Solution The displacement vector $\boldsymbol{u}$ at (3, 1, –2) is

$$\boldsymbol{u} = (3^2 + 1)\boldsymbol{i} + (3 - 2)\boldsymbol{j} + (3^2 + 2)\boldsymbol{k}$$
$$= 10\boldsymbol{i} + \boldsymbol{j} + 11\boldsymbol{k}$$

The initial position vector $\boldsymbol{r}$ of point P is

$$\boldsymbol{r} = 3\boldsymbol{i} + \boldsymbol{j} - 2\boldsymbol{k}$$

The final position vector $\boldsymbol{r}'$ of point P' is

$$\boldsymbol{r}' = \boldsymbol{r} + \boldsymbol{u} = 13\boldsymbol{i} + 2\boldsymbol{j} + 9\boldsymbol{k}$$

Example 2. 2 *Two points P and Q in the undeformed body have coordinates (0, 0, 1) and (2, 0, –1) respectively. Assuming that the displacement field given in Example 2.1 has been imposed on the body, what is the distance between points P and Q after deformation?*

Solution The displacement vector at point P is

$$\boldsymbol{u}(P) = (0 + 0)\boldsymbol{i} + (3 + 1)\boldsymbol{j} + (0 + 0)\boldsymbol{k} = 4\boldsymbol{j}$$

The displacement components at P are $u_x = 0$, $u_y = 4$, $u_z = 0$. Hence, the final coordinates of P after deformation are

$$P' : x + u_x = 0 + 0 = 0$$
$$y + u_y = 0 + 4 = 4$$
$$z + u_z = 1 + 0 = 1$$

or $$P': (0, 4, 1)$$

Similarly, the displacement components at point Q are,

$$u_x = 4, \quad u_y = 2, \quad u_z = 4$$

and the coordinates of Q' are (6, 2, 3).
The distance $P'Q'$ is therefore

$$d' = (6^2 + 2^2 + 2^2)^{1/2} = 2\sqrt{11}$$

2.3 DEFORMATION IN THE NEIGHBOURHOOD OF A POINT

Let P be a point in the body with coordinates (x, y, z). Consider a small region surrounding the point P. Let Q be a point in this region with coordinates $(x + \Delta x, y + \Delta y, z + \Delta z)$. When the body undergoes deformation, the points P and Q move to P' and Q'. Let the displacement vector $\boldsymbol{u}$ at P have components (u_x, u_y, u_z) (Fig. 2.3).

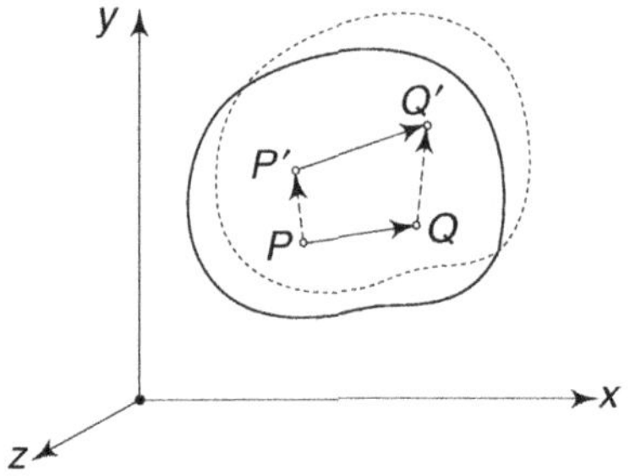

Fig. 2.3 *Displacements of two neighbouring points P and Q*

The coordinates of P, P' and Q are

P: (x, y, z)

P': $(x + u_x, y + u_y, z + u_z)$

Q: $(x + \Delta x, y + \Delta y, z + \Delta z)$

The displacement components at Q differ slightly from those at P since Q is away from P by Δx, Δy and Δz. Consequently, the displacements at Q are,

$$u_x + \Delta u_x,\ u_y + \Delta u_y,\ u_z + \Delta u_z.$$

If Q is very close to P, then to first-order approximation

$$\Delta u_x = \frac{\partial u_x}{\partial x}\Delta x + \frac{\partial u_x}{\partial y}\Delta y + \frac{\partial u_x}{\partial z}\Delta z \tag{2.7a}$$

The first term on the right-hand side is the rate of increase of u_x in x direction multiplied by the distance traversed, Δx. The second term is the rate of increase of u_x in y direction multiplied by the distance traversed in y direction, i.e. Δy. Similarly, we can also interpret the third term. For Δu_y and Δu_z too, we have

$$\Delta u_y = \frac{\partial u_y}{\partial x}\Delta x + \frac{\partial u_y}{\partial y}\Delta y + \frac{\partial u_y}{\partial z}\Delta z \tag{2.7b}$$

$$\Delta u_z = \frac{\partial u_z}{\partial x}\Delta x + \frac{\partial u_z}{\partial y}\Delta y + \frac{\partial u_z}{\partial z}\Delta z \tag{2.7c}$$

Therefore, the coordinates of Q' are,

$$Q' = (x + \Delta x + u_x + \Delta u_x,\ y + \Delta y + u_y + \Delta u_y,\ z + \Delta z + u_z + \Delta u_z) \tag{2.8}$$

Before deformation, the segment PQ had components Δx, Δy and Δz along the three axes. After deformation, the segment $P'Q'$ has components $\Delta x + \Delta u_x$, $\Delta y + \Delta u_y$, $\Delta z + \Delta u_z$ along the three axes. Terms like,

$$\frac{\partial u_x}{\partial x}, \frac{\partial u_x}{\partial y}, \frac{\partial u_x}{\partial z}, \text{ etc.}$$

are important in the analysis of strain. These are the gradients of the displacement components (at a point P) in x, y and z directions. One can represent these in the form of a matrix called the displacement-gradient matrix as

$$\left[\frac{\partial u_i}{\partial x_j}\right] = \begin{bmatrix} \frac{\partial u_x}{\partial x} & \frac{\partial u_x}{\partial y} & \frac{\partial u_x}{\partial z} \\ \frac{\partial u_y}{\partial x} & \frac{\partial u_y}{\partial y} & \frac{\partial u_y}{\partial z} \\ \frac{\partial u_z}{\partial x} & \frac{\partial u_z}{\partial y} & \frac{\partial u_z}{\partial z} \end{bmatrix}$$

Example 2.3 *The following displacement field is imposed on a body*

$$\boldsymbol{u} = (xy\boldsymbol{i} + 3x^2z\boldsymbol{j} + 4\boldsymbol{k})10^{-2}$$

Consider a point P and a neighbouring point Q where PQ has the following direction cosines

$$n_x = 0.200, \quad n_y = 0.800, \quad n_z = 0.555$$

Point P has coordinates (2, 1, 3). If PQ = Δs, find the components of **P′Q′** *after deformation.*

Solution Before deformation, the components of PQ are

$$\Delta x = n_x \Delta s = 0.2 \, \Delta s$$
$$\Delta y = n_y \Delta s = 0.8 \, \Delta \text{s}$$
$$\Delta z = n_z \Delta s = 0.555 \, \Delta s$$

Using Eqs (2.7a)–(2.7c), the values of Δu_x, Δu_y and Δu_z can be calculated. We are using $p = 10^{-2}$;

$$u_x = pxy \qquad u_y = 3px^2z \qquad u_z = 4p$$

$$\frac{\partial u_x}{\partial x} = py \qquad \frac{\partial u_y}{\partial x} = 6\,pxz \qquad \frac{\partial u_z}{\partial x} = 0$$

$$\frac{\partial u_x}{\partial y} = px \qquad \frac{\partial u_y}{\partial y} = 0 \qquad \frac{\partial u_z}{\partial y} = 0$$

$$\frac{\partial u_x}{\partial z} = 0 \qquad \frac{\partial u_y}{\partial z} = 3px^2 \qquad \frac{\partial u_z}{\partial z} = 0$$

At point $P(2, 1, 3)$ therefore,

$$\Delta u_x = (y\Delta x + x\Delta y)p = (\Delta x + 2\Delta y)p$$
$$\Delta u_y = (6xz\Delta x + 3x^2\Delta z)p = (36\Delta x + 12\Delta z)p$$
$$\Delta u_z = 0$$

Substituting for Δx, Δy and Δz, the components of $\Delta s' = |\boldsymbol{P'Q'}|$ are

$$\Delta x + \Delta u_x = 1.01 \, \Delta x + 0.02 \, \Delta y = (0.202 + 0.016) \, \Delta s = 0.218 \, \Delta s$$
$$\Delta y + \Delta u_y = (0.36 \, \Delta x + \Delta y + 0.12 \, \Delta z) = (0.072 + 0.8 + 0.067) \, \Delta s$$
$$= 0.939 \, \Delta s$$
$$\Delta z + \Delta u_z = \Delta z = 0.555 \, \Delta s$$

Hence, the new vector **P′Q′** can be written as

$$\boldsymbol{P'Q'} = (0.218\boldsymbol{i} + 0.939\boldsymbol{j} + 0.555\boldsymbol{k})\Delta s$$

2.4 CHANGE IN LENGTH OF A LINEAR ELEMENT

Deformation causes a point $P(x, y, z)$ in the solid body under consideration to be displaced to a new position P' with coordinates $(x + u_x, y + u_y, z + u_z)$ where u_x, u_y and u_z are the displacement components. A neighbouring point Q with coordinates $(x + \Delta x, y + \Delta y, z + \Delta z)$ gets displaced to Q' with new coordinates $(x + \Delta x + u_x + \Delta u_x, y + \Delta y + u_y + \Delta u_y, z + \Delta z + u_z + \Delta u_z)$. Hence, it is possible to determine the change in the length of the line element PQ caused by deformation. Let Δs be the length of the line element PQ. Its components are

$$\Delta s: (\Delta x, \Delta y, \Delta z)$$

$$\therefore \quad \Delta s^2: (PQ)^2 = \Delta x^2 + \Delta y^2 + \Delta z^2$$

Let $\Delta s'$ be the length of $P'Q'$. Its components are

$$\Delta s': (\Delta x' = \Delta x + \Delta u_x, \Delta y' = \Delta y + \Delta u_y, \Delta z' = \Delta z + \Delta u_z)$$

$$\therefore \quad \Delta s'^2: (P'Q')^2 = (\Delta x + \Delta u_x)^2 + (\Delta y + \Delta u_y)^2 + (\Delta z + \Delta u_z)^2$$

From Eqs (2.7a)–(2.7c),

$$\begin{aligned} \Delta x' &= \left(1 + \frac{\partial u_x}{\partial x}\right)\Delta x + \frac{\partial u_x}{\partial y}\Delta y + \frac{\partial u_x}{\partial z}\Delta z \\ \Delta y' &= \frac{\partial u_y}{\partial x}\Delta x + \left(1 + \frac{\partial u_y}{\partial y}\right)\Delta y + \frac{\partial u_y}{\partial z}\Delta z \\ \Delta z' &= \frac{\partial u_z}{\partial x}\Delta x + \frac{\partial u_z}{\partial y}\Delta y + \left(1 + \frac{\partial u_z}{\partial z}\right)\Delta z \end{aligned} \tag{2.9}$$

We take the difference between $\Delta s'^2$ and Δs^2

$$\begin{aligned} (P'Q')^2 - (PQ)^2 &= \Delta s'^2 - \Delta s^2 \\ &= (\Delta x'^2 + \Delta y'^2 + \Delta z'^2) - (\Delta x^2 + \Delta y^2 + \Delta z^2) \\ &= 2(E_{xx}\Delta x^2 + E_{yy}\Delta y^2 + E_{zz}\Delta z^2 + E_{xy}\Delta x\,\Delta y \\ &\quad + E_{yz}\Delta y\,\Delta z + E_{xz}\Delta x\,\Delta z) \end{aligned} \tag{2.10}$$

where

$$\begin{aligned} E_{xx} &= \frac{\partial u_x}{\partial x} + \frac{1}{2}\left[\left(\frac{\partial u_x}{\partial x}\right)^2 + \left(\frac{\partial u_y}{\partial x}\right)^2 + \left(\frac{\partial u_z}{\partial x}\right)^2\right] \\ E_{yy} &= \frac{\partial u_y}{\partial y} + \frac{1}{2}\left[\left(\frac{\partial u_x}{\partial y}\right)^2 + \left(\frac{\partial u_y}{\partial y}\right)^2 + \left(\frac{\partial u_z}{\partial y}\right)^2\right] \\ E_{zz} &= \frac{\partial u_z}{\partial z} + \frac{1}{2}\left[\left(\frac{\partial u_x}{\partial z}\right)^2 + \left(\frac{\partial u_y}{\partial z}\right)^2 + \left(\frac{\partial u_z}{\partial z}\right)^2\right] \\ E_{xy} &= \frac{\partial u_x}{\partial y} + \frac{\partial u_y}{\partial x} + \frac{\partial u_x}{\partial x}\frac{\partial u_x}{\partial y} + \frac{\partial u_y}{\partial x}\frac{\partial u_y}{\partial y} + \frac{\partial u_z}{\partial x}\frac{\partial u_z}{\partial y} \\ E_{yz} &= \frac{\partial u_y}{\partial z} + \frac{\partial u_z}{\partial y} + \frac{\partial u_x}{\partial y}\frac{\partial u_x}{\partial z} + \frac{\partial u_y}{\partial y}\frac{\partial u_y}{\partial z} + \frac{\partial u_z}{\partial y}\frac{\partial u_z}{\partial z} \\ E_{xz} &= \frac{\partial u_x}{\partial z} + \frac{\partial u_z}{\partial x} + \frac{\partial u_x}{\partial x}\frac{\partial u_x}{\partial z} + \frac{\partial u_y}{\partial x}\frac{\partial u_y}{\partial z} + \frac{\partial u_z}{\partial x}\frac{\partial u_z}{\partial z} \end{aligned} \tag{2.11}$$

It is observed that

$$E_{xy} = E_{yx}, \qquad E_{yz} = E_{zy}, \qquad E_{xz} = E_{zx}$$

We introduce the notation

$$E_{PQ} = \frac{\Delta s' - \Delta s}{\Delta s} \tag{2.12}$$

E_{PQ} is the ratio of the increase in distance between the points P and Q caused by the deformation to their initial distance. This quantity will be called the relative extension at point P in the direction of point Q. Now,

$$\frac{\Delta s'^2 - \Delta s^2}{2} = \left(\frac{\Delta s' - \Delta s}{\Delta s} + \frac{(\Delta s' - \Delta s)^2}{2\Delta s^2} \right) \Delta s^2$$

$$= \left(E_{PQ} + \frac{1}{2} E_{PQ}^2 \right) \Delta s^2 \tag{2.13}$$

$$= E_{PQ} \left(1 + \frac{1}{2} E_{PQ} \right) \Delta s^2$$

From Eq. (2.10), substituting for $(\Delta s'^2 - \Delta s^2)$

$$E_{PQ}\left(1 + \frac{1}{2} E_{PQ}\right) \Delta s^2 = E_{xx}\, \Delta x^2 + E_{yy}\, \Delta y^2 + E_{zz}\, \Delta z^2$$

$$+ E_{xy}\, \Delta x\, \Delta y + E_{yz}\, \Delta y\, \Delta z + E_{xz}\, \Delta x\, \Delta z$$

If n_x, n_y and n_z are the direction cosines of PQ, then

$$n_x = \frac{\Delta x}{\Delta s}, \qquad n_y = \frac{\Delta y}{\Delta s}, \qquad n_z = \frac{\Delta z}{\Delta s}$$

Substituting these in the above expression

$$E_{PQ}\left(1 + \frac{1}{2} E_{PQ}\right) = E_{xx} n_x^2 + E_{yy} n_y^2 + E_{zz} n_z^2 + E_{xy} n_x n_y$$

$$+ E_{yz} n_y n_z + E_{xz} n_x n_z \tag{2.14}$$

Equation (2.14) gives the value of the relative extension at point P in the direction PQ with direction cosines n_x, n_y and n_z.

If the line segment PQ is parallel to the x axis before deformation, then $n_x = 1$, $n_y = n_z = 0$ and

$$E_x \left(1 + \frac{1}{2} E_x\right) = E_{xx} \tag{2.15}$$

Hence,
$$E_x = [1 + 2E_{xx}]^{1/2} - 1 \tag{2.16}$$

This gives the relative extension of a line segment originally parallel to the x-axis. By analogy, we get

$$E_y = [1 + 2E_{yy}]^{1/2} - 1, \qquad E_z = [1 + 2E_{zz}]^{1/2} - 1 \tag{2.17}$$

2.5 CHANGE IN LENGTH OF A LINEAR ELEMENT—LINEAR COMPONENTS

Equation (2.11) in the previous section contains linear quantities like $\partial u_x/\partial x$, $\partial u_y/\partial y$, $\partial u_x/\partial y$, . . ., etc., as well as non-linear terms, like $(\partial u_x/\partial x)^2$, $(\partial u_x/\partial x \cdot \partial u_y/\partial y)$, ..., etc. If the deformation imposed on the body is small, the quantities like $\partial u_x/\partial x$, $\partial u_y/\partial y$, etc. are extremely small so that their squares and products can be neglected. Retaining only linear terms, the following equations are obtained

$$\varepsilon_{xx} = \frac{\partial u_x}{\partial x}, \qquad \varepsilon_{yy} = \frac{\partial u_y}{\partial y}, \qquad \varepsilon_{zz} = \frac{\partial u_z}{\partial z} \tag{2.18}$$

$$\gamma_{xy} = \frac{\partial u_x}{\partial y} + \frac{\partial u_y}{\partial x}, \qquad \gamma_{yz} = \frac{\partial u_y}{\partial z} + \frac{\partial u_z}{\partial y}, \qquad \gamma_{xz} = \frac{\partial u_x}{\partial z} + \frac{\partial u_z}{\partial x} \tag{2.19}$$

$$E_{PQ} \approx \varepsilon_{PQ} = \varepsilon_{xx} n_x^2 + \varepsilon_{yy} n_y^2 + \varepsilon_{zz} n_z^2 + \varepsilon_{xy} n_x n_y + \varepsilon_{yz} n_y n_z + \varepsilon_{xz} n_x n_z \tag{2.20}$$

Equation 2.20 directly gives the linear strain at point P in the direction PQ with direction cosines n_x, n_y, n_z. When $n_x = 1$, $n_y = n_z = 0$, the line element PQ is parallel to the x axis and the linear strain is

$$E_x \approx \varepsilon_{xx} = \frac{\partial u_x}{\partial x}$$

Similarly, $$E_y \approx \varepsilon_{yy} = \frac{\partial u_y}{\partial y} \quad \text{and} \quad E_z \approx \varepsilon_{zz} = \frac{\partial u_z}{\partial z}$$

are the linear strains in y and z directions respectively. In the subsequent analyses, we will use only the linear terms in strain components and neglect squares and products of strain components. The relations expressed by Eqs (2.18) and (2.19) are known as the strain displacement relations of Cauchy.

2.6 RECTANGULAR STRAIN COMPONENTS

ε_{xx}, ε_{yy} and ε_{zz} are the linear strains at a point in x, y and z directions. It will be shown later that γ_{xy}, γ_{yz} and γ_{xz} represent shear strains in xy, yz and xz planes respectively. Analogous to the rectangular stress components, these six strain components are called the rectangular strain components at a point.

2.7 THE STATE OF STRAIN AT A POINT

Knowing the six rectangular strain components at a point P, one can calculate the linear strain in any direction PQ, using Eq. (2.20). The totality of all linear strains in every possible direction PQ defines the state of strain at point P. This definition is similar to that of the state of stress at a point. Since all that is required to determine the state of strain are the six rectangular strain components, these six components are said to define the state of strain at a point. We can write this as

$$[\varepsilon_{ij}] = \begin{bmatrix} \varepsilon_{xx} & \gamma_{xy} & \gamma_{xz} \\ \gamma_{xy} & \varepsilon_{yy} & \gamma_{yz} \\ \gamma_{xz} & \gamma_{yz} & \varepsilon_{zz} \end{bmatrix} \tag{2.21}$$

To maintain consistency, we could have written

$$\varepsilon_{xy} = \gamma_{xy}, \qquad \varepsilon_{yz} = \gamma_{yz}, \qquad \varepsilon_{xz} = \gamma_{xz}$$

but as it is customary to represent the shear strain by γ, we have retained this notation. In the theory of elasticity, $1/2\gamma_{xy}$ is written as e_{xy}, i.e.

$$\frac{1}{2}\gamma_{xy} = \frac{1}{2}\left(\frac{\partial u_x}{\partial y} + \frac{\partial u_y}{\partial x}\right) = e_{xy} \tag{2.22}$$

If we follow the above notation and use

$$e_{xx} = \varepsilon_{xx}, \qquad e_{yy} = \varepsilon_{yy}, \qquad e_{zz} = \varepsilon_{zz}$$

then Eq. (2.20) can be written in a very short form as

$$\varepsilon_{PQ} = \sum_i \sum_j e_{ij}\, n_i n_j$$

where i and j are summed over x, y and z, Note that $e_{ij} = e_{ji}$

2.8 INTERPRETATION OF γ_{xy}, γ_{yz}, γ_{xz} AS SHEAR STRAIN COMPONENTS

It was shown in the previous section that

$$\varepsilon_{xx} = \frac{\partial u_x}{\partial x}, \qquad \varepsilon_{yy} = \frac{\partial u_y}{\partial y}, \qquad \varepsilon_{zz} = \frac{\partial u_z}{\partial z}$$

represent the linear strains of line elements parallel to the x, y and z axes respectively. It was also stated that

$$\gamma_{xy} = \frac{\partial u_x}{\partial y} + \frac{\partial u_y}{\partial x}, \qquad \gamma_{yz} = \frac{\partial u_y}{\partial z} + \frac{\partial u_z}{\partial y}, \qquad \gamma_{xz} = \frac{\partial u_x}{\partial z} + \frac{\partial u_z}{\partial x}$$

represent the shear strains in the xy, yz and xz planes respectively. This can be shown as follows.

Consider two line elements, PQ and PR, originally perpendicular to each other and parallel to the x and y axes respectively (Fig. 2.4a).

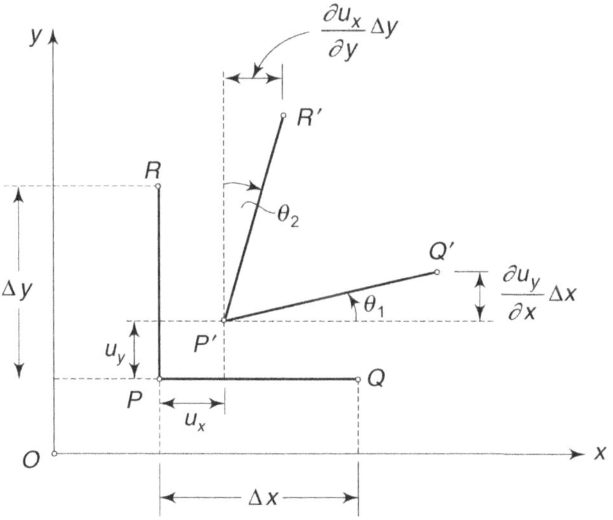

Fig. 2.4 *(a) Change in orientations of line segments PQ and PR-shear strain*

Let the coordinates of P be (x, y) before deformation and let the lengths of PQ and PR be Δx and Δy respectively. After deformation, point P moves to P', point Q to Q' and point R to R'.

Let u_x, u_y be the displacements of point P, so that the coordinates of P' are $(x + u_x,\ y + u_y)$. Since point Q is Δx distance away from P, the displacement components of $Q(x + \Delta x, y)$ are

$$u_x + \frac{\partial u_x}{\partial x}\,\Delta x \quad \text{and} \quad u_y + \frac{\partial u_y}{\partial x}\,\Delta x$$

Similarly, the displacement components of $R(x, y + \Delta y)$ are

$$u_x + \frac{\partial u_x}{\partial y}\Delta y \quad \text{and} \quad u_y + \frac{\partial u_y}{\partial y}\Delta y$$

From Fig. 2.4(a), it is seen that if θ_1 and θ_2 are small, then

$$\theta_1 \approx \tan \theta_1 = \frac{\partial u_y}{\partial x}$$

$$\theta_2 \approx \tan \theta_2 = \frac{\partial u_x}{\partial y} \tag{2.23}$$

so that the total change in the original right angle is

$$\theta_1 + \theta_2 = \frac{\partial u_x}{\partial y} + \frac{\partial u_y}{\partial x} = \gamma_{xy} \tag{2.24}$$

This is the shear strain in the xy plane at point P. Similarly, the shear strains γ_{yz} and γ_{zx} can be interpreted appropriately.

If the element PQR undergoes a pure rigid body rotation through a small angular displacement, then from Fig. 2.4(b) we note

$$\omega_{zo} = \frac{\partial u_y}{\partial x} = -\frac{\partial u_x}{\partial y}$$

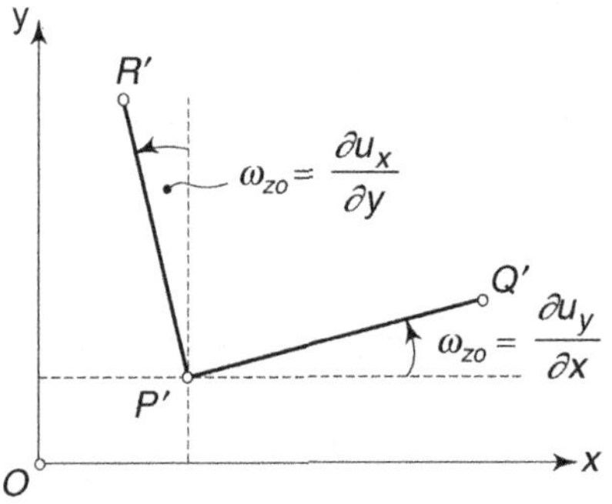

Fig. 2.4 *(b) Change in orientations of line segments PQ and PR-rigid body rotation*

taking the counter-clockwise rotation as positive. The negative sign in $(-\partial u_x/\partial y)$ comes since a positive $\partial u_x/\partial y$ will give a movement from the y to the x axis as shown in Fig. 2.4(a). No strain occurs during this rigid body displacement. We define

$$\omega_z = \frac{1}{2}\left(\frac{\partial u_y}{\partial x} - \frac{\partial u_x}{\partial y}\right) = \omega_{yx} \tag{2.25}$$

This represents the average of the sum of rotations of the x and y elements and is called the rotational component. Similarly, for rotations about the x and y axes, we get

$$\omega_x = \frac{1}{2}\left(\frac{\partial u_z}{\partial y} - \frac{\partial u_y}{\partial z}\right) = \omega_{zy} \tag{2.26}$$

$$\omega_y = \frac{1}{2}\left(\frac{\partial u_x}{\partial z} - \frac{\partial u_z}{\partial x}\right) = \omega_{xz} \tag{2.27}$$

Example 2.4 *Consider the displacement field*

$$\boldsymbol{u} = [y^2\boldsymbol{i} + 3yz\boldsymbol{j} + (4 + 6x^2)\boldsymbol{k}]10^{-2}$$

What are the rectangular strain components at the point P(1, 0, 2)? Use only linear terms.

Solution $\quad u_x = y^2 \cdot 10^{-2} \qquad u_y = 3yz \cdot 10^{-2} \qquad u_z = (4 + 6x^2) \cdot 10^{-2}$

$$\frac{\partial u_x}{\partial x} = 0 \qquad \frac{\partial u_y}{\partial x} = 0 \qquad \frac{\partial u_z}{\partial x} = 12x \cdot 10^{-2}$$

$$\frac{\partial u_x}{\partial y} = 2y \cdot 10^{-2} \qquad \frac{\partial u_y}{\partial y} = 3z \cdot 10^{-2} \qquad \frac{\partial u_z}{\partial y} = 0$$

$$\frac{\partial u_x}{\partial z} = 0 \qquad \frac{\partial u_y}{\partial z} = 3y \cdot 10^{-2} \qquad \frac{\partial u_z}{\partial z} = 0$$

The linear strains at (1, 0, 2) are

$$\varepsilon_{xx} = \frac{\partial u_x}{\partial x} = 0, \qquad \varepsilon_{yy} = \frac{\partial u_y}{\partial y} = 6 \times 10^{-2}, \qquad \varepsilon_{zz} = \frac{\partial u_z}{\partial z} = 0$$

The shear strains at (1, 0, 2) are

$$\gamma_{xy} = \frac{\partial u_x}{\partial y} + \frac{\partial u_y}{\partial x} = 0 + 0 = 0$$

$$\gamma_{yz} = \frac{\partial u_y}{\partial z} + \frac{\partial u_z}{\partial y} = 0 + 0 = 0$$

$$\gamma_{xz} = \frac{\partial u_x}{\partial z} + \frac{\partial u_z}{\partial x} = 0 + 12 \times 10^{-2} = 12 \times 10^{-2}$$

2.9 CHANGE IN DIRECTION OF A LINEAR ELEMENT

It is easy to calculate the change in the orientation of a linear element resulting from the deformation of the solid body. Let PQ be the element of length Δs, with direction cosines n_x, n_y and n_z. After deformation, the element becomes $P'Q'$ of length $\Delta s'$, with direction cosines n'_x, n'_y and n'_z. If u_x, u_y, u_z are the displacement components of point P, then the displacement components of point Q are.

$$u_x + \Delta u_x, \qquad u_y + \Delta u_y, \qquad u_z + \Delta u_z$$

where Δu_x, Δu_y and Δu_z are given by Eq. (2.7a)–(2.7c).
From Eq. (2.12), remembering that in the linear range $E_{PQ} = \varepsilon_{PQ}$,

$$\Delta s' = \Delta s\,(1 + \varepsilon_{PQ}) \tag{2.28}$$

The coordinates of P, Q, P' and Q' are as follows:

P: (x, y, z)

Q: $(x + \Delta x, y + \Delta y, z + \Delta z)$

P': $(x + u_x, y + u_y, z + u_z)$

Q': $(x + \Delta x + u_x + \Delta u_x, y + \Delta y + u_y + \Delta u_y, z + \Delta z + u_z + \Delta u_z)$

Hence,

$$n_x = \frac{\Delta x}{\Delta s}, \qquad n_y = \frac{\Delta y}{\Delta s}, \qquad n_z = \frac{\Delta z}{\Delta s}$$

$$n_x' = \frac{\Delta x + \Delta u_x}{\Delta s'}, \quad n_y' = \frac{\Delta y + \Delta u_y}{\Delta s'}, \quad n_z = \frac{\Delta z + \Delta u_z}{\Delta s'}$$

Substituting for $\Delta s'$ from Eq. (2.28) and for Δu_x, Δu_y, Δu_z from Eq. (2.7a)–(2.7c)

$$n_x' = \frac{1}{1+\varepsilon_{PQ}}\left[\left(1+\frac{\partial u_x}{\partial x}\right)n_x + \frac{\partial u_x}{\partial y}n_y + \frac{\partial u_x}{\partial z}n_z\right]$$

$$n_y' = \frac{1}{1+\varepsilon_{PQ}}\left[\frac{\partial u_y}{\partial x}n_x + \left(1+\frac{\partial u_y}{\partial y}\right)n_y + \frac{\partial u_y}{\partial z}n_z\right] \qquad (2.29)$$

$$n_z' = \frac{1}{1+\varepsilon_{PQ}}\left[\frac{\partial u_z}{\partial x}n_x + \frac{\partial u_z}{\partial y}n_y + \left(1+\frac{\partial u_z}{\partial z}\right)n_z\right]$$

The value of ε_{PQ} is obtained using Eq. (2.20).

2.10 CUBICAL DILATATION

Consider a point A with coordinates (x, y, z) and a neighbouring point B with coordinates $(x + \Delta x, y + \Delta y, z + \Delta z)$. After deformation, the points A and B move to A' and B' with coordinates

$$A' : (x + u_x, y + u_y, z + u_z)$$

$$B' : (x + \Delta x + u_x + \Delta u_x, y + \Delta y + u_y + \Delta u_y, z + \Delta z + u_z + \Delta u_z)$$

where u_x, u_y and u_z are the components of diplacement of point A, and from Eqs (2.7a)–(2.7c)

$$\Delta u_x = \frac{\partial u_x}{\partial x}\Delta x + \frac{\partial u_x}{\partial y}\Delta y + \frac{\partial u_x}{\partial z}\Delta z$$

$$\Delta u_y = \frac{\partial u_y}{\partial x}\Delta x + \frac{\partial u_y}{\partial y}\Delta y + \frac{\partial u_y}{\partial z}\Delta z$$

$$\Delta u_z = \frac{\partial u_z}{\partial x}\Delta x + \frac{\partial u_z}{\partial y}\Delta y + \frac{\partial u_z}{\partial z}\Delta z$$

The displaced segement $A'B'$ will have the following components along the x, y and z axes:

$$x \text{ axis: } \Delta x + \Delta u_x = \left(1+\frac{\partial u_x}{\partial x}\right)\Delta x + \frac{\partial u_x}{\partial y}\Delta y + \frac{\partial u_x}{\partial z}\Delta z$$

$$y \text{ axis: } \Delta y + \Delta u_y = \frac{\partial u_y}{\partial x}\Delta x + \left(1+\frac{\partial u_y}{\partial y}\right)\Delta y + \frac{\partial u_y}{\partial z}\Delta z \qquad (2.30)$$

$$z \text{ axis: } \Delta z + \Delta u_z = \frac{\partial u_z}{\partial x}\Delta x + \frac{\partial u_z}{\partial y}\Delta y + \left(1+\frac{\partial u_z}{\partial z}\right)\Delta z$$

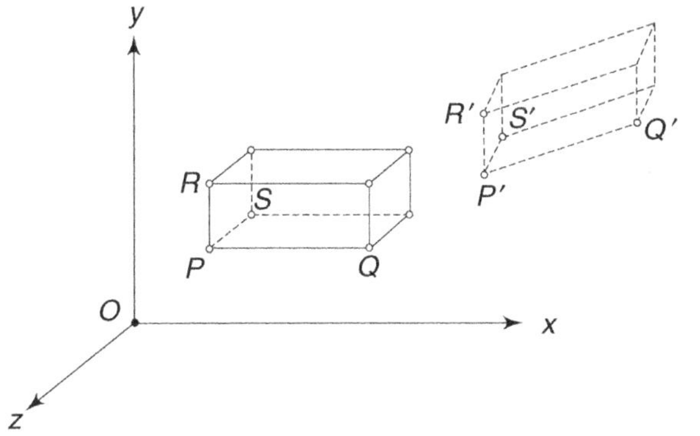

Fig. 2.5 *Deformation of right parallelepiped*

Consider now an infinitesimal rectangular parallelepiped with sides Δx, Δy and Δz (Fig. 2.5). When the body undergoes deformation, the right parallelepiped $PQRS$ becomes an oblique parallelepiped $P'Q'R'S'$.

Identifying PQ of Fig. 2.5 with AB of Eqs (2.30), one has $\Delta y = \Delta z = 0$. Then, from Eqs (2.30) the projections of $P'Q'$ will be

$$\text{along } x \text{ axis: } \left(1 + \frac{\partial u_x}{\partial x}\right)\Delta x$$

$$\text{along } y \text{ axis: } \frac{\partial u_y}{\partial x}\Delta x$$

$$\text{along } z \text{ axis: } \frac{\partial u_z}{\partial x}\Delta x$$

Hence, one can successively identify AB with PQ ($\Delta y = \Delta z = 0$), PR ($\Delta x = \Delta z = 0$), PS ($\Delta x = \Delta y = 0$) and get the components of $P'Q'$, $P'R'$ and $P'S'$ along the x, y and z axes as

	$P'Q'$	$P'R'$	$P'S'$
x axis:	$\left(1 + \frac{\partial u_x}{\partial x}\right)\Delta x$	$\frac{\partial u_x}{\partial y}\Delta y$	$\frac{\partial u_x}{\partial z}\Delta z$
y axis:	$\frac{\partial u_y}{\partial x}\Delta x$	$\left(1 + \frac{\partial u_y}{\partial y}\right)\Delta y$	$\frac{\partial u_y}{\partial z}\Delta z$
z axis:	$\frac{\partial u_z}{\partial x}\Delta x$	$\frac{\partial u_z}{\partial y}\Delta y$	$\left(1 + \frac{\partial u_z}{\partial z}\right)\Delta z$

The volume of the right parallelepiped before deformation is equal to $V = \Delta x\, \Delta y\, \Delta z$. The volume of the deformed parallelepiped is obtained from the well-known formula from analytic geometry as

$$V' = V + \Delta V = D \cdot \Delta x\, \Delta y\, \Delta z$$

where D is the following determinant:

$$D = \begin{vmatrix} \left(1+\dfrac{\partial u_x}{\partial x}\right) & \dfrac{\partial u_x}{\partial y} & \dfrac{\partial u_x}{\partial z} \\ \dfrac{\partial u_y}{\partial x} & \left(1+\dfrac{\partial u_y}{\partial y}\right) & \dfrac{\partial u_y}{\partial z} \\ \dfrac{\partial u_z}{\partial x} & \dfrac{\partial u_z}{\partial y} & \left(1+\dfrac{\partial u_z}{\partial z}\right) \end{vmatrix} \tag{2.31}$$

If we assume that the strains are small quantities such that their squares and products can be negelected, the above determinant becomes

$$\begin{aligned} D &\approx 1 + \frac{\partial u_x}{\partial x} + \frac{\partial u_y}{\partial y} + \frac{\partial u_z}{\partial z} \\ &= 1 + \varepsilon_{xx} + \varepsilon_{yy} + \varepsilon_{zz} \end{aligned} \tag{2.32}$$

Hence, the new volume according to the linear strain theory will be

$$V' = V + \Delta V = (1 + \varepsilon_{xx} + \varepsilon_{yy} + \varepsilon_{zz})\, \Delta x\, \Delta y\, \Delta z \tag{2.33}$$

The volumetric strain is defined as

$$\Delta = \frac{\Delta V}{V} = \varepsilon_{xx} + \varepsilon_{yy} + \varepsilon_{zz} \tag{2.34}$$

Therefore, according to the linear theory, the volumetric strain, also known as cubical dilatation, is equal to the sum of three linear strains.

Example 2.5 *The following state of strain exists at a point P*

$$[\varepsilon_{ij}] = \begin{bmatrix} 0.02 & -0.04 & 0 \\ -0.04 & 0.06 & -0.02 \\ 0 & -0.02 & 0 \end{bmatrix}$$

In the direction PQ having direction cosines $n_x = 0.6$, $n_y = 0$ *and* $n_z = 0.8$, *determine* ε_{PQ}.

Solution From Eq. (2.20)

$$\begin{aligned} \varepsilon_{PQ} &= 0.02\,(0.36) + 0.06\,(0) + 0\,(0.64) - 0.04\,(0) - 0.02\,(0) + 0\,(0.48) \\ &= 0.007 \end{aligned}$$

Example 2.6 *In Example 2.5, what is the cubical dilatation at point P?*

Solution From Eq. (2.34)

$$\begin{aligned} \Delta &= \varepsilon_{xx} + \varepsilon_{yy} + \varepsilon_{zz} \\ &= 0.02 + 0.06 + 0 = 0.08 \end{aligned}$$

2.11 CHANGE IN THE ANGLE BETWEEN TWO LINE ELEMENTS

Let PQ be a line element with direction cosines n_{x1}, n_{y1}, n_{z1} and PR be another line element with direction cosines n_{x2}, n_{y2}, n_{z2}, (Fig. 2.6). Let θ be the angle between the two line elements before deformation. After deformation, the line segments become $P'Q'$ and $P'R'$ with an included angle θ'. We can determine θ' easily from the results obtained in Sec. 2.9.

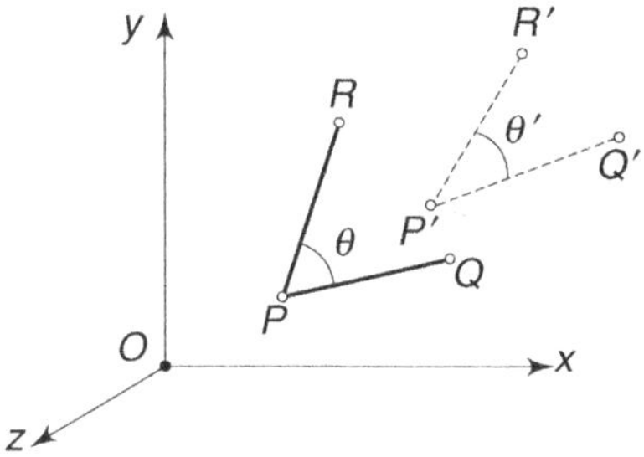

Fig. 2.6 *Change in angle between two line segments*

From analytical geometry

$$\cos\theta' = n'_{x1}\, n'_{x2} + n'_{y1}\, n'_{y2} + n'_{z1}\, n'_{z2}$$

The values of $n'_{x1}\, n'_{y1}, n'_{z1}, n'_{x2}, n'_{y2}$ and n'_{z2} can be substituted from Eq. (2.29). Neglecting squares and products of small strain components.

$$\cos\theta' = \frac{1}{(1+\varepsilon_{PQ})(1+\varepsilon_{PR})}[(1+2\varepsilon_{xx})\, n_{x1}n_{x2} + (1+2\varepsilon_{yy})\, n_{y1}n_{y2}$$
$$+ (1+2\varepsilon_{zz})\, n_{z1}n_{z2} + \gamma_{xy}(n_{x1}n_{y2} + n_{x2}n_{y1})$$
$$+ \gamma_{yz}(n_{y1}n_{z2} + n_{y2}n_{z1}) + \gamma_{zx}(n_{x1}n_{z2} + n_{x2}\, n_{z1})] \quad (2.35)$$

In particular, if the two line segments PQ and PR are at right angles to each other before strain, then after strain,

$$\cos\theta' = \frac{1}{(1+\varepsilon_{PQ})(1+\varepsilon_{PR})}[2\varepsilon_{xx}\, n_{x1}\, n_{x2} + 2\varepsilon_{yy}\, n_{y1}\, n_{y2} + 2\varepsilon_{zz}\, n_{z1}\, n_{z2}$$
$$+ \gamma_{xy}\,(n_{x1}n_{y2} + n_{x2}n_{y1}) + \gamma_{yz}(n_{y1}n_{z2} + n_{y2}n_{z1})$$
$$+ \gamma_{zx}(n_{x1}\, n_{z2} + n_{x2}\, n_{z1})] \quad (2.36a)$$

Now $(90° - \theta')$ represents the change in the initial right angle. If this is denoted by α, then

$$\theta' = 90° - \alpha \quad (2.36b)$$

or $$\cos\theta' = \cos(90° - \alpha) = \sin\alpha \approx \alpha \quad (2.36c)$$

since α is small. Therefore Eq. (2.36a) gives the shear strain α between PQ and PR. If we represent the directions of PQ and PR at P by x' and y' axes, then

$\gamma_{x'y'}$ at $P = \cos\theta'$ = expression given in Eqs (2.36a), (2.36b) and (2.36c). (2.36d)

Example 2.7 *The displacement field for a body is given by*

$$\boldsymbol{u} = k(x^2 + y)\boldsymbol{i} + k(y + z)\boldsymbol{j} + k(x^2 + 2z^2)\boldsymbol{k} \quad \text{where } k = 10^{-3}$$

At a point P(2, 2, 3), consider two line segments PQ and PR having the following direction cosines before deformation

$$PQ: n_{x1} = n_{y1} = n_{z1} = \frac{1}{\sqrt{3}}, \qquad PR: n_{x2} = n_{y2} = \frac{1}{\sqrt{2}}, \qquad n_{z2} = 0$$

Determine the angle between the two segments before and after deformation.

Solution Before deformation, the angle θ between PQ and PR is

$$\cos\,\theta = n_{x1}\,n_{x2} + n_{y1}n_{y2} + n_{z1}n_{z2} = \frac{1}{\sqrt{6}} + \frac{1}{\sqrt{6}} = 0.8165$$

$\therefore$ $\theta = 35.3°$

The strain components at P after deformation are

$$\varepsilon_{xx} = \frac{\partial u_x}{\partial x} = 2kx = 4k, \quad \varepsilon_{yy} = \frac{\partial u_y}{\partial y} = k, \quad \varepsilon_{zz} = \frac{\partial u_z}{\partial z} = 4kz = 12k$$

$$\gamma_{xy} = \frac{\partial u_x}{\partial y} + \frac{\partial u_y}{\partial x} = k, \quad \gamma_{yz} = \frac{\partial u_y}{\partial z} + \frac{\partial u_z}{\partial y} = k, \quad \gamma_{zx} = \frac{\partial u_z}{\partial x} + \frac{\partial u_x}{\partial z} = 4k$$

The linear strains in directions PQ and PR are from Eq. (2.20)

$$\varepsilon_{PQ} = k\left[\left(4\times\frac{1}{3}\right) + \frac{1}{3} + \left(12\times\frac{1}{3}\right) + \left(1\times\frac{1}{3}\right) + \left(1\times\frac{1}{3}\right) + \left(4\times\frac{1}{3}\right)\right] = \frac{23}{3}k$$

$$\varepsilon_{PR} = k\left[\left(4\times\frac{1}{2}\right) + \left(1\times\frac{1}{2}\right) + (12\times 0) + \left(1\times\frac{1}{2}\right) + 0 + 0\right] = 3k$$

After deformation, the angle beteween $P'Q'$ and $P'R'$ is from Eq. (2.35)

$$\cos\,\theta' = \frac{1}{(1+23/3k)\,(1+3k)}\left[(1+8k)\,\frac{1}{\sqrt{6}} + (1+2k)\,\frac{1}{\sqrt{6}} + 0\right.$$

$$\left. + \left(\frac{1}{\sqrt{6}} + \frac{1}{\sqrt{6}}\right)k + \left(0 + \frac{1}{\sqrt{6}}\right)k + \left(0 + \frac{1}{\sqrt{6}}\right)4k\right]$$

$$= 0.8144 \quad \text{and} \quad \theta = 35.5°$$

2.12 PRINCIPAL AXES OF STRAIN AND PRINCIPAL STRAINS

It was shown in Sec. 2.5 that when a displacement field is defined at a point P, the relative extension (i.e. strain) at P in the direction PQ is given by Eq. (2.20) as

$$\varepsilon_{PQ} = \varepsilon_{xx}n_x^2 + \varepsilon_{yy}n_y^2 + \varepsilon_{zz}n_z^2 + \gamma_{xy}n_xn_y + \gamma_{yz}n_yn_z + \gamma_{xz}n_xn_z$$

As the values of n_x, n_y and n_z change, we get different values of strain ε_{PQ}. Now we ask ourselves the following questions:

What is the direction (n_x, n_y, n_z) along which the strain is an extremum (i.e. maximum or minimum) and what is the corresponding extremum value?

According to calculus, in order to find the maximum or the minimum, we would have to equate,

$$\partial\varepsilon_{PQ}/\partial n_x, \quad \partial\varepsilon_{PQ}/\partial n_y, \quad \partial\varepsilon_{PQ}/\partial n_z,$$

to zero, if n_x, n_y and n_z were all independent. However, n_x, n_y and n_z are not all independent since they are related by the condition

$$n_x^2 + n_y^2 + n_z^2 = 1 \tag{2.37}$$

Taking n_x and n_y as independent and differentiating Eq. (2.37) with respect to n_x and n_y we get

$$
\begin{aligned}
2n_x + 2n_z \frac{\partial n_z}{\partial n_x} &= 0 \\
2n_y + 2n_z \frac{\partial n_z}{\partial n_y} &= 0
\end{aligned}
\tag{2.38}
$$

Differentiating ε_{PQ} with respect to n_x and n_y and equating them to zero for extremum

$$0 = 2n_x\varepsilon_{xx} + n_y\gamma_{xy} + n_z\gamma_{zx} + \frac{\partial n_z}{\partial n_x}(n_x\gamma_{zx} + n_y\gamma_{zy} + 2n_z\varepsilon_{zz})$$

$$0 = 2n_y\varepsilon_{yy} + n_x\gamma_{xy} + n_z\gamma_{yz} + \frac{\partial n_z}{\partial n_y}(n_x\gamma_{zx} + n_y\gamma_{zy} + 2n_z\varepsilon_{zz})$$

Substituting for $\partial n_z/\partial n_x$ and $\partial n_z/\partial n_y$ from Eqs (2.38),

$$\frac{2n_x\varepsilon_{xx} + n_y\gamma_{xy} + n_z\gamma_{zx}}{n_x} = \frac{n_x\gamma_{zx} + n_y\gamma_{zy} + 2n_z\varepsilon_{zz}}{n_z}$$

$$\frac{2n_y\varepsilon_{yy} + n_x\gamma_{xy} + n_z\gamma_{yz}}{n_y} = \frac{n_x\gamma_{zx} + n_y\gamma_{zy} + 2n_z\varepsilon_{zz}}{n_z}$$

Denoting the right-hand side expression in the above two equations by 2ε and rearranging,

$$2\varepsilon_{xx}n_x + \gamma_{xy}n_y + \gamma_{xz}n_z - 2\varepsilon n_x = 0 \tag{2.39a}$$

$$\gamma_{xy}n_x + 2\varepsilon_{yy}n_y + \gamma_{yz}n_z - 2\,\varepsilon n_y = 0 \tag{2.39b}$$

and

$$\gamma_{zx}n_x + \gamma_{zy}n_y + 2\varepsilon_{zz}n_z - 2\varepsilon n_z = 0 \tag{2.39c}$$

One can solve Eqs (2.39a)–(2.39c) to get the values of n_x, n_y and n_z, which determine the direction along which the relative extension is an extremum. Let us assume that this direction has been determined. Multiplying the first equation by n_x, second by n_y and the third by n_z and adding them, we get

$$2(\varepsilon_{xx}n_x^2 + \varepsilon_{yy}n_y^2 + \varepsilon_{zz}n_z^2 + \gamma_{xy}n_xn_y + \gamma_{yz}n_yn_z + \gamma_{zx}n_zn_x) = 2\varepsilon(n_x^2 + n_y^2 + n_z^2)$$

If we impose the condition $n_x^2 + n_y^2 + n_z^2 = 1$, the right-hand side becomes equal to 2ε. From Eq. (2.20), the left-hand side is the expression for $2\varepsilon_{PQ}$. Therefore

$$\varepsilon_{PQ} = \varepsilon$$

This means that in Eqs (2.39a)–(2.39c) the values of n_x, n_y and n_z determine the direction along which the relative extension is an extremum and further, the value of ε is equal to this extremum. Equations (2.39a)–(2.39c) can be written as

$$
\begin{aligned}
(\varepsilon_{xx} - \varepsilon)n_x + \frac{1}{2}\gamma_{xy}n_y + \frac{1}{2}\gamma_{xz}n_z &= 0 \\
\frac{1}{2}\gamma_{yx}n_x + (\varepsilon_{yy} - \varepsilon)n_y + \frac{1}{2}\gamma_{yz}n_z &= 0 \\
\frac{1}{2}\gamma_{zx}n_x + \frac{1}{2}\gamma_{zy}n_y + (\varepsilon_{zz} - \varepsilon)n_z &= 0
\end{aligned}
\tag{2.40a}
$$

If we adopt the notation given in Eq. (2.22), i.e. put

$$\frac{1}{2}\gamma_{xy} = e_{xy}, \quad \frac{1}{2}\gamma_{yz} = e_{yz}, \quad \frac{1}{2}\gamma_{zx} = e_{zx}$$

then Eqs (2.40a) can be written as

$$\begin{aligned} (\varepsilon_{xx} - \varepsilon)n_x + e_{xy}n_y + e_{xz}n_z &= 0 \\ e_{yx}n_x + (\varepsilon_{yy} - \varepsilon)n_y + e_{yz}n_z &= 0 \\ e_{zx}n_x + e_{zy}n_y + (\varepsilon_{zz} - \varepsilon)n_z &= 0 \end{aligned} \tag{2.40b}$$

The above set of equations is homogeneous in n_x, n_y and n_z. For the existence of a non-trivial solution, the determinant of its coefficient must be equal to zero, i.e.

$$\begin{vmatrix} (\varepsilon_{xx} - \varepsilon) & e_{xy} & e_{xz} \\ e_{yx} & (\varepsilon_{yy} - \varepsilon) & e_{yz} \\ e_{zx} & e_{zy} & (\varepsilon_{zz} - \varepsilon) \end{vmatrix} = 0 \tag{2.41}$$

Expanding the determinant, we get

$$\varepsilon^3 - J_1\varepsilon^2 + J_2\varepsilon - J_3 = 0 \tag{2.42}$$

where

$$J_1 = \varepsilon_{xx} + \varepsilon_{yy} + \varepsilon_{zz} \tag{2.43}$$

$$J_2 = \begin{vmatrix} \varepsilon_{xx} & e_{xy} \\ e_{yx} & \varepsilon_{yy} \end{vmatrix} + \begin{vmatrix} \varepsilon_{yy} & e_{yz} \\ e_{zy} & \varepsilon_{zz} \end{vmatrix} + \begin{vmatrix} \varepsilon_{xx} & e_{xz} \\ e_{zx} & \varepsilon_{zz} \end{vmatrix} \tag{2.44}$$

$$J_3 = \begin{vmatrix} \varepsilon_{xx} & e_{xy} & e_{xz} \\ e_{yx} & \varepsilon_{yy} & e_{yz} \\ e_{zx} & e_{zy} & \varepsilon_{zz} \end{vmatrix} \tag{2.45}$$

It is important to observe that J_2 and J_3 involve e_{xy}, e_{yz} and e_{zx} not γ_{xy}, γ_{yz} and γ_{zx}. Equations (2.41)–(2.45) are all similar to Eqs (1.8), (1.9), (1.12), (1.13) and (1.14). The problem posed and its analysis are similar to the analysis of principal stress axes and principal stresses. The results of Sec. 1.10–1.15 can be applied to the case of strain.

For a given state of strain at point P, if the relative extension (i.e. strain) ε is an extremum in a direction $\boldsymbol{n}$, then ε is the principal strain at P and $\boldsymbol{n}$ is the principal strain direction associated with ε.

In every state of strain there exist at least three mutually perpendicular principal axes and at most three distinct principal strains. The principal strains ε_1, ε_2 and ε_3, are the roots of the cubic equation.

$$\varepsilon^3 - J_1\varepsilon^2 + J_2\varepsilon - J_3 = 0 \tag{2.46}$$

where J_1, J_2, J_3 are the first, second and third invariants of strain. The principal directions associated with ε_1, ε_2 and ε_3 are obtained by substituting ε_i $(i = 1, 2, 3)$ in the following equations and solving for n_x, n_y and n_z.

$$\begin{aligned} (\varepsilon_{xx} - \varepsilon_i)n_x + e_{xy}n_y + e_{xz}n_z &= 0 \\ e_{xy}n_x + (\varepsilon_{yy} - \varepsilon_i)n_y + e_{yz}n_z &= 0 \\ n_x^2 + n_y^2 + n_z^2 &= 1 \end{aligned} \tag{2.47}$$

If ε_1, ε_2 and ε_3 are distinct, then the axes of $\boldsymbol{n}_1$, $\boldsymbol{n}_2$ and $\boldsymbol{n}_3$ are unique and mutually peprendicular. If, say $\varepsilon_1 = \varepsilon_2 \neq \varepsilon_3$, then the axis of $\boldsymbol{n}_3$ is unique and every direction perpendicular to $\boldsymbol{n}_3$ is a principal direction associated with $\varepsilon_1 = \varepsilon_2$.

If $\varepsilon_1 = \varepsilon_2 = \varepsilon_3$, then every direction is a principal direction.

Example 2.8 *The displacement field in micro units for a body is given by*

$$\boldsymbol{u} = (x^2 + y)\boldsymbol{i} + (3 + z)\boldsymbol{j} + (x^2 + 2y)\boldsymbol{k}$$

Determine the principal strains at (3, 1, –2) and the direction of the minimum principal strain.

Solution The displacement components in micro units are,

$$u_x = x^2 + y, \quad u_y = 3 + z, \quad u_z = x^2 + 2y.$$

The rectangular strain components are

$$\varepsilon_{xx} = \frac{\partial u_x}{\partial x} = 2x, \quad \varepsilon_{yy} = \frac{\partial u_y}{\partial y} = 0, \quad \varepsilon_{zz} = \frac{\partial u_z}{\partial z} = 0$$

$$\gamma_{xy} = \frac{\partial u_x}{\partial y} + \frac{\partial u_y}{\partial x} = 1, \quad \gamma_{yz} = \frac{\partial u_y}{\partial z} + \frac{\partial u_z}{\partial y} = 3, \quad \gamma_{zx} = \frac{\partial u_z}{\partial x} + \frac{\partial u_x}{\partial z} = 2x$$

At point (3, 1, –2) the strain components are therefore,

$$\varepsilon_{xx} = 6, \qquad \varepsilon_{yy} = 0, \qquad \varepsilon_{zz} = 0$$
$$\gamma_{xy} = 1, \qquad \gamma_{yz} = 3, \qquad \gamma_{zx} = 6$$

The strain invariants from Eqs (2.43) – (2.45) are

$$J_1 = \varepsilon_{xx} + \varepsilon_{yy} + \varepsilon_{zz} = 6$$

$$J_2 = \begin{vmatrix} 6 & \frac{1}{2} \\ \frac{1}{2} & 0 \end{vmatrix} + \begin{vmatrix} 0 & \frac{3}{2} \\ \frac{3}{2} & 0 \end{vmatrix} + \begin{vmatrix} 6 & 3 \\ 3 & 0 \end{vmatrix} = -\frac{23}{2}$$

Note that J_2 and J_3 involve $e_{xy} = \frac{1}{2}\gamma_{xy}, \quad e_{yz} = \frac{1}{2}\gamma_{yz}, \quad e_{zx} = \frac{1}{2}\gamma_{zx}$

$$J_3 = \begin{vmatrix} 6 & \frac{1}{2} & 3 \\ \frac{1}{2} & 0 & \frac{3}{2} \\ 3 & \frac{3}{2} & 0 \end{vmatrix} = -9$$

The cubic from Eq. (2.46) is

$$\varepsilon^3 - 6\varepsilon^2 - \frac{23}{2}\varepsilon + 9 = 0$$

Following the standard method suggested in Sec. 1.15

$$a = \frac{1}{3}\left(-\frac{69}{2} - 36\right) = -\frac{47}{2}$$

$$b = \frac{1}{27}(-432 - 621 + 243) = -30$$

$$\cos\phi = -\frac{-30}{2\times\sqrt{-a^3/27}} = 0.684$$

$$\therefore \qquad \phi = 46°48'$$

$$g = 2\sqrt{-a/3} = 5.6$$

The principal strains in micro units are

$$\varepsilon_1 = g\cos\phi/3 + 2 = +7.39$$

$$\varepsilon_2 = g\cos(\phi/3 + 120°) + 2 = -2$$

$$\varepsilon_3 = g\cos(\phi/3 + 240°) + 2 = +0.61$$

As a check, the first invariant J_1 is

$$\varepsilon_{xx} + \varepsilon_{yy} + \varepsilon_{zz} = \varepsilon_1 + \varepsilon_2 + \varepsilon_3 = 7.39 - 2 + 0.61 = 6$$

The second invariant J_2 is

$$\varepsilon_1\varepsilon_2 + \varepsilon_2\varepsilon_3 + \varepsilon_3\varepsilon_1 = -14.78 - 1.22 + 4.51 = -11.49$$

The third invariant J_3 is

$$\varepsilon_1\varepsilon_2\varepsilon_3 = 7.39\times 2\times 0.61 = -9$$

These agree with the earlier values.
The minimum principal strain is –2. For this, from Eq. (2.47)

$$(6+2)\,n_x + \frac{1}{2}n_y + 3n_z = 0$$

$$\frac{1}{2}n_x + 2n_y + \frac{3}{2}n_z = 0$$

$$n_x^2 + n_y^2 + n_z^2 = 1$$

The solutions are $n_x = 0.267$, $n_y = 0.534$ and $n_z = -0.801$.

Example 2.9 *For the state of strain given in Example 2.5, determine the principal strains and the directions of the maximum and minimum principal strains.*

Solution From the strain matrix given, the invariants are

$$J_1 = \varepsilon_{xx} + \varepsilon_{yy} + \varepsilon_{zz} = 0.02 + 0.06 + 0 = 0.08$$

$$J_2 = \begin{vmatrix} 0.02 & -0.02 \\ -0.02 & 0.06 \end{vmatrix} + \begin{vmatrix} 0.06 & -0.01 \\ -0.01 & 0 \end{vmatrix} + \begin{vmatrix} 0.02 & 0 \\ 0 & 0 \end{vmatrix}$$

$$= (0.0012 - 0.0004) + (-0.0001) + 0 = 0.0007$$

$$J_3 = \begin{vmatrix} 0.02 & -0.02 & 0 \\ -0.02 & 0.06 & -0.01 \\ 0 & -0.01 & 0 \end{vmatrix} = 0.02\,(-0.0001) + 0 + 0 = -\,0.000002$$

The cubic equation is

$$\varepsilon^3 - 0.08\varepsilon^2 + 0.0007\varepsilon + 0.000002 = 0$$

Following the standard procedure described in Sec. 1.15, one can determine the principal strains. However, observing that the constant J_3 in the cubic is very small, one can ignore it and write the cubic as

$$\varepsilon^2 - 0.08\varepsilon^2 + 0.0007\varepsilon = 0$$

One of the solutions obviously is $\varepsilon = 0$. For the other two solutions (ε not equal to zero), dividing by ε

$$\varepsilon^2 - 0.08\,\varepsilon + 0.0007 = 0$$

The solutions of this quadratic equation are

$$\varepsilon = 0.4 \pm 0.035, \text{ i.e. } 0.075 \quad \text{and} \quad 0.005$$

Rearranging such that $\varepsilon_1 \geq \varepsilon_2 \geq \varepsilon_3$, the principal strains are

$$\varepsilon_1 = 0.07, \quad \varepsilon_2 = 0.01, \quad \varepsilon_3 = 0$$

As a check:

$$J_1 = \varepsilon_1 + \varepsilon_2 + \varepsilon_3 = 0.07 + 0.01 = 0.08$$

$$J_2 = \varepsilon_1\varepsilon_2 + \varepsilon_2\varepsilon_2 + \varepsilon_3\varepsilon_1 = (0.07 \times 0.01) = 0.0007$$

$$J_3 = \varepsilon_1\varepsilon_2\varepsilon_3 = 0 \qquad \text{(This was assumed as zero)}$$

Hence, these values agree with their previous values. To determine the direction of $\varepsilon_1 = 0.07$, from Eqs (2.47)

$$(0.02 - 0.07)\,n_x - 0.02n_y = 0$$

$$-0.02n_x + (0.06 - 0.07)\,n_y - 0.01n_z = 0$$

$$n_x^2 + n_y^2 + n_z^2 = 1$$

The solutions are $n_x = 0.44$, $n_y = -0.176$ and $n_z = 0.88$.
Similarly, for $\varepsilon_3 = 0$, from Eqs (2.47)

$$0.02n_x - 0.02n_y = 0$$

$$-0.02n_x + 0.06n_y - 0.01n_z = 0$$

$$n_x^2 + n_y^2 + n_z^2 = 1$$

The solutions are $n_x = n_y = 0.236$ and $n_z = 0.944$.

2.13 PLANE STATE OF STRAIN

If, in a given state of strain, there exists a coordinate system $Oxyz$, such that for this system

$$\varepsilon_{zz} = 0, \quad \gamma_{yz} = 0, \quad \gamma_{zx} = 0 \tag{2.48}$$

then the state is said to have a plane state of strain parallel to the xy plane. The non-vanishing strain components are ε_{xx}, ε_{yy} and γ_{xy}.

If PQ is a line element in this xy plane, with direction cosines n_x, n_y, then the relative extension or the strain ε_{PQ} is obtained from Eq. (2.20) as

$$\varepsilon_{PQ} = \varepsilon_{xx}\,n_x^2 + \varepsilon_{yy}\,n_y^2 + \gamma_{xy}\,n_x\,n_y$$

or if PQ makes an angle θ with the x axis, then

$$\varepsilon_{PQ} = \varepsilon_{xx}\cos^2\theta + \varepsilon_{yy}\sin^2\theta + \frac{1}{2}\gamma_{xy}\sin 2\theta \tag{2.49}$$

If ε_1 and ε_2 are the principal strains, then

$$\varepsilon_1, \varepsilon_2, = \frac{\varepsilon_{xx} + \varepsilon_{yy}}{2} \pm \left[\left(\frac{\varepsilon_{xx} - \varepsilon_{yy}}{2} \right)^2 + \left(\frac{\gamma_{xy}}{2} \right)^2 \right]^{1/2} \tag{2.50}$$

Note that $\varepsilon_3 = \varepsilon_{zz}$ is also a principal strain. The principal strain axes make angles ϕ and $\phi + 90°$ with the x axis, such that

$$\tan 2\phi = \frac{\gamma_{xy}}{\varepsilon_{xx} - \varepsilon_{yy}} \tag{2.51}$$

The discussions and conclusions will be identical with the analysis of stress if we use ε_{xx}, ε_{yy}, and ε_{zz} in place of σ_x, σ_y and σ_z respectively, and $e_{xy} = \frac{1}{2}\gamma_{xy}$, $e_{yz} = \frac{1}{2}\gamma_{yz}$, $e_{zx} = \frac{1}{2}\gamma_{zx}$ in place of τ_{xy}, τ_{yz} and τ_{zx} respectively.

2.14 THE PRINCIPAL AXES OF STRAIN REMAIN ORTHOGONAL AFTER STRAIN

Let PQ be one of the principal extensions or strain axes with direction cosines n_{x1}, n_{y1} and n_{z1}. Then according to Eqs (2.40b)

$$(\varepsilon_{xx} - \varepsilon_1)n_{x1} + e_{xy}n_{y1} + e_{xz}n_{z1} = 0$$
$$e_{xy}n_{x1} + (\varepsilon_{yy} - \varepsilon_1)n_{y1} + e_{yz}n_{z1} = 0$$
$$e_{xz}n_{x1} + e_{yz}n_{y1} + (\varepsilon_{zz} - \varepsilon_1)n_{z1} = 0$$

Let n_{x2}, n_{y2} and n_{z2} be the direction cosines of a line PR, perpendicular to PQ before strain. Therefore,

$$n_{x1}n_{x2} + n_{y1}n_{y2} + n_{z1}n_{z2} = 0$$

Multiplying Eq. (2.40b), given above, by n_{x2}, n_{y2} and n_{z2} respectively and adding, we get,

$$\varepsilon_{xx}n_{x1}n_{x2} + \varepsilon_{yy}n_{y1}n_{y2} + \varepsilon_{zz}n_{z1}n_{z2} + e_{xy}(n_{x1}n_{y2} + n_{y1}n_{x2}) + e_{yz}(n_{y1}n_{z2} + n_{y2}n_{z1}) + e_{zx}(n_{x1}n_{z2} + n_{x2}n_{z1}) = 0$$

Multiplying by 2 and putting

$$2e_{xy} = \gamma_{xy}, \qquad 2e_{yz} = \gamma_{yz}, \qquad 2e_{zx} = \gamma_{zx}$$

we get

$$2\varepsilon_{xx}n_{x1}n_{x2} + 2\varepsilon_{yy}n_{y1}n_{y2} + 2\varepsilon_{zz}n_{z1}nz_2 + \gamma_{xy}(n_{x1}n_{y2} + n_{y1}n_{x2}) + \gamma_{yz}(n_{y1}n_{z2} + n_{y2}n_{z1}) + \gamma_{zx}(n_{x1}n_{z2} + n_{z1}n_{x2}) = 0$$

Comparing the above with Eq. (2.36a), we get

$$\cos\theta' (1 + \varepsilon_{PQ})(1 + \varepsilon_{PR}) = 0$$

where θ' is the new angle between PQ and PR after strain.

Since ε_{PQ} and ε_{PR} are quite general, to satisfy the equation, $\theta' = 90°$, i.e. the line segments remain perpendicular after strain also. Since PR is an arbitrary perpendicular line to the principal axis PQ, every line perpendicular to PQ before strain remains perpendicular after strain. In particular, PR can be the second principal axis of strain.

Repeating the above steps, if PS is the third principal axis of strain perpendicular to PQ and PR, it remains perpendicular after strain also. Therefore, at point P, we can identify a small rectangular element, with faces normal to the principal axes of strain, that will remain rectangular after strain also.

2.15 PLANE STRAINS IN POLAR COORDINATES

We now consider displacements and deformations of a two-dimensional radial element in polar coordinates. The polar coordinates of a point a are r and θ. The radial and circumferential displacements are denoted by u_r and u_θ. Consider an elementary radial element $abcd$, as shown in Fig. 2.7.

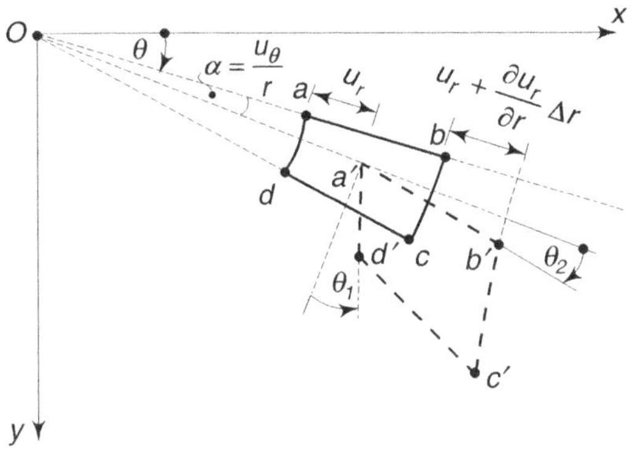

Fig. 2.7 *Displacement components of a radial element*

Point a with coordinates (r, θ) gets displaced after deformation to position a' with coordinates $(r + u_r, \theta + \alpha)$. The neighbouring point $b(r + \Delta r, \theta)$ gets moved to b' with coordinates

$$\left(r + \Delta r + u_r + \frac{\partial u_r}{\partial r}\Delta r, \theta + \alpha + \frac{\partial \alpha}{\partial r}\Delta r\right)$$

The length of $a'b'$ is therefore

$$\Delta r + \frac{\partial u_r}{\partial r}\Delta r$$

The radial strain ε_r is therefore

$$\varepsilon_r = \frac{\partial u_r}{\partial r} \tag{2.52}$$

The circumferential strain ε_θ is caused in two ways. If the element $abcd$ undergoes a purely radial displacement, then the length $ad = r\,\Delta\theta$ changes to $(r + u_r)\Delta\theta$. The strain due to this radial movement alone is

$$\frac{u_r \Delta\theta}{r\Delta\theta} = \frac{u_r}{r}$$

In addition to this, the point d moves circumferentially to d' through the distance

$$u_\theta + \frac{\partial u_\theta}{\partial \theta}\Delta\theta$$

Since point a moves circumferentially through u_θ, the change in ad is $\frac{\partial u_\theta}{\partial \theta}\Delta\theta$. The strain due to this part is

$$\frac{\partial u_\theta}{\partial \theta}\frac{\Delta\theta}{r\Delta\theta} = \frac{1}{r}\frac{\partial u_\theta}{\partial \theta}$$

The total circumferential strain is therefore

$$\varepsilon_\theta = \frac{u_r}{r} + \frac{1}{r}\frac{\partial u_\theta}{\partial \theta} \tag{2.53}$$

To determine the shear strain we observe the following:

The circumferential displacement of a is u_θ, whereas that of b is

$u_\theta + \frac{\partial u_\theta}{\partial r}\Delta r$. The magnitude of θ_2 is

$$\left[\left(u_\theta + \frac{\partial u_\theta}{\partial r}\Delta r\right) - \alpha\left(r + \Delta r\right)\right]\frac{1}{\Delta r}$$

But $$\alpha = \frac{u_\theta}{r}.$$

Hence, $$\theta_2 = \left(u_\theta + \frac{\partial u_\theta}{\partial r}\Delta r - u_\theta - \frac{u_\theta}{r}\Delta r\right)\frac{1}{\Delta r}$$

$$= \frac{\partial u_\theta}{\partial r} - \frac{u_\theta}{r}$$

Similarly, the radial displacement of a is u_r, whereas that of d is $u_r + \frac{\partial u_r}{\partial \theta}\Delta\theta$.

Hence,

$$\theta_1 = \frac{1}{r\Delta\theta}\left[\left(u_r + \frac{\partial u_r}{\partial \theta}\Delta_\theta\right) - u_r\right]$$

$$= \frac{1}{r}\frac{\partial u_r}{\partial \theta}$$

Hence, the shear strain $\gamma_{r\theta}$ is

$$\gamma_{r\theta} = \theta_1 + \theta_2 = \frac{1}{r}\frac{\partial u_r}{\partial \theta} + \frac{\partial u_\theta}{\partial r} - \frac{u_\theta}{r} \tag{2.54}$$

2.16 COMPATIBILITY CONDITIONS

It was observed that the displacement of a point in a solid body can be represented by a displacement vector $\boldsymbol{u}$, which has components,

$$u_x, u_y, u_z.$$

along the three axes x, y and z respectively. The deformation at a point is specified by the six strain components,

$$\varepsilon_{xx}, \varepsilon_{yy}, \varepsilon_{zz}, \gamma_{xy}, \gamma_{yz} \text{ and } \gamma_{zx}.$$

The three displacement components and the six rectangular strain components are related by the six strain displacement relations of Cauchy, given by Eqs (2.18) and (2.19). The determination of the six strain components from the three displacement functions is straightforward since it involves only differentiation. However, the reverse operation, i.e. determination of the three displacement functions from the six strain components is more complicated since it involves integrating six equations to obtain three functions. One may expect, therefore, that all the six strain components cannot be prescribed arbitrarily and there must exist certain relations among these. The total number of these relations are six and they fall into two groups.

First group: We have

$$\varepsilon_{xx} = \frac{\partial u_x}{\partial x}, \qquad \varepsilon_{yy} = \frac{\partial u_y}{\partial y}, \qquad \gamma_{xy} = \frac{\partial u_x}{\partial y} + \frac{\partial u_y}{\partial x}$$

Differentiate the first two of the above equations as follows:

$$\frac{\partial^2 \varepsilon_{xx}}{\partial y^2} = \frac{\partial^3 u_x}{\partial x \partial y^2} = \frac{\partial^2}{\partial x \partial y}\left(\frac{\partial u_x}{\partial y}\right)$$

$$\frac{\partial^2 \varepsilon_{yy}}{\partial x^2} = \frac{\partial^3 u_y}{\partial y \partial x^2} = \frac{\partial^2}{\partial x \partial y}\left(\frac{\partial u_y}{\partial x}\right)$$

Adding these two, we get

$$\frac{\partial^2}{\partial x \partial y}\left(\frac{\partial u_x}{\partial y} + \frac{\partial u_y}{\partial x}\right) = \frac{\partial^2 \gamma_{xy}}{\partial x \partial y}$$

i.e.

$$\frac{\partial^2 \varepsilon_{xx}}{\partial y^2} + \frac{\partial^2 \varepsilon_{yy}}{\partial x^2} = \frac{\partial^2 \gamma_{xy}}{\partial x \partial y}$$

Similarly, by considering ε_{yy}, ε_{zz} and γ_{yz}, and ε_{zz}, ε_{xx} and γ_{zx}, we get two more conditions. This leads us to the first group of conditions.

$$\begin{aligned}
\frac{\partial^2 \varepsilon_{xx}}{\partial y^2} + \frac{\partial^2 \varepsilon_{yy}}{\partial x^2} &= \frac{\partial^2 \gamma_{xy}}{\partial x \partial y} \\
\frac{\partial^2 \varepsilon_{yy}}{\partial z^2} + \frac{\partial^2 \varepsilon_{zz}}{\partial y^2} &= \frac{\partial^2 \gamma_{yz}}{\partial y \partial z} \\
\frac{\partial^2 \varepsilon_{zz}}{\partial x^2} + \frac{\partial^2 \varepsilon_{xx}}{\partial z^2} &= \frac{\partial^2 \gamma_{zx}}{\partial z \partial x}
\end{aligned} \tag{2.55}$$

Second group: This group establishes the conditions among the shear strains. We have

$$\gamma_{xy} = \frac{\partial u_x}{\partial y} + \frac{\partial u_y}{\partial x}$$

$$\gamma_{yz} = \frac{\partial u_y}{\partial z} + \frac{\partial u_z}{\partial y}$$

$$\gamma_{xz} = \frac{\partial u_z}{\partial x} + \frac{\partial u_x}{\partial z}$$

Differentiating

$$\frac{\partial \gamma_{xy}}{\partial z} = \frac{\partial^2 u_x}{\partial z \partial y} + \frac{\partial^2 u_y}{\partial z \partial x}$$

$$\frac{\partial \gamma_{yz}}{\partial x} = \frac{\partial^2 u_y}{\partial x \partial z} + \frac{\partial^2 u_z}{\partial x \partial y}$$

$$\frac{\partial \gamma_{zx}}{\partial y} = \frac{\partial^2 u_z}{\partial x \partial y} + \frac{\partial^2 u_x}{\partial y \partial z}$$

Adding the last two equations and subtracting the first

$$\frac{\partial \gamma_{yz}}{\partial x} + \frac{\partial \gamma_{zx}}{\partial y} - \frac{\partial \gamma_{xy}}{\partial z} = 2\frac{\partial^2 u_z}{\partial x\, \partial y}$$

Differentiating the above equation once more with respect to z and observing that

$$\frac{\partial^3 u_z}{\partial x \partial y \partial z} = \frac{\partial^2 \varepsilon_{zz}}{\partial x\, \partial y}$$

we get,

$$\frac{\partial}{\partial z}\left(\frac{\partial \gamma_{yz}}{\partial x} + \frac{\partial \gamma_{zx}}{\partial y} - \frac{\partial \gamma_{xy}}{\partial z}\right) = 2\frac{\partial^3 u_z}{\partial x\, \partial y\, \partial z} = 2\frac{\partial^2 \varepsilon_{zz}}{\partial x\, \partial y}$$

This is one of the required relations of the second group. By a cyclic change of the letters we get the other two equations. Collecting all equations, the six strain compatibility relations are

$$\frac{\partial^2 \varepsilon_{xx}}{\partial y^2} + \frac{\partial^2 \varepsilon_{yy}}{\partial x^2} = \frac{\partial^2 \gamma_{xy}}{\partial x\, \partial y} \tag{2.56a}$$

$$\frac{\partial^2 \varepsilon_{yy}}{\partial z^2} + \frac{\partial^2 \varepsilon_{zz}}{\partial y^2} = \frac{\partial^2 \gamma_{yz}}{\partial y\, \partial z} \tag{2.56b}$$

$$\frac{\partial^2 \varepsilon_{zz}}{\partial x^2} + \frac{\partial^2 \varepsilon_{xx}}{\partial z^2} = \frac{\partial^2 \gamma_{zx}}{\partial z\, \partial x} \tag{2.56c}$$

$$\frac{\partial}{\partial z}\left(\frac{\partial \gamma_{yz}}{\partial x} + \frac{\partial \gamma_{zx}}{\partial y} - \frac{\partial \gamma_{xy}}{\partial z}\right) = 2\frac{\partial^2 \varepsilon_{zz}}{\partial x\, \partial y} \tag{2.56d}$$

$$\frac{\partial}{\partial x}\left(\frac{\partial \gamma_{zx}}{\partial y} + \frac{\partial \gamma_{xy}}{\partial z} - \frac{\partial \gamma_{yz}}{\partial x}\right) = 2\frac{\partial^2 \varepsilon_{xx}}{\partial y\, \partial z} \tag{2.56e}$$

$$\frac{\partial}{\partial y}\left(\frac{\partial \gamma_{xy}}{\partial z} + \frac{\partial \gamma_{yz}}{\partial x} - \frac{\partial \gamma_{zx}}{\partial y}\right) = 2\frac{\partial^2 \varepsilon_{yy}}{\partial x\, \partial z} \tag{2.56f}$$

The above six equations are called Saint-Venant's equations of compatibility. We can give a geometrical interpretation to the above equations. For this purpose, imagine an elastic body cut into small parallelepipeds and give each of them the deformation defined by the six strain components. It is easy to conceive that if the components of strain are not connected by certain relations, it is impossible to make a continuous deformed solid from individual deformed parallelepipeds. Saint-Venant's compatibility relations furnish these conditions. Hence, these equations are also known as continuity equations.

Example 2.10 *For a circular rod subjected to a torque (Fig. 2.8), the displacement components at any point (x, y, z) are obtained as*

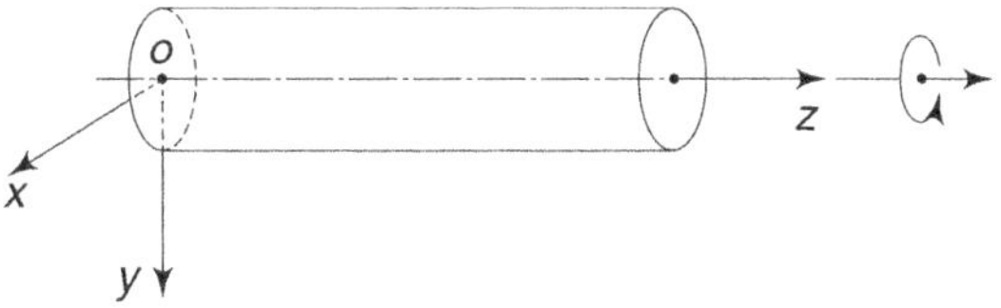

Fig. 2.8 *Example 2.8*

$$u_x = -\tau yz + ay + bz + c$$
$$u_y = \tau xz - ax + ez + f$$
$$u_z = -bx - ey + k$$

where a, b, c, e, f and k are constants, and τ is the shear stress.

(i) Select the constants a, b, c, e, f, k such that the end section z = 0 is fixed in the following manner:
  (a) Point o has no displacement.
  (b) The element Δz of the axis rotates neither in the plane xoz nor in the plane yoz
  (c) The element Δy of the axis does not rotate in the plane xoy.
(ii) Determine the strain components.
(iii) Verify whether these strain components satisfy the compatibility conditions.

Solution

(i) Since point '*o*' does not have any displacement

$$u_x = c = 0, \qquad u_y = f = 0, \qquad u_z = k = 0$$

The displacements of a point Δz from '*o*' are

$$\frac{\partial u_x}{\partial z}\Delta z, \quad \frac{\partial u_y}{\partial z}\Delta z \quad \text{and} \quad \frac{\partial u_z}{\partial z}\Delta z$$

Similarly, the displacements of a point Δy from '*o*' are

$$\frac{\partial u_x}{\partial y}\Delta y, \quad \frac{\partial u_y}{\partial y}\Delta y \quad \text{and} \quad \frac{\partial u_z}{\partial y}\Delta y$$

Hence, according to condition (b)

$$\frac{\partial u_y}{\partial z}\Delta z = 0 \quad \text{and,} \quad \frac{\partial u_x}{\partial z}\Delta z = 0$$

and according to condition (c)

$$\frac{\partial u_x}{\partial y}\Delta y = 0$$

Applying these requirements

$\frac{\partial u_y}{\partial z}$ at '*o*' is e and hence, $e = 0$

$\frac{\partial u_x}{\partial z}$ at '*o*' is b and hence, $b = 0$

$\frac{\partial u_x}{\partial y}$ at '*o*' is a and hence, $a = 0$

Consequently, the displacement components are

$$u_x = -\tau yz, \qquad u_y = \tau xz \qquad \text{and} \qquad u_z = 0$$

(ii) The strain components are

$$\varepsilon_{xx} = \frac{\partial u_x}{\partial x} = 0, \quad \varepsilon_{yy} = \frac{\partial u_y}{\partial y} = 0, \quad \varepsilon_{zz} = 0;$$

$$\gamma_{xy} = \frac{\partial u_x}{\partial y} + \frac{\partial u_y}{\partial x} = -\tau z + \tau z = 0$$

$$\gamma_{yz} = \frac{\partial u_y}{\partial z} + \frac{\partial u_z}{\partial x} = \tau x$$

$$\gamma_{zx} = \frac{\partial u_z}{\partial x} + \frac{\partial u_x}{\partial z} = -\tau y$$

(iii) Since the strain components are linear in x, y and z, the Saint-Venant's compatibility requirements are automatically satisfied.

2.17 STRAIN DEVIATOR AND ITS INVARIANTS

Similar to the analysis of stress, we can resolve the e_{ij} matrix into a spherical part (i.e. isotoropic) and a deviatoric part. The e_{ij} matrix is

$$\left[e_{ij}\right] = \begin{bmatrix} \varepsilon_{xx} & e_{xy} & e_{xz} \\ e_{xy} & \varepsilon_{yy} & e_{yz} \\ e_{xz} & e_{yz} & \varepsilon_{zz} \end{bmatrix}$$

This can be resolved into two parts as

$$\left[e_{ij}\right] = \begin{bmatrix} \varepsilon_{xx} - e & e_{xy} & e_{xz} \\ e_{xy} & \varepsilon_{yy} - e & e_{yz} \\ e_{xz} & e_{yz} & \varepsilon_{zz} - e \end{bmatrix} + \begin{bmatrix} e & 0 & 0 \\ 0 & e & 0 \\ 0 & 0 & e \end{bmatrix} \tag{2.57}$$

where
$$e = \frac{1}{3}\left(\varepsilon_{xx} + \varepsilon_{yy} + \varepsilon_{zz}\right) \tag{2.58}$$

represents the mean elongation at a given point. The second matrix on the right-hand side of Eq. (2.57) is the spherical part of the strain matrix. The first matrix represents the deviatoric part or the strain deviator. If an isolated element of the body is subjected to the strain deviator only, then according to Eq. (2.34), the volumetric strain is equal to

$$\begin{aligned} \frac{\Delta V}{V} &= (\varepsilon_{xx} - e) + (\varepsilon_{yy} - e) + (\varepsilon_{zz} - e) \\ &= \varepsilon_{xx} + \varepsilon_{yy} + \varepsilon_{zz} - 3e \\ &= 0 \end{aligned} \tag{2.59}$$

This means that an element subjected to deviatoric strain undergoes pure deformation without a change in volume. Hence, this part is also known as the pure shear part of the strain matrix. This discussion is analogous to that made in Sec. 1.22. The spherical part of the strain matrix, i.e. the second matrix on the right-hand side of Eq. (2.57) is an isotropic state of strain. It is called isotropic because when a body is subjected to this particular state of strain, then every direction is a principal strain direction, with a strain of magnitude e, according to Eq. (2.20). A sphere subjected to this state of strain will uniformally expand or contract and remain spherical.

Consider the invariants of the strain deviator. These are constructed in the same way as the invariants of the stress and strain matrices with an appropriate replacement of notations.

(i) Linear invariant is zero since

$$J_1' = (\varepsilon_{xx} - e) + (\varepsilon_{yy} - e) + (\varepsilon_{zz} - e) = 0 \tag{2.60}$$

(ii) Quadratic invariant is

$$J_2' = \begin{bmatrix} \varepsilon_{xx} - e & e_{xy} \\ e_{xy} & \varepsilon_{yy} - e \end{bmatrix} + \begin{bmatrix} \varepsilon_{yy} - e & e_{yz} \\ e_{yz} & \varepsilon_{zz} - e \end{bmatrix} + \begin{bmatrix} \varepsilon_{xx} - e & e_{xz} \\ e_{xz} & \varepsilon_{zz} - e \end{bmatrix}$$

$$= -\frac{1}{6}\Big[(\varepsilon_{xx} - \varepsilon_{yy})^2 + (\varepsilon_{yy} - \varepsilon_{zz})^2 + (\varepsilon_{zz} - \varepsilon_{xx})^2 \tag{2.61}$$

$$+ 6(e_{xy} + e_{yx} + e_{zx})^2\Big]$$

(iii) Cubic invariant is

$$J_3' = \begin{bmatrix} \varepsilon_{xx} - e & e_{xy} & e_{xz} \\ e_{xy} & \varepsilon_{yy} - e & e_{yz} \\ e_{xz} & e_{zy} & \varepsilon_{zz} - e \end{bmatrix} \tag{2.62}$$

The second and third invariants of the deviatoric strain matrix describe the two types of distortions that an isolated element undergoes when subjected to the given strain matrix e_{ij}.

Problems

2.1 The displacement field for a body is given by

$$\boldsymbol{u} = (x^2 + y)\boldsymbol{i} + (3 + z)\boldsymbol{j} + (x^2 + 2y)\boldsymbol{k}$$

Write down the displacement gradient matrix at point (2, 3, 1).

$$\left[Ans. \quad \begin{bmatrix} 4 & 1 & 0 \\ 0 & 0 & 1 \\ 4 & 2 & 0 \end{bmatrix} \right]$$

2.2 The displacement field for a body is given by

$$\boldsymbol{u} = [(x^2 + y^2 + 2)\boldsymbol{i} + (3x + 4y^2)\boldsymbol{j} + (2x^3 + 4z)\boldsymbol{k}]10^{-4}$$

What is the displaced position of a point originally at (1, 2, 3)?

[*Ans.* (1.0007, 2.0019, 3.0014)]

2.3 For the displacement field given in Problem 2.2, what are the strain components at (1, 2, 3). Use only linear terms.

$$\left[\begin{array}{l} Ans. \quad \varepsilon_{xx} = 0.0002,\ \varepsilon_{yy} = 0.0016,\ \varepsilon_{zz} = 0.0004 \\ \qquad\quad \gamma_{xy} = 0.0007,\ \gamma_{yz} = 0,\ \gamma_{zx} = 0.0006 \end{array}\right]$$

2.4 What are the strain acomponents for Problem 2.3, if non-linear terms are also included?

$$\left[\begin{array}{ll} Ans.\ E_{xx} = 2p + 24.5p^2, & E_{yy} = 16p + 136p^2,\ E_{zz} = 4p + 8p^2 \\ \qquad E_{xy} = 7p + 56p^2, & E_{yz} = 0,\ E_{zx} = 6p + 24p^2 \text{ where } p = 10^{-4} \end{array}\right]$$

2.5 If the displacement field is given by

$$u_x = kxy, \qquad u_y = kxy, \qquad u_z = 2k(x + y)z$$

where k is a constant small enough to ensure applicability of the small deformation theory,

(a) write down the strain matrix

(b) what is the strain in the direction $n_x = n_y = n_z = 1/\sqrt{3}$?

$$\left[\begin{array}{l} Ans.\ (a) \left[\varepsilon_{ij}\right] = k\begin{bmatrix} y & x+y & 2z \\ x+y & x & 2z \\ 2z & 2z & 2(x+y) \end{bmatrix} \\ \qquad (b)\ \varepsilon_{PQ} = \dfrac{4k}{3}(x + y + z) \end{array}\right]$$

2.6 The displacement field is given by

$$u_x = k(x^2 + 2z), \qquad u_y = k(4x + 2y^2 + z), \qquad u_z = 4kz^2$$

k is a very small constant. What are the strains at (2, 2, 3) in directions

(a) $n_x = 0,\ n_y = 1/\sqrt{2},\ n_z = 1/\sqrt{2}$

(b) $n_x = 1,\ n_y = n_z = 0$

(c) $n_x = 0.6,\ n_y = 0,\ n_z = 0.8$

$$\left[Ans.\ \text{(a)}\ \frac{33}{2}k,\ \text{(b)}\ 4k,\ \text{(c)}\ 17.76k\right]$$

2.7 For the displacement field given in Problem 2.6, with $k = 0.001$, determine the change in angle between two line segments PQ and PR at $P(2, 2, 3)$ having direction cosines before deformation as

(a) PQ: $\quad n_{x1} = 0,\ n_{y1} = n_{z1} = \dfrac{1}{\sqrt{2}}$

PR: $\quad n_{x2} = 1,\ n_{y2} = n_{z2} = 0$

(b) PQ: $\quad n_{x1} = 0,\ n_{y1} = n_{z1} = \dfrac{1}{\sqrt{2}}$

PR: $\quad n_{x2} = 0.6,\ n_{y2} = 0,\ n_{z2} = 0.8$

$$\left[\begin{array}{l} Ans.\ \text{(a)}\ 90° - 89.8° = 0.2° \\ \qquad\ \ \text{(b)}\ 55.5° - 50.7° = 4.8° \end{array}\right]$$

2.8. The rectangular components of a small strain at a point is given by the following matrix. Determine the principal strains and the direction of the maximum unit strain (i.e. ε_{max}).

$$\left[\varepsilon_{ij}\right] = p\begin{bmatrix} 1 & 0 & 0 \\ 0 & 0 & -4 \\ 0 & -4 & 3 \end{bmatrix} \text{ where } p = 10^{-4}$$

$$\left[\begin{array}{l} Ans.\ \varepsilon_1 = 4p,\ \varepsilon_2 = p,\ \varepsilon_3 = -p \\ \text{for } \varepsilon_1 : n_x = 0,\ n_y = 0.447,\ n_z = 0.894 \\ \text{for } \varepsilon_2 : n_x = 1,\ n_y = n_z = 0 \\ \text{for } \varepsilon_3 : n_x = 0,\ n_y = 0.894,\ n_z = 0.447 \end{array}\right]$$

2.9 For the following plane strain distribution, verify whether the compatibility condition is satisfied:

$$\varepsilon_{xx} = 3x^2y, \qquad \varepsilon_{yy} = 4y^2x + 10^{-2}, \qquad \gamma_{xy} = 2xy + 2x^3$$

[*Ans.* Not satisfied]

2.10 Verify whether the following strain field satisfies the equations of compatibility. p is a constant:

$$\varepsilon_{xx} = py, \qquad \varepsilon_{yy} = px, \qquad \varepsilon_{zz} = 2p(x + y)$$

$$\gamma_{xy} = p(x + y), \qquad \varepsilon_{yz} = 2pz, \qquad \varepsilon_{zx} = 2pz$$

[*Ans.* Yes]

2.11 State the conditions under which the following is a possible system of strains:

$$\varepsilon_{xx} = a + b(x^2 + y^2)\,x^4 + y^4, \qquad \gamma_{yz} = 0$$

$$\varepsilon_{yy} = \alpha + \beta(x^2 + y^2) + x^4 + y^4, \qquad \gamma_{zx} = 0$$

$$\gamma_{xy} = A + Bxy\,(x^2 + y^2 - c^2), \qquad \varepsilon_{zz} = 0$$

[*Ans.* $B = 4;\ b + \beta + 2c^2 = 0$]

2.12 Given the following system of strains

$$\varepsilon_{xx} = 5 + x^2 + y^2 + x^4 + y^4$$

$$\varepsilon_{yy} = 6 + 3x^2 + 3y^2 + x^4 + y^4$$

$$\gamma_{xy} = 10 + 4xy\,(x^2 + y^2 + 2)$$

$$\varepsilon_{zz} = \gamma_{yz} = \gamma_{zx} = 0$$

determine whether the above strain field is possible. If it is possible, determine the displacement components in terms of x and y, assuming that $u_x = u_y = 0$ and $\omega_{xy} = 0$ at the origin.

$$\left[\begin{array}{l} Ans.\ \text{It is possible.}\quad u_x = 5x + \frac{1}{3}x^3 + xy^2 + \frac{1}{5}x^5 + xy^4 + cy \\ \qquad\qquad\qquad\quad u_y = 6y + 3x^2y + y^3 + x^4y + \frac{1}{5}y^5 + cx \end{array}\right]$$

2.13 For the state of strain given in Problem 2.12, write down the spherical part and the deviatoric part and determine the volumetric strain.

$$\left[\begin{array}{l} Ans.\ \text{Components of spherical part are} \\ e = \frac{1}{3}\,[11 + 4(x^2 + y^2) + 2(x^4 + y^4)] \\ \text{Volumetric strain} = 11 + 4(x^2 + y^2) + 2(x^4 + y^4) \end{array}\right]$$

Appendix

On Compatibility Conditions

It was stated in Sec. 2.16 that the six strain components e_{ij} (i.e., $e_{xx} = \varepsilon_{xx}$, $e_{yy} = \varepsilon_{yy}$, $e_{zz} = \varepsilon_{zz}$, $e_{xy} = \frac{1}{2}\gamma_{xy}$, $e_{yz} = \frac{1}{2}\gamma_{zy}$, $e_{zx} = \frac{1}{2}\gamma_{zx}$) should satisfy certain necessary conditions for the existence of single-valued, continuous displacement functions, and these were called compatibility conditions. In a two-dimensional case, these conditions reduce to

$$\frac{\partial^2 e_{xx}}{\partial y^2} + \frac{\partial^2 e_{yy}}{\partial x^2} = 2\frac{\partial^2 e_{xy}}{\partial x \partial y}$$

Generally, these equations are obtained by differentiating the expressions for e_{xx}, e_{yy}, e_{xy}, and showing their equivalence in the above manner. However, their requirement for the existence of single-value displacement is not shown. In this section, this aspect will be treated for the plane case.

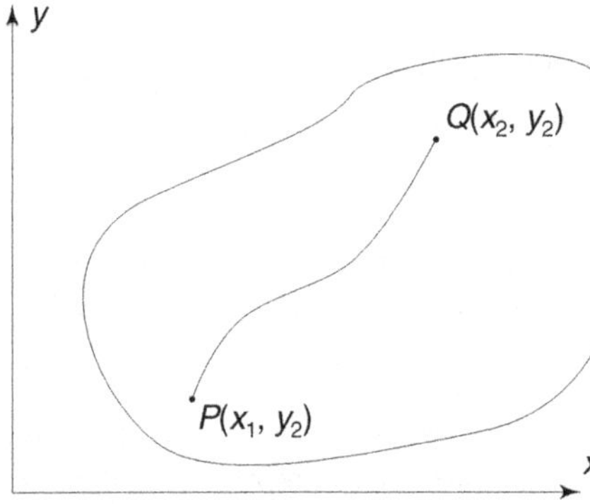

Fig. A.1 *Continuous curve connecting P and Q in a simply connected body.*

Let $P(x_1, x_2)$ be some point in a simply connected region at which the displacement (u_x°, u_y°) are known. We try to determine the displacements (u_x, u_y) at another point Q in terms of the known functions e_{xx}, e_{yy}, e_{xy}, ω_{xy} by means of a line integral over a simple continuous curve C joining the points P and Q.

Consider the displacement u_x

$$u_x(x_2, y_2) = u_x^\circ + \int_P^Q du_x \tag{A.1}$$

Since,

$$du_x = \frac{\partial u_x}{\partial x}dx + \frac{\partial u_x}{\partial y}dy$$

$$u_x(x_2, y_2) = u_x^\circ + \int_P^Q \frac{\partial u_x}{\partial x}dx + \int_P^Q \frac{\partial u_x}{\partial y}dy$$

$$= u_x^\circ + \int_P^Q e_{xx}d_x + \int_P^Q \frac{\partial u_x}{\partial y}dy$$

Now, $$\frac{\partial u_x}{\partial y} = \frac{1}{2}\left(\frac{\partial u_x}{\partial y} + \frac{\partial u_y}{\partial x}\right) + \frac{1}{2}\left(\frac{\partial u_x}{\partial y} - \frac{\partial u_y}{\partial x}\right)$$

$$= e_{xy} - \omega_{yx} \text{ from equations (2.22) and (2.25).}$$

$$\therefore \quad u_x(x_2, y_2) = u_x^{\circ} + \int_P^Q e_{xx} dx + \int_P^Q e_{xy} dy - \int_P^Q \omega_{yx} dy \tag{A.2}$$

Integrating by parts, the last integral on the right-hand side

$$\int_P^Q \omega_{yx} dy = (y\omega_{yx})\Big|_P^Q - \int_P^Q y d(\omega_{yx})$$

$$= (y\omega_{yx})\Big|_P^Q - \int_P^Q y\left(\frac{\partial \omega_{yx}}{\partial x} dx + \frac{\partial \omega_{yx}}{\partial y} dy\right) \tag{A.3}$$

Substituting, Eq. (A.2) becomes

$$u_x(x_2, y_2) = u_x^{\circ} + \int_P^Q e_{xx} dx + \int_P^Q e_{xy} dx - (y\omega_{yx})\Big|_P^Q - \int_P^Q y\left(\frac{\partial \omega_{yx}}{\partial x} dx + \frac{\partial \omega_{yx}}{\partial y} dy\right) \tag{A.4}$$

Now consider the terms in the last integral on the right-hand side.

$$\frac{\partial \omega_{yx}}{\partial x} = \frac{1}{2}\frac{\partial}{\partial x}\left(\frac{\partial u_x}{\partial y} - \frac{\partial u_y}{\partial x}\right)$$

$$= \frac{1}{2}\frac{\partial}{\partial x}\left(\frac{\partial u_x}{\partial y} - \frac{\partial u_y}{\partial x}\right) + \frac{1}{2}\frac{\partial}{\partial y}\left(\frac{\partial u_x}{\partial x} - \frac{\partial u_x}{\partial x}\right)$$

adding and subtracting $\frac{1}{2}\frac{\partial}{\partial y}\left(\frac{\partial u_x}{\partial x}\right)$.

Since the order of differentiation is immaterial.

$$\frac{\partial \omega_{yx}}{\partial x} = \frac{1}{2}\frac{\partial}{\partial y}\left(\frac{\partial u_x}{\partial x} + \frac{\partial u_x}{\partial x}\right) - \frac{1}{2}\frac{\partial}{\partial x}\left(\frac{\partial u_x}{\partial y} + \frac{\partial u_y}{\partial x}\right)$$

$$= \frac{\partial}{\partial y} e_{xx} - \frac{\partial}{\partial x} e_{xy} \tag{A.5}$$

Similarly, $$\frac{\partial \omega_{xy}}{\partial y} = \frac{1}{2}\frac{\partial}{\partial y}\left(\frac{\partial u_x}{\partial y} - \frac{\partial u_y}{\partial x}\right)$$

$$= \frac{1}{2}\frac{\partial}{\partial y}\left(\frac{\partial u_x}{\partial y} - \frac{\partial u_y}{\partial x}\right) + \frac{1}{2}\frac{\partial}{\partial x}\left(\frac{\partial u_y}{\partial y} - \frac{\partial u_y}{\partial y}\right)$$

$$= \frac{1}{2}\frac{\partial}{\partial y}\left(\frac{\partial u_x}{\partial y} + \frac{\partial u_y}{\partial x}\right) - \frac{1}{2}\frac{\partial}{\partial x}\left(\frac{\partial u_y}{\partial y} + \frac{\partial u_y}{\partial y}\right) \tag{A.6}$$

$$= \frac{\partial}{\partial y} e_{xy} - \frac{\partial}{\partial x} e_{yy}$$

Substituting (A.5) and (A.6) in (A.4)

$$u_x(x_2, y_2) = u_x^\circ - (y\omega_{yx})\Big|_P^Q + \int_P^Q e_{xx}dx + \int_P^Q e_{xy}dy$$

$$-\int y\left[\left(\frac{\partial}{\partial y}e_{xx} - \frac{\partial}{\partial x}e_{xy}\right)dx + \left(\frac{\partial}{\partial y}e_{yx} - \frac{\partial}{\partial x}e_{yy}\right)dy\right]$$

Regrouping,

$$u_x(x_2, y_2) = u_x^\circ - (y\omega_{yx})\Big|_P^Q + \int_P^Q\left[e_{xx} - y\frac{\partial e_{xx}}{\partial y} + y\frac{\partial e_{xy}}{\partial x}\right]dx$$

$$+\int_P^Q\left[e_{xy} - y\frac{\partial e_{yx}}{\partial y} + y\frac{\partial e_{yy}}{\partial x}\right]dy \qquad \text{(A.7)}$$

Since the displacement is single–valued, the integral should be independent of the path of integration. This means that the integral is a perfect differential. This means

$$\frac{\partial}{\partial y}\left[e_{xx} - y\frac{\partial e_{xx}}{\partial y} + y\frac{\partial e_{xy}}{\partial x}\right] = \frac{\partial}{\partial x}\left[e_{xy} - y\frac{\partial e_{yx}}{\partial y} + y\frac{\partial e_{yy}}{\partial x}\right]$$

i.e.,
$$\frac{\partial e_{xx}}{\partial y} - \frac{\partial e_{xx}}{\partial y} - y\frac{\partial^2 e_{xx}}{\partial y^2} + \frac{\partial e_{xy}}{\partial x} + y\frac{\partial^2 e_{xy}}{\partial x\partial y} = \frac{\partial e_{xy}}{\partial x} - y\frac{\partial^2 e_{xy}}{\partial x\partial y} + y\frac{\partial^2 e_{yy}}{\partial x^2}$$

Since $e_{xy} = e_{yx}$, the above equation reduces to

$$\frac{\partial^2 e_{xx}}{\partial y^2} + \frac{\partial^2 e_{yy}}{\partial x^2} = 2\frac{\partial^2 e_{xy}}{\partial x\partial y} \qquad \text{(A.8)}$$

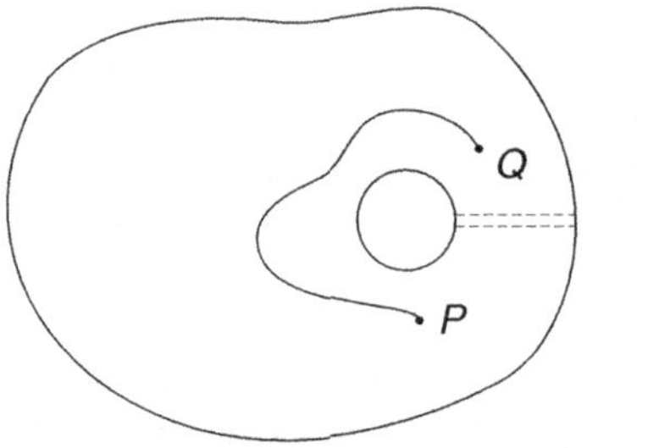

Fig. A.2 *Continuous curve connecting P and Q but not passing through the cut of multiply connected body*

An identical expression is obtained while considering the displacement u_y (x_2, y_2). Hence, the compatibility condition is a necessary and sufficient condition for the existence of single-valued displacement functions in simply connected bodies. For a multiply connected body, it is a necessary but not a sufficient condition. A multiply connected body can be made simply connected by a suitable cut. The displacement functions will then become single-valued when the path of integration does not pass through the cut.

Stress–Strain Relations for Linearly Elastic Solids

CHAPTER

3

3.1 INTRODUCTION

In the preceding two chapters we dealt with the state of stress at a point and the state of strain at a point. The strain components were related to the displacement components through six of Cauchy's strain-displacement relationships. In this chapter, the relationships between the stress and strain components will be established. Such equations are termed constitutive equations. They depend on the manner in which the material resists deformation.

The constitutive equations are mathematical descriptions of the physical phenomena based on experimental observations and established principles. Consequently, they are approximations of the true behavioural pattern, since an accurate mathematical representation of the physical phenomena would be too complicated and unworkable.

The constitutive equations describe the behaviour of a material, not the behaviour of a body. Therefore, the equations relate the state of stress at a point to the state of strain at the point.

3.2 GENERALISED STATEMENT OF HOOKE'S LAW

Consider a uniform cylindrical rod of diameter d subjected to a tensile force P. As is well known from experimental observations, when P is gradually increased from zero to some positive value, the length of the rod also increases. Based on experimental observations, it is postulated in elementary strength of materials that the axial stress σ is proportional to the axial strain ε up to a limit called the proportionality limit. The constant of proportionality is the Young's Modulus E, i.e.

$$\varepsilon = \frac{\sigma}{E} \quad \text{or} \quad \sigma = E\varepsilon \tag{3.1}$$

It is also well known that when the uniform rod elongates, its lateral dimensions, i.e. its diameter, decreases. In elementary strength of materials, the ratio of lateral strain to longitudinal strain was termed as Poisson's ratio ν. We now extend this information or knowledge to relate the six rectangular components of stress to the six rectangular components of strain. We assume that each of the six independent components of stress may be expressed as a linear function of the six components of strain and vice versa.

The mathematical expressions of this statement are the six stress–strain equations:

$$
\begin{aligned}
\sigma_x &= a_{11}\varepsilon_{xx} + a_{12}\varepsilon_{yy} + a_{13}\varepsilon_{zz} + a_{14}\gamma_{xy} + a_{15}\gamma_{yz} + a_{16}\gamma_{zx} \\
\sigma_y &= a_{21}\varepsilon_{xx} + a_{22}\varepsilon_{yy} + a_{23}\varepsilon_{zz} + a_{24}\gamma_{xy} + a_{25}\gamma_{yz} + a_{26}\gamma_{zx} \\
\sigma_z &= a_{31}\varepsilon_{xx} + a_{32}\varepsilon_{yy} + a_{33}\varepsilon_{zz} + a_{34}\gamma_{xy} + a_{35}\gamma_{yz} + a_{36}\gamma_{zx} \\
\tau_{xy} &= a_{41}\varepsilon_{xx} + a_{42}\varepsilon_{yy} + a_{43}\varepsilon_{zz} + a_{44}\gamma_{xy} + a_{45}\gamma_{yz} + a_{46}\gamma_{zx} \\
\tau_{yz} &= a_{51}\varepsilon_{xx} + a_{52}\varepsilon_{yy} + a_{53}\varepsilon_{zz} + a_{54}\gamma_{xy} + a_{55}\gamma_{yz} + a_{56}\gamma_{zx} \\
\tau_{zx} &= a_{61}\varepsilon_{xx} + a_{62}\varepsilon_{yy} + a_{63}\varepsilon_{zz} + a_{64}\gamma_{xy} + a_{65}\gamma_{yz} + a_{66}\gamma_{zx}
\end{aligned}
\tag{3.2}
$$

Or conversely, six strain-stress equations of the type:

$$
\varepsilon_{xx} = b_{11}\sigma_x + b_{12}\sigma_y + b_{13}\sigma_z + b_{14}\tau_{xy} + b_{15}\tau_{yz} + b_{16}\tau_{zx} \tag{3.3}
$$

$$
\varepsilon_{yy} = \ldots \text{etc}
$$

where a_{11}, a_{12}, b_{11}, b_{12}, . . . , are constants for a given material. Solving Eq. (3.2) as six simultaneous equations, one can get Eq. (3.3), and vice versa. For homogeneous, linearly elastic material, the six Eqs (3.2) or (3.3) are known as Generalised Hooke's Law. Whether we use the set given by Eq. (3.2) or that given by Eq. (3.3), 36 elastic constants are apparently involved.

3.3 STRESS–STRAIN RELATIONS FOR ISOTROPIC MATERIALS

We now make a further assumption that the ideal material we are dealing with has the same properties in all directions so far as the stress-strain relations are concerned. This means that the material we are dealing with is isotropic, i.e. it has no directional property.

Care must be taken to distinguish between the assumption of isotropy, which is a particular statement regarding the stress-strain properties at a given point, and that of homogeneity, which is a statement that the stress-strain properties, whatever they may be, are the same at all points. For example, timber of regular grain is homogeneous but not isotropic.

Assuming that the material is isotropic, one can show that only two independent elastic constants are involved in the generalised statement of Hooke's law. In Chapter 1, it was shown that at any point there are three faces (or planes) on which the resultant stresses are wholly normal, i.e. there are no shear stresses on these planes. These planes were termed the principal planes and the stresses on these planes the principal stresses. In Sec. 2.14, it was shown that at any point one can identify before strain, a small rectangular parallelepiped or a box which remains rectangular after strain. The normals to the faces of this box were called the principal axes of strain. Since in an isotropic material, a small rectangular box the faces of which are subjected to pure normal stresses, will remain rectangular after deformation (no asymmetrical deformation), the normal to these faces coincide with the principal strain axes. Hence, for an isotropic material, one can relate the principal stresses σ_1, σ_2, σ_3 with the three principal strains ε_1, ε_2 and ε_3 through suitable elastic constants. Let the axes x, y and z coincide with the principal stress and principal strain directions. For the principal stress σ_1 the equation becomes

$$\sigma_1 = a\varepsilon_1 + b\varepsilon_2 + c\varepsilon_3$$

where a, b and c are constants. But we observe that b and c should be equal since the effect of σ_1 in the directions of ε_2 and ε_3, which are both at right angles to σ_1, must be the same for an isotropic material. In other words, the effect of σ_1 in any direction transverse to it is the same in an isotropic material. Hence, for σ_1 the equation becomes

$$\begin{aligned}\sigma_1 &= a\varepsilon_1 + b(\varepsilon_2 + \varepsilon_3)\\ &= (a - b)\varepsilon_1 + b(\varepsilon_1 + \varepsilon_2 + \varepsilon_3)\end{aligned}$$

by adding and subtracting $b\varepsilon_1$. But $(\varepsilon_1 + \varepsilon_2 + \varepsilon_3)$ is the first invariant of strain $\boldsymbol{J}_1$ or the cubical dilatation Δ. Denoting b by λ and $(a - b)$ by 2μ, the equation for σ_1 becomes

$$\sigma_1 = \lambda\Delta + 2\mu\varepsilon_1 \tag{3.4a}$$

Similarly, for σ_2 and σ_3 we get

$$\sigma_2 = \lambda\Delta + 2\mu\varepsilon_2 \tag{3.4b}$$

$$\sigma_3 = \lambda\Delta + 2\mu\varepsilon_3 \tag{3.4c}$$

The constants λ and μ are called Lame's coefficients. Thus, there are only two elastic constants involved in the relations between the principal stresses and principal strains for an isotropic material. As the next sections show, this can be extended to the relations between rectangular stress and strain components also.

3.4 MODULUS OF RIGIDITY

Let the co-ordinate axes Ox, Oy, Oz coincide with the principal stress axes. For an isotropic body, the principal strain axes will also be along Ox, Oy, Oz. Consider another orthogonal frame of reference Ox', Oy', Oz', such that the direction cosines of Ox' are n_{x1}, n_{y1}, n_{z1} and those of Oy' are n_{x2}, n_{y2}, n_{z2}. Since Ox' and Oy' are at right angles to each other.

$$n_{x1}n_{x2} + n_{y1}n_{y2} + n_{z1}n_{z2} = 0 \tag{3.5}$$

The normal stress $\sigma_{x'}$ and the shear stress $\tau_{x'y'}$ are obtained from Cauchy's formula, Eqs. (1.9). The resultant stress vector on the x' plane will have components as

$$\overset{x'}{T}_x = n_{x1}\sigma_1, \qquad \overset{x'}{T}_y = n_{y1}\sigma_2, \qquad \overset{x'}{T}_z = n_{z1}\sigma_3$$

These are the components in x, y and z directions. The normal stress on this x' plane is obtained as the sum of the projections of the components along the normal, i.e.

$$\sigma_n = \sigma_{x'} = n_{x1}^2\sigma_1 + n_{y1}^2\sigma_2 + n_{z1}^2\sigma_3 \tag{3.6a}$$

Similarly, the shear stress component on this x' plane in y' direction is obtained as the sum of the projections of the components in y' direction, which has direction cosines n_{x2}, n_{y2}, n_{z2}. Thus

$$\tau_{x'y'} = n_{x1}n_{x2}\sigma_1 + n_{y1}n_{y2}\sigma_2 + n_{z1}n_{z2}\sigma_3 \tag{3.6b}$$

On the same lines, if ε_1, ε_2 and ε_3 are the principal strains, which are also along x, y, z directions, the normal strain in x' direction, from Eq. (2.20), is

$$\varepsilon_{x'x'} = n_{x1}^2\varepsilon_1 + n_{y1}^2\varepsilon_2 + n_{z1}^2\varepsilon_3 \tag{3.7a}$$

The shear strain $\gamma_{x'y'}$ is obtained from Eqs. (2.36a) and (2.36d)

$$\gamma_{x'y'} = \frac{1}{(1+\varepsilon_{x'})(1+\varepsilon_{y'})}\Big[2\big(n_{x1}n_{x2}\varepsilon_1 + n_{y1}n_{y2}\varepsilon_2 + n_{z1}n_{z2}\varepsilon_3\big)$$
$$+ n_{x1}n_{x2} + n_{y1}n_{y2} + n_{z1}n_{z2}\Big]$$

Using Eq. (3.5), and observing that $\varepsilon_{x'}$ and $\varepsilon_{y'}$ are small compared to unity in the denominator,

$$\gamma_{x'y'} = 2(n_{x1}n_{x2}\varepsilon_1 + n_{y1}n_{y2}\varepsilon_2 + n_{z1}n_{z2}\varepsilon_3) \tag{3.7b}$$

Substituting the values of σ_1, σ_2 and σ_3 from Eqs (3.4a)–(3.4c) into Eq. (3.6b)

$$\tau_{x'y'} = n_{x1}n_{x2}(\lambda\Delta + 2\mu\varepsilon_1) + n_{y1}n_{y2}\,(\lambda\Delta + 2\mu\varepsilon_2) + n_{z1}n_{z2}\,(\lambda\Delta + 2\mu\varepsilon_3)$$
$$= \lambda\Delta(n_{x1}n_{x2} + n_{y1}n_{y2} + n_{z1}n_{z2}) + 2\mu\,(n_{x1}n_{x2}\,\varepsilon_1 + n_{y1}\,n_{y2}\,\varepsilon_2 + n_{z1}\,n_{z2}\,\varepsilon_3)$$

Hence, from Eqs (3.5) and (3.7b)

$$\tau_{x'y'} = \mu\gamma_{x'y'} \tag{3.8}$$

Equation (3.8) relates the rectangular shear stress component $\tau_{x'y'}$ with the rectangular shear strain component $\gamma_{x'y'}$. Comparing this with the relation used in elementary strength of materials, one observes that μ is the modulus of rigidity, usually denoted by G.

By taking another axis Oz' with direction cosines n_{x3}, n_{y3} and n_{z3} and at right angles to Ox' and Oy' (so that $Ox'y'z'$ forms an orthogonal set of axes), one can get equations similar to (3.6a) and (3.6b) for the other rectangular stress components. Thus,

$$\sigma_{y'} = n_{x2}^2\sigma_1 + n_{y2}^2\sigma_2 + n_{z2}^2\,\sigma_3 \tag{3.9a}$$

$$\sigma_{z'} = n_{x3}^2\sigma_1 + n_{y3}^2\sigma_2 + n_{z3}^2\sigma_3 \tag{3.9b}$$

$$\tau_{y'z'} = n_{x2}n_{x3}\sigma_1 + n_{y2}n_{y3}\sigma_2 + n_{z2}n_{z3}\sigma_3 \tag{3.9c}$$

$$\tau_{z'x'} = n_{x3}n_{x1}\sigma_1 + n_{y3}n_{y1}\sigma_2 + n_{z3}n_{z1}\sigma_3 \tag{3.9d}$$

Similarly, following Eqs (3.7a) and (3.7b) for the other rectangular strain components, one gets

$$\varepsilon_{y'y'} = n_{x2}^2\varepsilon_1 + n_{y2}^2\varepsilon_2 + n_{z2}^2\varepsilon_3 \tag{3.10a}$$

$$\varepsilon_{z'z'} = n_{x3}^2\,\varepsilon_1 + n_{y3}^2\,\varepsilon_2 + n_{z3}^2\,\varepsilon_3 \tag{3.10b}$$

$$\gamma_{y'z'} = 2(n_{x2}n_{x3}\varepsilon_1 + n_{y2}n_{y3}\varepsilon_2 + n_{z2}n_{z3}\varepsilon_3) \tag{3.10c}$$

$$\gamma_{z'x'} = 2(n_{x3}n_{x1}\varepsilon_1 + n_{y3}n_{y1}\,\varepsilon_2 + n_{z3}n_{z1}\varepsilon_3) \tag{3.10d}$$

From Eqs (3.6a), (3.4a)–(3.4c) and (3.7a)

$$\sigma_{x'} = n_{x1}^2\sigma_1 + n_{y1}^2\sigma_2 + n_{z1}^2\sigma_3$$

$$= \lambda \Delta\left(n_{x1}^2 + n_{y1}^2 + n_{z1}^2\right) + 2\mu\left(\varepsilon_1 n_{x1}^2 + \varepsilon_2 n_{y1}^2 + \varepsilon_3 n_{z1}^2\right)$$

$$= \lambda \Delta + 2\mu \varepsilon_{x'x'} \tag{3.11a}$$

Similarly, one gets

$$\sigma_{y'} = \lambda \Delta + 2\mu \varepsilon_{y'y'} \tag{3.11b}$$

$$\sigma_{z'} = \lambda \Delta + 2\mu \varepsilon_{z'z'} \tag{3.11c}$$

Similar to Eq. (3.8),

$$\tau_{y'z'} = \mu \gamma_{y'z'} \tag{3.12a}$$

$$\tau_{x'z'} = \mu \gamma_{z'x'} \tag{3.12b}$$

Equations (3.11a)–(3.11c), (3.8) and (3.12a) and (3.12b) relate the six rectangular stress components to six rectangular strain components and in these only two elastic constants are involved. Therefore, the Hooke's law for an isotropic material will involve two independent elastic constants λ and μ (or G).

3.5 BULK MODULUS

Adding equations (3.11a)–(3.11c)

$$\sigma_{x'} + \sigma_{y'} + \sigma_{z'} = 3\lambda\Delta + 2\mu\left(\varepsilon_{x'x'} + \varepsilon_{y'y'} + \varepsilon_{z'z'}\right) \tag{3.13a}$$

Observing that

$$\sigma_{x'} + \sigma_{y'} + \sigma_{z'} = I_1 = \sigma_1 + \sigma_2 + \sigma_3 \qquad \text{(first invariant of stress),}$$

and

$$\varepsilon_{x'x'} + \varepsilon_{y'y'} + \varepsilon_{z'z'} = J_1 = \varepsilon_1 + \varepsilon_2 + \varepsilon_3 \qquad \text{(first invariant of strain),}$$

Eq. (3.13a) can be written in several alternative forms as

$$\sigma_1 + \sigma_2 + \sigma_3 = (3\lambda + 2\mu)\Delta \tag{3.13b}$$

$$\sigma_{x'} + \sigma_{y'} + \sigma_{z'} = (3\lambda + 2\mu)\Delta \tag{3.13c}$$

$$I_1 = (3\lambda + 2\mu)J_1 \tag{3.13d}$$

Noting from Eq. (2.34) that Δ is the volumetric strain, the definition of bulk modulus K is

$$K = \frac{\text{pressure}}{\text{volumetric strain}} = \frac{p}{\Delta} \tag{3.14a}$$

If $\sigma_1 = \sigma_2 = \sigma_3 = p$, then from Eq. (3.13b)

$$3p = (3\lambda + 2\mu)\Delta$$

or

$$3\,\frac{p}{\Delta} = (3\lambda + 2\mu)$$

and from Eq. (3.14a)

$$K = \frac{1}{3}(3\lambda + 2\mu) \tag{3.14b}$$

Thus, the bulk modulus for an isotropic solid is related to Lame's constants through Eq. (3.14b).

3.6 YOUNG'S MODULUS AND POISSON'S RATIO

From Eq. (3.13b), we have

$$\Delta = \frac{\sigma_1 + \sigma_2 + \sigma_3}{(3\lambda + 2\mu)}$$

Substituting this in Eq. (3.4a)

$$\sigma_1 = \frac{\lambda}{(3\lambda + 2\mu)}(\sigma_1 + \sigma_2 + \sigma_3) + 2\mu\varepsilon_1$$

or

$$\varepsilon_1 = \frac{\lambda + \mu}{\mu(3\lambda + 2\mu)}\left[\sigma_1 - \frac{\lambda}{2(\lambda + \mu)}(\sigma_2 + \sigma_3)\right] \tag{3.15}$$

From elementary strength of materials

$$\varepsilon_1 = \frac{1}{E}[\sigma_1 - \nu(\sigma_2 + \sigma_3)]$$

where E is Young's modulus, and ν is Poisson's ratio. Comparing this with Eq. (3.15),

$$E = \frac{\mu(3\lambda + 2\mu)}{(\lambda + \mu)}; \qquad \nu = \frac{\lambda}{2(\lambda + \mu)} \tag{3.16}$$

3.7 RELATIONS BETWEEN THE ELASTIC CONSTANTS

In elementary strength of materials, we are familiar with Young's modulus E, Poisson's ratio ν, shear modulus or modulus of rigidity G and bulk modulus K. Among these, only two are independent, and E and ν are generally taken as the independent constants. The other two, namely, G and K, are expressed as

$$G = \frac{E}{2(1 + \nu)}, \qquad K = \frac{E}{3(1 - 2\nu)} \tag{3.17}$$

It has been shown in this chapter, that for an isotropic material, the 36 elastic constants involved in the Generalised Hooke's law, can be reduced to two independent elastic constants. These two elastic constants are Lame's coefficients λ and μ. The second coefficient μ is the same as the rigidity modulus G. In terms of these, the other elastic constants can be expressed as

$$E = \frac{\mu(3\lambda + 2\mu)}{(\lambda + \mu)}, \qquad \nu = \frac{\lambda}{2(\lambda + \mu)} \tag{3.18}$$

$$K = \frac{(3\lambda + 2\mu)}{3}, \qquad G \equiv \mu, \qquad \lambda = \frac{\nu E}{(1 + \nu)(1 - 2\nu)},$$

It should be observed from Eq. (3.17) that for the bulk modulus to be positive, the value of Poisson's ratio ν cannot exceed 1/2. This is the upper limit for ν. For $\nu = 1/2$,

$$3G = E \quad \text{and} \quad K = \infty$$

A material having Poisson's ratio equal to 1/2 is known as an incompressible material, since the volumetric strain for such an isotropic material is zero.

For easy reference one can collect the equations relating stresses and strains that have been obtained so far.

(i) In terms of principal stresses and principal strains:

$$\begin{aligned}\sigma_1 &= \lambda\Delta + 2\mu\varepsilon_1\\ \sigma_2 &= \lambda\Delta + 2\mu\varepsilon_2\\ \sigma_3 &= \lambda\Delta + 2\mu\varepsilon_3\end{aligned} \tag{3.19}$$

where $\Delta = \varepsilon_1 + \varepsilon_2 + \varepsilon_3 = J_1$.

$$\begin{aligned}\varepsilon_1 &= \frac{\lambda+\mu}{\mu(3\lambda+2\mu)}\left[\sigma_1 - \frac{\lambda}{2(\lambda+\mu)}(\sigma_2+\sigma_3)\right]\\ \varepsilon_2 &= \frac{\lambda+\mu}{\mu(3\lambda+2\mu)}\left[\sigma_2 - \frac{\lambda}{2(\lambda+\mu)}(\sigma_3+\sigma_1)\right]\\ \varepsilon_3 &= \frac{\lambda+\mu}{\mu(3\lambda+2\mu)}\left[\sigma_3 - \frac{\lambda}{2(\lambda+\mu)}(\sigma_1+\sigma_2)\right]\end{aligned} \tag{3.20}$$

(ii) In terms of rectangular stress and strain components referred to an orthogonal coordinate system *Oxyz*:

$$\begin{aligned}\sigma_x &= \lambda\Delta + 2\mu\varepsilon_{xx}\\ \sigma_y &= \lambda\Delta + 2\mu\varepsilon_{yy}\\ \sigma_z &= \lambda\Delta + 2\mu\varepsilon_{zz}\end{aligned} \tag{3.21a}$$

where $\Delta = \varepsilon_{xx} + \varepsilon_{yy} + \varepsilon_{zz} = J_1$.

$$\tau_{xy} = \mu\gamma_{xy}, \qquad \tau_{yz} = \mu\gamma_{yz}, \quad \tau_{zx} = \mu\gamma_{zx} \tag{3.21b}$$

$$\begin{aligned}\varepsilon_{xx} &= \frac{\lambda+\mu}{\mu(3\lambda+2\mu)}\left[\sigma_x - \frac{\lambda}{2(\lambda+\mu)}(\sigma_y+\sigma_z)\right]\\ \varepsilon_{yy} &= \frac{\lambda+\mu}{\mu(3\lambda+2\mu)}\left[\sigma_y - \frac{\lambda}{2(\lambda+\mu)}(\sigma_z+\sigma_x)\right]\\ \varepsilon_{zz} &= \frac{\lambda+\mu}{\mu(3\lambda+2\mu)}\left[\sigma_z - \frac{\lambda}{2(\lambda+\mu)}(\sigma_x+\sigma_y)\right]\end{aligned} \tag{3.22a}$$

$$\gamma_{xy} = \frac{1}{\mu}\tau_{xy}, \qquad \gamma_{yz} = \frac{1}{\mu}\tau_{yz}, \qquad \gamma_{zx} = \frac{1}{\mu}\tau_{zx} \tag{3.22b}$$

In the preceeding sets of equations, λ and μ are Lame's constants. In terms of the more familiar elastic constants E and ν, the stress-strain relations are:

(iii) with $\varepsilon_{xx} + \varepsilon_{yy} + \varepsilon_{zz} = J_1 = \Delta$,

$$\begin{aligned}\sigma_x &= \frac{E}{(1+\nu)}\left[\frac{\nu}{(1-2\nu)}\Delta + \varepsilon_{xx}\right]\\ &= \lambda J_1 + 2G\varepsilon_{xx}\end{aligned}$$

$$\sigma_y = \frac{E}{(1+\nu)}\left[\frac{\nu}{(1-2\nu)}\Delta + \varepsilon_{yy}\right] \tag{3.23a}$$

$$= \lambda J_1 + 2G\varepsilon_{yy}$$

$$\sigma_z = \frac{E}{(1+\nu)}\left[\frac{\nu}{(1-2\nu)}\Delta + \varepsilon_{zz}\right]$$

$$= \lambda J_1 + 2G\varepsilon_{zz}$$

$$\tau_{xy} = G\gamma_{xy}, \qquad \tau_{yz} = G\gamma_{yz}, \qquad \tau_{xx} = G\gamma_{zx} \tag{3.23b}$$

$$\varepsilon_{xx} = \frac{1}{E}\left[\sigma_x - \nu(\sigma_y + \sigma_z)\right]$$

$$\varepsilon_{yy} = \frac{1}{E}\left[\sigma_y - \nu(\sigma_z + \sigma_x)\right] \tag{3.24a}$$

$$\varepsilon_{zz} = \frac{1}{E}\left[\sigma_z - \nu(\sigma_x + \sigma_y)\right]$$

$$\gamma_{xy} = \frac{1}{G}\tau_{xy}, \qquad \gamma_{yz} = \frac{1}{G}\tau_{yz}, \qquad \gamma_{zx} = \frac{1}{G}\tau_{zx} \tag{3.24b}$$

3.8 DISPLACEMENT EQUATIONS OF EQUILIBRIUM

In Chapter 1, it was shown that if a solid body is in equilibrium, the six rectangular stress components have to satisfy the three equations of equilibrium. In this chapter, we have shown how to relate the stress components to the strain components using the stress-strain relations. Hence, stress equations of equilibrium can be converted to strain equations of equilibrium. Further, in Chapter 2, the strain components were related to the displacement components. Therefore, the strain equations of equilibrium can be converted to displacement equations of equilibrium. In this section, this result will be derived.

The first equation from Eq. (1.65) is

$$\frac{\partial \sigma_x}{\partial x} + \frac{\partial \tau_{xy}}{\partial y} + \frac{\partial \tau_{zx}}{\partial z} = 0$$

For an isotropic material

$$\sigma_x = \lambda\Delta + 2\mu\varepsilon_{xx}; \qquad \tau_{xy} = \mu\gamma_{xy}; \qquad \tau_{xz} = \mu\gamma_{xz}$$

Hence, the above equation becomes

$$\lambda\frac{\partial\Delta}{\partial x} + \mu\left(2\frac{\partial\varepsilon_{xx}}{\partial x} + \frac{\partial\gamma_{xy}}{\partial y} + \frac{\partial\gamma_{xz}}{\partial z}\right) = 0$$

From Cauchy's strain-displacement relations

$$\varepsilon_{xx} = \frac{\partial u_x}{\partial x}, \qquad \gamma_{xy} = \frac{\partial u_x}{\partial y} + \frac{\partial u_y}{\partial x}, \qquad \gamma_{zx} = \frac{\partial u_x}{\partial z} + \frac{\partial u_z}{\partial x}$$

Substituting these

$$\lambda \frac{\partial \Delta}{\partial x} + \mu \left(2\frac{\partial^2 u_x}{\partial x^2} + \frac{\partial^2 u_x}{\partial y^2} + \frac{\partial^2 u_y}{\partial x \partial y} + \frac{\partial^2 u_x}{\partial z^2} + \frac{\partial^2 u_z}{\partial x \partial z} \right) = 0$$

or $$\lambda \frac{\partial \Delta}{\partial x} + \mu \left(\frac{\partial^2 u_x}{\partial x^2} + \frac{\partial^2 u_x}{\partial y^2} + \frac{\partial^2 u_x}{\partial z^2} \right) + \mu \left(\frac{\partial^2 u_x}{\partial x^2} + \frac{\partial^2 u_y}{\partial x \partial y} + \frac{\partial^2 u_z}{\partial x \partial z} \right) = 0$$

or $$\lambda \frac{\partial \Delta}{\partial x} + \mu \left(\frac{\partial^2 u_x}{\partial x^2} + \frac{\partial^2 u_x}{\partial y^2} + \frac{\partial^2 u_x}{\partial z^2} \right) + \mu \frac{\partial}{\partial x} \left(\frac{\partial u_x}{\partial x} + \frac{\partial u_y}{\partial y} + \frac{\partial u_z}{\partial z} \right) = 0$$

Observing that

$$\Delta = \varepsilon_{xx} + \varepsilon_{yy} + \varepsilon_{zz} = \frac{\partial u_x}{\partial x} + \frac{\partial u_y}{\partial y} + \frac{\partial u_z}{\partial z}$$

$$(\lambda + \mu) \frac{\partial}{\partial x} \left(\frac{\partial u_x}{\partial x} + \frac{\partial u_y}{\partial y} + \frac{\partial u_z}{\partial z} \right) + \mu \left(\frac{\partial^2 u_x}{\partial x^2} + \frac{\partial^2 u_x}{\partial y^2} + \frac{\partial^2 u_x}{\partial z^2} \right) = 0$$

This is one of the displacement equations of equilibrium. Using the notation

$$\nabla^2 = \frac{\partial^2}{\partial x^2} + \frac{\partial^2}{\partial y^2} + \frac{\partial^2}{\partial z^2}$$

the displacement equation of equilibrium becomes

$$(\lambda + \mu) \frac{\partial \Delta}{\partial x} + \mu \nabla^2 u_x = 0 \tag{3.25a}$$

Similarly, from the second and third equations of equilibrium, one gets

$$(\lambda + \mu) \frac{\partial \Delta}{\partial y} + \mu \nabla^2 u_y = 0 \tag{3.25b}$$

$$(\lambda + \mu) \frac{\partial \Delta}{\partial z} + \mu \nabla^2 u_z = 0$$

These are known as Lame's displacement equations of equilibrium. They involve a synthesis of the analysis of stress, analysis of strain and the relations between stresses and strains. These equations represent the mechanical, geometrical and physical characteristics of an elastic solid. Consequently, Lame's equations play a very prominent role in the solutions of problems.

Example 3.1 *A rubber cube is inserted in a cavity of the same form and size in a steel block and the top of the cube is pressed by a steel block with a pressure of p pascals. Considering the steel to be absolutely hard and assuming that there is no friction between steel and rubber, find (i) the pressure of rubber against the box walls, and (ii) the extremum shear stresses in rubber.*

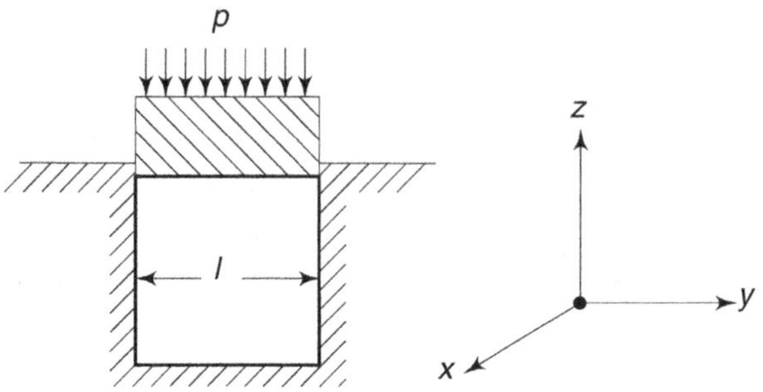

Fig. 3.1 *Example 3.1*

Solution

(i) Let l be the dimension of the cube. Since the cube is constrained in x and y directions

$$\varepsilon_{xx} = 0 \quad \text{and} \quad \varepsilon_{yy} = 0$$

and $\quad \sigma_z = -p$

Therefore

$$\varepsilon_{xx} = \frac{1}{E}\left[\sigma_x - \nu\left(\sigma_y + \sigma_z\right)\right] = 0$$

$$\varepsilon_{yy} = \frac{1}{E}\left[\sigma_y - \nu\left(\sigma_x + \sigma_z\right)\right] = 0$$

Solving

$$\sigma_x = \sigma_y = \frac{\nu}{1-\nu}\sigma_z = -\frac{\nu}{1-\nu}p$$

If Poisson's ratio = 0.5, then

$$\sigma_x = \sigma_y = \sigma_z = -p$$

(ii) The extremum shear stresses are

$$\tau_2 = \frac{\sigma_1 - \sigma_3}{2}, \qquad \tau_3 = \frac{\sigma_1 - \sigma_2}{2}, \qquad \tau_1 = \frac{\sigma_2 - \sigma_3}{2}$$

If $\nu \le 0.5$, then σ_x and σ_y are numerically less than or equal to σ_z. Since σ_x, σ_y and σ_z are all compressive

$$\sigma_1 = \sigma_x = -\frac{\nu}{1-\nu}p$$

$$\sigma_2 = \sigma_y = -\frac{\nu}{1-\nu}p$$

$$\sigma_3 = \sigma_z = -p$$

$$\therefore \qquad \tau_1 = p\left(1 - \frac{\nu}{1-\nu}\right) = \frac{1-2\nu}{1-\nu}p, \quad \tau_2 = \frac{1-2\nu}{1-\nu}p, \quad \tau_3 = 0$$

If $\nu = 0.5$, the shear stresses are zero.

Example 3.2 *A cubical element is subjected to the following state of stress.*

$$\sigma_x = 100 \text{ MPa}, \quad \sigma_y = -20 \text{ MPa}, \quad \sigma_z = -40 \text{ Mpa}, \quad \tau_{xy} = \tau_{yz} = \tau_{zx} = 0$$

Assuming the material to be homogeneous and isotropic, determine the principal shear strains and the octahedral shear strain, if $E = 2 \times 10^5$ MPa and $\nu = 0.25$.

Solution Since the shear stresses on x, y and z planes are zero, the given stresses are principal stresses. Arranging such that $\sigma_1 \geq \sigma_2 \geq \sigma_3$

$$\sigma_1 = 100 \text{ MPa}, \quad \sigma_2 = -20 \text{ MPa}, \quad \sigma_3 = -40 \text{ MPa}$$

The extremal shear stresses are

$$\tau_1 = \frac{1}{2}(\sigma_2 - \sigma_3) = \frac{1}{2}(-20 + 40) = 10 \text{ Mpa}$$

$$\tau_2 = \frac{1}{2}(\sigma_3 - \sigma_1) = \frac{1}{2}(-40 - 100) = -70 \text{ Mpa}$$

$$\tau_3 = \frac{1}{2}(\sigma_1 - \sigma_2) = \frac{1}{2}(100 + 20) = 60 \text{ Mpa}$$

The modulus of rigidity G is

$$G = \frac{E}{2(1+\nu)} = \frac{2 \times 10^5}{2 \times 1.25} = 8 \times 10^4 \text{ MPa}$$

The principal shear strains are therefore

$$\gamma_1 = \frac{\tau_1}{G} = \frac{10}{8 \times 10^4} = 1.25 \times 10^{-4}$$

$$\gamma_2 = \frac{\tau_2}{G} = -\frac{70}{8 \times 10^4} = -8.75 \times 10^{-4}$$

$$\gamma_3 = \frac{\tau_3}{G} = \frac{60}{8 \times 10^4} = 7.5 \times 10^{-4}$$

From Eq. (1.44a), the octahedral shear stress is

$$\tau_0 = \frac{1}{3}[(\sigma_1 - \sigma_2)^2 + (\sigma_2 - \sigma_3)^2 + (\sigma_3 - \sigma_1)^2]^{1/2}$$

$$= \frac{1}{3}[120^2 + 20^2 + 140^2]^{1/2} = 61.8 \text{ MPa}$$

The octahedral shear strain is therefore

$$\gamma_0 = \frac{\tau_0}{G} = \frac{61.8}{8 \times 10^4} = 7.73 \times 10^{-4}$$

Problems

3.1 Compute Lame's coefficients λ and μ for
  (a) steel having $E = 207 \times 10^6$ kPa (2.1×10^6 kgf/cm^2) and $\nu = 0.3$.
  (b) concrete having $E = 28 \times 10^6$ kPa (2.85×10^5 kgf/cm^2) and $\nu = 0.2$.

[*Ans.* (a) 120×10^6 kPa (1.22×10^6 kgf/cm^2), 80×10^6 kPa (8.1680×10^5 kgf/cm^2)
(b) 7.8×10^6 kPa (7.96×10^4 kgf/cm^2), 11.7×10^6 kPa (1.2×10^5 kgf/cm^2)]

3.2 For steel, the following data is applicable:

$$E = 207 \times 10^6 \text{ kPa } (2.1 \times 10^6 \text{ kgf/cm}^2),$$

and $$G = 80 \times 10^6 \text{ kPa } (0.82 \times 10^6 \text{ kgf/cm}^2)$$

For the given strain matrix at a point, determine the stress matrix.

$$[\varepsilon_{ij}] = \begin{bmatrix} 0.001 & 0 & -0.002 \\ 0 & -0.003 & 0.0003 \\ -0.002 & 0.003 & 0 \end{bmatrix}$$

$$\left[Ans. \; [\tau_{ij}] = \begin{bmatrix} -68.4 & 0 & -160 \\ 0 & -708.4 & 24 \\ -160 & 24 & -228.4 \end{bmatrix} \times 10^3 \text{ kPa} \right]$$

3.3 A thin rubber sheet is enclosed between two fixed hard steel plates (see Fig. 3.2). Friction between the rubber and steel faces is negligible. If the rubber plate is subjected to stresses σ_x and σ_y as shown, determine the strains ε_{xx} and ε_{yy}, and also the stress ε_{zz}

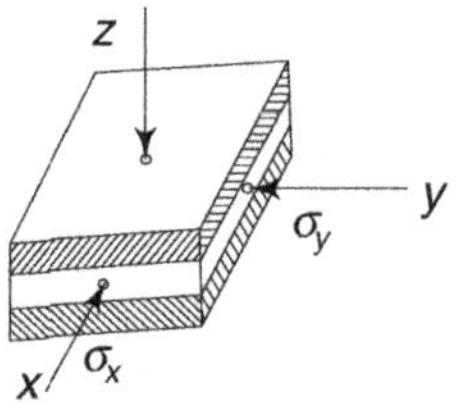

Fig. 3.2 *Example 3.2*

[*Ans.* $\sigma_z = +\nu(\sigma_x + \sigma_y)$

$$\varepsilon_{xx} = +\frac{1+\nu}{E}[(1-\nu)\sigma_x - \nu\sigma_y]$$

$$\varepsilon_{yy} = +\frac{1+\nu}{E}[(1-\nu)\sigma_y \; \nu\sigma_x]$$]

Theories of Failure or Yield Criteria and Introduction to Ideally Plastic Solid

CHAPTER 4

4.1 INTRODUCTION

It is known from the results of material testing that when bars of ductile materials are subjected to uniform tension, the stress-strain curves show a linear range within which the materials behave in an elastic manner and a definite yield zone where the materials undergo permanent deformation. In the case of the so-called brittle materials, there is no yield zone. However, a brittle material, under suitable conditions, can be brought to a plastic state before fracture occurs. In general, the results of material testing reveal that the behaviour of various materials under similar test conditions, e.g. under simple tension, compression or torsion, varies considerably.

In the process of designing a machine element or a structural member, the designer has to take precautions to see that the member under consideration does not fail under service conditions. The word 'failure' used in this context may mean either fracture or permanent deformation beyond the operational range due to the yielding of the member. In Chapter 1, it was stated that the state of stress at any point can be characterised by the six rectangular stress components—three normal stresses and three shear stresses. Similarly, in Chapter 2, it was shown that the state of strain at a point can be characterised by the six rectangular strain components. When failure occurs, the question that arises is: what causes the failure? Is it a particular state of stress, or a particular state of strain or some other quantity associated with stress and strain? Further, the cause of failure of a ductile material need not be the same as that for a brittle material.

Consider, for example, a uniform rod made of a ductile material subject to tension. When yielding occurs,

(i) The principal stress σ at a point will have reached a definite value, usually denoted by σ_y;

(ii) The maximum shearing stress at the point will have reached a value equal to $\tau = \frac{1}{2}\sigma_y$;

(iii) The principal extension will have become $\varepsilon = \sigma_y/E$;

(iv) The octahedral shearing stress will have attained a value equal to $(\sqrt{2}/3)\,\sigma_y$;

and so on.

Any one of the above or some other factors might have caused the yielding. Further, as pointed out earlier, the factor that causes a ductile material to yield might be quite different from the factor that causes fracture in a brittle material under the same loading conditions. Consequently, there will be many criteria or theories of failure. It is necessary to remember that failure may mean fracture or yielding. Whatever may be the theory adopted, the information regarding it will have to be obtained from a simple test, like that of a uniaxial tension or a pure torsion test. This is so because the state of stress or strain which causes the failure of the material concerned can easily be calculated. The critical value obtained from this test will have to be applied for the stress or strain at a point in a general machine or a structural member so as not to initiate failure at that point.

There are six main theories of failure and these are discussed in the next section. Another theory, called Mohr's theory, is slightly different in its approach and will be discussed separately.

4.2 THEORIES OF FAILURE

Maximum Principal Stress Theory

This theory is generally associated with the name of Rankine. According to this theory, the maximum principal stress in the material determines failure regardless of what the other two principal stresses are, so long as they are algebraically smaller. This theory is not much supported by experimental results. Most solid materials can withstand very high hydrostatic pressures without fracture or without much permanent deformation if the pressure acts uniformly from all sides as is the case when a solid material is subjected to high fluid pressure. Materials with a loose or porous structure such as wood, however, undergo considerable permanent deformation when subjected to high hydrostatic pressures.On the other hand, metals and other crystalline solids (including consolidated natural rocks) which are impervious, are elastically compressed and can withstand very high hydrostatic pressures. In less compact solid materials, a marked evidence of failure has been observed when these solids are subjected to hydrostatic pressures. Further, it has been observed that even brittle materials, like glass bulbs, which are subject to high hydrostatic pressure do not fail when the pressure is acting, but fail either during the period the pressure is being reduced or later when the pressure is rapidly released. It is stated that the liquid could have penentrated through the fine invisible surface cracks and when the pressure was released, the entrapped liquid may not have been able to escape fast enough. Consequently, high pressure gradients are caused on the surface of the material which tend to burst or explode the glass. As Karman pointed out, this penentration and the consequent failure of the material can be prevented if the latter is covered by a thin flexible metal foil and then subjected to high hydrostatic pressures. Further noteworthy observations on the bursting action of a liquid which is used to transmit pressure were made by Bridgman who found that cylinders of hardened chrome-nickel steel were not able to withstand an internal pressure well if the liquid transmitting the pressure was mercury instead of viscous oil. It appears that small atoms of mercury are able to penentrate the cracks, whereas the large molecules of oil are not able to penentrate so easily.

From these observations, we draw the conclusion that a pure state of hydrostatic pressure [$\sigma_1 = \sigma_2 = \sigma_3 = -p$ ($p > 0$)] cannot produce permanent deformation in compact crystalline or amorphous solid materials but produces only a small elastic contraction, provided the liquid is prevented from entering the fine surface cracks or crevices of the solid. This contradicts the maximum principal stress theory. Further evidence to show that the maximum principal stress theory cannot be a good criterion for failure can be demonstrated in the following manner:

Consider the block shown in Fig. 4.1, subjected to stresses σ_1 and σ_2, where σ_1 is tensile and σ_2 is compressive.

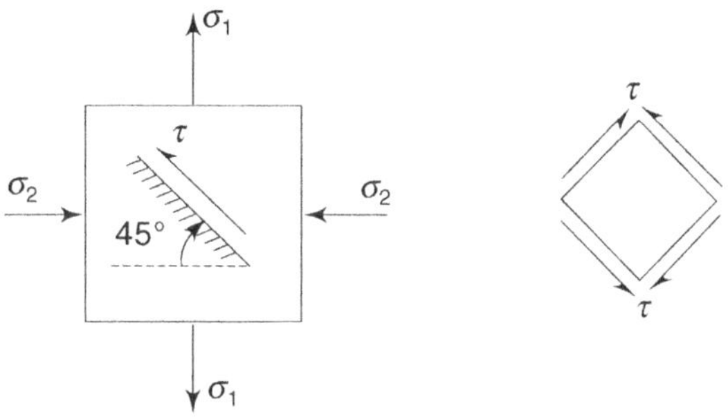

Fig. 4.1 *Rectangular element with 45° plane*

If σ_1 is equal to σ_2 in magnitude, then on a 45° plane, from Eq. (1.63b), the shearing stress will have a magnitude equal to σ_1. Such a state of stress occurs in a cylindrical bar subjected to pure torsion. If the maximum principal stress theory was valid, σ_1 would have been the limiting value. However, for ductile materials subjected to pure torsion, experiments reveal that the shear stress limit causing yield is much less than σ_1 in magnitude.

Notwithstanding all these, the maximum principal stress theory, because of its simplicity, is considered to be reasonably satisfactory for brittle materials which do not fail by yielding. Using information from a uniaxial tension (or compression) test, we say that failure occurs when the maximum principal stress at any point reaches a value equal to the tensile (or compressive) elastic limit or yield strength of the material obtained from the uniaxial test. Thus, if $\sigma_1 > \sigma_2 > \sigma_3$ are the principal stresses at a point and σ_y the yield stress or tensile elastic limit for the material under a uniaxial test, then failure occurs when

$$\sigma_1 \geq \sigma_y \tag{4.1}$$

Maximum Shearing Stress Theory

Observations made in the course of extrusion tests on the flow of soft metals through orifices lend support to the assumption that the plastic state in such metals is created when the maximum shearing stress just reaches the value of the resistance of the metal against shear. Assuming $\sigma_1 > \sigma_2 > \sigma_3$, yielding, according to this theory, occurs when the maximum shearing stress

$$\tau_{max} = \frac{\sigma_1 - \sigma_3}{2}$$

reaches a critical value. The maximum shearing stress theory is accepted to be fairly well justified for ductile materials. In a bar subject to uniaxial tension or compression, the maximum shear stress occurs on a plane at 45° to the load axis. Tension tests conducted on mild steel bars show that at the time of yielding, the so-called slip lines occur approximately at 45°, thus supporting the theory. On the other hand, for brittle crystalline materials which cannot be brought into the plastic state under tension but which may yield a little before fracture under compression, the angle of the slip planes or of the shear fracture surfaces, which usually develop along these planes, differs considerably from the planes of maximum shear. Further, in these brittle materials, the values of the maximum shear in tension and compression are not equal. Failure of material under triaxial tension (of equal magnitude) also does not support this theory, since equal triaxial tensions cannot produce any shear.

However, as remarked earlier, for ductile load carrying members where large shears occur and which are subject to unequal triaxial tensions, the maximum shearing stress theory is used because of its simplicity.

If $\sigma_1 > \sigma_2 > \sigma_3$ are the three principal stresses at a point, failure occurs when

$$\tau_{\max} = \frac{\sigma_1 - \sigma_3}{2} \geq \frac{\sigma_y}{2} \tag{4.2}$$

where $\sigma_y/2$ is the shear stress at yield point in a uniaxial test.

Maximum Elastic Strain Theory

According to this theory, failure occurs at a point in a body when the maximum strain at that point exceeds the value of the maximum strain in a uniaxial test of the material at yield point. Thus, if σ_1, σ_2 and σ_3 are the principal stresses at a point, failure occurs when

$$\varepsilon_1 = \frac{1}{E}\left[\sigma_1 - \nu(\sigma_2 + \sigma_3)\right] \geq \frac{\sigma_y}{E} \tag{4.3}$$

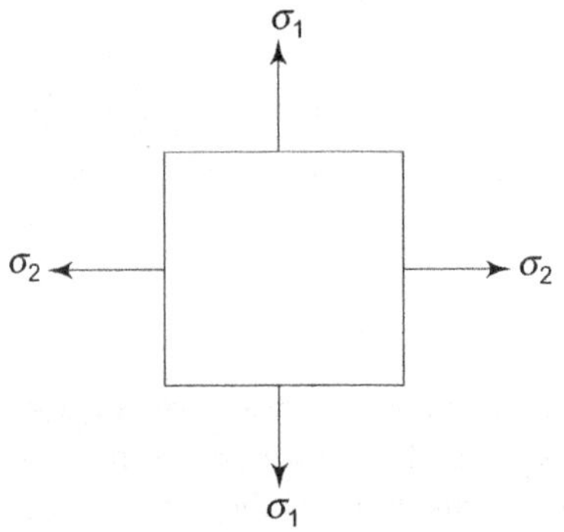

Fig. 4.2 *Biaxial state of stress*

We have observed that a material subjected to triaxial compression does not suffer failure, thus contradicting this theory. Also, in a block subjected to a biaxial tension, as shown in Fig. 4.2, the principal strain ε_1 is

$$\varepsilon_1 = \frac{1}{E}(\sigma_1 - \nu\sigma_2)$$

and is smaller than σ_1/E because of σ_2. Therefore, according to this theory, σ_1 can be increased more than σ_y without causing failure, whereas, if σ_2 were compressive, the magnitude of σ_1 to cause failure would be less than σ_y. However, this is not supported by experiments.

While the maximum strain theory is an improvement over the maximum stress theory, it is not a good theory for ductile materials. For materials which fail by brittle fracture, one may prefer the maximum strain theory to the maximum stress theory.

Octahedral Shearing Stress Theory

According to this theory, the critical quantity is the shearing stress on the octahedral plane. The plane which is equally inclined to all the three principal axes *Ox*, *Oy* and *Oz* is called the octahedral plane. The normal to this plane has direction cosines n_x, n_y and $n_z = 1/\sqrt{3}$. The tangential stress on this plane is the octahedral shearing stress. If σ_1, σ_2 and σ_3 are the principal stresses at a point, then from Eqs (1.44a) and (1.44c)

$$\tau_{\text{oct}} = \frac{1}{3}\left[(\sigma_1 - \sigma_2)^2 + (\sigma_2 - \sigma_3)^2 + (\sigma_3 - \sigma_1)^2\right]^{1/2}$$

$$= \frac{\sqrt{2}}{3}\left(I_1^2 - 3I_2\right)^{1/2}$$

In a uniaxial test, at yield point, the octahedral stress $(\sqrt{2}/3)\ \sigma_y = 0.47\sigma_y$. Hence, according to the present theory, failure occurs at a point where the values of principal stresses are such that

$$\tau_{\text{oct}} = \frac{1}{3}\left[(\sigma_1 - \sigma_2)^2 + (\sigma_2 - \sigma_3)^2 + (\sigma_3 - \sigma_1)^2\right]^{1/2} \geq \frac{\sqrt{2}}{3}\sigma_y \qquad (4.4a)$$

or

$$\left(I_1^2 - 3I_2\right) \geq \sigma_y^2 \qquad (4.4b)$$

This theory is supported quite well by experimental evidences. Further, when a material is subjected to hydrostatic pressure, $\sigma_1 = \sigma_2 = \sigma_3 = -p$, and τ_{oct} is equal to zero. Consequently, according to this theory, failure cannot occur and this, as stated earlier, is supported by experimental results. This theory is equivalent to the maximum distortion energy theory, which will be discussed subsequently.

Maximum Elastic Energy Theory

This theory is associated with the names of Beltrami and Haigh. According to this theory, failure at any point in a body subject to a state of stress begins only when the energy per unit volume absorbed at the point is equal to the energy absorbed per unit volume by the material when subjected to the elastic limit under a uniaxial state of stress. To calculate the energy absorbed per unit volume we proceed as follows:

Let σ_1, σ_2 and σ_3 be the principal stresses and let their magnitudes increase uniformly from zero to their final magnitudes. If ε_1, ε_2 and ε_3 are the corresponding principal strains, then the work done by the forces, from Fig. 4.3(b), is

$$\Delta W = \frac{1}{2}\sigma_1\ \Delta y\ \Delta z\ (\delta\Delta x) + \frac{1}{2}\sigma_2\ \Delta x \Delta z\ (\delta\Delta y) + \frac{1}{2}\sigma_3\ \Delta x\ \Delta y\ (\delta\Delta z)$$

where $\delta\Delta x$, $\delta\Delta y$ and $\delta\Delta z$ are extensions in *x*, *y* and *z* directions respectively.

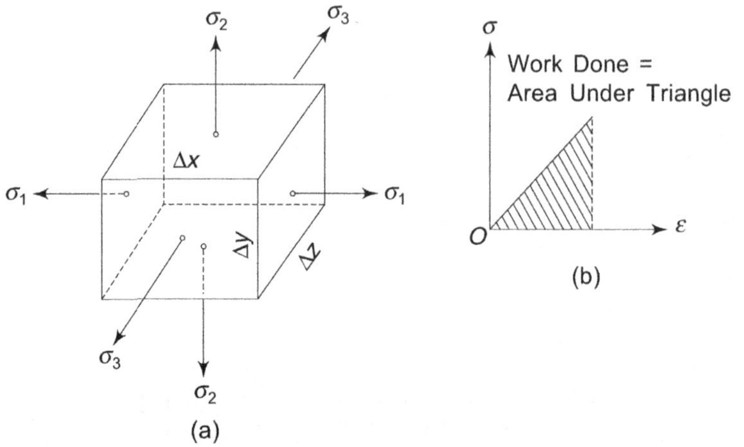

Fig. 4.3 *(a) Principal stresses on a rectangular block*
(b) Area representing work done

From Hooke' s law

$$\delta\Delta x = \varepsilon_1\ \Delta x = \frac{1}{E}\left[\sigma_1 - \nu\left(\sigma_2 + \sigma_3\right)\right]\Delta x$$

$$\delta\Delta y = \varepsilon_2\ \Delta y = \frac{1}{E}\left[\sigma_2 - \nu\left(\sigma_1 + \sigma_3\right)\right]\Delta y$$

$$\delta\Delta z = \varepsilon_3\ \Delta z = \frac{1}{E}\left[\sigma_3 - \nu\left(\sigma_1 + \sigma_2\right)\right]\Delta z$$

Substituting these

$$\Delta W = \frac{1}{2E}\left[\sigma_1^2 + \sigma_2^2 + \sigma_3^2 - 2\nu\left(\sigma_1\sigma_2 + \sigma_2\sigma_3 + \sigma_3\sigma_1\right)\right]\Delta x\ \Delta y\ \Delta z$$

The above work is stored as internal energy if the rate of deformation is small. Consequently, the energy U per unit volume is

$$\frac{1}{2E}\left[\sigma_1^2 + \sigma_2^2 + \sigma_3^2 - 2\nu\left(\sigma_1\sigma_2 + \sigma_2\sigma_3 + \sigma_3\sigma_1\right)\right] \tag{4.5}$$

In a uniaxial test, the energy stored per unit volume at yield point or elastic limit is $1/2E\ \sigma_y^2$. Hence, failure occurs when

$$\sigma_1^2 + \sigma_2^2 + \sigma_3^2 - 2\nu\left(\sigma_1\sigma_2 + \sigma_2\sigma_3 + \sigma_3\sigma_1\right) \geq \sigma_y^2 \tag{4.6}$$

This theory does not have much significance since it is possible for a material to absorb considerable amount of energy without failure or permanent deformation when it is subjected to hydrostatic pressure.

Energy of Distortion Theory

This theory is based on the work of Huber, von Mises and Hencky. According to this theory, it is not the total energy which is the criterion for failure; in fact the energy absorbed during the distortion of an element is responsible for failure. The energy of distortion can be obtained by subtracting the energy of volumetric expansion from the total energy. It was shown in the Analysis of Stress (Sec. 1.22) that any given state of stress can be uniquely resolved into an isotropic state and a pure shear (or deviatoric) state. If σ_1, σ_2 and σ_3 are the principal stresses at a point then

$$\begin{bmatrix} \sigma_1 & 0 & 0 \\ 0 & \sigma_2 & 0 \\ 0 & 0 & \sigma_3 \end{bmatrix} = \begin{bmatrix} p & 0 & 0 \\ 0 & p & 0 \\ 0 & 0 & P \end{bmatrix} = \begin{bmatrix} \sigma_1 - p & 0 & 0 \\ 0 & \sigma_2 - p & 0 \\ 0 & 0 & \sigma_3 - p \end{bmatrix} \tag{4.7}$$

where $p = \frac{1}{3}(\sigma_1 + \sigma_2 + \sigma_3)$.

The first matrix on the right-hand side represents the isotropic state and the second matrix the pure shear state. Also, recall that the necessary and sufficient condition for a state to be a pure shear state is that its first invariant must be equal to zero. Similarly, in the Analysis of Strain (Section 2.17), it was shown that any given state of strain can be resolved uniquely into an isotropic and a deviatoric state of strain. If ε_1, ε_2 and ε_3 are the principal strains at the point, we have

$$\begin{bmatrix} \varepsilon_1 & 0 & 0 \\ 0 & \varepsilon_2 & 0 \\ 0 & 0 & \varepsilon_3 \end{bmatrix} = \begin{bmatrix} e & 0 & 0 \\ 0 & e & 0 \\ 0 & 0 & e \end{bmatrix} = \begin{bmatrix} \varepsilon_1 - e & 0 & 0 \\ 0 & \varepsilon_2 - e & 0 \\ 0 & 0 & \varepsilon_3 - e \end{bmatrix} \tag{4.8}$$

where $e = \frac{1}{3}(\varepsilon_1 + \varepsilon_2 + \varepsilon_3)$.

It was also shown that the volumetric strain corresponding to the deviatoric state of strain is zero since its first invariant is zero.

It is easy to see from Eqs (4.7) and (4.8) that, by Hooke's law, the isotropic state of strain is related to the isotropic state of stress because

$$\varepsilon_1 = \frac{1}{E}\left[\sigma_1 - \nu(\sigma_2 + \sigma_3)\right]$$

$$\varepsilon_2 = \frac{1}{E}\left[\sigma_2 - \nu(\sigma_3 + \sigma_1)\right]$$

$$\varepsilon_3 = \frac{1}{E}\left[\sigma_3 - \nu(\sigma_2 + \sigma_1)\right]$$

Adding and taking the mean

$$\frac{1}{3}(\varepsilon_1 + \varepsilon_2 + \varepsilon_3) = e$$

$$= \frac{1}{3E}\left[(\sigma_1 + \sigma_2 + \sigma_3) - 2\nu(\sigma_1 + \sigma_2 + \sigma_3)\right]$$

or

$$e = \frac{1}{E}\left[(1 - 2\nu)\,p\right] \tag{4.9}$$

i.e. e is connected to p by Hooke's law. This states that the volumetric strain $3e$ is proportional to the pressure p, the proportionality constant being equal to $\frac{3}{E}(1 - 2\nu) = K$, the bulk modulus, Eq. (3.14).

Consequently, the work done or the energy stored during volumetric change is

$$U' = \frac{1}{2}pe + \frac{1}{2}pe + \frac{1}{2}pe = \frac{3}{2}pe$$

Substituting for e from Eq. (4.9)

$$U' = \frac{3}{2E}(1-2\nu)\,p^2 \qquad (4.10)$$

$$= \frac{1-2\nu}{6E}(\sigma_1 + \sigma_2 + \sigma_3)^2$$

The total elastic strain energy density is given by Eq. (4.5). Hence, subtracting U' from U

$$U^* = \frac{1}{2E}\left(\sigma_1^2 + \sigma_2^2 + \sigma_3^2\right) - \frac{\nu}{E}(\sigma_1\sigma_2 + \sigma_2\sigma_3 + \sigma_3\sigma_1)$$

$$-\frac{1-2\nu}{6E}(\sigma_1 + \sigma_2 + \sigma_3)^2 \qquad (4.11a)$$

$$= \frac{2(1+\nu)}{6E}\left(\sigma_1^2 + \sigma_2^2 + \sigma_3^2 - \sigma_1\sigma_2 - \sigma_2\sigma_3 - \sigma_3\sigma_1\right) \qquad (4.11b)$$

$$= \frac{(1+\nu)}{6E}\left[(\sigma_1 - \sigma_2)^2 + (\sigma_2 - \sigma_3)^2 + (\sigma_3 - \sigma_1)^2\right] \qquad (4.11c)$$

Substituting $G = \dfrac{E}{2(1+\nu)}$ for the shear modulus,

$$U^* = \frac{1}{6G}\left(\sigma_1^2 + \sigma_2^2 + \sigma_3^2 - \sigma_1\sigma_2 - \sigma_2\sigma_3 - \sigma_3\sigma_1\right) \qquad (4.12a)$$

or

$$U^* = \frac{1}{12G}\left[(\sigma_1 - \sigma_2)^2 + (\sigma_2 - \sigma_3)^2 + (\sigma_3 - \sigma_1)^2\right] \qquad (4.12b)$$

This is the expression for the energy of distortion. In a uniaxial test, the energy of distortion is equal to $\frac{1}{6G}\sigma_y^2$. This is obtained by simply putting $\sigma_1 = \sigma_y$ and $\sigma_2 = \sigma_3 = 0$ in Eq. (4.12). This is also equal to $\frac{(1+\nu)}{3E}\sigma_y^2$ from Eq. (4.11c).

Hence, according to the distortion energy theory, failure occurs at that point where σ_1, σ_2 and σ_3 are such that

$$(\sigma_1 - \sigma_2)^2 + (\sigma_2 - \sigma_3)^2 + (\sigma_3 - \sigma_1)^2 \geq 2\sigma_y^2 \qquad (4.13)$$

But we notice that the expression for the octahedral shearing stress from Eq. (1.22) is

$$\tau_{oct} = \frac{1}{3}\left[(\sigma_1 - \sigma_2)^2 + (\sigma_2 - \sigma_3)^2 + (\sigma_3 - \sigma_1)^2\right]^{1/2}$$

Hence, the distortion energy theory states that failure occurs when

$$9\tau_{\text{oct}}^2 = \geq 2\sigma_y^2$$

or

$$\tau_{\text{oct}} = \geq \frac{\sqrt{2}}{3}\sigma_y \tag{4.14}$$

This is identical to Eq. (4.4). Therefore, the octahedral shearing stress theory and the distortion energy theory are identical. Experiments made on the flow of ductile metals under biaxial states of stress have shown that Eq. (4.14) or equivalently, Eq. (4.13) expresses well the condition under which the ductile metals at normal temperatures start to yield. Further, as remarked earlier, the purely elastic deformation of a body under hydrostatic pressure ($\tau_{\text{oct}} = 0$) is also supported by this theory.

4.3 SIGNIFICANCE OF THE THEORIES OF FAILURE

The mode of failure of a member and the factor that is responsible for failure depend on a large number of factors such as the nature and properties of the material, type of loading, shape and temperature of the member, etc. We have observed, for example, that the mode of failure of a ductile material differs from that of a brittle material. While yielding or permanent deformation is the characteristic feature of ductile materials, fracture without permanent deformation is the characteristic feature of brittle materials. Further, if the loading conditions are suitably altered, a brittle material may be made to yield before failure. Even ductile materials fail in a different manner when subjected to repeated loadings (such as fatigue) than when subjected to static loadings. All these factors indicate that any rational procedure of design of a member requires the determination of the mode of failure (either yielding or fracture), and the factor (such as stress, strain and energy) associated with it. If tests could be performed on the actual member, subjecting it to all the possible conditions of loading that the member would be subjected to during operation, then one could determine the maximum loading condition that does not cause failure. But this may not be possible except in very simple cases. Consequently, in complex loading conditions, one has to identify the factor associated with the failure of a member and take precautions to see that this factor does not exceed the maximum allowable value. This information is obtained by performing a suitable test (uniform tension or torsion) on the material in the laboratory.

In discussing the various theories of failure, we have expressed the critical value associated with each theory in terms of the yield point stress σ_y obtained from a uniaxial tensile stress. This was done since it is easy to perform a uniaxial tensile stress and obtain the yield point stress value. It is equally easy to perform a pure torsion test on a round specimen and obtain the value of the maximum shear stress τ_y at the point of yielding. Consequently, one can also express the critical value associated with each theory of failure in terms of the yield point shear stress τ_y. In a sense, using σ_y or τ_y is equivalent because during a uniaxial tension, the maximum shear stress τ at a point is equal to $\frac{1}{2}\sigma$; and in the case of pure shear, the normal stresses on a 45° element are σ and $-\sigma$, where σ is numerically equivalent to τ. These are shown in Fig. 4.4.

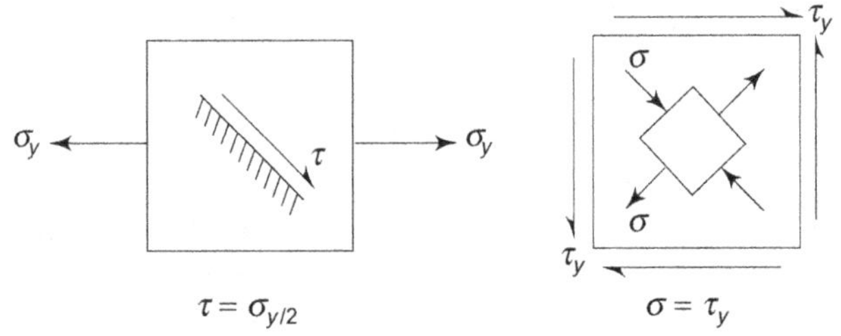

Fig. 4.4 *Uniaxial and pure shear state of stress*

If one uses the yield point shear stress τ_y obtained from a pure torsion test, then the critical value associated with each theory of failure is as follows:

(i) Maximum Normal Stress Theory According to this theory, failure occurs when the normal stress *s* at any point in the stressed member reaches a value

$$\sigma \geq \tau_y$$

This is because, in a pure torsion test when yielding occurs, the maximum normal stress *s* is numerically equivalent to t_y.

(ii) Maximum Shear Stress Theory According to this theory, failure occurs when the shear stress *t* at a point in the member reaches a value

$$\tau \geq \tau_y$$

(iii) Maximum Strain Theory According to this theory, failure occurs when the maximum strain at any point in the member reaches a value

$$\varepsilon = \frac{1}{E}[\sigma_1 - \nu(\sigma_2 + \sigma_3)]$$

From Fig. 4.4, in the case of pure shear

$$\sigma_1 = \sigma = \tau, \quad \sigma_2 = 0, \ \sigma_3 = -\sigma = -\tau$$

Hence, failure occurs when the strain *e* at any point in the member reaches a value

$$\varepsilon = \frac{1}{E}(\tau_y + \nu\tau_y) = \frac{1}{E}(1 + \nu)\tau_y$$

(iv) Octahedral Shear Stress Theory When an element is subjected to pure shear, the maximum and minimum normal stresses at a point are *s* and –*s* (each numerically equal to the shear stress *t*), as shown in Fig. 4.4. Corresponding to this, from Eq. (1.44a), the octahedral shear stress is

$$\tau_{oct} = \frac{1}{3}\left[(\sigma_1 - \sigma_2)^2 + (\sigma_2 - \sigma_3)^2 + (\sigma_3 - \sigma_1)^2\right]^{1/2}$$

Observing that $\sigma_1 = \sigma = \tau$, $\sigma_2 = 0$, $\sigma_3 = -\sigma = -\tau$

$$\tau_{oct} = \frac{1}{3}(\sigma^2 + \sigma^2 + 4\sigma^2)^{1/2}$$

$$= \frac{\sqrt{6}}{3}\sigma = \sqrt{\frac{2}{3}}\tau$$

So, failure occurs when the octahedral shear stress at any point is

$$\tau_{oct} = \sqrt{\frac{2}{3}}\,\tau_y$$

(v) Maximum Elastic Energy Theory The elastic energy per unit volume stored at a point in a stressed body is, from Eq. (4.5),

$$U = \frac{1}{E}\left[\sigma_1^2 + \sigma_2^2 + \sigma_3^2 - 2\nu\left(\sigma_1\sigma_2 + \sigma_2\sigma_3 + \sigma_3\sigma_1\right)\right]$$

In the case of pure shear, from Fig. 4.4,

$$\sigma_1 = \tau, \qquad \sigma_2 = 0, \qquad \sigma_3 = -\tau$$

Hence,
$$U = \frac{1}{2E}\left[\tau^2 + \tau^2 - 2\nu\left(-\tau^2\right)\right]$$
$$= \frac{1}{E}(1+\nu)\,\tau^2$$

So, failure occurs when the elastic energy density at any point in a stressed body is such that

$$U = \frac{1}{E}(1+\nu)\,\tau_y^2$$

(vi) Distortion Energy Theory The distortion energy density at a point in a stressed body is, from Eq. (4.12),

$$U^* = \frac{1}{12G}\left[(\sigma_1 - \sigma_2)^2 + (\sigma_2 - \sigma_3)^2 + (\sigma_3 - \sigma_1)^2\right]$$

Once again, by observing that in the case of pure shear

$$\sigma_1 = \tau, \qquad \sigma_2 = 0, \qquad \sigma_3 = -\tau$$

$$U^* = \frac{1}{12G}\left[\tau^2 + \tau^2 + 4\tau^2\right]$$

$$= \frac{1}{2G}\tau^2$$

So, failure occurs when the distortion energy density at any point is equal to

$$U^* = \frac{1}{2G}\tau_y^2 = \frac{1}{2}\cdot\frac{2(1+\nu)}{E}\tau_y^2$$

$$= \frac{(1+\nu)}{E}\tau_y^2$$

The foregoing results show that one can express the critical value associated with each theory of failure either in terms of σ_y or in terms of τ_y. Assuming that a particular theory of failure is correct for a given material, then the values of σ_y and τ_y obtained from tests conducted on the material should be related by the corresponding expressions. For example, if the distortion energy is a valid theory for a material, then the value of the energy in terms of σ_y and that in terms of τ_y should be equal. Thus,

$$U^* = \frac{(1+\nu)}{E}\tau_y^2 = \frac{(1+\nu)}{3E}\sigma_y^2$$

or

$$\tau_y = \frac{1}{\sqrt{3}}\sigma_y = 0.577\,\sigma_y$$

This means that the value of τ_y obtained from pure torsion test should be equal to 0.577 times the value of σ_y obtained from a uniaxial tension test conducted on the same material.

Table 4.1 summarizes these theories and the corresponding expressions. The first column lists the six theories of failure. The second column lists the critical value associated with each theory in terms of σ_y, the yield point stress in uniaxial tension test. For example, according to the octahedral shear stress theory, failure occurs when the octahedral shear stress at a point assumes a value equal to $\sqrt{2}/3\,\sigma_y$. The third column lists the critical value associated with each theory in terms of τ_y, the yield point shear stress value in pure torsion. For example, according to octahedral shear stress theory, failure occurs at a point when the octahedral shear stress equals a value $\sqrt{2/3}\,\tau_y$. The fourth column gives the relationship that should exist between τ_y and σ_y in each case if each theory is valid. Assuming octahedral shear stress theory is correct, then the value of τ_y obtained from pure torsion test should be equal to 0.577 times the yield point stress σ_y obtained from a uniaxial tension test.

Tests conducted on many ductile materials reveal that the values of τ_y lie between 0.50 and 0.60 of the tensile yield strength σ_y, the average value being about 0.57. This result agrees well with the octahedral shear stress theory and the

Table 4.1

Failure theory	*Tension*	*Shear*	*Relationship*
Max. normal stress	σ_y	$\sigma_y = \tau_y$	$\tau_y = \sigma_y$
Max. shear stress	$\tau = \frac{1}{2}\sigma_y$	τ_y	$\tau_y = 0.5\,\sigma_y$
Max. strain $\left(\nu = \frac{1}{4}\right)$	$\varepsilon = \frac{1}{E}\sigma_y$	$\varepsilon = \frac{5}{4}\frac{\tau_y}{E}$	$\tau_y = 0.8\,\sigma_y$
Octahedral shear	$\tau_{oct} = \frac{\sqrt{2}}{3}\sigma_y$	$\tau_{oct} = \sqrt{\frac{2}{3}}\,\tau_y$	$\tau_y = 0.577\,\sigma_y$
Max. energy $\left(\nu = \frac{1}{4}\right)$,	$U = \frac{1}{2E}\sigma_y^2$	$U = \frac{5}{4}\frac{1}{E}\tau_y^2$	$\tau_y = 0.632\,\sigma_y$
Distortion energy	$U^* = \frac{1+\nu}{3}\frac{\sigma_y^2}{E}$	$U^* = (1+\nu)\frac{\tau_y^2}{E}$	$\tau_y = 0.577\,\sigma_y$

distortion energy theory. The maximum shear stress theory predicts that shear yield value τ_y is 0.5 times the tensile yield value. This is about 15% less than the value predicted by the distortion energy (or the octahedral shear) theory. The maximum shear stress theory gives values for design on the safe side. Also, because of its simplicity, this theory is widely used in machine design dealing with ductile materials.

4.4 USE OF FACTOR OF SAFETY IN DESIGN

In designing a member to carry a given load without failure, usually a factor of safety N is used. The purpose is to design the member in such a way that it can carry N times the actual working load without failure. It has been observed that one can associate different factors for failure according to the particular theory of failure adopted. Consequently, one can use a factor appropriately reduced during the design process. Let X be a factor associated with failure and let F be the load. If X is directly proportional to F, then designing the member to safely carry a load equal to NF is equivalent to designing the member for a critical factor equal to X/N. However, if X is not directly proportional to F, but is, say, proportional to F^2, then designing the member to safely carry a load to equal to NF is equivalent to limiting the critical factor to $\sqrt{X/N}$. Hence, in using the factor of safety, care must be taken to see that the critical factor associated with failure is not reduced by N, but rather the load-carrying capacity is increased by N. This point will be made clear in the following example.

Example 4.1 *Determine the diameter d of a circular shaft subjected to a bending moment M and a torque T, according to the several theories of failure. Use a factor of safety N.*

Solution Consider a point P on the periphery of the shaft. If d is the diameter, then owing to the bending moment M, the maximum normal stress σ at P on a plane normal to the axis of the shaft is, from elementary strength of materials,

$$\sigma = \frac{My}{I} = M\frac{d}{2}\frac{64}{\pi d^4} \tag{4.15}$$

$$= \frac{32M}{\pi d^3}$$

The shearing stress on a transverse plane at P due to torsion T is

$$\tau = \frac{Td}{2I_P} = \frac{Td \cdot 32}{2\pi d^4} \tag{4.16}$$

$$= \frac{16T}{\pi d^3}$$

Therefore, the principal stresses at P are

$$\sigma_{1,3} = \frac{1}{2}\sigma \pm \frac{1}{2}\sqrt{\left(\sigma^2 + 4\tau^2\right)}, \quad \sigma_2 = 0 \tag{4.17}$$

(i) Maximum Normal Stress Theory At point P, the maximum normal stress should not exceed σ_y, the yield point stress in tension. With a factor of safety N, when the load is increased N times, the normal and shearing stresses are Ns and $N\tau$. Equating the maximum normal stress to σ_y,

$$\sigma_{\max} = \sigma_1 = N\left[\frac{\sigma}{2} + \frac{1}{2}\left(\sigma^2 + 4\tau^2\right)^{1/2}\right] = \sigma_y$$

or
$$\sigma + \left(\sigma^2 + 4\tau^2\right)^{1/2} = \frac{2\sigma_y}{N}$$

i.e.,
$$\frac{32M}{\pi d^3} + \frac{1}{\pi d^3} \times 32\left(M^2 + T^2\right)^{1/2} = \frac{2\,\sigma_y}{N}$$

i.e.,
$$16M + 16\left(M^2 + T^2\right)^{1/2} = \frac{\pi d^3\,\sigma_y}{N}$$

From this, the value of d can be determined with the known values of M, T and s_y.

(ii) Maximum Shear Stress Theory At point P, the maximum shearing stress from Eq. (4.17) is

$$\tau_{\max} = \frac{1}{2}\left(\sigma_1 - \sigma_3\right) = \frac{1}{2}\left(\sigma^2 + 4\tau^2\right)^{1/2}$$

When the load is increased N times, the shear stress becomes Nt.

Hence,

$$N\tau_{\max} = \frac{1}{2}N\left(\sigma^2 + 4\tau^2\right)^{1/2} = \frac{\sigma_y}{2}$$

or,
$$\left(\sigma^2 + 4\tau^2\right)^{1/2} = \frac{\sigma_y}{N}$$

Substituting for σ and τ

$$\frac{32}{\pi d^3}\left(M^2 + T^2\right)^{1/2} = \frac{\sigma_y}{N}$$

or,
$$32\left(M^2 + T^2\right)^{1/2} = \frac{\pi d^3 \sigma_y}{N}$$

(iii) Maximum Strain Theory The maximum elastic strain at point P with a factor of safety N is

$$\varepsilon_{\max} = \frac{N}{E}\left[\sigma_1 - \nu\left(\sigma_2 + \sigma_3\right)\right]$$

From Eq. (4.3)

$$\sigma_1 - \nu\left(\sigma_2 + \sigma_3\right) = \frac{\sigma_y}{N}$$

Since $\sigma_2 = 0$, we have $\sigma_1 - \nu\,\sigma_3 = \dfrac{\sigma_y}{N}$

or $$\frac{\sigma}{2}+\frac{1}{2}\left(\sigma^2+4\tau^2\right)^{1/2}-\nu\frac{\sigma}{2}+\frac{\nu}{2}\left(\sigma^2+4\tau^2\right)^{1/2}=\frac{\sigma_y}{N}$$

Substituting for σ and τ

$$(1-\nu)\frac{16M}{\pi d^3}+(1+\nu)\frac{16}{\pi d^3}\left(M^2+T^2\right)^{1/2}=\frac{\sigma_y}{N}$$

or $$(1-\nu)16M+(1+\nu)16\left(M^2+T^2\right)^{1/2}=\frac{\pi d^3\sigma_y}{N}$$

(iv) Octahedral Shear Stress Theory The octahedral shearing stress at point P from Eq. (4.4a), and using a factor of safety N, is

$$N\tau_{\text{oct}}=\frac{N}{3}\left[(\sigma_1-\sigma_2)^2+(\sigma_2-\sigma_3)^2+(\sigma_3-\sigma_1)^2\right]^{1/2}=\frac{\sqrt{2}}{3}\sigma_y$$

or $$\left[(\sigma_1-\sigma_2)^2+(\sigma_2-\sigma_3)^2+(\sigma_3-\sigma_1)^2\right]^{1/2}=\frac{\sqrt{2}}{N}\sigma_y$$

With $\sigma_2=0$

$$\left[2\sigma_1^2+2\sigma_3^2-2\sigma_1\sigma_3\right]^{1/2}=\frac{\sqrt{2}}{N}\sigma_y$$

or $$\left[\sigma_1^2+\sigma_3^2-\sigma_1\sigma_3\right]^{1/2}=\frac{\sigma_y}{N}$$

Substituting for σ_1 and σ_3

$$\left[\frac{1}{4}\sigma^2+\frac{1}{4}\left(\sigma^2+4\tau^2\right)+\frac{1}{2}\sigma\left(\sigma^2+4\tau^2\right)^{1/2}+\frac{1}{4}\sigma^2+\frac{1}{4}\left(\sigma^2+4\tau^2\right)\right.$$

$$\left.-\frac{1}{2}\sigma\left(\sigma^2+4\tau^2\right)^{1/2}-\frac{1}{4}\sigma^2+\frac{1}{4}\left(\sigma^2+4\tau^2\right)\right]^{1/2}=\frac{\sigma_y}{N}$$

or $$\left(\sigma^2+3\tau^2\right)^{1/2}=\frac{\sigma_y}{N}$$

Substituting for σ and τ

$$\frac{16}{\pi d^3}\left(4M^2+3T^2\right)^{1/2}=\frac{\sigma_y}{N}$$

or $$16\,(4M^2+3T^2)^{1/2}=\frac{\pi d^3\sigma_y}{N}$$

(v) Maximum Energy Theory The maximum elastic energy at P from Eq. (4.6) and with a factor of safety N is

$$U=\frac{N^2}{2E}\left[\sigma_1^2+\sigma_2^2+\sigma_3^2-2\nu\left(\sigma_1\sigma_2+\sigma_2\sigma_3+\sigma_3\sigma_1\right)\right]=\frac{\sigma_y^2}{2E}$$

Note: Since the stresses for design are $N\sigma_1$, $N\sigma_2$ and $N\sigma_3$, the factor N^2 appears in the expression for U. In the previous four cases, only N appeared because of the particular form of the expression.

With $\sigma_2 = 0$,

$$(\sigma_1^2 + \sigma_3^2 - 2\nu\, \sigma_1\, \sigma_3) = \frac{\sigma_y^2}{N^2}$$

Substituting for σ_1 and σ_3

$$\left[\frac{1}{4}\sigma^2 + \frac{1}{4}\left(\sigma^2 + 4\tau^2\right) + \frac{1}{2}\sigma\left(\sigma^2 + 4\tau^2\right)^{1/2} + \frac{1}{4}\sigma^2\right.$$

$$\left. + \frac{1}{4}\left(\sigma^2 + 4\tau^2\right) - \frac{1}{2}\sigma\left(\sigma^2 + 4\tau^2\right)^{1/2} + 2\nu\tau^2\right] = \frac{\sigma_y^2}{N^2}$$

or $$\sigma^2 + (2 + 2\nu)\,\tau^2 = \frac{\sigma_y^2}{N^2}$$

i.e. $$\left[\sigma^2 + (2 + 2\nu)\,\tau^2\right]^{1/2} = \frac{\sigma_y}{N}$$

i.e. $$\frac{16}{\pi d^3}\left[4M^2 + (2 + 2\nu)\,T^2\right]^{1/2} = \frac{\sigma_y}{N}$$

or $$\left[4M^2 + 2\,(1 + \nu)\,T^2\right]^{1/2} = \frac{\pi d^3 \sigma_y}{16\,N}$$

(vi) Maximum Distortion Energy Theory The distortion energy associated with Ns_1, Ns_2 and Ns_3 at P is given by Eq. (4.11c). Equating this to distortion energy in terms of s_y

$$U_d = \frac{N^2(1+\nu)}{6E}\left[(\sigma_1 - \sigma_2)^2 + (\sigma_2 - \sigma_3)^2 + (\sigma_3 - \sigma_1)^2\right]$$

$$= \frac{1+\nu}{3E}\,\sigma_y^2$$

With $\sigma_2 = 0$,

$$(2\sigma_1^2 + 2\sigma_3^2 - 2\sigma_1\,\sigma_3) = \frac{2\,\sigma_y^2}{N^2}$$

or $$(\sigma_1^2 + \sigma_3^2 - \sigma_1\,\sigma_3)^{1/2} = \frac{\sigma_y}{N}$$

This yields the same result as the octahedral shear stress theory.

4.5 A NOTE ON THE USE OF FACTOR OF SAFETY

As remarked earlier, when a factor of safety N is prescribed, we may consider two ways of introducing it in design:

(i) Design the member so that it safely carries a load NF.

(ii) If the factor associated with failure is X, then see that this factor at any point in the member does not exceed X/N.

But the second method of using N is not correct, since by the definition of the factor of safety, the member is to be designed for N times the load. So long as X is directly proportional to F, whether one uses NF or X/N for design analysis, the result will be identical. If X is not directly proportional to F, method (ii) may give wrong results. For example, if we adopt method (ii) with the maximum energy theory, the result will be

$$U = \frac{1}{2E}\left[\sigma_1^2 + \sigma_2^2 + \sigma_3^2 - 2\nu\left(\sigma_1\,\sigma_2 + \sigma_2\,\sigma_3 + \sigma_3\,\sigma_1\right)\right] = \frac{1}{N}\frac{\sigma_y^2}{2E}$$

where X, the factor associated with failure, is $\frac{1}{2}\frac{\sigma_y^2}{E}$. But method (i) gives

$$U = \frac{N^2}{2E}\left[\sigma_1^2 + \sigma_2^2 + \sigma_3^2 - 2\nu\left(\sigma_1\,\sigma_2 + \sigma_2\,\sigma_3 + \sigma_3\,\sigma_1\right)\right] = \frac{\sigma_y^2}{2E}$$

The result obtained from method (i) is correct, since $N\sigma_1$, $N\sigma_2$ and $N\sigma_3$ are the principal stresses corresponding to the load NF. As one an see, the results are not the same. The result given by method (ii) is not the right one.

Example 4.2 *A force F = 45,000 N is necessary to rotate the shaft shown in Fig. 4.5 at uniform speed. The crank shaft is made of ductile steel whose elastic limit is 207,000 kPa, both in tension and compression. With $E = 207 \times 10^6$ kPa, $\nu = 0.25$, determine the diameter of the shaft, using the octahedral shear stress theory and the maximum shear stress theory. Use a factor of safety N = 2. Consider a point on the periphery at section A for analysis.*

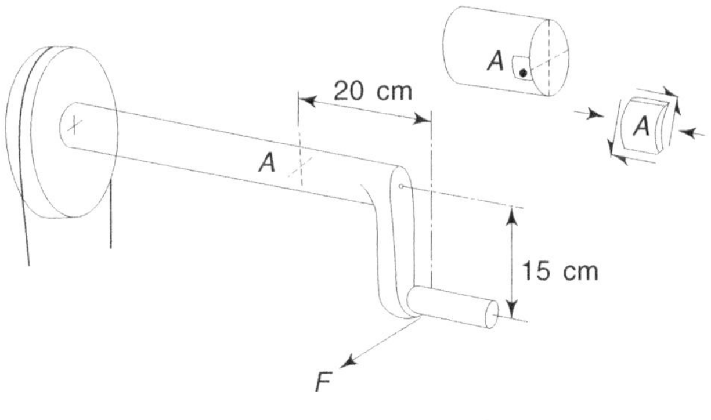

Fig. 4.5 *Example 4.2*

Solution The moment at section A is

$$M = 45{,}000 \times 0.2 = 9000 \text{ Nm}$$

and the torque on the shaft is

$$T = 45{,}000 \times 0.15 = 6750 \text{ Nm}$$

The normal stress due to M at A is

$$\sigma = -\frac{64Md}{2\pi d^4} = -\frac{32M}{\pi d^3}$$

and the maximum shear stress due to T at A is

$$\tau = \frac{32Td}{2\pi d^4} = \frac{16T}{\pi d^3}$$

The shear stress due to the shear force F is zero at A. The principal stresses from Eq. (1.61) are

$$\sigma_{1,3} = \frac{1}{2}\sigma \pm \frac{1}{2}\left(\sigma^2 + 4\tau^2\right)^{1/2}, \qquad \sigma_2 = 0$$

*(i) **Maximum Shear Stress Theory***

$$\tau_{max} = \frac{1}{2}(\sigma_1 - \sigma_3)$$

$$= \frac{1}{2}(\sigma^2 + 4\tau^2)^{1/2}$$

$$= \frac{1}{2}\frac{32}{\pi d^3}(M^2 + T^2)^{1/2}$$

$$= \frac{16}{\pi d^3}\left(9000^2 + 6750^2\right)^{1/2} = \frac{57295.8}{d^3} \text{ Pa}$$

With a factor of safety $N = 2$, the value of τ_{max} becomes

$$N\tau_{max} = \frac{114591.6}{d^3} \text{ Pa}$$

This should not exceed the maximum shear stress value at yielding in uniaxial tension test. Thus,

$$\frac{1}{d^3}(114591.6) = \frac{\sigma_y}{2} = \frac{207}{2} \times 10^6$$

$$\therefore \qquad d^3 = 1107 \times 10^{-6} \text{ m}^3$$

$$\text{or} \qquad d = 10.35 \times 10^{-2} \text{ m} = 10.4 \text{ cm}$$

(ii) Octahedral Shear Stress Theory

$$\tau_{oct} = \frac{1}{3}\left[(\sigma_1 - \sigma_2)^2 + (\sigma_2 - \sigma_3)^2 + (\sigma_3 - \sigma_1)^2\right]^{1/2}$$

With $\sigma_2 = 0$,

$$\tau_{oct} = \frac{1}{3}\left[2\sigma_1^2 + 2\sigma_3^2 - 2\sigma_1\sigma_3\right]^{1/2}$$

Substituting for σ_1 and σ_3 and simplifying

$$\tau_{oct} = \frac{\sqrt{2}}{3}\left(\sigma^2 + 3\tau^2\right)^{1/2}$$

$$= \frac{\sqrt{2}}{3\pi d^3}\left[(32M)^2 + 3(16T)^2\right]^{1/2}$$

$$= \frac{16\sqrt{2}}{3\pi d^3}\left(4M^2 + 3T^2\right)^{1/2}$$

$$= \frac{16\sqrt{2}}{3\pi d^3}\left[4\,(9000)^2 + 3(6750)^2\right]^{1/2}$$

$$= \frac{\sqrt{2}}{3\pi d^3} \times 343418$$

Equating this to octahedral shear stress at yielding of a uniaxial tension bar, and using a factor of safety $N = 2$,

$$\frac{\sqrt{2}}{3\pi d^3} \times 2 \times 343418 = \frac{\sqrt{2}}{3}\,\sigma_y$$

or $\quad 2 \times 343418 = \pi d^3\,\sigma_y = \pi d^3 \times 207 \times 10^6$

$\therefore \quad d^3 = 1.056 \times 10^{-3}$

or $\quad d = 0.1018\text{ m} = 10.18\text{ cm}$

Example 4.3 *A cylindrical bar of 7 cm diameter is subjected to a torque equal to 3400 Nm, and a bending moment M. If the bar is at the point of failing in accordance with the maximum principal stress theory, determine the maximum bending moment it can support in addition to the torque. The tensile elastic limit for the material is 207 MPa, and the factor of safety to be used is 3.*

Solution From Example 4.1(i)

$$16\,M + 16\,(M^2 + T^2)^{1/2} = \frac{\pi d^3}{N}\,\sigma_y$$

i.e. $\quad 16\,M + 16\,(M^2 + 3400^2)^{1/2} = \dfrac{\pi \times 7^3 \times 10^{-6} \times 207 \times 10^6}{3}$

or $\quad (M^2 + 3400^2)^{1/2} = 4647 - M$

or $\quad M^2 + 3400^2 = 4647^2 + M^2 - 9294\,M$

$\therefore \quad M = 1080\text{ Nm}$

Example 4.4 *In Example 4.3, if failure is governed by the maximum strain theory, determine the diameter of the bar if it is subjected to a torque $T = 3400$ Nm and a bending moment $M = 1080$ Nm. The elastic modulus for the material is $E = 103 \times 10^6$ kPa, $\nu = 0.25$, factor of safety $N = 3$ and $\sigma_y = 207$ MPa.*

Solution According to the maximum strain theory and Example 4.1(iii)

$$16(1-\nu)\,M + 16(1+\nu)\,(M^2 + T^2)^{1/2} = \frac{\pi d^3}{N}\,\sigma_y$$

$$(16 \times 0.75 \times 1080) + (16 \times 1.25)\,(1080^2 + 3400^2)^{1/2} = \frac{\pi d^3}{3} \times 207 \times 10^6$$

i.e., $12960 + 71348 = 216.77 \times 10^6 d^3$

or $d^3 = 389 \times 10^{-6}$

or $d = 7.3 \times 10^{-2}\,\text{m} = 7.3$ cm

Example 4.5 *An equipment used in deep sea investigation is immersed at a depth H. The weight of the equipment in water is W. The rope attached to the instrument has a specific weight γ_r and the water has a specific weight γ. Analyse the strength of the rope. The rope has a cross-sectional area A. (Refer to Fig. 4.6.)*

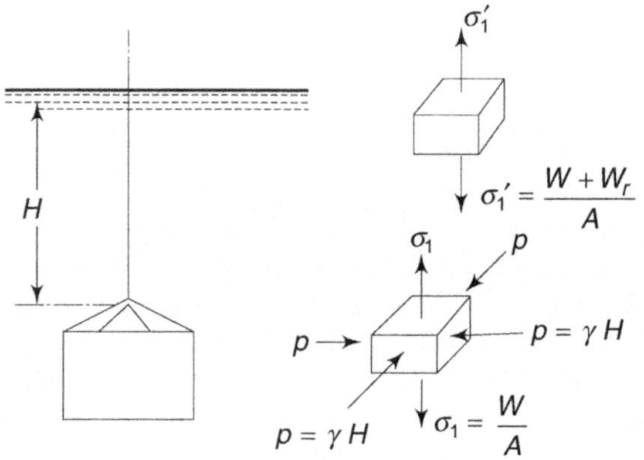

Fig. 4.6 *Example 4.5*

Solution The lower end of the rope is subjected to a triaxial state of stress. There is a tensile stress σ_1 due to the weight of the equipment and two hydrostatic compressions each equal to p, where

$$\sigma_1 = \frac{W}{A}, \qquad \sigma_2 = \sigma_3 = -\gamma H \text{ (compression)}$$

At the upper section there is only a uniaxial tension σ_1' due to the weight of the equipment and rope immersed in water.

$$\sigma_1' = \frac{W}{A} + (\gamma_r - \gamma) H; \quad \sigma_2' = \sigma_3' = 0$$

Therefore, according to the maximum shear stress theory, at lower section

$$\tau_{\max} = \frac{\sigma_1 - \sigma_3}{2} = \frac{1}{2}\left(\frac{W}{A} + \gamma H\right)$$

and at the upper section

$$\tau_{\max} = \frac{\sigma_1' - \sigma_3'}{2} = \frac{1}{2}\left(\frac{W}{A} - \gamma H + \gamma_r H\right)$$

If the specific weight of the rope is more than twice that of water, then the upper section is the critical section. When the equipment is above the surface of the water, near the hoist, the stress is

$$\sigma_1 = \frac{W'}{A} \quad \text{and} \quad \sigma_2 = \sigma_3 = 0$$

$$\tau_{max} = \frac{1}{2}\frac{W'}{A}$$

W' is the weight of equipment in air and is more than W. It is also necessary to check the strength of the rope for this stress.

4.6 MOHR'S THEORY OF FAILURE

In the previous discussions on failure, all the theories had one common feature. This was that the criterion of failure is unaltered by a reversal of sign of the stress. While the yield point stress σ_y for a ductile material is more or less the same in tension and compression, this is not true for a brittle material. In such a case, according to the maximum shear stress theory, we would get two different values for the critical shear stress. Mohr's theory is an attempt to extend the maximum shear stress theory (also known as the stress-difference theory) so as to avoid this objection.

To explain the basis of Mohr's theory, consider Mohr's circles, shown in Fig. 4.7, for a general state of stress.

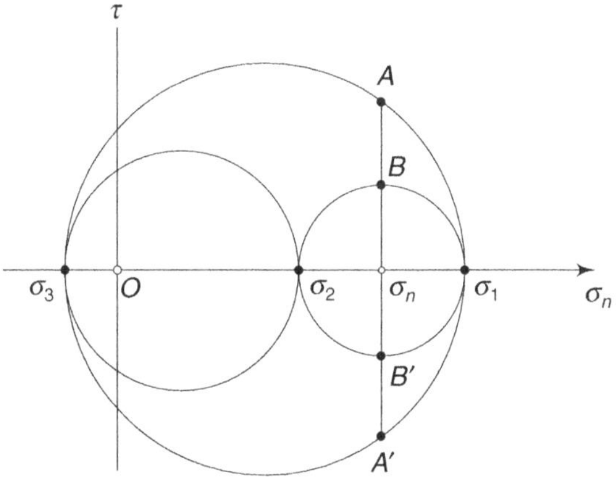

Fig. 4.7 *Mohr's circles*

σ_1, σ_2 and σ_3 are the principal stresses at the point. Consider the line $ABB'A'$. The points lying on BA and $B'A'$ represent a series of planes on which the normal stresses have the same magnitude σ_n but different shear stresses. The maximum shear stress associated with this normal stress value is τ, represented by point A or A'. The fundamental assumption is that if failure is associated with a given normal stress value, then the plane having this normal stress and a maximum shear stress accompanying it, will be the critical plane. Hence, the critical point for the normal stress σ_n will be the point A. From Mohr's circle diagram, the planes having maximum shear stresses for given normal stresses, have their representative points on the outer circle. Consequently, as far as failure is concerned, the critical circle is the outermost circle in Mohr's circle diagram, with diameter $(\sigma_1 - \sigma_3)$.

Now, on a given material, we conduct three experiments in the laboratory, relating to simple tension, pure shear and simple compression. In each case, the test is conducted until failure occurs. In simple tension, $\sigma_1 = \sigma_{yt}$, $\sigma_2 = \sigma_3 = 0$. The outermost circle in the circle diagram (there is only one circle) corresponding to this state is shown as T in Fig. 4.8. The plane on which failure occurs will have its representative point on this outer circle. For pure shear, $\tau_{ys} = \sigma_1 = -\sigma_3$ and $\sigma_2 = 0$. The outermost circle for this state is indicated by S. In simple compression, $\sigma_1 = \sigma_2 = 0$ and $\sigma_3 = -\sigma_{yc}$. In general, for a brittle material, σ_{yc} will be greater than σ_{yt} numerically. The outermost circle in the circle diagram for this case is represented by C.

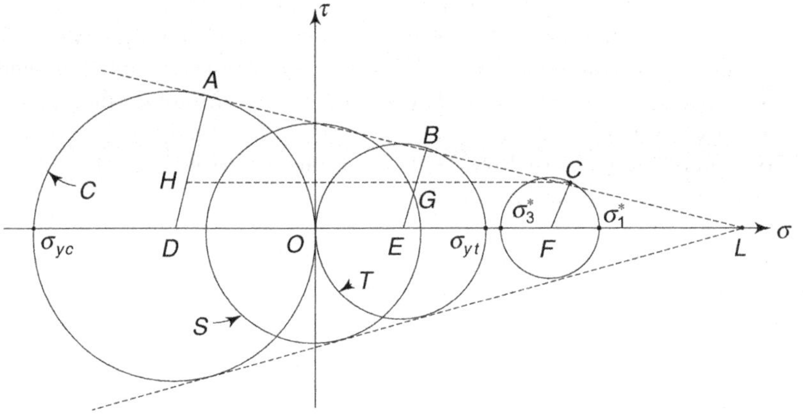

Fig. 4.8 *Diagram representing Mohr's failure theory*

In addition to the three simple tests, we can perform many more tests (like combined tension and torsion) until failure occurs in each case, and correspondingly for each state of stress, we can construct the outermost circle. For all these circles, we can draw an envelope. The point of contact of the outermost circle for a given state with this envelope determines the combination of σ and τ, causing failure. Obviously, a large number of tests will have to be performed on a single material to determine the envelope for it.

If the yield point stress in simple tension is small, compared to the yield point stress in simple compression, as shown in Fig. 4.8, then the envelope will cut the horizontal axis at point L, representing a finite limit for 'hydrostatic tension'. Similarly, on the left-hand side, the envelope rises indefinitely, indicating no elastic limit under hydrostatic compression.

For practical application of this theory, one assumes the envelopes to be straight lines, i.e. tangents to the circles as shown in Fig. 4.8. When a member is subjected to a general state of stress, for no failure to take place, the Mohr's circle with $(\sigma_1 - \sigma_3)$ as diameter should lie within the envelope. In the limit, the circle can touch the envelope. If one uses a factor of safety N, then the circle with $N(\sigma_1 - \sigma_3)$ as diameter can touch the envelopes. Figure 4.8 shows this limiting state of stress, where $\sigma_1^* = N\sigma_1$ and $\sigma_3^* = N\sigma_3$.

The envelopes being common tangents to the circles, triangles *LCF, LBE* and *LAD* are similar. Draw *CH* parallel to *LO* (the σ axis), making *CBG* and *CAH* similar. Then,

$$\frac{BG}{CG} = \frac{AH}{CH} \tag{a}$$

Now, $$BG = BE - GE = BE - CF = \frac{1}{2}\sigma_{yt} - \frac{1}{2}(\sigma_1^* - \sigma_3^*)$$

$$CG = FE = FO - EO = \frac{1}{2}(\sigma_1^* + \sigma_3^*) - \frac{1}{2}\sigma_{yt}$$

$$AH = AD - HD = AD - CF = \frac{1}{2}\sigma_{yc} - \frac{1}{2}(\sigma_1^* - \sigma_3^*)$$

$$CH = FD = FO + OD = \frac{1}{2}(\sigma_1^* + \sigma_3^*) + \frac{1}{2}\sigma_{yc}$$

Substituting these in Eq. (a), and after simplification,

$$\sigma_{yt} = \sigma_1^* - \frac{\sigma_{yt}}{\sigma_{yc}}\sigma_3^*$$

$$= N(\sigma_1 - k\sigma_3) \tag{4.18a}$$

where $$k = \frac{\sigma_{yt}}{\sigma_{yc}} \tag{4.18b}$$

Equation (4.18a) states that for a general state of stress where σ_1 and σ_3 are the maximum and minimum principal stresses, to avoid failure according to Mohr's theory, the condition is

$$\sigma_1 - k\sigma_3 \le \frac{\sigma_{yt}}{N} = \sigma_{eq}$$

where N is the factor of safety used for design, and k is the ratio of σ_{yt} to σ_{yc} for the material. For a brittle material with no yield stress value, k is the ratio of σ ultimate in tension to σ ultimate in compression, i.e.

$$k = \frac{\sigma_{ut}}{\sigma_{uc}} \tag{4.18c}$$

σ_{yt}/N is sometimes called the equivalent stress σ_{eq} in uniaxial tension corresponding to Mohr's theory of failure. When $\sigma_{yt} = \sigma_{yc}$, k will become equal to 1 and Eq. (4.18a) becomes identical to the maximum shear stress theory, Eq. (4.2).

Example 4.6 *Consider the problem discussed in Example 4.2. Let the crank-shaft material have* $\sigma_{yt} = 150$ *MPa and* $\sigma_{yc} = 330$ *MPa. If the diameter of the shaft is 10 cm, determine the allowable force F according to Mohr's theory of failure. Let the factor of safety be 2. Consider a point on the surface of the shaft where the stress due to bending is maximum.*

Solution Bending moment at section $A = (20 \times 10^{-2}\ F)$ Nm

$$\text{Torque} = (15 \times 10^{-2}\ F)\ \text{Nm}$$

$$\therefore \qquad \sigma\,(\text{bending}) = \frac{64Md}{2\pi d^4} = \frac{32M}{\pi d^3}\ \text{Pa}$$

$$\tau\,(\text{torsion}) = \frac{32Td}{2\pi d^4} = \frac{16T}{\pi d^3}\ \text{Pa}$$

$$\sigma_{1,3} = \frac{1}{2}\sigma \pm \frac{1}{2}(\sigma^2 + 4\tau^2)^{1/2}, \qquad \sigma_2 = 0$$

$$\sigma_{1,3} = \frac{16M}{\pi d^3} \pm \frac{8}{\pi d^4}(4M^2 + T^2)^{1/2}$$

$$= \frac{8F}{\pi \times 10^{-3}}\left[2\,(20 \times 10^{-2}) \pm 10^{-2}\,(1600 + 22F)^{1/2}\right]$$

$$= \frac{80F}{\pi}(40 \pm 42.7) = 2106F; \qquad -68.75F$$

$$k = \frac{\sigma_{yt}}{\sigma_{yc}} = \frac{150}{330} = 0.4545$$

$$\therefore \qquad N(\sigma_1 - k\sigma_3) = 2F(2106 + 31.25) = 4274.5F$$

From Eq. (4.18a),

$$4274.5F = \sigma_{yt} = 150 \times 10^6\ \text{Pa}$$

or $\qquad F = 34092\text{N}$

4.7 IDEALLY PLASTIC SOLID

If a rod of a ductile metal, such as mild steel, is tested under a simple uniaxial tension, the stress–strain diagram would be like the one shown in Fig. 4.9(a). As can be observed, the curve has several distinct regions. Part *OA* is linear, signifying that in this region, the strain is proportional to the stress. If a specimen is loaded within this limit and gradually unloaded, it returns to its original length

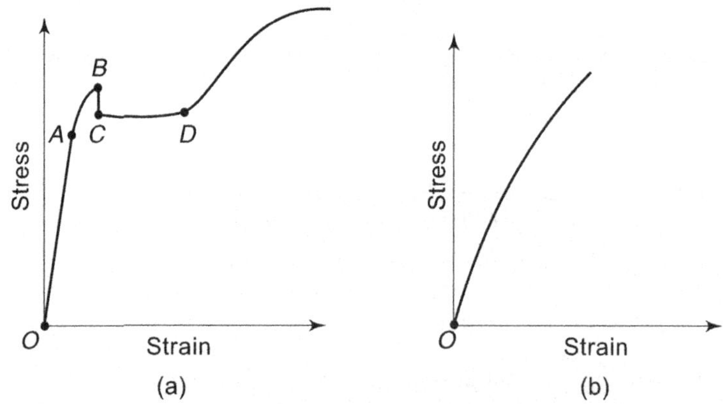

Fig. 4.9 *Stress–strain diagram for (a) Ductile material (b) Brittle material*

without any permanent deformation. This is the linear elastic region and point A denotes the limit of proportionality. Beyond A, the curve becomes slightly non-linear. However, the strain upto point B is still elastic. Point B, therefore, represents the elastic limit.

If the specimen is strained further, the stress drops suddenly (represented by point C) and thereafter the material yields at constant stress. After D, further straining is accompanied by increased stress, indicating work hardening. In the figure, the elastic region is shown exaggerated for clarity.

Most metals and alloys do not have a distinct yield point. The change from the purely elastic to the elastic-plastic state is gradual. Brittle materials, such as cast iron, titanium carbide or rock material, allow very little plastic deformation before reaching the breaking point. The stress-strain diagram for such a material would look like the one shown in Fig. 4.9(b).

In order to develop stress-strain relations during plastic deformation, the actual stress-strain diagrams are replaced by less complicated ones. These are shown in Fig. 4.10. In these, Fig. 4.10(a) represents a linearly elastic material, while Fig. 4.10(b) represents a material which is rigid (i.e. has no deformation) for stresses below σ_y and yields without limit when the stress level reaches the value σ_y. Such a material is called a rigid perfectly plastic material. Figure 4.10(c) shows the behaviour of a material which is rigid for stresses below σ_y and for stress levels above σ_y a linear work hardening characteristics is exhibited. A material exhibiting this characteristic behaviour is designated as rigid linear work hardening. Figure 4.10(d) and (e) represent respectively linearly elastic, perfectly plastic and linearly elastic–linear work hardening.

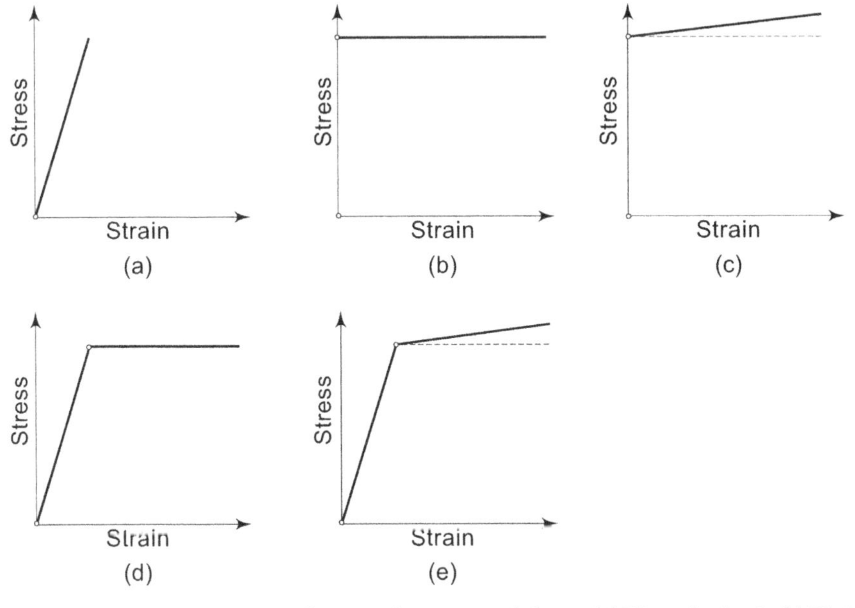

Fig. 4.10 *Ideal stress-strain diagram for a material that is (a) Linearly elastic (b) Rigid-perfectly plastic (c) Rigid-linear work hardening (d) Linearly elastic-perfectly plastic (e) Linearly elastic-linear work hardening*

In the following sections, we shall very briefly discuss certain elementary aspects of the stress-strain relations for an ideally plastic solid. It is assumed that the material behaviour in tension or compression is identical.

4.8 STRESS SPACE AND STRAIN SPACE

The state of stress at a point can be represented by the six rectangular stress components τ_{ij} $(i, j = 1, 2, 3)$. One can imagine a six-dimensional space called the stress space, in which the state of stress can be represented by a point. Similarly, the state of strain at a point can be represented by a point in a six- dimensional strain space. In particular, a state of plastic strain $\varepsilon_{ij}^{(p)}$ can be so represented. A history of loading can be represented by a path in the stress space and the corresponding deformation or strain history as a path in the strain space.

A basic assumption that is now made is that there exists a scalar function called a stress function or loading function, represented by $f(\tau_{ij}, \varepsilon_{ij}, K)$, which depends on the states of stress and strain, and the history of loading. The function $f = 0$ represents a closed surface in the stress space. The function f characterises the yielding of the material as follows:

As long as $f < 0$ no plastic deformation or yielding takes place; $f > 0$ has no meaning. Yielding occurs when $f = 0$. For materials with no work hardening characteristics, the parameter $K = 0$.

In the previous sections of this chapter, several yield criteria have been considered. These criteria were expressed in terms of the principal stresses $(\sigma_1, \sigma_2, \sigma_3)$ and the principal strains $(\varepsilon_1, \varepsilon_2, \varepsilon_3)$. We have also observed that a material is said to be isotropic if the material properties do not depend on the particular coordinate axes chosen. Similarly, the plastic characteristics of the material are said to be isotropic if the yield function f depends only on the invariants of stress, strain and strain history. The isotropic stress theory of plasticity gives function f as an isotropic function of stresses alone. For such theories, the yield function can be expressed as $f(I_1, I_2, I_3)$ where I_1, I_2 and I_3 are the stress invariants. Equivalently, one may express the function as $f(\sigma_1, \sigma_2, \sigma_3)$. It is, therefore, possible to represent the yield surface in a three-dimensional space with coordinate axes σ_1, σ_2 and σ_3.

The Deviatoric Plane or the π Plane

In Section 4.2(a), it was stated that most metals can withstand considerable hydrostatic pressure without any permanent deformation. It has also been observed that a given state of stress can be uniquely resolved into a hydrostatic (or isotropic) state and a deviatoric (i.e. pure shear) state, i.e.

$$\begin{bmatrix} \sigma_1 & 0 & 0 \\ 0 & \sigma_2 & 0 \\ 0 & 0 & \sigma_3 \end{bmatrix} = \begin{bmatrix} p & 0 & 0 \\ 0 & p & 0 \\ 0 & 0 & P \end{bmatrix} = \begin{bmatrix} \sigma_1 - p & 0 & 0 \\ 0 & \sigma_2 - p & 0 \\ 0 & 0 & \sigma_3 - p \end{bmatrix}$$

or

$$[\sigma_i] = [p] + [\sigma_i^*], \qquad (i = 1, 2, 3) \tag{4.19}$$

where $p = \frac{1}{3}(\sigma_1 + \sigma_2 + \sigma_3)$

is the mean normal stress, and

$$\sigma_i^* = \sigma_i - p, \qquad (i = 1, 2, 3)$$

Consequently, the yield function will be independent of the hydrostatic state. For the deviatoric state, $I_1^* = 0$. According to the isotropic stress theory, therefore, the yield function will be a function of the second and third invariants of the devatoric state, i.e. $f(I_2^*, I_3^*)$. The equation

$$\sigma_1^* + \sigma_2^* + \sigma_3^* = 0 \tag{4.20}$$

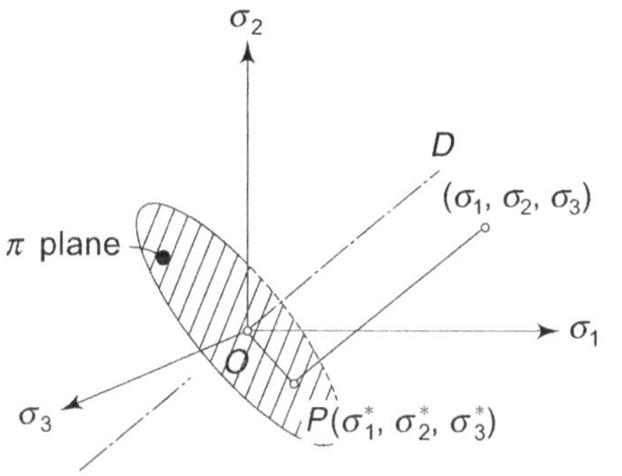

Fig. 4.11 *The π Plane*

represents a plane passing through the origin, whose normal OD is equally inclined (with direction cosines $1/\sqrt{3}$) to the axes σ_1, σ_2 and σ_3. This plane is called the deviatoric plane or the π plane. If the stress state $(\sigma_1^*, \sigma_2^*, \sigma_3^*)$ causes yielding, the point representing this state will lie in the π plane. This is shown by point P in Fig. 4.11. Since the addition or subtraction of an isotropic state does not affect the yielding process, point P can be moved parallel to OD. Hence, the yield function will represent a cylinder perpendicular to the π plane. The trace of this surface on the π plane is the yield locus.

4.9 GENERAL NATURE OF THE YIELD LOCUS

Since the yield surface is a cylinder perpendicular to the π plane, we can discuss its characteristics with reference to its trace on the π plane, i.e. with reference to the yield locus. Figure 4.12 shows the π plane and the projections of the σ_1, σ_2 and σ_3 axes on this plane as σ'_1, σ'_2 and σ'_3. These projections make an angle of 120° with each other.

Let us assume that the state (6, 0, 0) lies on the yield surface, i.e. the state $\sigma_1 = 6$, $\sigma_2 = 0$, $\sigma_3 = 0$, causes yielding. Since we have assumed isotropy, the states (0, 6, 0) and (0, 0, 6) also should cause yielding. Further, as we have assumed that the material behaviour in tension is identical to that in compression, the states (–6, 0, 0), (0, –6, 0) and (0, 0, –6) also cause yielding. Thus, appealing to isotropy and the property of the material in tension and compression, one point on the yield surface locates five other points. If we choose a general point (a, b, c) on the yield surface, this will generate 11 other (or a total of 12) points on the surface. These are (a, b, c) (c, a, b), (b, c, a), (a, c, b), (c, b, a) (b, a, c) and the remaining six are obtained by multiplying these by –1. Therefore, the yield locus is a symmetrical curve.

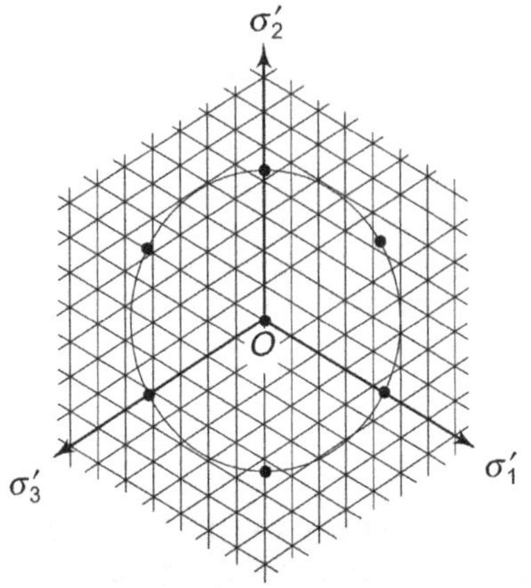

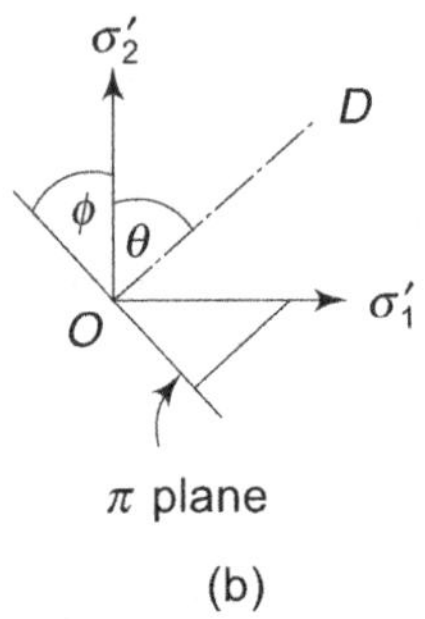

Fig. 4.12 *(a) The yield locus (b) Projection of π plane*

4.10 YIELD SURFACES OF TRESCA AND VON MISES

One of the yield conditions studied in Section 4.2 was stated by the maximum shear stress theory. According to this theory, if $\sigma_1 > \sigma_2 > \sigma_3$, the yielding starts when the maximum shear $\frac{1}{2}(\sigma_1 - \sigma_3)$ becomes equal to the maximum shear $\sigma_y/2$ in uniaxial tension yielding. In other words, yielding begins when $\sigma_1 - \sigma_3 = \sigma_y$. This condition is generally named after Tresca.

Let us assume that only σ_1 is acting. Then, yielding occurs when $\sigma_1 = \sigma_y$. The σ_1 axis is inclined at an angle of ϕ to its projection σ'_1 axis on the π plane, and $\sin\phi = \cos\theta = 1/\sqrt{3}$, [Fig. 4.12(b)]. Hence, the point $\sigma_1 = \sigma_y$ will have its projection on the π plane as $\sigma_y \cos\phi = \sqrt{2/3}\,\sigma_y$ along the σ'_1 axis. Similarly, other points on the π plane will be at distances of $\pm\sqrt{2/3}\,\sigma_y$ along the projections of σ_1, σ_2 and σ_3 axes on the π plane, i.e., along σ'_1, σ'_2, σ'_3 axes in Fig. 4.13. If σ_1, σ_2 and σ_3 are all acting (with $\sigma_1 > \sigma_2 > \sigma_3$), yielding occurs when $\sigma_1 - \sigma_3 = \sigma_y$. This defines a straight line joining points at a distance of σ_y along σ_1 and $-\sigma_3$ axes. The projection of this line on the π plane will be a straight line joining points at a distance of $\sqrt{2/3}\,\sigma_y$ along the σ'_1 and $-\sigma'_3$ axes, as shown by *AB* in Fig. 4.13. Consequently, the yield locus is a hexagon.

Another yield criterion discussed in Section 4.2 was the octahedral shearing stress or the distortion energy theory. According to this criterion, Eq. (4.4b), yielding occurs when

$$f(l_1, l_2, l_3) = f(l_1^2 - 3l_2) = \sigma_y^2 \tag{4.21}$$

Since a hydrostatic state of stress does not have any effect on yielding, one can deal with the deviatoric state (for which $l_1^* = 0$) and write the above condition as

$$f(l_2^*, l_3^*) = f(l_2^*) = -3\,l_2^* = \sigma_y^2 \tag{4.22}$$

The yield function can, therefore, be written as

$$f = l_2^* + \frac{1}{3}\sigma_y^2 = l_2^* + s^2 \tag{4.23}$$

where s is a constant. This yield criterion is known as the von Mises condition for yielding. The yield surface is defined by

$$I_2^* + s^2 = 0$$

or $$\sigma_1\sigma_2 + \sigma_2\sigma_3 + \sigma_3\sigma_1 - 3p^2 = -s^2 \tag{4.24}$$

The other alternative forms of the above expression are

$$(\sigma_1 - p)^2 + (\sigma_2 - p)^2 + (\sigma_3 - p)^2 = 2s^2 \tag{4.25}$$

$$(\sigma_1 - \sigma_2)^2 + (\sigma_2 - \sigma_3)^2 + (\sigma_3 - \sigma_1)^2 = 6s^2 \tag{4.26}$$

Equation (4.25) can also be written as

$$\sigma_1^* + \sigma_2^* + \sigma_3^* = 2s^2 \tag{4.27}$$

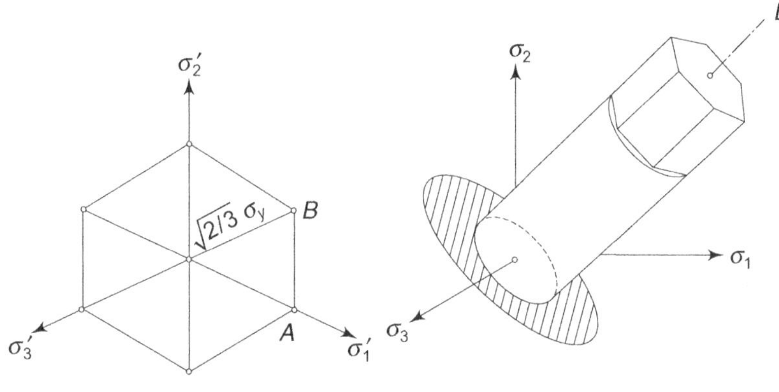

Fig. 4.13 *Yield surfaces of Tresca and von Mises*

This is the curve of intersection between the sphere $\sigma_1^2 + \sigma_2^2 + \sigma_3^2 = 2s^2$ and the π plane defined by $\sigma_1^* + \sigma_2^* + \sigma_3^* = 0$. This curve is, therefore, a circle with radius $\sqrt{2}s$ in the π plane. The yield surface according to the von Mises criterion is, therefore, a right circular cylinder. From Eq. (4.23)

$$s^2 = \frac{1}{3}\sigma_y^2, \quad \text{or,} \quad s = \frac{1}{\sqrt{3}}\sigma_y \tag{4.28}$$

Hence, the radius of the cylinder is $\sqrt{2/3}\,\sigma_y$ i.e. the cylinder of von Mises circumscribes Tresca's hexagonal cylinder. This is shown in Fig. 4.13.

4.11 STRESS–STRAIN RELATIONS (PLASTIC FLOW)

The yield locus that has been discussed so far defines the boundary of the elastic zone in the stress space. When a stress point reaches this boundary, plastic deformation takes place. In this context, one can speak of only the change in the plastic strain rather than the total plastic strain because the latter is the sum total of all plastic strains that have taken place during the previous strain history of the specimen. Consequently, the stress–strain relations for plastic flow relate the strain increments. Another way of explaining this is to realise that the process of plastic flow is irreversible; that most of the deformation work is transformed into heat and that the stresses in the final state depend on the strain path. Consequently, the equations governing plastic deformation cannot, in principle, be finite relations concerning stress and strain components as in the case of Hooke's law, but must be differential relations.

The following assumptions are made:

(i) The body is isotropic

(ii) The volumetric strain is an elastic strain and is proportional to the mean pressure ($\sigma_m = p = \sigma$)

$$\varepsilon = 3k\sigma$$

or
$$d\varepsilon = 3kd\sigma \tag{4.29}$$

(iii) The total strain increments $d\varepsilon_{ij}$ are made up of the elastic strain increments $d\varepsilon_{ij}^e$ and plastic strain increaments $d\varepsilon_{ij}^p$

$$d\varepsilon_{ij} = d\varepsilon_{ij}^e + d\varepsilon_{ij}^p \tag{4.30}$$

(iv) The elastic strain increments are related to stress components σ_{ij} through Hooke's law

$$\begin{aligned}
d\varepsilon_{xx}^e &= \frac{1}{E}[\sigma_x - \nu(\sigma_y + \sigma_z)] \\
d\varepsilon_{yy}^e &= \frac{1}{E}[\sigma_y - \nu(\sigma_x + \sigma_z)] \\
d\varepsilon_{zz}^e &= \frac{1}{E}[\sigma_z - \nu(\sigma_x + \sigma_y)] \\
d\varepsilon_{xy}^e &= d\gamma_{xy}^e = \frac{1}{G}\tau_{xy} \\
d\varepsilon_{yz}^e &= d\gamma_{yz}^e = \frac{1}{G}\tau_{yz} \\
d\varepsilon_{zx}^e &= d\gamma_{zx}^e = \frac{1}{G}\tau_{zx}
\end{aligned} \tag{4.31}$$

(v) The deviatoric components of the plastic strain increments are proportional to the components of the deviatoric state of stress

$$d\left[\varepsilon_{xx}^p - \frac{1}{3}(\varepsilon_{xx}^p + \varepsilon_{yy}^p + \varepsilon_{zz}^p)\right] = \left[\sigma_x - \frac{1}{3}(\sigma_x + \sigma_y + \sigma_z)\right] d\lambda \tag{4.32}$$

where $d\lambda$ is the instantaneous constant of proportionality.

From (ii), the volumetric strain is purely elastic and hence

$$\varepsilon = \varepsilon_{xx}^e + \varepsilon_{yy}^e + \varepsilon_{zz}^e$$

But
$$\varepsilon = \varepsilon_{xx}^e + \varepsilon_{yy}^e + \varepsilon_{zz}^e + (\varepsilon_{xx}^p + \varepsilon_{yy}^p + \varepsilon_{zz}^p)$$

Hence,

$$\varepsilon_{xx}^p + \varepsilon_{yy}^p + \varepsilon_{zz}^p = 0 \tag{4.33}$$

Using this in Eq. (4.32)

$$d\varepsilon_{xx}^{p} = d\lambda\left[\sigma_x - \frac{1}{3}(\sigma_x + \sigma_y + \sigma_z)\right]$$

Denoting the components of stress deviator by s_{ij}, the above equations and the remaining ones are

$$\begin{aligned} d\varepsilon_{xx}^{p} &= d\lambda\, s_{xx} \\ d\varepsilon_{yy}^{p} &= d\lambda\, s_{yy} \\ d\varepsilon_{zz}^{p} &= d\lambda\, s_{zz} \\ d\gamma_{xy}^{p} &= d\lambda\, s_{xy} \\ d\gamma_{yz}^{p} &= d\lambda\, s_{yz} \\ d\gamma_{zx}^{p} &= d\lambda\, s_{zx} \end{aligned} \tag{4.34}$$

Equivalently

$$d\varepsilon_{ij}^{p} = d\lambda\, s_{ij} \tag{4.35}$$

4.12 PRANDTL–REUSS EQUATIONS

Combining Eqs (4.30), (4.31) and (4.35)

$$d\varepsilon_{ij} = d\varepsilon_{ij}^{(e)} + d\lambda\, s_{ij} \tag{4.36}$$

where $d\varepsilon_{ij}^{(e)}$ is related to stress components through Hooke's law, as given in Eq. (4.31). Equations (4.30), (4.31) and (4.35) constitute the Prandtl–Reuss equations. It is also observed that the principal axes of stress and plastic strain increments coincide. It is easy to show that $d\lambda$ is non-negative. For this, consider the work done during the plastic strain increment

$$\begin{aligned} dW_p &= \sigma_x d\varepsilon_{xx}^{p} + \sigma_y d\varepsilon_{yy}^{p} + \sigma_z d\varepsilon_{zz}^{p} + \tau_{xy} d\gamma_{xy}^{p} + \tau_{yz} d\gamma_{yz}^{p} + \tau_{zx} d\gamma_{zx}^{p} \\ &= d\lambda\left(\sigma_x s_{xx} + \sigma_y s_{yy} + \sigma_z s_{zz} + \tau_{xy} s_{xy} + \tau_{yz} s_{yz} + \tau_{zx} s_{zx}\right) \\ &= d\lambda\left[\sigma_x(\sigma_x - p) + \sigma_y(\sigma_y - p) + \sigma_z(\sigma_z - p) + \tau_{xy}^2 + \tau_{yz}^2 + \tau_{zx}^2\right] \end{aligned}$$

or $\quad dW_p = d\lambda\left[(\sigma_x - p)^2 + (\sigma_y - p)^2 + (\sigma_z - p)^2 + \tau_{xy}^2 + \tau_{yz}^2 + \tau_{zx}^2\right]$

i.e. $\quad dW_p = d\lambda T^2 \qquad (4.37)$

Since $\quad dW_p \geq 0$

we have $\quad d\lambda \geq 0$

If the von Mises condition is applied, from Eqs (4.23) and (4.35)

$$dW_p = d\lambda 2s^2$$

or $$d\lambda = \frac{dW_p}{2s^2} \tag{4.38}$$

i.e $d\lambda$ is proportional to the increment of plastic work.

4.13 SAINT VENANT–VON MISES EQUATIONS

In a fully developed plastic deformation, the elastic components of strain are very small compared to plastic components. In such a case

$$d\varepsilon_{ij} \approx d\varepsilon_{ij}^p$$

and this gives the equations of the Saint Venant–von Mises theory of plasticity in the form

$$d\varepsilon_{ij} = d\lambda\, s_{ij} \tag{4.39}$$

Expanding this

$$d\varepsilon_{xx} = \frac{2}{3} d\lambda \left[\sigma_x - \frac{1}{2}\left(\sigma_y + \sigma_z\right)\right]$$

$$d\varepsilon_{yy} = \frac{2}{3} d\lambda \left[\sigma_y - \frac{1}{2}\left(\sigma_z + \sigma_x\right)\right]$$

$$d\varepsilon_{zz} = \frac{2}{3} d\lambda \left[\sigma_z - \frac{1}{2}\left(\sigma_x + \sigma_y\right)\right] \tag{4.40}$$

$$d\gamma_{xy} = d\lambda \tau_{xy}$$
$$d\gamma_{yz} = d\lambda \tau_{yz}$$
$$d\gamma_{zx} = d\lambda \tau_{zx}$$

The above equations are also called Levy–Mises equations. In this case, it should be observed that the principal axes of strain increments coincide with the axes of the principal stresses.

Problems

4.1 Figure 4.14 shows three elements *a, b, c* subjected to different states of stress. Which one of these three, do you think, will yield first according to
(i) the maximum stress theory?
(ii) the maximum strain theory?

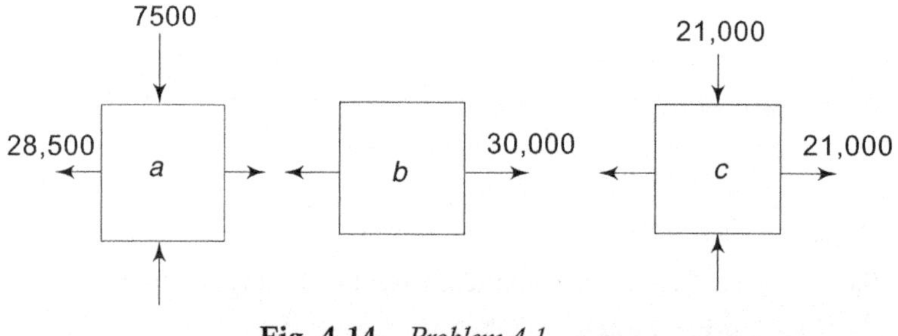

Fig. 4.14 *Problem 4.1*

(iii) the maximum shear stress theory?
Poisson's ratio $\nu = 0.25$ [*Ans.* (i) *b*, (ii) *a*, (iii) *c*]

4.2 Determine the diameter of a cold-rolled steel shaft, 0.6 m long, used to transmit 50 hp at 600 rpm. The shaft is simply supported at its ends in bearings. The shaft experiences bending owing to its own weight also. Use a factor of safety 2. The tensile yield limit is 280×10^3 kPa (2.86×10^3 kgf/cm^2) and the shear yield limit is 140×10^3 kPa (1.43×10^3 kgf/cm^2). Use the maximum shear stress theory. [*Ans.* $d = 3.6$ cm]

4.3 Determine the diameter of a ductile steel bar (Fig 4.15) if the tensile load F is 35,000 N and the torsional moment T is 1800 Nm. Use a factor of safety $N = 1.5$.

$E = 207 \times 10^6$ kPa $\left(2.1 \times 10^6 \text{ kgf/cm}^2\right)$ and σ_{yp} is 207,000 kPa (2100 kgf/cm^2).

Use the maximum shear stress theory. [*Ans.* $d = 4.1$ cm]

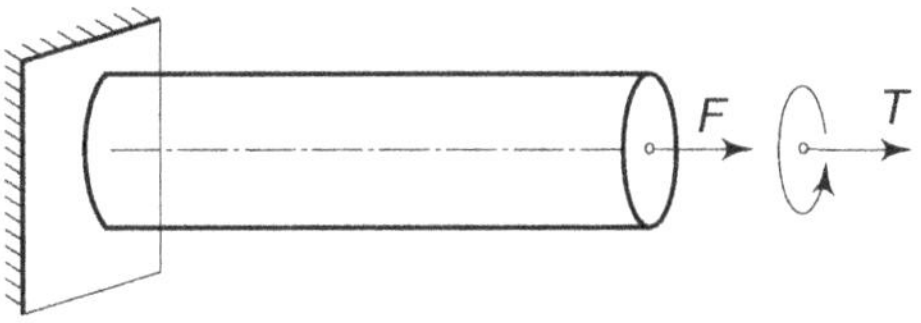

Fig. 4.15 *Problem 4.3*

4.4 For the problem discussed in Problem 4.3, determine the diameter according to Mohr's theory if $\sigma_{yt} = 207$ MPa, $\sigma_{yc} = 310$ MPa. The factor of safety $N = 1.5$; $F = 35{,}000$ N and $T = 1800$ Nm. [*Ans.* $d = 4.2$ cm]

4.5 At a point in a steel member, the state of stress is as shown in Fig. 4.16. The tensile elastic limit is 413.7 kPa. If the shearing stress at the point is 206.85 kPa, when yielding starts, what is the tensile stress σ at the point (a) according to the maximum shearing stress theory, and (b) according to the octahedral shearing stress theory?

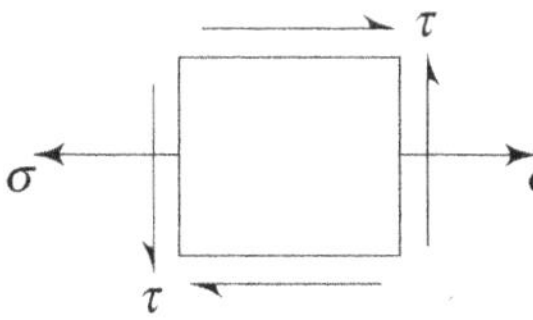

Fig. 4.16 *Problem 4.5*

[*Ans.* (a) zero; (b) 206.85 kPa (2.1 kgf/cm^2)]

4.6 A torque T is transmitted by means of a system of gears to the shaft shown in Fig. 4.17. If $T = 2500$Nm (25,510 kgf cm), $R = 0.08$ m, $a = 0.8$ m and $b = 0.1$ m, determine the diameter of the shaft, using the maximum shear stress theory. $\sigma_y = 290000$ kPa. The factor of safety is 2. Note that when a torque is being transmitted, in addition to the tangential force, there occurs a radial force equal to $0.4F$, where F is the tangential force. This is shown in Fig. 4.17(b).

Hint: The forces F and $0.4F$ acting on the gear A are shown in Fig. 4.17(b). The reactions at the bearings are also shown. There are two bending moments—one in the vertical plane and the other in the horizontal plane. In the vertical plane, the maximum moment is $\dfrac{(0.4Fab)}{(a+b)}$;

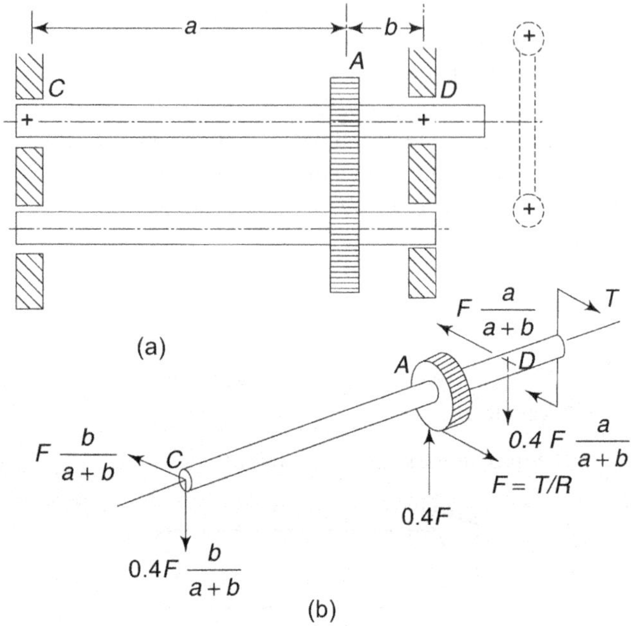

Fig. 4.17 *Problem 4.6*

in the horizontal plane the maximum moment is $\dfrac{(Fab)}{(a+b)}$; both these maximums occur at the gear section A. The resultant bending is

$$(M)_{max} = \left[\left(\frac{0.4\,Fab}{a+b}\right)^2 + \left(\frac{Fab}{a+b}\right)^2\right]^{1/2}$$

$$= 1.08F\ \frac{ab}{a+b}$$

The critical point to be considered is the circumferential point on the shaft subjected to this maximum moment. [*Ans.* $d \approx 65$ mm]

4.7 If the material of the bar in Problem 4.4 has $\sigma_{yt} = 207 \times 10^6$ Pa and $\sigma_{yc} = 517 \times 10^6$ Pa determine the diameter of the bar according to Mohr's theory of failure. The other conditions are as given in Problem 4.4. [*Ans.* $d = 4.6$ cm]

CHAPTER

5

Energy Methods

5.1 INTRODUCTION

In Chapters 1 and 2, attention was focussed on the analysis of stress and strain at a point. Except for the condition that the material we considered was a continuum, the shape or size of the body as a whole was not considered. In Chapter 3, the stresses and strains at a point were related through the material or the constitutive equations. Here too, the material properties rather than the behaviour of the body as such was not considered. Chapter 4, on the theory of failure, also discussed the critical conditions to impend failure at a point. In this chapter, we shall consider the entire body or structural member or machine element, along with the forces acting on it. Hooke's law will relate the force acting on the body to the displacement. When the body deforms under the action of the externally applied forces, the work done by these forces is stored as strain energy inside the body, which can be recovered when the latter is elastic in nature. It is assumed that the forces are applied gradually.

The strain energy methods are extremely important for the solution of many problems in the mechanics of solids and in structural analysis. Many of the theorems developed in this chapter can be used with great advantage to solve displacement problems and statically indeterminate structures and frameworks.

5.2 HOOKE'S LAW AND THE PRINCIPLE OF SUPERPOSITION

We have observed in Chapter 3 that the rectangular stress components at a point can be related to the rectangular strain components at the same point through a set of linear equations that were designated as the generalised Hooke's Law. In this chapter, however, we shall state Hooke's law as applicable to the elastic body as a whole, i.e. relate the complete system of forces acting on the body to the deformation of the body as a whole. The law asserts that 'deflections are proportional to the forces which produce them'. This is a very general assertion without any restriction as to the shape or size of the loaded body.

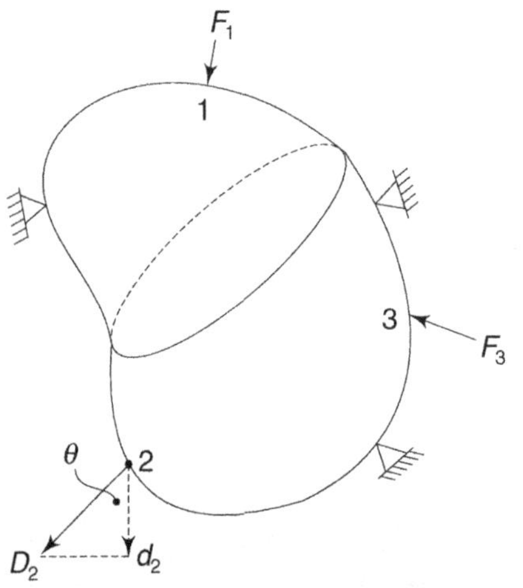

Fig. 5.1 *Elastic solid and Hooke's law*

In Fig. 5.1, a force F_1 is applied at point 1 of a body in equilibrium, and in consequence, point 2 undergoes a displacement or a deflection, which according to Hooke's law, is proportionate to F_1. This deflection of point 2 may take place in a direction which is quite different from that of F_1. If D_2 is the actual deflection, we have

$$D_2 = k_{21}F_1$$

where k_{21} is some proportionality constant.

When F_1 is increased, D_2 also increases proportionately. Let d_2 be the component of D_2 in a specified direction. If θ is the angle between D_2 and d_2

$$d_2 = D_2 \cos\theta = k_{21} \cos\theta \, F_1$$

If we keep θ constant, i.e. if we fix our attention on the deflection in a specified direction, then

$$d_2 = a_{21}F_1$$

where a_{21} is a constant. Therefore, one can consider the displacement of point 2 in any specified direction and apply Hooke's Law. Let us consider the vertical component of the deflection of point 2. If d_2 is the vertical component, then Hooke's law asserts that

$$d_2 = a_{21}F_1 \tag{5.1}$$

where a_{21} is a constant called the 'influence coefficient' for vertical deflection at point 2 due to a force applied in the specified direction (that of F_1) at point 1. If F_1 is a unit force, then a_{21} is the actual value of the vertical deflection at 2. If a force equal and opposite to F_1 is applied at 1, then a deflection equal and opposite to the earlier deflection takes place. If several forces, all having the direction of F_1, are applied simultaneously at 1, the resultant vertical deflection which they produce at 2 will be the resultant of the deflections which they would have produced if applied separately. This is the principle of superposition for small deformations.

Consider a force F_3 acting alone at point 3, and let d_2' be the vertical component of the deflection of 2. Then, according to Hooke's Law, as stated by Eq. (5.1)

$$d_2' = a_{23}F_3 \tag{5.2}$$

where a_{23} is the influence coefficient for vertical deflection at point 2 due to a force applied in the specified direction (that of F_3) at point 3. The question that we now examine is whether the principle of superposition holds true to two or more forces, such as F_1 and F_3, which act in different directions and at different points.

Let F_1 be applied first, and then F_3. The vertical deflection at 2 is

$$d_2 = a_{21}F_1 + a_{23}'F_3 \tag{5.3}$$

where a'_{23} may be different from a_{23}. This difference, if it exists, is due to the presence of F_1 when F_3 is applied. Now apply $-F_1$. Then

$$d''_2 = a_{21}F_1 + a'_{23}F_3 - a'_{21}F_1$$

a'_{21} may be different from a_{21}, since F_3 is acting when $-F_1$ is applied. Only F_3 is acting now. If we apply $-F_3$, the deflection finally becomes

$$d'''_2 = a_{21}F_1 + a'_{23}F_3 - a'_{21}F_1 - a_{23}F_3 \tag{5.4}$$

Since the elastic body is not subjected to any force now, the final deflection given by Eq. (5.4) must be zero. Hence,

$$a_{21}F_1 + a'_{23}F_3 - a'_{21}F_1 - a_{23}F_3 = 0$$

i.e. $$(a_{21} - a'_{21})F_1 = (a_{23} - a'_{23})F_3$$

or $$\frac{a_{21} - a'_{21}}{F_3} = \frac{a_{23} - a'_{23}}{F_1} \tag{5.5}$$

The difference $a_{21} - a'_{21}$, if it exists, must be due to the action of F_3. Hence, the left-hand side is a function of F_3 alone. Similarly, if the difference $a_{23} - a'_{23}$ exists, it must be due to the action of F_1 and, therefore, the right-hand side must be a function F_1 alone. Consequently, Eq. (5.5) becomes

$$\frac{a_{21} - a'_{21}}{F_3} = \frac{a_{23} - a'_{23}}{F_1} = k \tag{5.6}$$

where k is a constant independent of F_1 and F_3. Hence

$$a'_{23} = a_{23} - kF_1$$

Substituting this in Eq. (5.3)

$$d_2 = a_{21}F_1 + a_{23}F_3 - kF_1F_3$$

The last term on the right-hand side in the above equation is non-linear, which is contradictory to Hooke's law, unless k vanishes. Hence, $k = 0$, and

$$a_{23} = a'_{23} \quad \text{and} \quad a_{21} = a'_{21}$$

The principle of superposition is, therefore, valid for two different forces acting at two different points. This can be extended by induction to include a third or any number of other forces. This means that the deflection at 2 due to any number of forces, including force F_2 at 2 is

$$d_2 = a_{21}F_1 + a_{22}F_2 + a_{23}F_3 + \ldots \tag{5.7}$$

5.3 CORRESPONDING FORCE AND DISPLACEMENT OR WORK-ABSORBING COMPONENT OF DISPLACEMENT

Consider an elastic body which is in equilibrium under the action of forces F_1, F_2, F_3, . . . The forces of reaction at the points of support will also be considered as applied forces. This is shown in Fig. 5.2.

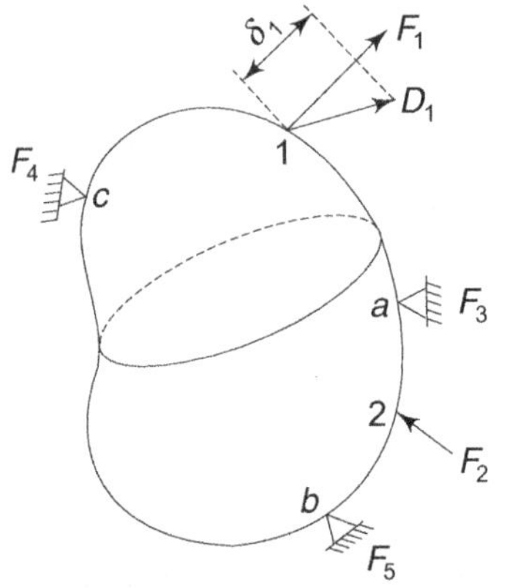

Fig. 5.2 *Corresponding forces and displacements*

The displacement d_1 in a specified direction at point 1 is given by Eq. (5.7). If the actual displacement is D_1 and takes place in a direction as shown in Fig. (5.2), then the component of this displacement in the direction of force F_1 is called the corresponding displacement at point 1. This corresponding displacement is denoted by δ_1. At every loaded point, a corresponding displacement can be identified. If the points of support a, b and c do not yield, then at these points the corresponding displacements are zero. One can apply Hooke's law to these corresponding displacements and obtain from Eq. (5.7)

$$\delta_1 = a_{11}F_1 + a_{12}F_2 + a_{13}F_3 + \ldots$$
$$\delta_2 = a_{21}F_1 + a_{22}F_2 + a_{23}F_3 + \ldots \text{ etc.} \quad (5.8)$$

where $a_{11}, a_{12}, a_{13}, \ldots$, are the influence coefficients of the kind discussed earlier. The corresponding displacement is also called the work-absorbing component of the displacement.

5.4 WORK DONE BY FORCES AND ELASTIC STRAIN ENERGY STORED

Equations (5.8) show that the displacements $\delta_1, \delta_2, \ldots$ .etc., depend on all the forces $F_1, F_2, \ldots$, etc. If we slowly increase the forces $F_1, F_2, \ldots$, etc. from zero to their full magnitudes, the deflections also increase similarly. For example, when the forces $F_1, F_2, \ldots$, etc. are one-half of their full magnitudes, the deflections are

$$\frac{1}{2}\delta_1 = a_{11}\left(\frac{1}{2}F_1\right) + a_{12}\left(\frac{1}{2}F_2\right) + \ldots,$$

$$\frac{1}{2}\delta_2 = a_{21}\left(\frac{1}{2}F_1\right) + a_{22}\left(\frac{1}{2}F_2\right) + \ldots, \text{ etc.,}$$

i.e. the deflections reached are also equal to half their full magnitudes. Similarly, when $F_1, F_2, \ldots$, etc. reach two-thirds of their full magnitudes, the deflections reached are also equal to two-thirds of their full magnitudes. Assuming that the forces are increased in constant proportion and the increase is gradual, the work done by F_1 at its point of application will be

$$W_1 = \frac{1}{2}F_1\delta_1$$
$$= \frac{1}{2}F_1(a_{11}F_1 + a_{12}F_2 + a_{13}F_3 + \ldots) \quad (5.9)$$

Similar expressions hold good for other forces also. The total work done by external forces is, therefore, given by

$$W_1 + W_2 + W_3 + \ldots = \frac{1}{2}(F_1\delta_1 + F_2\delta_2 + F_3\delta_3 + \ldots)$$

If the supports are rigid, then no work is done by the support reactions. When the forces are gradually reduced to zero, keeping their ratios constant, negative work will be done and the total work will be recovered. This shows that the work done is stored as potential energy and its magnitude should be independent of the order in which the forces are applied. If it were not so, it would be possible to store or extract energy by merely changing the order of loading and unloading. This would be contradictory to the principle of conservation of energy.

The potential energy that is stored as a consequence of the deformation of any elastic body is termed elastic strain energy. If F_1, F_2, F_3 are the forces in a particular configuration and δ_1, δ_2, δ_3 are the corresponding displacements then the elastic strain energy stored is

$$U = \frac{1}{2}(F_1\delta_1 + F_2\delta_2 + F_3\delta_3 + \ldots) \tag{5.10}$$

It must be noted that though this expression has been obtained on the assumption that the forces F_1, F_2, F_3 . . ., are increased in constant proportion, the conservation of energy principle and the superposition principle dictate that this expression for U must hold without restriction on the manner or order of the application of these forces.

5.5 RECIPROCAL RELATION

It is easy to show that the influence coefficient a_{12} in Eq. (5.8) is equal to the influence coefficient a_{21}. In general, $a_{ij} = a_{ji}$. To show this, consider a force F_1 applied at point 1, and let δ_1 be the corresponding displacement. The energy stored is

$$U_1 = \frac{1}{2}F_1\delta_1 = \frac{1}{2}a_{11}F_1^2$$

since $\qquad \delta_1 = a_{11}F_1$

Next, apply force F_2 at point 2. The corresponding deflection at point 2 is $a_{22}F_2$ and that at point 1 is $a_{12}F_2$. During this displacement, force F_1 is fully acting and hence, the additional energy stored is

$$U_2 = \frac{1}{2}F_2(a_{22}F_2) + F_1(a_{12}F_2)$$

The total elastic energy stored is therefore

$$U = U_1 + U_2 = \frac{1}{2}a_{11}F_1^2 + \frac{1}{2}a_{22}F_2^2 + a_{12}F_1F_2$$

Now, if F_2 is applied before F_1, the elastic energy stored is

$$U' = \frac{1}{2}a_{22}F_2^2 + \frac{1}{2}a_{11}F_1^2 + a_{21}F_1F_2$$

Since the elastic energy stored is independent of the order of application of F_1 and F_2, U and U' must be equal. Consequently,

$$a_{12} = a_{21} \tag{5.11a}$$

or in general

$$a_{ij} = a_{ji} \tag{5.11b}$$

The result expressed in Eq. (5.11b) has great importance in the mechanics of solids, as shown in the next section.

One can obtain an expression for the elastic strain energy in terms of the applied forces, using the above reciprocal relationship. From Eq. (5.10)

$$U = \frac{1}{2}(F_1\delta_1 + F_2\delta_2 + \ldots + F_n\delta_n)$$

$$= \frac{1}{2}F_1(a_{11}F_1 + a_{12}F_2 + \ldots + a_{1n}F_n)$$

$$+ \ldots \frac{1}{2}F_n(a_{n1}F_1 + a_{n2}F_2 + \ldots + a_{nn}F_n)$$

$$U = \frac{1}{2}(a_{11}F_1^2 + a_{22}F_2^2 + \ldots + a_{nn}F_n^2)$$

$$+ \frac{1}{2}(a_{12}F_1F_2 + a_{13}F_1F_3 + \ldots + a_{1n}F_1F_n + \ldots)$$

$$= \frac{1}{2}\Sigma(a_{11}F_1^2) + \Sigma(a_{12}F_1F_2) \qquad (5.12)$$

5.6 MAXWELL–BETTI–RAYLEIGH RECIPROCAL THEOREM

Consider two systems of forces $F_1, F_2, \ldots$, and $F_1', F_2', \ldots$, both systems having the same points of application and the same directions. Let $\delta_1, \delta_2, \ldots$, be the corresponding displacements caused by $F_1, F_2, \ldots$, and $\delta_1', \delta_2', \ldots$, the corresponding displacements caused by $F_1', F_2', \ldots$, Then, making use of the reciprocal relation given by Eq. (5.11) we have

$$F_1'\delta_1 + F_2'\delta_2 + \ldots + F_n'\delta_n$$

$$= F_1'(a_{11}F_1 + a_{12}F_2 + \ldots + a_{1n}F_n)$$

$$+ F_2'(a_{21}F_1 + a_{22}F_2 + \ldots + a_{2n}F_n)$$

$$+ \ldots + F_n'(a_{n1}F_1 + a_{n2}F_2 + \ldots + a_{nn}F_n)$$

$$= a_{11}F_1F_1' + a_{22}F_2F_2' + a_{nn}F_nF_n'$$

$$+ a_{12}(F_1'F_2 + F_2'F_1) + a_{13}(F_1'F_3 + F_3'F_1)$$

$$+ \ldots + a_{1n}(F_1'F_n + F_n'F_1) \qquad (5.13)$$

The symmetry of the expressions between the primed and unprimed quantities in the above expression shows that it is equal to

$$F_1\delta_1' + F_2\delta_2' + \ldots + F_n\delta_n'$$

i.e. $$F_1\delta_1' + F_2\delta_2' + \ldots = F_1'\delta_1 + F_2'\delta_2 + \ldots \qquad (5.14)$$

In words:

'The forces of the first system ($F_1, F_2, \ldots$, etc.) acting through the corresponding displacements produced by any second system ($F_1', F_2', \ldots$, etc.) do the same amount of work as that done by the second system of forces acting through the corresponding displacements produced by the first system of forces'.

This is the reciprocal theorem of Maxwell, Betti and Rayleigh.

5.7 GENERALISED FORCES AND DISPLACEMENTS

In the above discussions, F_1 F_2, . . . , etc. represented concentrated forces and δ_1, δ_2, . . . , etc. the corresponding linear displacements. It is possible to extend the term 'force' to include not only a concentrated force but also a bending moment or a torque. Similarly, the term 'displacement' may mean linear or angular displacement. Consider, for example, the elastic body shown in Fig. 5.3, subjected to a concentrated force F_1 at point 1 and a couple $F_2 = M$ at point 2. δ_1 will now stand for the corresponding linear displacement of point 1 and δ_2 for the corresponding angular rotation of point 2. If F_1 is a unit force acting alone, then a_{11}, the influence coefficient, gives the linear displacement of point 1 corresponding to the direction of F_1. Similarly, a_{12} stands for the corresponding linear displacement of point 1 caused by a unit couple F_2 applied at point 2. a_{21} gives the corresponding angular rotation of point 2 caused by a unit concentrated force F_1 at point 1.

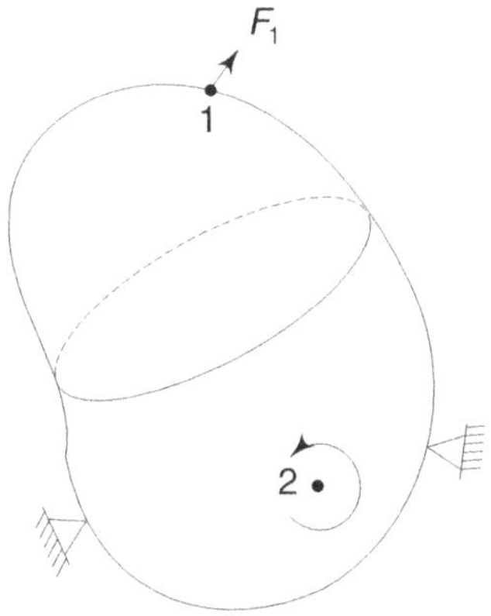

Fig. 5.3 *Generalised forces and displacements*

The reciprocal relation $a_{12} = a_{21}$ can also be interpreted appropriately. For example, making reference to Fig. 5.3, the above relation reveals that the linear displacement at point 1 in the direction of F_1 caused by a unit couple acting alone at point 2, is equal to the angular rotation at point 2 in the direction of the moment F_2 caused by a unit load acting alone at point 1. This fact will be demonstrated in the next few examples.

With the above generalised definitions for forces and displacements, the work done when the forces are gradually increased from zero to their full magnitudes is given by

$$W = \frac{1}{2}\left(F_1\delta_1 + F_2\delta_2 + \ldots + F_n\delta_n\right)$$

The reciprocal theorem of Maxwell, Betti and Rayleigh can also be given wider meaning with these extended definitions.

Example 5.1 *Consider a cantilever loaded by unit concentrated forces, as shown in Figs. 5.4(a) and (b). Check the deflections at points 1 and 2.*

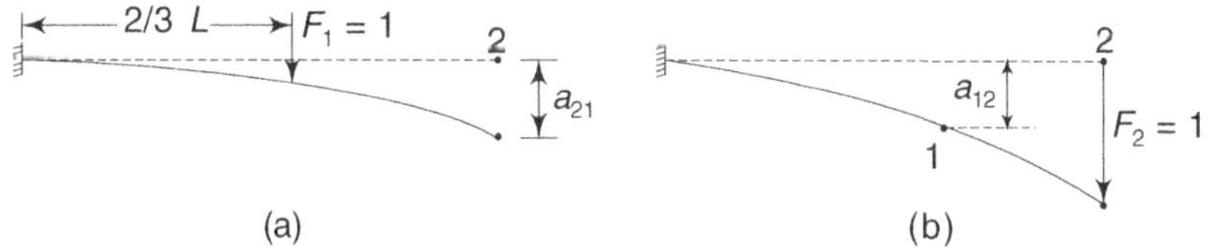

Fig. 5.4 *Example 5.1*

Solution In Fig. 5.4(a), the unit load F_1 acts at point 1. As a result, the deflection of point 2 is a_{21}. In Fig. 5.4(b) the unit load F_2 acts at point 2 and as a result, the deflection of point 1 is a_{12}. The reciprocal relation conveys that these two deflections are equal. If L is the length of the cantilever and if point 1 is at a distance of $\frac{2}{3}L$ from the fixed end, we have from elementary strength of materials

δ_2 due to F_1= deflection at 1 due to F_1 + deflection due to slope

$$= \frac{8L^3}{81\,EI} + \frac{4L^3}{54\,EI}$$

δ_1 due to F_2 = deflection at 1 due to a unit load at 1 + deflection at 1 due to a moment (L/3) at 1

$$= \frac{8L^3}{81\,EI} + \frac{4L^3}{54\,EI}$$

Example 5.2 *Consider a cantilever beam subjected to a concentrated force F at point 1 (Fig 5.5). Let us determine the curve of deflection for the beam.*

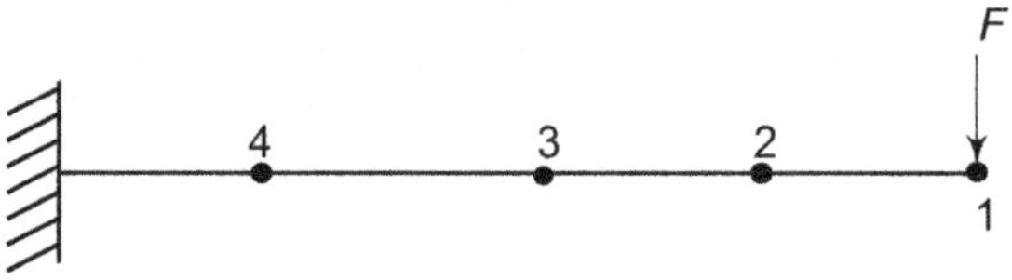

Fig. 5.5 *Example 5.2*

Solution One obvious method would be to use a travelling microscope and take readings at points 2, 3, 4, etc. These readings would be very small and consequently, errors would creep in. On the other hand, the reciprocal relation can be used to obtain this curve of deflection more accurately. The deflection at 2 due to F at 1 is the same as the deflection at 1 due to F at 2, i.e. $a_{21} = a_{12}$. Similarly, the deflection at 3 due to F at 1 is the same as the deflection at 1 due to F at 3, i.e. $a_{31} = a_{13}$. Hence, one observes the deflections at 1 as F is moved along the beam to get the required information.

Example 5.3 *The cantilever beam shown in Fig. 5.6(a) is subjected to a bending moment $M = F_1$ at point 1, and in Fig. 5.6(b), it is subjected to a concentrated load $P = F_2$ at point 2. Point 2 is 2/3 L from the fixed end. Verify the reciprocal theorem.*

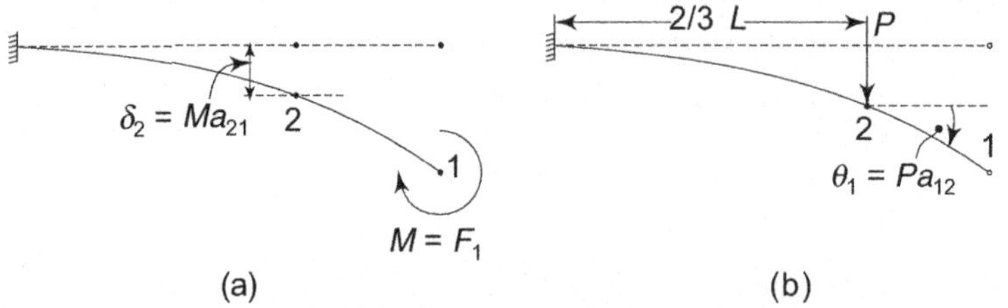

Fig. 5.6 *Example 5.3*

Solution From elementary strength of materials the deflection at point 2 due to the moment M at point 1 is

$$\delta_2 = M\left(\frac{2}{3}L\right)^2 \frac{1}{2EI} = \frac{2ML^2}{9EI}$$

The slope (angular displacement) at point 1 due to the concentrated force P at point 2 is

$$\theta_1 = P\left(\frac{2}{3}L\right)^2 \frac{1}{2EI} = \frac{2PL^2}{9EI}$$

Hence, the work done by M through the displacement (angular displacement) produced by P is equal to

$$M\theta_1 = \frac{2MPL^2}{9EI}$$

This is equal to the work done by P acting through the displacement produced by the moment M.

Example 5.4 *Determine the change in volume of an elastic body subjected to two equal and opposite forces, as shown. The distance between the points of application is h and the elastic constants for the material are E and ν, (Fig. 5.7).*

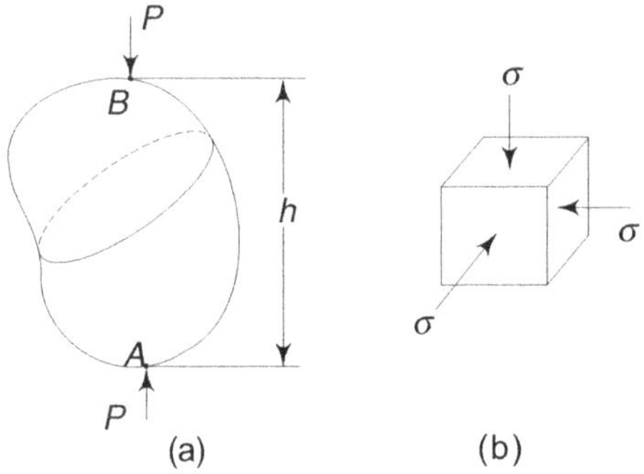

Fig. 5.7 *Example 5.4*

Solution This is a very general problem, the solution of which is apparently difficult. However, we can get a solution very easily by applying the reciprocal theorem. Let the elastic body be subjected to a hydrostatic pressure of value σ. Every volume element will be in a state of hydrostatic (isotropic) stress. Consequently, the unit contraction in any direction from Fig. 5.7(b) is

$$\varepsilon = \frac{\sigma}{E} - 2\nu\frac{\sigma}{E} = (1-2\nu)\frac{\sigma}{E}$$

The two points of application A and B, therefore, move towards each other by a distance.

$$\Delta h = h(1-2\nu)\frac{\sigma}{E}$$

Now we have two systems of forces:

System 1	Force	P
	Volume change	ΔV

System 2	Force	σ
	Distance change	Δh

From the reciprocal theorem

$$P\,\Delta h = \sigma\,\Delta V$$

or

$$\Delta V = \frac{P}{\sigma}\Delta h$$

$$= \frac{Ph}{E}(1-2\nu)$$

If ν is equal to 0.5, the change in volume is zero.

5.8 BEGG'S DEFORMETER

In this section, we shall demonstrate the application of the reciprocal theorem to a problem in experimental mechanics. Figure 5.8 shows a structural member subjected to a force P at point E. It is required to determine the forces of reaction at point B. The reaction forces are V, H and M and these make the displacements (vertical, horizontal and angular) at B equal to zero. A theoretical analysis is quite difficult for an odd structure like the one shown. The reactions at the other supports also are such that the displacement at these supports are zero. To determine V at B we proceed as follows.

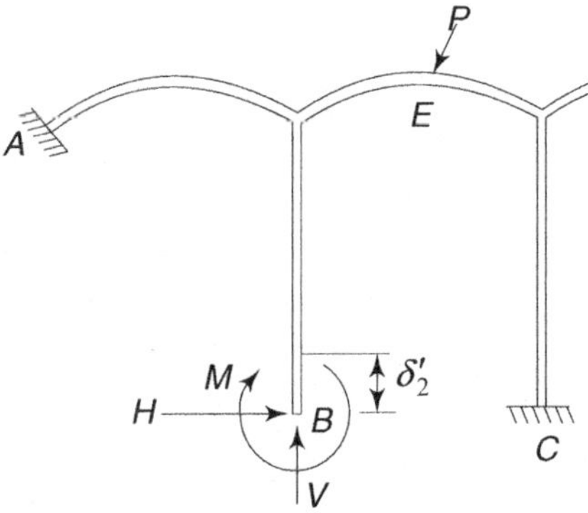

Fig. 5.8 *Reactions due to force P*

A known vertical displacement δ_2' is imposed at B, keeping A, C, D fixed and preventing angular rotation and horizontal displacement at B. The corresponding displacement at E (i.e. displacement in the direction of P) is measured. Let this be δ_1'. During the vertical displacement of B, the forces V', M' and H' that are induced at B are not measured. The two systems involved in the reciprocal theorem are as follows:

System 1 Specified

Forces V, H, M at B (unknown) and other reactive forces at A, C, D (also unknown), P at E (known)

Corresponding displacements 0, 0, 0 at B; 0, 0, 0 at A, C and D; δ_1 (unknown) at E.

System 2 Experimental

Forces V', H', M' at B (unknown) and other reactive forces at A, C, D (all unknown); 0 at E (i.e. point E not loaded)

Corresponding displacements δ_2', 0, 0 at B; 0, 0, 0 at A, C and D; δ_1' at E
Applying the reciprocal theorem

$$(V \cdot \delta_2') + (H \cdot 0) + (M \cdot 0) + 0 + (P \cdot \delta_1')$$
$$= (V' \cdot 0) + (H' \cdot 0) + (M' \cdot 0) + 0 + (0 \cdot \delta_1)$$

i.e.-
$$V = -P \frac{\delta_1'}{\delta_2'} \tag{5.15}$$

Since δ_2' is the known displacement imposed at B and δ_1' is the corresponding displacement at E that is experimentally measured, the value of V can be determined. It is necessary to remember that the corresponding displacement δ_1' at E is positive when it is in the direction of P.

To determine H at B, we proceed as above. A known horizontal displacement δ_2' is imposed at B, with all other displacements being kept zero. The corresponding displacement δ_1' at E is measured. The result is

$$H = -P \frac{\delta_1'}{\delta_2'}$$

To determine M at B, a known amount of small rotation θ' is imposed at B, keeping all other displacements zero. The corresponding displacement δ_1' resulting at E is measured. The reciprocal theorem again gives

$$M = -P \frac{\delta_1'}{\theta'}$$

5.9 FIRST THEOREM OF CASTIGLIANO

From Eq. (5.12), the expression for the elastic strain energy is

$$U = \frac{1}{2}\left(a_{11}F_1^2 + a_{22}F_2^2 + \ldots + a_{nn}F_n^2\right)$$
$$+ \left(a_{12}F_1F_2 + a_{13}F_1F_3 + \ldots + a_{1n}F_1F_n\right) + \ldots$$

In the above expression, F_1, F_2, etc. are the generalised forces, i.e. concentrated loads, moments or torques. a_{11}, a_{12}, . . . , etc. are the corresponding influence coefficients. The rate at which U increases with F_1 is given by $\frac{\partial U}{\partial F_1}$. From the above expression for U,

$$\frac{\partial U}{\partial F_1} = a_{11}F_1 + a_{12}F_2 + a_{13}F_3 + \ldots + a_{1n}F_n$$

This is nothing but the corresponding displacement at F_1, Eq. (5.8). Hence, if δ_1 stands for the generalised displacement (linear or angular) corresponding to the generalised force F_1, then

$$\frac{\partial U}{\partial F_1} = \delta_1 \tag{5.16}$$

In exactly the same way, one can show that

$$\frac{\partial U}{\partial F_2} = \delta_2, \quad \frac{\partial U}{\partial F_3} = \delta_3, \ldots, \text{etc.}$$

That is to say, 'the partial differential coefficient of the strain energy function with respect to F_r gives the displacement corresponding with F_r'. This is Castigliano's first theorem. In the form derived in Eq. (5.16), the theorem is applicable to only linearly elastic bodies, i.e. bodies satisfying Hooke's Law (see Sec. 5.15).

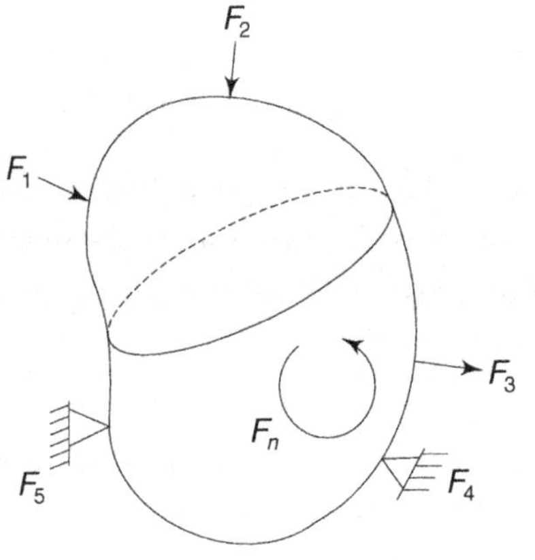

Fig. 5.9 *Elastic body in equilibrium under forces F_1, F_2, etc.*

This theorem is extremely useful in determining the displacements of structures as well as in the solutions of many statically indeterminate structures. Several examples will illustrate these subsequently. We can give an alternative proof for this theorem as follows:

Consider an elastic system in equilibrium under the force F_1, F_2, . . . F_n, etc. (Fig. 5.9). Some of these are concentrated loads and some are couples and torques. Let the strain energy stored be U. Now increase one of the forces, say F_n, by ΔF_n and as a result the strain energy increases to $U + \Delta U$, where

$$\Delta U = \frac{\Delta U}{\Delta F_n} \Delta F_n$$

Now we calculate the strain energy in a different manner. Let the elastic system be free of all forces. Let ΔF_n be applied first. The energy stored is

$$\frac{1}{2} \Delta F_n \, \Delta \delta_n$$

where $\Delta\delta_n$ is the elementary displacement corresponding to ΔF_n. This is a quantity of the second order which can be neglected since ΔF_n will be made to tend to zero in the limit. Next, we put all the other forces, F_1, F_2, . . . ,etc. These forces by themselves do an amount of work equal to U. But while these displacements are taking place, the elementary force ΔF_n is acting all the time with full magnitude at the point n which is undergoing a displacement δ_n. Hence, this elementary force does work equal to $\Delta F_n \delta_n$. The total energy stored is therefore

$$U + \Delta F_n \, \delta_n + \frac{1}{2} \Delta F_n \, \Delta \delta_n$$

Equating this to the previous expression, we get

$$U + \frac{\Delta U}{\Delta F_n} \Delta F_n = U + \Delta F_n \, \delta_n + \frac{1}{2} \Delta F_n \, \Delta \delta_n$$

In the limit, when $\Delta F_n \rightarrow 0$

$$\frac{\partial U}{\partial F_n} = \delta_n$$

it is important to note that δ_n is a linear displacement if F_n is a concentrated load, or an angular displacement if F_n is a couple or a torque. Further, we must express the strain energy in terms of the forces (including moments and couples) since it is the partial derivative with respect to a particular force that gives the corresponding displacement. In the next section, expressions for strain energies in terms of forces will be obtained.

5.10 EXPRESSIONS FOR STRAIN ENERGY

In this section we shall develop expressions for strain energy when an elastic member is subjected to axial force, shear force, bending moment and torsion. Figure 5.10(a) shows an elastic member subjected to several forces. Consider a section of the member at C. In general, this section will be subjected to three forces F_x, F_y and F_z and three moments M_x, M_y and M_z (Fig. 5.10(b)). The force F_x is the axial force and forces F_y and F_z are the shear forces across the section. Moment M_x is the torque T and moments M_y and M_z are the bending moments about the y and z axes respectively. Let Δs be an elementary length of the member; then when Δs is very small, we can assume that these forces and moments remain constant over Δs. At the left-hand section of this elementary member, the forces and moments have opposite signs. During the deformation caused by the axial force F_x alone, the remaining forces and moments do no work. Similarly, during the twist caused by the torque $T = M_x$, no work is assumed to be done (since the deformations are extremely small) by the other forces and moments.

Consequently, the work done by each of these forces and moments can be determined individually and added together to determine the total elastic strain energy stored by Δs while it undergoes deformation. We shall make use of the formulas available from elementary strength of materials.

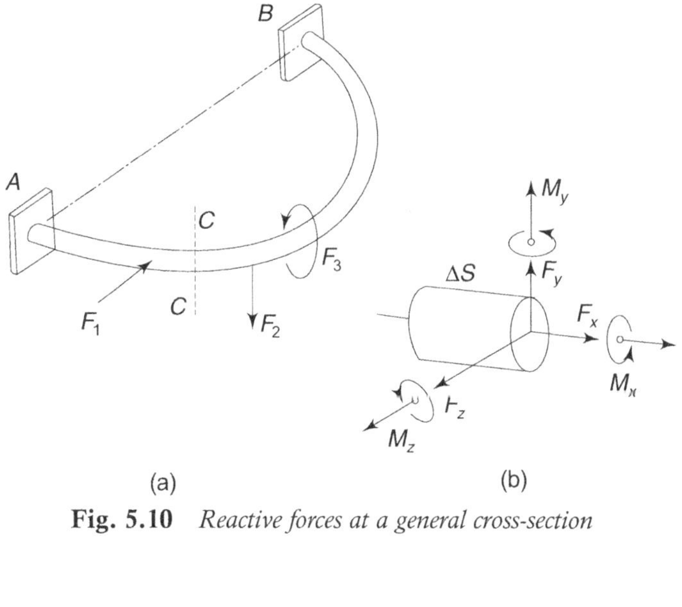

Fig. 5.10 *Reactive forces at a general cross-section*

(i) Elastic energy due to axial force: If δ_x is the axial extension due to F_x, then

$$\Delta U = \frac{1}{2} F_x \delta_x$$

$$= \frac{1}{2} F_x \cdot \frac{F_x}{AE} \Delta s$$

using Hooke's law.

$$\therefore \quad \Delta U = \frac{F_x^2}{2AE} \Delta s \tag{5.17}$$

A is the cross-sectional area and E is Young's modulus.

(ii) Elastic energy due to shear force: The shear force F_y (or F_z) is distributed across the section in a complicated manner depending on the shape of the cross-section. If we assume that the shear force is distributed uniformly across the section (which is not strictly correct), the shear displacement will be (from Fig. 5.11) $\Delta s\ \Delta\gamma$ and the work done by F_y will be

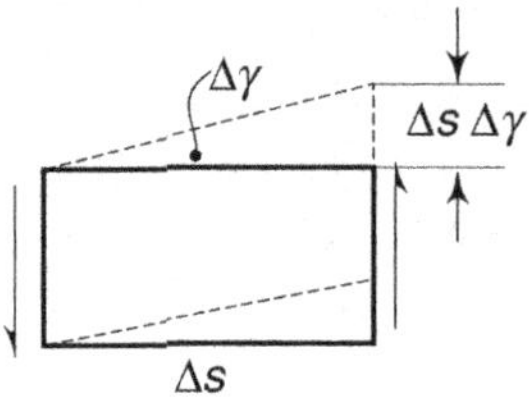

Fig. 5.11 *Displacement due to shear force*

$$\Delta U = \frac{1}{2} F_y\ \Delta s\ \Delta\gamma$$

From Hooke's law,

$$\Delta\gamma = \frac{F_y}{AG}$$

where A is the cross-sectional area and G is the shear modulus. Substituting this

$$\Delta U = \frac{1}{2} F_y\ \Delta s \frac{F_y}{AG}$$

or $$\Delta U = \frac{F_y^2}{2AG} \Delta s$$

It will be shown that the strain energy due to shear deformation is extremely small, which is often ignored. Hence, the error caused in assuming uniform distribution of the shear force across the section will be very small. However, to take into account the different cross-sections and non-uniform distribution, a factor k is introduced. With this

$$\Delta U = \frac{k F_y^2}{2AG} \Delta s \tag{5.18}$$

A similar expression is obtained for the shear force F_z.

(iii) Elastic energy due to bending moment: Making reference to Fig. 5.12, if $\Delta\phi$ is the angle of rotation due to the moment M_z(or M_y), the work done is

$$\Delta U = \frac{1}{2} M_z\ \Delta\phi$$

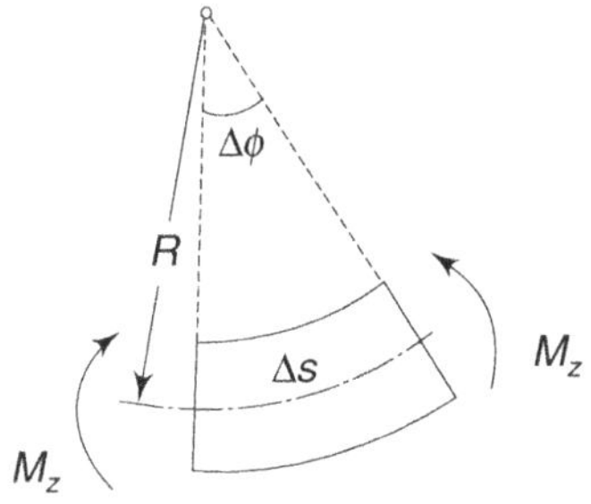

Fig. 5.12 *Displacement due to bending moment*

From the elementary flexure formula, we have

$$\frac{M_z}{I_z} = \frac{E}{R}$$

or

$$\frac{1}{R} = \frac{M_z}{EI_z}$$

where R is the radius of curvature and I_z is the area moment of inertia about the z axis. Hence

$$\Delta\phi = \frac{\Delta s}{R} = \frac{M_z}{EI_z}\Delta s$$

Substituting this

$$\Delta U = \frac{M_z^2}{2EI_z}\Delta s \tag{5.19}$$

A similar expression can be obtained for the moment M_y.

(iv) Elastic energy due to torque : Because of the torque T, the elementary member rotates through an angle $\Delta\theta$ according to the formula for a circular section

$$\frac{T}{I_p} = \frac{G\Delta\theta}{\Delta s}$$

i.e.

$$\Delta\theta = \frac{T}{GI_p}\Delta s$$

I_p is the polar moment of inertia. The work done due to this twist is,

$$\Delta U = \frac{1}{2}T\Delta\theta$$

$$= \frac{T^2}{2GI_p}\Delta s \tag{5.20}$$

Equations (5.17)–(5.20) give important expressions for the strain energy stored in the elementary length Δs of the elastic member. The elastic energy for the entire member is therefore

(i) Due to axial force

$$U_1 = \int_0^S \frac{F_x^2}{2AE} ds \tag{5.21}$$

(ii) Due to shear force

$$U_2 = \int_0^S \frac{k_y F_y^2}{2AG} ds \tag{5.22}$$

$$U_3 = \int_0^S \frac{k_z F_z^2}{2AG} ds \tag{5.23}$$

(iii) Due to bending moment

$$U_4 = \int_0^S \frac{M_y^2}{2EI_y} ds \tag{5.24}$$

$$U_5 = \int_0^S \frac{M_z^2}{2EI_z} ds \tag{5.25}$$

(iv) Due to torque

$$U_6 = \int_0^S \frac{T^2}{2GI_p} ds \tag{5.26}$$

Example 5.5 *Determine the deflection at end A of the cantilever beam shown in Fig. 5.13.*

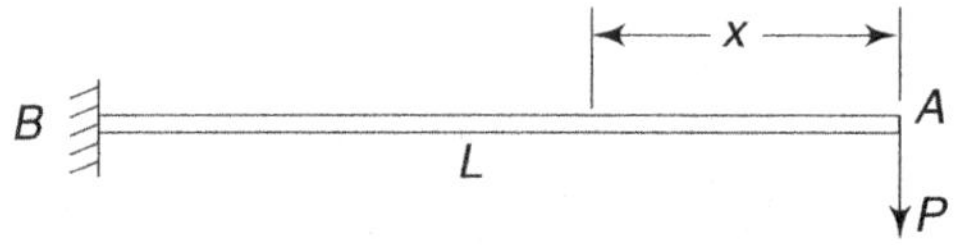

Fig. 5.13 *Example 5.5*

Solution The bending moment at any section x is

$$M = Px$$

The elastic energy due to bending moment is, therefore, from Eq. (5.24)

$$U_1 = \int_0^L \frac{(Px)^2\, dx}{2EI} = \frac{P^2 L^3}{6EI}$$

The elastic energy due to shear from Eq. (5.22) is (putting $k_1 = 1$)

$$U_2 = \int_0^L \frac{P^2}{2AG} dx = \frac{P^2 L}{2AG}$$

One can now show that U_2 is small as compared to U_1. If the beam is of a rectangular section

$$A = bd, \quad I = \frac{1}{12} bd^3$$

and $\qquad 2G \approx E$

Substituting these

$$\frac{U_2}{U_1} = \frac{P^2 L}{2bdG} \cdot \frac{6bd^3}{12P^2 L^3} \cdot 2G$$

$$= \frac{d^2}{2L^2}$$

For a member to be designated as beam, the length must be fairly large compared to the cross-sectional dimension. Hence, $L > d$ and the above ratio is extremely small. Consequently, one can neglect shear energy as compared to bending energy. With

$$U = \frac{P^2 L^3}{6EI}$$

we get

$$\frac{\partial U}{\partial P} = \frac{PL^3}{3EI} = \delta_A$$

which agrees with the solution from elementary strength of materials.

Example 5.6 *For the cantilever of total length L shown in Fig. 5.14, determine the deflection at end A. Neglect shear energy.*

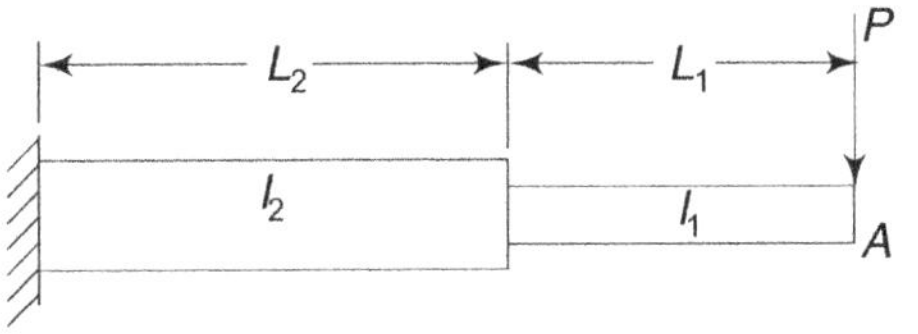

Fig. 5.14 *Example 5.6*

Solution The bending energy is

$$U = \int_0^{L_1} \frac{(Px)^2}{2EI_1} dx + \int_{L_1}^{L} \frac{(Px)^2}{2EI_2} dx$$

$$= \frac{P^2 L_1^3}{6EI_1} + \frac{P^2}{6EI_2}\left(L^3 - L_1^3\right)$$

$$\delta_A = \frac{\partial U}{\partial P} = \frac{P L_1^3}{3EI_1} + \frac{P}{3EI_2}\left(L^3 - L_1^3\right)$$

Example 5.7 *Determine the support reaction for the propped cantilever (Fig. 5.15.)*

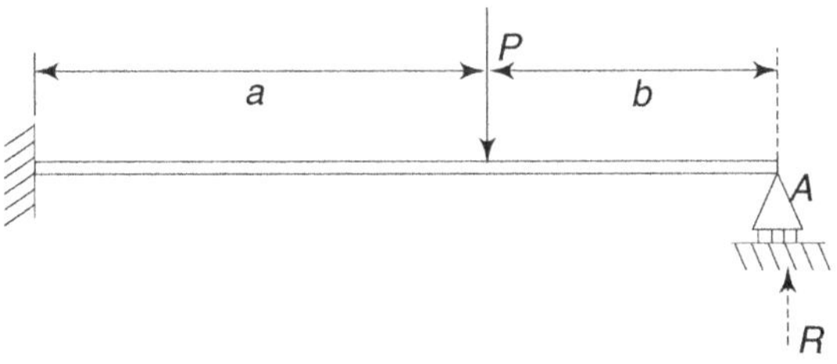

Fig. 5.15 *Example 5.7*

Solution The reaction R at A is such that the deflection there is zero. The energy is

$$U = \int_0^b \frac{(-Rx)^2}{2EI} dx + \int_0^a \frac{\left[-R(b+x) + Px\right]^2}{2EI} dx$$

$$U = \frac{1}{EI}\left(\frac{R^2 b^3}{6} + \frac{R^2 b^2 a}{2} + \frac{R^2 a^3}{6} + \frac{R^2 ba^2}{2} + \frac{P^2 a^3}{6} - \frac{PRb a^2}{2} - \frac{2PRa^3}{6}\right)$$

$$\frac{\partial U}{\partial R} = \frac{1}{EI}\left(\frac{Rb^3}{3} + Rb^2 a + \frac{Ra^3}{3} + Rba^2 - \frac{Pba^2}{2} - \frac{Pa^3}{3}\right)$$

Equating this to zero and solving for R,

$$R = \frac{Pa^2}{2} \frac{3b + 2a}{(b+a)^3}$$

Remembering that $a + b = L$, the length of cantilever,

$$R = P\left(\frac{a}{L}\right)^2 \left(\frac{3}{2} - \frac{a}{2L}\right)$$

Example 5.8 *For the structure shown in Fig. 5.16, what is the vertical deflection at end* A?

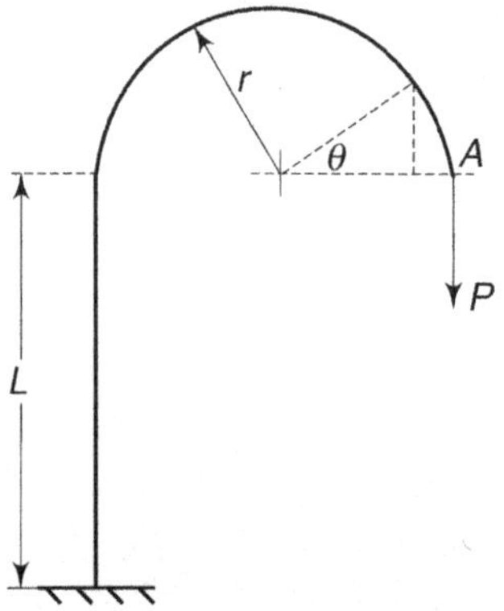

Fig. 5.16 *Example 5.8*

Solution The moment at any section θ of the curved part is $Pr\,(1 - \cos\theta)$. The bending moment for the vertical part of the structure is a constant equal to $2Pr$. The bending energy therefore is

$$\int_0^{\pi} \frac{\left[Pr\,(1-\cos\theta)\right]^2}{2EI}\, r\,d\theta + \int_0^{L} \frac{(2Pr)^2}{2EI}\,dx$$

We neglect the energy due to the axial force. Then

$$U = \frac{3}{4}\frac{\pi P^2 r^3}{EI} + \frac{2P^2 r^2 L}{EI}$$

$$\therefore\ \delta_A = \frac{\partial U}{\partial P} = \left(\frac{3}{2}\pi r + 4L\right)\frac{\mathrm{Pr}^2}{EI}$$

Example 5.9 *The end of the semi-circular member shown in Fig. 5.17, is subjected to torque T. What is the twist of end A? The member is circular in section.*

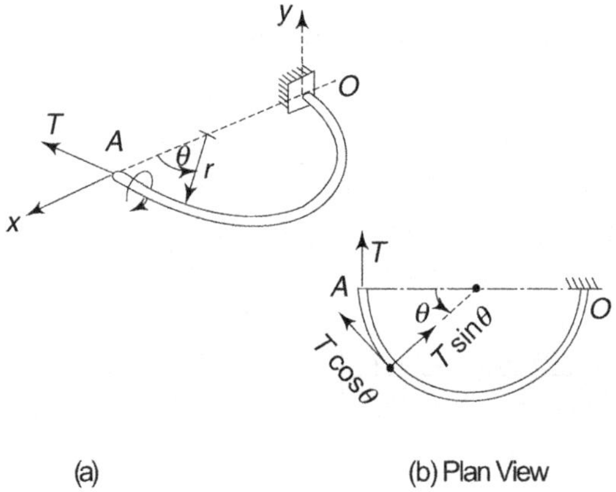

Fig. 5.17 *Example 5.9*

Solution The torque is a moment in the xy plane and can be represented by vector $\boldsymbol{T}$, as shown. At any section θ, this vector can be resolved into two components $\boldsymbol{T}\cos\theta$ and $\boldsymbol{T}\sin\theta$. The component $\boldsymbol{T}\cos\theta$ acts as torque and the component $\boldsymbol{T}\sin\theta$ as a moment.

The energy due to torque is, from Eq. (5.26),

$$U_1 = \int_0^{\pi} \frac{(T\cos\theta)^2}{2GI_P}\, r\,d\theta = \frac{\pi r T^2}{4GI_P}$$

The energy due to bending is, from Eq. (5.24),

$$U_2 = \int_0^{\pi} \frac{(T\sin\theta)^2}{2EI}\, r\,d\theta = \frac{\pi r T^2}{4EI}$$

I_p is the polar moment of inertia. For a circular member

$$I_p = 2I = \frac{\pi r^4}{2}$$

Substituting, the total energy is

$$U = U_1 + U_2 = \frac{\pi r T^2}{4}\left(\frac{1}{GI_p} + \frac{1}{EI}\right)$$

Hence, the twist is

$$\theta = \frac{\partial U}{\partial T} = \frac{\pi r T}{2}\left(\frac{1}{2G} + \frac{1}{E}\right)\frac{2}{\pi r^4}$$

$$= \frac{1}{r^3}\left(\frac{1}{2G} + \frac{1}{E}\right)T$$

5.11 FICTITIOUS LOAD METHOD

Castigliano's first theorem described above helps us to determine the displacement at a point corresponding to the force acting there. Situations arise where it may be desirable to determine the displacement (either linear or angular) at a point where there is no force (concentrated load or a couple) acting. In such situations, we assume a small fictitious or dummy load to be acting at the point where the displacement is required. Castigliano's theorem is then applied, and in the final result, the fictitious load is put equal to zero. The following example will describe the technique.

Example 5.10 *Determine the slope at end A of the cantilever in Fig. 5.18 which is subjected to load P.*

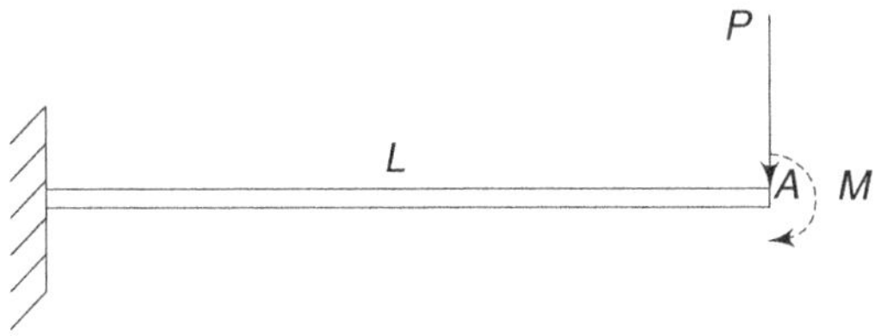

Fig. 5.18 *Example 5.10*

Solution To determine the slope by Castigliano's method we have to determine U and take its partial derivative with respect to the corresponding force, i.e. a moment. But no moment is acting at A. So, we assume a fictitious moment M to be acting at A and determine the slope caused by P and M. Since the magnitude of M is actually zero, in the final result, M is equated to zero.

The energy due to P and M is,

$$U = \int_0^L \frac{(Px + M)^2}{2EI}\,dx$$

$$= \frac{P^2 L^3}{6EI} + \frac{M^2 L}{2EI} + \frac{MPL^2}{2EI}$$

$$\theta = \frac{\partial U}{\partial M} = \frac{ML}{EI} + \frac{PL^2}{2EI}$$

This gives the slope when M and P are both acting. If M is zero, the slope due to P alone is

$$\theta = \frac{PL^2}{2EI}$$

If on the other hand, P is zero and M alone is acting the slope is

$$\theta = \frac{ML}{EI}$$

Example 5.11 *For the member shown in Fig. 5.16, Example 5.8, determine the ratio of L to r if the horizontal and vertical deflections of the loaded end A are equal. P is the only force acting.*

Solution In addition to the vertical for P at A, apply a horizontal fictitious force F to the right. The bending moment at section θ of the semi-circular part is

$$M_1 = Pr\,(1 - \cos\,\theta) - Fr\,\sin\,\theta)$$

At any section x in the vertical part, the moment is

$$M_2 = 2Pr + Fx$$

Hence,

$$U = \frac{1}{2EI}\int_0^{\pi}\left[Pr(1-\cos\theta) - Fr\sin\theta\right]^2 r\,d\theta + \frac{1}{2EI}\int_0^{L}(2Pr+Fx)^2\,dx$$

$$\therefore \quad \frac{\partial U}{\partial F} = -\frac{r^2}{EI}\int_0^{\pi}\left[\mathrm{Pr}(1-\cos\theta) - Fr\sin\theta\right]\sin\theta\,d\theta + \frac{1}{EI}\int_0^{L}(2Pr+Fx)\,x\,dx$$

and

$$\left.\frac{\partial U}{\partial F}\right|_{F=0} = \delta_h = -\frac{r^2}{EI}\int_0^{\pi}\left[Pr(1-\cos\theta)\sin\theta\right]d\theta + \frac{1}{EI}\int_0^{L}2Pr\,xdx$$

$$= -\frac{2Pr^3}{EI} + \frac{PrL^2}{EI} = \frac{Pr}{EI}\left(-2r^2 + L^2\right)$$

From Example 5.8

$$\delta_v = \frac{Pr^2}{EI}\left(\frac{3}{2}\pi r + 4L\right)$$

Equating δ_v to δ_h

$$\frac{Pr^2}{EI}\left(\frac{3}{2}\pi r + 4L\right) = \frac{Pr}{EI}\left(-2r^2 + L^2\right)$$

or

$$L^2 - 4Lr - r^2\left(\frac{3\pi}{2} + 2\right) = 0$$

Dividing by r^2 and putting $\frac{L}{r} = \rho$

$$\rho^2 - 4\rho - \left(\frac{3\pi}{2} + 2\right) = 0$$

Solving, $$\rho = \frac{4 \pm \sqrt{[16 + 4(3\pi/2 + 2)]}}{2}$$

or $$\rho = 2 + \sqrt{6 + \frac{3}{2}\pi}$$

5.12 SUPERPOSITION OF ELASTIC ENERGIES

When an elastic body is subjected to several forces, one cannot obtain the total elastic energy by adding the energies caused by individual forces. In other words, the sum of individual energies is not equal to the total energy of the system. The reason for this is simple. Consider an elastic body subjected to two forces F_1 and F_2. When F_1 is applied first, let the energy stored be U_1. When F_2 is applied next (with F_1 continuing to act), the additional energy stored is equal to U_2 due to F_2 alone, plus the work done by F_1 during the displacement caused by F_2. Hence, the total energy stored when both F_1 and F_2 are acting is equal to $(U_1 + U_2 + U_3)$, where U_1 is the work energy caused by F_1 alone, U_2 is the work energy caused by F_2 alone, and U_3 is the energy due to the work done by F_1 during the displacement caused by F_2. Another way of observing this is to note that the strain energy functions are not linear functions. Hence, individual energies cannot be added to get the total energy. As a specific example, consider the cantilever shown in Fig. 5.18, Example 5.10. Let P and M be actual forces acting on the cantilever, i.e M is not a fictitious force as was assumed in that example. The elastic energy stored due to P and M is given by (a), i.e.

$$U = \frac{P^2L^3}{6EI} + \frac{M^2L}{2EI} + \frac{MPL^2}{2EI}$$

The energy due to P alone is

$$U_1 = \frac{1}{2EI}\int_0^L (Px)^2\,dx = \frac{P^2L^3}{6EI}$$

Similarly, the energy due to M alone is

$$U_2 = \frac{1}{2EI}\int_0^L M^2\,dx = \frac{M^2L}{2EI}$$

Obviously, $U_1 + U_2$ is not equal to U. However, if P is applied first and then M, the total energy is given by $U_1 + U_2$ + work done by P during the displacement caused by M.

The deflection at the end of the cantilever (where P is acting with full magnitude) caused by M is

$$\delta_A^* = \frac{ML^2}{2EI}$$

During this deflection, the work done by P is

$$U_3 = P\left(\frac{ML^2}{2EI}\right)$$

If this additional energy is added to $U_1 + U_2$, then one gets the previous expression for U. It is immaterial whether P is applied first or M is applied. The order of loading is immaterial. Thus, one should be careful in applying the superposition principle to the energies. However, the individual energies caused by axial force, bending moment and torsion can be added since the force causing one kind of deformation will not do any work during a different kind of deformation caused by another force. For example, an axial force causing linear deformation will not do work during an angular deformation (or twist) caused by a torque. This is true in the case of small deformation as we have been assuming throughout our discussions. Similarly, a bending moment will not do any work during axial or linear displacement caused by an axial force.

5.13 STATICALLY INDETERMINATE STRUCTURE

Many statically indeterminate structural problems can be conveniently solved, using Castigliano's theorem. The technique is to determine the forces and moments to produce the required displacement. Example 5.7 was one such problem. The following example will further illustrate this method.

Example 5.12 *A rectangular frame with all four sides of equal cross section is subjected to forces P, as shown in Fig. 5.19. Determine the moment at section C and also the increase in the distance between the two points of application of force P.*

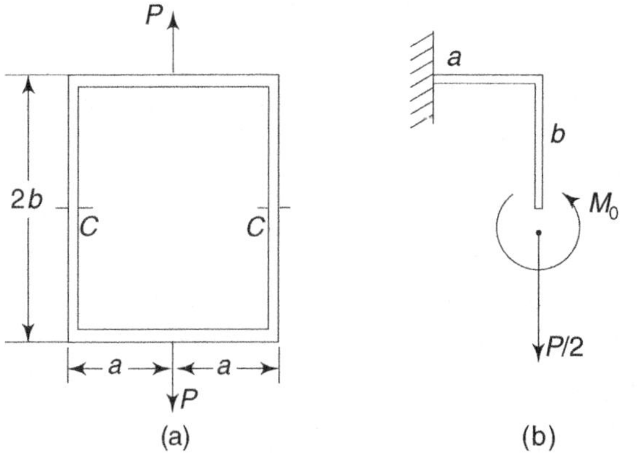

Fig. 5.19 *Example 5.12*

Solution The symmetry conditions indicate that the top and bottom members deform in such a manner that the tangents at the points of loading remain horizontal. Also, there is no Change in slopes at sections C-C. Hence, one can consider only a quarter part of the frame, as shown in (b).

Considering only the bending energy and neglecting the energies due to direct tension and shear force, we get

$$U' = \int_0^b \frac{M_0^2}{2EI}\,dx + \int_0^a \frac{(M_0 - P/2\,x)^2}{2EI}\,dx$$

$$= \frac{1}{2EI}\left(M_0^2\,b + M_0^2\,a - M_0 P\,\frac{a^2}{2} + \frac{1}{12}\,P^2 a^3\right)$$

Because of symmetry, the change in slope at section C is zero. Hence

$$\frac{\partial U'}{\partial M_0} = \frac{1}{2EI}\left[2M_0\,(a+b) - \frac{1}{2}\,Pa^2\right]$$

Equating this to zero,

$$M_0 = \frac{Pa^2}{4\,(a+b)}$$

To determine the increase in distance between the two load points, we determine the partial derivative of $4U'$ with respect to P (assuming that the bottom loaded point is held fixed).

$$U = 4U' = \frac{4}{2EI}\left[\frac{P^2a^4}{16(a+b)^2}(a+b) - \frac{P^2a^4}{8(a+b)} + \frac{P^2a^3}{12}\right]$$

$$\therefore \qquad \frac{\partial U}{\partial P} = \frac{Pa^3}{12EI}\frac{(a+4b)}{(a+b)}$$

Example 5.13 *A thin circular ring of radius r is subjected to two diametrically opposite loads P in its own plane as shown in Fig. 5.20(a). Obtain an expression for the bending moment at any section. Also, determine the change in the vertical diameter.*

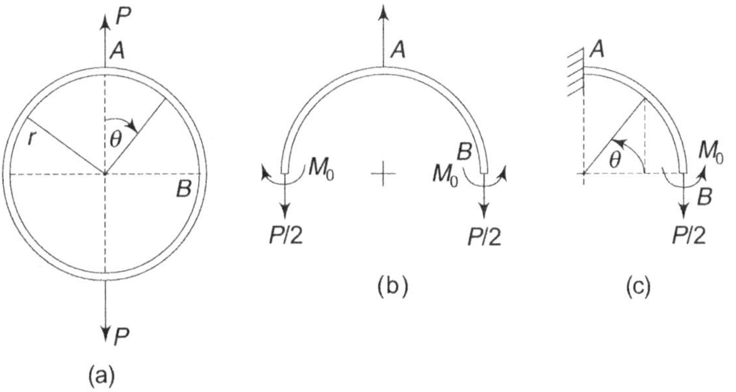

Fig. 5.20 *Example 5.13*

Solution Because of symmetry, during deformation there is no change in the slopes at A and B. So, one can consider only a quarter of the ring for calculation as shown in Fig. 5.20(c). The value of M_0 is such as to cause no change in slope at B. Section at A can be considered as built-in.

Moment at $\qquad \theta = M = \frac{P}{2}r(1-\cos\theta) - M_0$

$$U - \frac{1}{2EI}\int_0^{\pi/2}\left[\frac{P}{2}r(1-\cos\theta) - M_0\right]^2 r\,d\theta$$

Since there is no change in slope at B

$$\frac{\partial U}{\partial M_0} = -\frac{r}{2EI}\int_0^{\pi/2} 2\left[\frac{P}{2}r(1-\cos\theta) - M_0\right]d\theta = 0$$

i.e. $$\int_0^{\pi/2} \left[\frac{P}{2} r (1 - \cos\theta) - M_0\right] d\theta = 0$$

i.e. $$\frac{P}{2} r \left(\frac{\pi}{2} - 1\right) - M_0 \frac{\pi}{2} = 0$$

or $$M_0 = \frac{Pr}{2}\left(1 - \frac{2}{\pi}\right)$$

$\therefore$ $$M \text{ at } \theta = \frac{P}{2} r (1 - \cos\theta) - \frac{P}{2} r \left(1 - \frac{2}{\pi}\right) = \frac{Pr}{2}\left(\frac{2}{\pi} - \cos\theta\right)$$

To determine the increase in the diameter along the loads, one has to determine the elastic energy and take the differential. If one considers the quarter ring, Fig. 5.20(c), the elastic energy is

$$U^* = \int_0^{\pi/2} \frac{1}{2EI}\left[\frac{Pr}{2}\left(\frac{2}{\pi} - \cos\theta\right)\right]^2 r\, d\theta$$

The differential of this with respect to (P/2) will give the vertical deflection of the end B with reference to A. Observe that in order to determine the deflection at B, one has to take the differential with respect to the particular load that is acting at that point, which is (P/2). Putting (P/2) = Q.

$$U^* = \frac{1}{2EI}\int_0^{\pi/2} \left[Qr\left(\frac{2}{\pi} - \cos\theta\right)\right]^2 r\, d\theta$$

$$= \frac{Q^2 r^3}{2EI}\int_0^{\pi/2} \left(\frac{2}{\pi} - \cos\theta\right)^2 d\theta$$

$\therefore$ $$\frac{\partial U^*}{\partial Q} = \frac{Qr^3}{EI}\int_0^{\pi/2} \left(\frac{4}{\pi^2} + \cos^2\theta - \frac{4}{\pi}\cos\theta\right) d\theta$$

$$= \frac{Qr^3}{EI}\left(\frac{4}{\pi^2}\frac{\pi}{2} + \frac{\pi}{4} - \frac{4}{\pi}\right)$$

$$= \frac{Qr^3}{EI}\left(\frac{\pi}{4} - \frac{2}{\pi}\right) = \frac{Pr^3}{2EI}\left(\frac{\pi}{4} - \frac{2}{\pi}\right)$$

As this gives only the increase in the radius, the increase in the diameter is twice this quantity, i.e.

$$\delta_v = \frac{Pr^3}{EI}\left(\frac{\pi}{4} - \frac{2}{\pi}\right)$$

5.14 THEOREM OF VIRTUAL WORK

Consider an elastic system subjected to a number of forces (including moments) $F_1, F_2, \ldots$, etc. Let $\delta_1, \delta_2, \ldots$, etc. be the corresponding displacements. Remember that these are the work absorbing components (linear and angular displacements) in the corresponding directions of the forces (Fig. 5.21).

Let one of the displacements δ_1 be increased by a small quantity $\Delta\delta_1$. During this additional displacement, all other displacements where forces are acting are

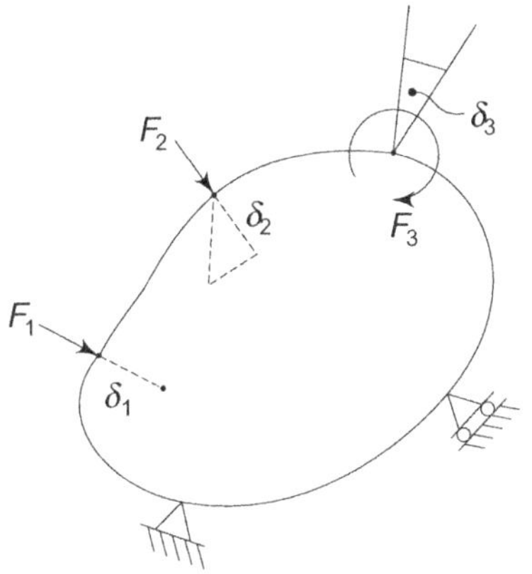

Fig. 5.21 *Generalised forces and displacements*

held fixed, which means that additional forces may be necessary to maintain such a condition. Further, the small displacement $\Delta\delta_1$ that is imposed must be consistent with the constraints acting. For example, if point l is constrained in such a manner that it can move only in a particular direction, then $\Delta\delta_1$ must be consistent with such a constraint. A hypothetical displacement of such a kind is called a virtual displacement. In applying this virtual displacement, the forces F_1, F_2, . . ., etc. (except F_1) do no work at all because their points of application do not move (at least in the work-absorbing direction). The only force doing work is F_1 by an amount $F_1\,\Delta\delta_1$ plus a fraction of $\Delta F_1\,\Delta\delta_1$, caused by the change in F_1. This additional work is stored as strain energy ΔU. Hence

$$\Delta U = F_1\,\Delta\delta_1 + k\,\Delta F_1\,\Delta\delta_1$$

or
$$\frac{\Delta U}{\Delta\delta_1} = F_1 + k\,\Delta F_1$$

and
$$\underset{\Delta\delta_1 \to 0}{\text{Lt}} \frac{\Delta U}{\Delta\delta_1} = \frac{\partial U}{\partial\delta_1} = F_1 \tag{5.27}$$

This is the theorem of virtual work. Note that in this case, the strain energy must be expressed in terms of δ_1, δ_2, . . ., etc. whereas in the application of Castigliano's theorem U had to be expressed in terms of F_1, F_2, . . ., etc.

It is important to observe that in obtaining the above equation, we have not assumed that the material is linearly elastic, i.e. that it obeys Hooke's law. The theorem is applicable to any elastic body, linear or nonlinear, whereas Castigliano's first theorem, as derived in Eq. (5.16), is strictly applicable to linear elastic or Hookean materials. This aspect will be discussed further in Sec. 5.15.

Example 5.14 *Three elastic members AD, BD and CD are connected by smooth pins, as shown in Fig. 5.22. All the members have the same cross-sectional areas and are of the same material. BD is 100 cm long and members AD and CD are each 200 cm long. What is the deflection of D under load W?*

Solution Under the action of load W, it is possible for D to move vertically and horizontally. If δ_1 and δ_2 are the vertical and horizontal displacements, then according to the principle of virtual work.

$$\frac{\partial U}{\partial\delta_1} = W, \quad \frac{\partial U}{\partial\delta_2} = 0$$

where U is the total strain energy of the system.

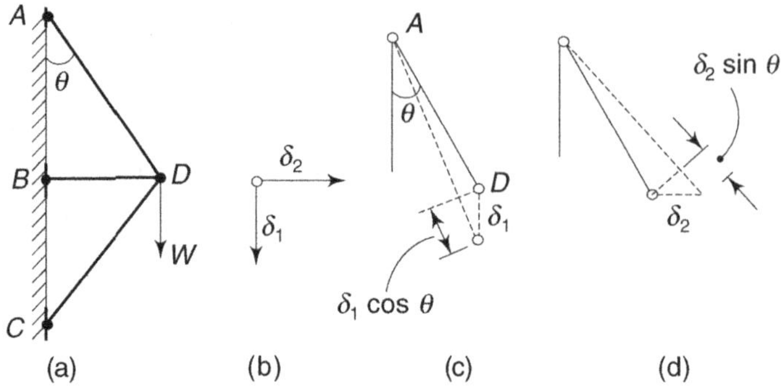

Fig. 5.22 *Example 5.14*

Becacse of δ_1, *BD* will not undergo any changes in length but *AD* will extend by $\delta_1 \cos\theta$ and *CD* will contract by the same amount, From Fig. (a),

$$\cos\ \theta = \frac{\sqrt{3}}{2}$$

Because of δ_2, *BD* will extend by δ_2 and *AD* and *CD* each will extend by $\frac{1}{2}\delta_2$. Hence, the total extension of each member is

$$AD \text{ extends by } \frac{1}{2}\left(\sqrt{3}\,\delta_1 + \delta_2\right) \text{ cm}$$

$$BD \text{ extends by } \delta_2 \text{ cm}$$

$$CD \text{ extends by } \frac{1}{2}\left(-\sqrt{3}\,\delta_1 + \delta_2\right) \text{ cm}$$

To calculate the strain energy, one needs to know the force-deformation equation for the non-Hookean members. This aspect will be taken up in Sec. 5.17, and Example 5.17. For the present example, assuming Hooke's law, the forces in the members are (with δ corresponding extensions and a to cross-sections)

$$\text{in } AD: \frac{aE\delta}{L} = aE\,\frac{1}{2}\left(\sqrt{3}\,\delta_1 + \delta_2\right)\frac{1}{200}$$

$$\text{in } BD: \frac{aE\delta}{L} = aE\delta_2\,\frac{1}{100}$$

$$\text{in } CD: \frac{aE\delta}{L} = aE\,\frac{1}{2}\left(-\sqrt{3}\,\delta_1 + \delta_2\right)\frac{1}{200}$$

The total elastic strain energy taking only axial forces into account is

$$U = \Sigma\frac{P^2L}{2aE} = \frac{aE}{2}\left[\frac{1}{800}\left(\sqrt{3}\,\delta_1 + \delta_2\right)^2 + \frac{1}{100}\delta_2^2\right.$$

$$\left. + \frac{1}{800}\left(-\sqrt{3}\,\delta_1 + \delta_2\right)^2\right]$$

$$= aE\left(\frac{3}{800}\delta_1^2 + \frac{1}{160}\delta_2^2\right)$$

$$\therefore \qquad W = \frac{\partial U}{\partial \delta_1} = \frac{3aE}{400}\delta_1$$

and $$0 = \frac{\partial U}{\partial \delta_2} = \frac{aE}{80}\delta_2$$

Hence, δ_2 is zero, which means that D moves only vertically under W and the value of this vertical deflection δ_1 is

$$\delta_1 = \frac{400}{3aE}W$$

5.15 KIRCHHOFF'S THEOREM

In this section, we shall prove an important theorem dealing with the uniqueness of solution. First, we observe that the applied forces taken as a whole work on the body upon which they act. This means that some of the products $F_n\ \delta_n$ etc. may be negative but the sum of these products taken as a whole is positive. When the body is elastic, this work is stored as elastic strain energy. This amounts to the statement that U is an essentially positive quantity. If this were not so, it would have been possible to extract energy by applying an appropriate system of forces. Hence, every portion of the body must store positive energy or no energy at all. Accordingly, U will vanish only when every part of the body is undeformed. On the basis of this and the superposition principle, we can prove Kirchhoff's uniqueness theorem, which states the following:

An elastic body for which displacements are specified at some points and forces at others, will have a unique equilibrium configuration.

Let the specified displacements be $\delta_1, \delta_2, \ldots, \delta_r$ and the specified forces be F_s, $F_t, \ldots, F_n$. It is necessary to observe that it is not possible to prescribe simultaneously both force and displacement for one and the same point. Consequently, at those points where displacements are prescribed, the corresponding forces are $F_1', F_2', \ldots, F_r'$ and at those points where forces are prescribed, the corresponding displacement are $\delta_s', \delta_t', \ldots, \delta_n'$. Let this be the equilibrium configuration. If this system is not unique, then there should be another equilibrium configuration in which the forces corresponding to the displacements $\delta_1, \delta_2, \ldots, \delta_r$ have the values $F_1'', F_2'', \ldots, F_r''$ and the displacements corresponding to the forces $F_s, F_t, \ldots, F_n$ have the values $\delta_s'', \delta_t'', \ldots, \delta_n''$. We therefore have two distinct systems.

First System	Forces	$F_1', F_2', \ldots, F_r'$,		F_s,	$F_t, \ldots,$	F_n
	Corresponding displacements	$\delta_1, \delta_2, \ldots,$	δ_r	δ_s',	$\delta_t', \ldots,$	δ_n'
Second System	Forces	$F_1'', F_2'', \ldots,$	F_r''	F_s,	$F_t, \ldots,$	F_n
	Corresponding displacements	$\delta_1, \delta_2, \ldots,$	δ_r	δ_s'',	$\delta_t'', \ldots,$	δ_n''

We have assumed that these are possible equilibrium configurations. Hence, by the principle of superposition the difference between these two systems must also be an equilibrium configuration. Subtracting the second system from the first, we get the third equilibrium configuration as

Forces	$(F_1'-F_1''), (F_2'-F_2''), \ldots, (F_r'-F_r'')$;		0,	0,	...,	0
Corresponding displacements	0, ...,	0	0	$(\delta_s'-\delta_s'')$,	$(\delta_t'-\delta_t'')$, ...,	$(\delta_n'-\delta_n'')$

The strain energy corresponding to the third system is $U = 0$. Consequently the body remains completely undeformed. This means that the first and second systems are identical, i.e. there is a unique equilibrium configuration.

5.16 SECOND THEOREM OF CASTIGLIANO OR MENABREA'S THEOREM

This theorem is of great importance in the solution of redundant structures or frames. Let a framework consist of m number of members and j number of joints. Then, if

$$m > 3j - 6$$

the frame is termed a redundant frame. The reason is as follows. For each joint, we can write three force equilibrium equations (in a general three-dimensional case), thus giving a total of $3j$ number of equations. However, all these equation are not independent, since all the external forces by themselves are in equilibrium and, therefore, satisfy the three force equilibrium equations and the three moment equilibrium equations. Hence, the number of independent equations are $3j - 6$ and if the number of members exceed $3j - 6$, the frame is redundant. The number

$$N = m - 3j + 6$$

is termed the order of redundancy of the framework. If the skeleton diagram lies wholly in one plane, the framework is termed a plane frame. For a plane framework, the degree of redundancy is given by the number

$$N = m - 2j + 3$$

Castigliano's second theorem (also known as Menabrea's theorem) can be stated as follows:

The forces developed in a redundant framework are such that the total elastic strain energy is a minimum.

Thus, if F_1, F_2 and F_r are the forces in the redundant members of a framework and U is the elastic strain energy, then

$$\frac{\partial U}{\partial F_1} = 0, \quad \frac{\partial U}{\partial F_2} = 0, \ldots, \quad \frac{\partial U}{\partial F_r} = 0$$

This is also called the principle of least work and can be proven as follows:

Let r be the number of redundant members. Remove the latter and replace their actions by their respective forces, as shown in Fig. 5.23(b). Assuming that the values of these redundant forces F_1, F_2, . . ., F_r are known, the framework will have become statically determinate and the elastic strain energy of the remaining members can be determined. Let U_s be the strain energy of these members. Then by Castigliano's first theorem, the 'increase' in the distance between the joints a and b is given as

$$\delta'_{ab} = -\frac{\partial U_s}{\partial F_i} \tag{5.28}$$

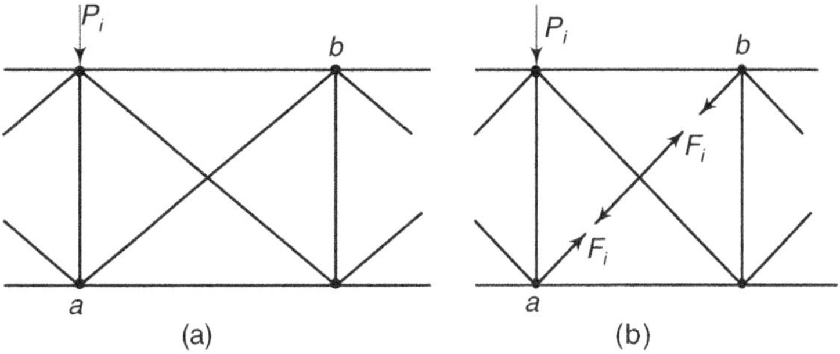

Fig. 5.23 *(a) Redundant structure (b) Structure with redundant member removed*

The negative appears because of the direction of F_i. The reactive force on the redundant member *ab* being F_i, its length will increase by

$$\delta_{ab} = \frac{F_i\, l_i}{A_i\, E_i} \tag{5.29}$$

where l_i is the length and A_i is the sectional area of the member. The increase in the distance given by Eq. (5.28) must be equal to the increase in the length of the member *ab*, given by Eq. (5.29). Hence

$$-\frac{\partial U_s}{\partial F_i} = \frac{F_i\, l_i}{A_i\, E_i} \tag{5.30}$$

The elastic strain energies of the redundant members are

$$U_1 = \frac{F_1^2\, l_1}{2A_1\, E_1},\quad U_2 = \frac{F_2^2\, l_2}{2A_2\, E_2}, \ldots,\quad U_r = \frac{F_r^2\, l_r}{2A_r\, E_r}$$

Hence, the total elastic energy of all redundant members is

$$U_1 + U_2 + \ldots U_r = \frac{F_1^2\, l_1}{2A_1\, E_1} + \frac{F_2^2\, l_2}{2A_2\, E_2} + \ldots + \frac{F_r^2\, l_r}{2A_r\, E_r}$$

$$\therefore \quad \frac{\partial}{\partial F_i}\left(U_1 + U_2 + \ldots + U_r\right) = \frac{F_i\, l_i}{A_i\, E_i}$$

since all terms, other than the *i*th term on the right-hand side, will vanish when differentiated with respect to F_i. Substituting this in Eq. (5.30)

$$-\frac{\partial U_s}{\partial F_i} = \frac{\partial}{\partial F_i}\left(U_1 + U_2 + \ldots + U_r\right) = 0$$

or $$\frac{\partial}{\partial F_i}\left(U_1 + U_2 + \ldots + U_r + U_s\right) = 0$$

The sum of the terms inside the parentheses is the total energy of the entire framework including the redundant members. If U is this total energy

$$\frac{\partial U}{\partial F_i} = 0$$

Similarly, by considering the redundant members one-by-one, we get

$$\frac{\partial U}{\partial F_1} = 0, \ \frac{\partial U}{\partial F_2} = 0, \ldots, \frac{\partial U}{\partial F_r} = 0 \tag{5.31}$$

This is the principle of least work.

Example 5.15 *The framework shown in Fig. 5.24 contains a redundant bar. All the members are of the same section and material. Determine the force in the horizontal redundant member.*

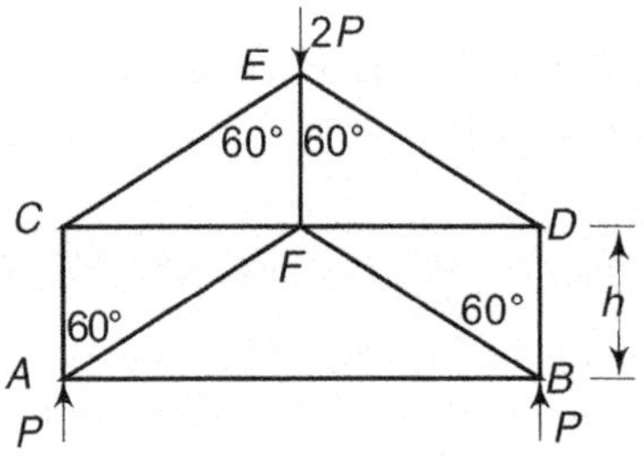

Fig. 5.24 *Example 5.15*

Solution Let T be the tension in the member AB. The forces in the members are

Members	*Length*	*Force*
AB	$2\sqrt{3}\,h$	$+T$
AC, BD	h	$T/\sqrt{3} - P$
AF, BF	$2h$	$-2T/\sqrt{3} + 0$
CF, DF	$\sqrt{3}\,h$	$-T + P\sqrt{3}$
CE, DE	$2h$	$2T/\sqrt{3} - 2P$
FE	h	$-2T/\sqrt{3} + 0$

The total strain energy is

$$U = \frac{h}{2EA}\left[2\sqrt{3}\,T^2 + 2\left(P^2 + \frac{T^2}{3} - \frac{2PT}{\sqrt{3}}\right) + \frac{16T^2}{3}\right.$$

$$+ 2\sqrt{3}\left(T^2 + 3P^2 - 2PT\sqrt{3}\right)$$

$$\left. + 16\left(\frac{T^2}{3} + P^2 - \frac{2PT}{\sqrt{3}}\right) + \frac{4T^2}{3}\right]$$

The condition for minimum strain energy or least work is

$$\frac{\partial U}{\partial T} = 0 = \frac{h}{2EA}\left[4\sqrt{3}T + \frac{4T}{3} - \frac{4P}{\sqrt{3}} + \frac{32T}{3} + 4\sqrt{3}\,T\right.$$

$$\left. -12P + \frac{32T}{3} - \frac{32}{\sqrt{3}}P + \frac{8T}{3}\right]$$

$$\therefore \quad T\left(4\sqrt{3} + \frac{4}{3} + \frac{32}{3} + 4\sqrt{3} + \frac{32}{3} + \frac{8}{3}\right) = P\left(\frac{4}{\sqrt{3}} + 12 + \frac{32}{\sqrt{3}}\right)$$

or

$$T = \frac{9\left(\sqrt{3}+1\right)}{6\sqrt{3}+19}P$$

Example 5.16 *A cantilever is supported at the free end by an elastic spring of spring constant k. Determine the reaction at A (Fig. 5.25). The cantilever beam is uniformly loaded. The intensity of loading is W.*

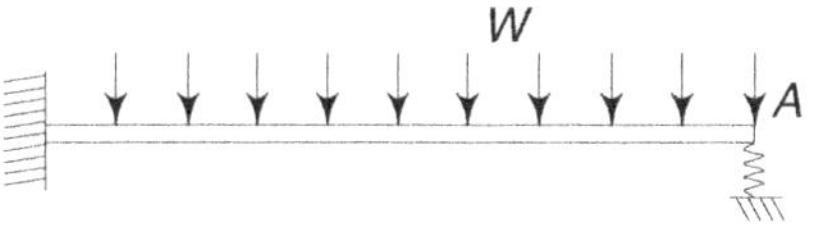

Fig. 5.25 *Example 5.16*

Solution Let R be the unknown reaction at A, i.e. R is the force on the spring. The strain energy in the spring is

$$U_1 = \frac{1}{2} R\delta = \frac{1}{2} R \frac{R}{k} = \frac{R^2}{2k}$$

where δ is the deflection of the spring. The strain energy in the beam is

$$U_2 = \int_0^L \frac{M^2\, dx}{2EI}$$

$$= \int_0^L \frac{\left(Rx - wx^2/2\right)^2 dx}{2EI}$$

$$= \frac{1}{EI}\left(\frac{1}{6} R^2 L^3 + \frac{1}{40} w^2 L^5 - \frac{1}{8} RwL^4\right)$$

Hence, the total strain energy for the system is

$$U = U_1 + U_2 = \frac{R^2}{2k} + \frac{1}{EI}\left(\frac{1}{6} R^2 L^3 + \frac{1}{40} w^2 L^5 - \frac{1}{8} RwL^4\right)$$

From Castigliano's second theorem

$$\frac{\partial U}{\partial R} = \frac{R}{k} + \frac{1}{EI}\left(\frac{1}{3} RL^3 - \frac{1}{8} wL^4\right) = 0$$

$$\therefore \qquad R = \frac{3kwL^4}{8\left(3EI + kL^3\right)}$$

5.17 GENERALISATION OF CASTIGLIANO'S THEOREM OR ENGESSER'S THEOREM

It is necessary to observe that in developing the first and second theorems of Castigliano, we have explicitly assumed that the elastic body satisfies Hooke's law, i.e. the body is linearly elastic. However, situations exist where the deformation is not proportional to load, though the body may be elastic. Consider the spring showns in Fig. 5.26(a), whose load–displacement curve is as given in Fig. 5.26(b).

The spring is a non-linear spring. Consider the area of OBC which is the strain energy. It is represented by

$$U = \int_0^x F\, dx \tag{5.32}$$

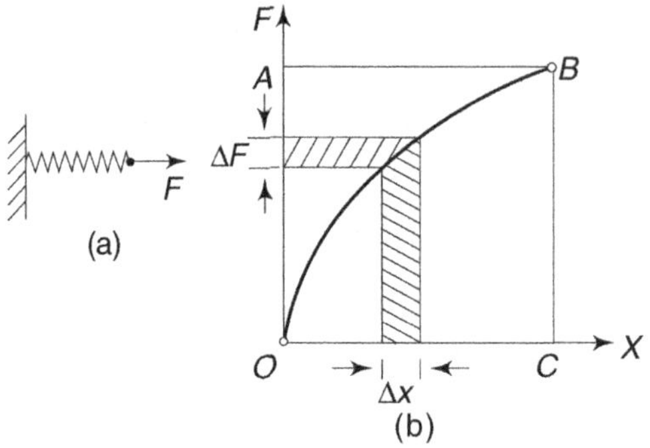

Fig. 5.26 *(a) Non-linear spring; (b) Non-linear load-displacement curve*

Hence $\frac{dU}{dx} = F$

This is the principle of virtual work, discussed in Sec. 5.14, and is applicable whether the elastic member is linear or non-linear. Now consider the area *OAB*. It is represented by

$$U^* = \int_0^F x\, dF \tag{5.33}$$

This is termed as a complementary energy. Differentiating the complementary energy with respect to *F* yields

$$\frac{dU^*}{dF} = x \tag{5.34}$$

This gives the deflection in the direction of *F*. If we compare with Castigliano's first theorem (Eq. 5.16), we notice that to obtain the corresponding deflection, we must take the derivative of the complementary energy and not that of the strain energy. When a material obeys Hooke's law, the curve *OB* is a straight line and consequently, the strain energy and the complementary strain energy are equal and it becomes immaterial which one we use in Castigliano's first theorem. The expression given by Eq. (5.34) represents Engesser's theorem.

Consider as an example an elastic spring the force deflection characteristic of which is represented by

$$F = ax^n$$

where a and n are constants.

The strain energy is

$$U = \int_0^x F\, dx = \int_0^x a(x')^n\, dx' = \frac{1}{n+1} ax^{n+1}$$

The complimentary strain energy is

$$U^* = \int_0^F x\, dF = \int_0^F \left(\frac{F}{a}\right)^{1/n} dF$$

$$= \frac{1}{a^{1/n}} \cdot \frac{n}{n+1} F^{(1+1/n)}$$

From these $\frac{dU}{dx} = ax^n = F$

$$\frac{dU^*}{dF} = \frac{1}{a^{1/n}} \cdot F^{1/n} = x$$

Further, expressing U in terms of F, we get

$$U = \frac{1}{n+1} \cdot a\left[\frac{1}{a^{1/n}} \cdot F^{1/n}\right]^{n+1}$$

$$\therefore \qquad \frac{dU}{dF} = \frac{1}{n}\left(\frac{F}{a}\right)^{1/n} = \frac{1}{n}x$$

and this does not agree with the correct result. Hence the principle of virtual work is valid both for linear and non-linear elastic material, whereas to obtain deflection using Castigliano's first theorem, we have to use the complementary energy U^* if the material is non-linear. If it is linearly elastic, it is immaterial wheather we use U or U^*, since both are equal.

Example 5.17 *Consider Fig. 5.27, which shows two identical bars hinged together, carrying a load W. Check Castigliano's first theorem, using the elastic and complementary strain energy.*

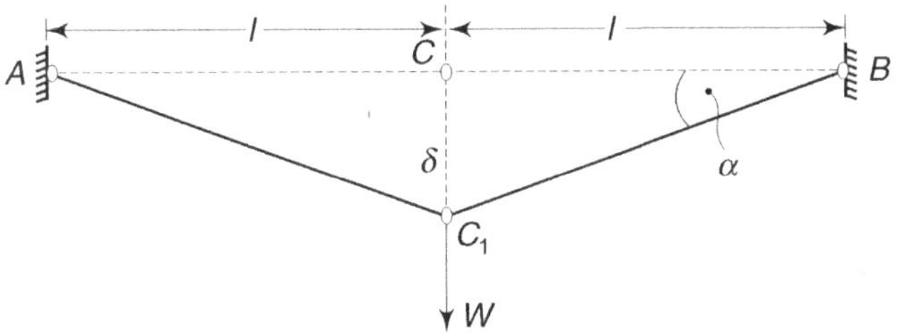

Fig. 5.27 *Example 5.17*

Solution When C has displacement $CC_1 = \delta$, we have from the figure for small α,

$$\tan\alpha \approx \sin\alpha \approx \delta/l$$

If F is the force in each member, a the cross-sectional area and ε the strain, then

$$F = \frac{W}{2\sin\alpha} \approx \frac{Wl}{2\delta}$$

and $$\varepsilon = \frac{\sqrt{l^2+\delta^2}-l}{l} \approx \frac{1}{2}\frac{\delta^2}{l^2}$$

Also $$\varepsilon = \frac{F}{aE} = \frac{Wl}{2\delta aE}$$

Equating the two strains

$$\frac{Wl}{2\delta aE} = \frac{\delta^2}{2l^2}$$

or $$\delta = l\left(\frac{W}{Ea}\right)^{1/3}$$

i.e. the deflection is not linearly related to the load.

The strain energy is

$$U = \int_0^{\delta} W d\delta = \frac{lW^{4/3}}{(aE)^{1/3}}$$

$$\therefore \qquad \frac{\partial U}{\partial W} = \frac{4lW^{1/3}}{3(aE)^{1/3}}$$

Hence, Castigliano's first theorem applied to the strain energy, does not yield the deflection δ. This is so because the load defection equation is not linearly related. If we consider the complementary energy,

$$U^* = \int_0^W \delta dW = \frac{l}{(Ea)^{1/3}} \int_0^W W^{1/3}\, dW$$

$$= \frac{3lW^{4/3}}{4(Ea)^{1/3}}$$

$$\frac{\partial U^*}{\partial W} = l\left(\frac{W}{Ea}\right)^{1/3} = \delta$$

Hence, Engesser's theorem gives the correct result.

5.18 MAXWELL–MOHR INTEGRALS

Castigliano's first theorem gives the displacement of points in the directions of the external forces where they are acting. When a displacement is required at a point where no external force is acting, a fictitious force in the direction of the required displacement is assumed at the point, and in the final result, the value of the fictitious load is considered equal to zero. This technique was discussed in Sec. 5.11. In this section, we shall develop certain integrals, which are based on the fictitious load techniques.

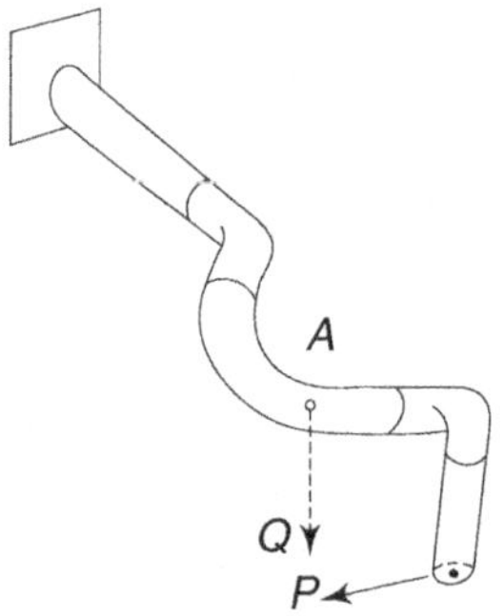

Fig. 5.28 *A general structure under load P*

Consider the determination of the vertical displacement of point A of a structure which is loaded by a force P, as shown in Fig. 5.28. Since no external force is acting at A in the corresponding direction, we apply a fictitious force Q in the corresponding direction at A. In order to calculate the strain energy in the elastic member, we need to determine the moments and forces across a general section. This is shown in Fig. 5.29.

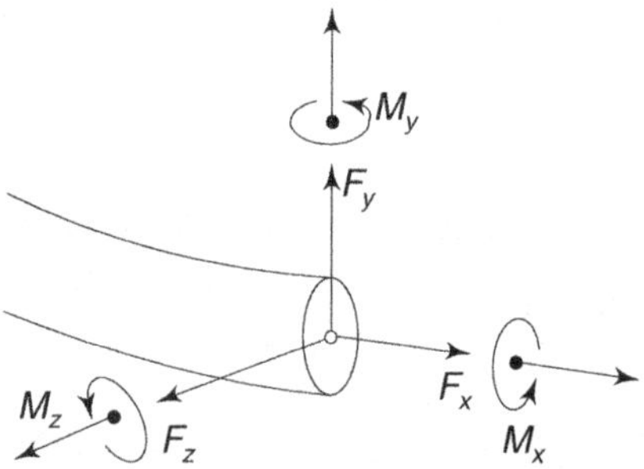

Fig. 5.29 *Moments and forces across a general section*

At any section, the moments and forces of reaction are caused by the actual external forces plus the fictitious load Q. For example, about the x axis we have

$$F_x = F_{xP} + F_{xQ},$$
$$M_x = M_{xP} + M_{xQ}$$

where F_{xP} is caused by the actual external forces, such as P, and F_{xQ} is due to the fictitious load Q. It is essential to observe that the additional force factors, such as F_{xQ}, M_{xQ}, etc. are directly proportional to Q. If Q is doubled, these factors also get doubled. Hence, one can write these as $F_{x1}Q$, $M_{x1}Q$, etc. where F_{x1}, M_{x1}, etc. are the force factors caused by a unit fictitious generalised force. Consequently, the force factors due to the actual loads and fictitious force are

$$\begin{aligned} F_x &= F_{xP} + F_{x1}Q, \qquad & M_x &= M_{xP} + M_{x1}Q \\ F_y &= F_{yP} + F_{y1}Q, \qquad & M_y &= M_{yP} + M_{y1}Q \\ F_z &= F_{zP} + F_{z1}Q, \qquad & M_z &= M_{zP} + M_{z1}Q \end{aligned} \tag{5.35}$$

Note that in Fig. 5.29 while M_x acts as a torque, M_y and M_z act as bending moments. These force factors vary from section to section. The total elastic energy is

$$U = \int_l \frac{(M_{xP} + M_{x1}Q)^2\, ds}{2GI_x} + \int_l \frac{(M_{yP} + M_{y1}Q)^2\, ds}{2EI_y} + \int_l \frac{(M_{zP} + M_{z1}Q)^2\, ds}{2EI_z} + \int_l \frac{(F_{xP} + F_{x1}Q)^2\, ds}{2EA} + \int_l \frac{k_y (F_{yP} + F_{y1}Q)^2\, ds}{2GA} + \int_l \frac{k_z (F_{zP} + F_{z1}Q)^2\, ds}{2GA}$$

Differentiating the above expression with respect to Q and putting $Q = 0$

$$\delta_A = \left.\frac{\partial U}{\partial Q}\right|_{Q=0} = \int_l \frac{M_{xP}\, M_{x1}\, ds}{GI_x} + \int_l \frac{M_{yP}\, M_{y1}\, ds}{EI_y} + \int_l \frac{M_{zP}\, M_{z1}\, ds}{EI_z} + \int_l \frac{F_{xP}\, F_{x1}\, ds}{EA} + \int_l \frac{k_y\, F_{yP}\, F_{y1}\, ds}{GA} + \int_l \frac{k_z\, F_{zP}\, F_{z1}\, ds}{GA} \tag{5.36}$$

If the fictitious force Q is replaced by a fictitious moment or torque, we get the corresponding deflection θ_A.

These sets of integrals are known as Maxwell–Mohr integrals. The above method is sometimes known as the unit load method. These integrals can be used to solve not only problems of finding displacements but also to solve problems connected with plane thin-walled rings. The above set of equations is generally written as

$$\delta_A = \int_l \frac{M_x \overline{M}_x}{GI_x} ds + \int_l \frac{M_y \overline{M}_y}{EI_y} ds + \int_l \frac{M_z \overline{M}_z}{EI_z} ds + \int_l \frac{F_x \overline{F}_x}{EA} ds + \int_l \frac{k_y F_y \overline{F}_y}{GA} ds + \int_l \frac{k_z F_z \overline{F}_z}{GA} ds \tag{5.37}$$

where $\bar{M}_x$, $\bar{M}_y$, . . . , $\bar{M}_z$ are the force factors caused by a generalised unit fictitious force applied where the appropriate displacement is needed.

Example 5.18 *Determine by what amount the straight portions of the ring are bought closer together when it is loaded, as shown in Fig. 5.30 consider only the bending energy.*

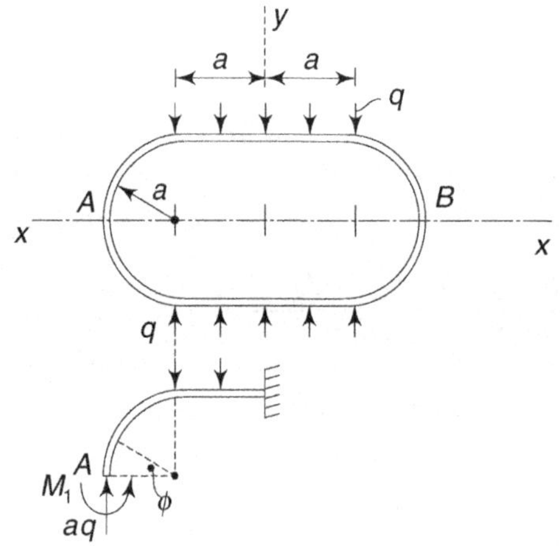

Fig. 5.30 *Example 5.18*

Solution Consider one quarter of the ring. The unknown moment M_1 is the redundant unknown generalised force. Owing to symmetry, there is no rotation of the section at point A. To determine the rotation, we assume a unit moment in the same direction as M_1. The moment due to this fictitious unit moment at any section is $\bar{M}$.

M at any section in quadrant = $aq \,.\, a\,(1 - \cos\phi) - M_1$

$\bar{M}$ at any section in quadrant = -1

M at any section in the top horizontal member = $aq\,(a + x) \;- qx^2/2 - M_1$ $\bar{M}$ at any section in the top horizontal member = -1

$$\therefore \qquad \theta_A = \int_0^{\pi/2} \frac{-a^2q(1-\cos\phi)+M_1}{EI}\,a\,d\phi - \int_0^{a} \frac{a\,q(a+x)-qx^2/2-M_1}{EI}\,dx$$

or
$$EI\theta_A = -a^3q\left(\frac{\pi}{2}+\frac{1}{3}\right)+M_1a\left(\frac{\pi}{2}+1\right)=0$$

$$\therefore \qquad M_1 = a^2q\,\frac{3\pi+2}{3(\pi+2)} \approx 0.74\,a^2q$$

This is the value of the redundant unknown moment. To determine the vertical displacements of the midpoints of the horizontal members, we apply a fictitious force $P_f = 1$ in an upward direction at point A of the quarter ring. Because of this

$\bar{M}$ at any section in quadrant = $-a\,(1 - \cos\phi)$

$\bar{M}$ at any section in top horizontal part = $-(a + x)$

Hence, the vertically upward displacement of point A is

$$\delta_A = \int_0^{\pi/2} -\frac{a^4q(1-\cos\phi-0.74)(1-\cos\phi)}{EI}\,d\phi$$

$$+\int_0^{a} -\frac{\left[aq\,(a+x)-\frac{1}{2}qx^2-0.74a^2q\right](a+x)}{EI}\,dx$$

$$= \frac{0.86\,a^4 q}{EI}$$

Hence, the two horizontal members approach each other by a distance equal to

$$\frac{2\,(0.86)\,a^4 q}{EI} = 1.72\,\frac{a^4 q}{EI}$$

Example 5.19 *A thin walled circular ring is loaded as shown in Fig. 5.31. Determine the vertical displacement of point A. Take only the bending energy.*

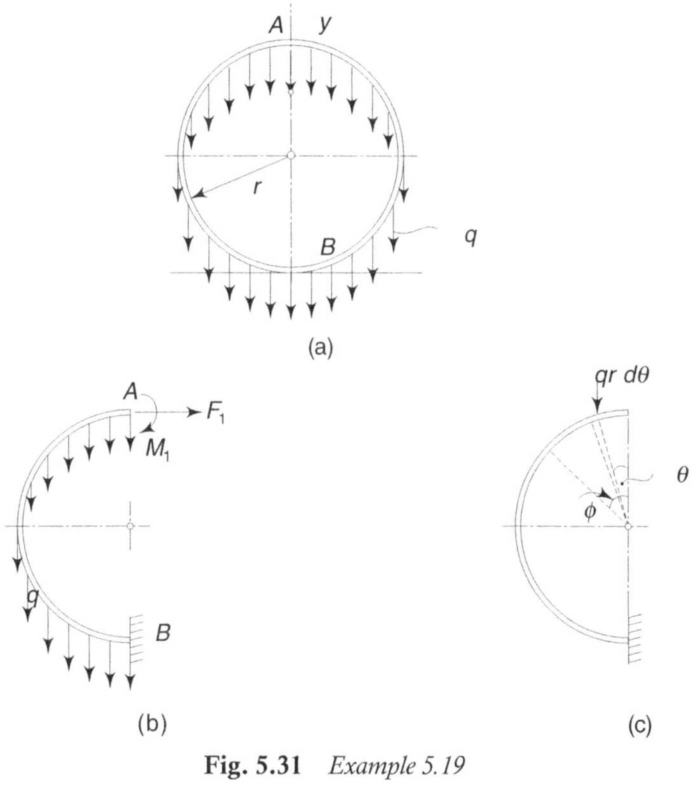

Fig. 5.31 *Example 5.19*

Solution Because of symmetry, we may consider one half of the ring. The reactive forces at section A are F_1 and M_1. Because of symmetry, section A does not rotate and also does not have a horizontal displacement. Hence in addition to M_1 and F_1, we assume a fictitious moment and a fictitious horizontal force, each of unit magnitude at section A.

The moment at any section ϕ due to the distributed loading q is

$$M_q = \int_0^{\phi} qr\,d\theta\, r\,(\sin\phi - \sin\theta) = qr^2(\phi\sin\phi + \cos\phi - 1)$$

M at any section ϕ with distributed loading F_1 and M_1 is

$$M = qr^2\,(\phi\sin\phi + \cos\phi - 1) + M_1 + F_1 r(1 - \cos\phi)$$

$\bar{M}$ at any section ϕ due to the unit fictitious horizontal force is

$$\bar{M} = r(1 - \cos\phi)$$

$$\therefore \quad \delta_A = \frac{I}{EI}\int_0^{\pi} r^2\left[qr^2(\phi\sin\phi + \cos\phi - 1)\right.$$

$$\left. + M_1 + F_1 r(1-\cos\phi)\right](1-\cos\phi)\,d\phi$$

$$= \frac{r^2}{EI}\left(-qr^2\frac{\pi}{4} + \pi M_1 + F_1 r\frac{3\pi}{2}\right)$$

Since this is equal to zero, we have

$$M_1 + \frac{3}{2}F_1 r = \frac{1}{4}qr^2 \tag{5.38}$$

$\bar{M}$ at any section ϕ due to unit fictitious moment is

$$\bar{M} = 1$$

$$\therefore \quad \theta_A = \frac{I}{EI}\int_0^{\pi} r\left[qr^2(\phi\sin\phi + \cos\phi - 1) + M_1 + F_1 r(1-\cos\phi)\right]d\phi$$

$$= \frac{r}{EI}(\pi M_1 + F_1 r\pi)$$

Since this is also equal to zero, we have

$$M_1 + F_1 r = 0 \tag{5.39}$$

Solving Eqs (5.38) and (5.39)

$$M_1 = -\frac{qr^2}{2} \quad \text{and} \quad F_1 = \frac{qr}{2}$$

To determine the vertical displacement of A we apply a fictitious unit force $P_f = 1$ at A in the downward direction.

$\bar{M}$ at any section ϕ due to $P_f = 1$ is $r\sin\phi$

$$\therefore \quad \delta_v = \int_0^{\pi} r^2\left[qr^2(\phi\sin\phi + \cos\phi - 1) + M_1 + F_1 r(1-\cos\phi)\right]\sin\phi\,d\phi$$

$$= \left(\frac{\pi^2}{4} - 2\right)\frac{qr^4}{EI} \approx 0.467\,\frac{qr^4}{EI}$$

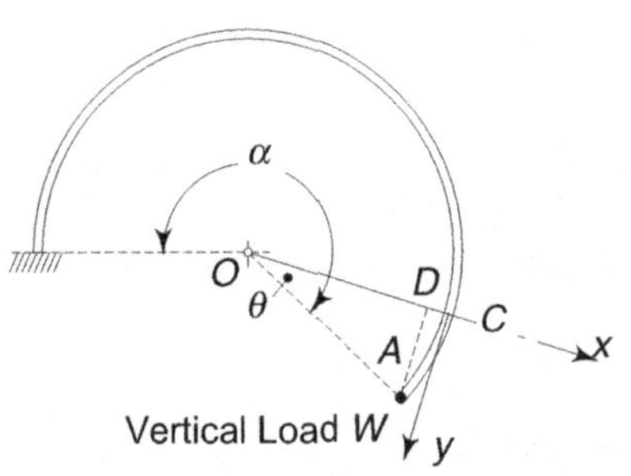

Fig. 5.32 *Example 5.20*

Example 5.20 *Figure 5.32 shows a circular member in its plan view. It carries a vertical load W at A perpendicular to the plane of the paper. Taking only bendng and torsional energies into account, determine the vertical deflection of the loaded end A. The radius of the member is R and the member subtends an angle α at the centre.*

Solution At section C, the moment of the force about x axis acts as bendng moment M and the moment about y axis acts as torque T. Hence,

$$M = W \times AD = WR \sin\theta$$

$$T = W \times DC = WR\,(1 - \cos\theta)$$

$$\therefore \qquad U = \int_0^{\alpha} \frac{1}{2EI}(WR\sin\theta)^2 R\,d\theta + \int_0^{\alpha} \frac{1}{2GJ}\left[WR(1-\cos\theta)\right]^2 R\,d\theta$$

When the load W is gradually applied, the work done by W during its vertical deflection is $\frac{1}{2}\, W\,\delta_V$ and this is stored as the elastic energy U. Thus,

$$\frac{1}{2}W\delta_V = U = \int_0^{\alpha} \frac{1}{2EI}(WR\sin\theta)^2 R\,d\theta + \int_0^{\alpha} \frac{1}{2GJ}\left[WR(1-\cos\theta)\right]^2 R\,d\theta$$

or
$$\delta_V = WR^3\left[\frac{1}{2EI}\left(\alpha - \frac{1}{2}\sin 2\alpha\right) + \frac{1}{GJ}\left(\frac{3}{2}\alpha + \frac{1}{4}\sin 2\alpha - 2\sin\alpha\right)\right]$$

This is the same as $\partial U/\partial W$.

if
$$\alpha = \frac{\pi}{2}, \quad \delta_V = WR^3\left[\frac{\pi}{4EI} - \frac{1}{GJ}\left(\frac{3\pi - 8}{4}\right)\right]$$

if
$$\alpha = \pi, \quad \delta_V = WR^3\,\pi\left(\frac{1}{EI} - \frac{3}{GJ}\right)$$

Problems

5.1 *A* load $P = 6000$ N acting at point R of a beam shown in Fig. 5.33 produces vertical deflections at three points A, B, and C of the beam as

$$\delta_A = 3 \text{ cm} \qquad \delta_B = 8 \text{ cm} \qquad \delta_C = 5 \text{ cm}$$

Find the deflection of point R when the beam is loaded at points, A, B and C by

$$P_A = 7500 \text{ N},\ P_B = 3500 \text{ N and } P_C = 5000 \text{ N}.$$

[*Ans.* 12.6 cm (approx.)]

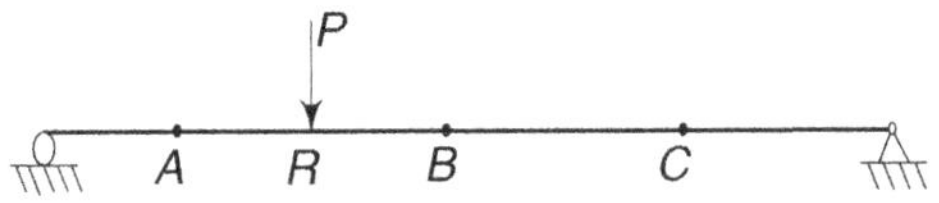

Fig. 5.33 *Problem 5.1*

5.2 For the horizontal beam shown in Fig. 5.34, a vertical displacement of 0.6 cm of support B causes a reaction $R_a = 10{,}000$ N at A. Determine the reaction R_b at B due to a vertical displacement of 0.8 cm at support A. [*Ans.* $R_b = 13{,}333$ N]

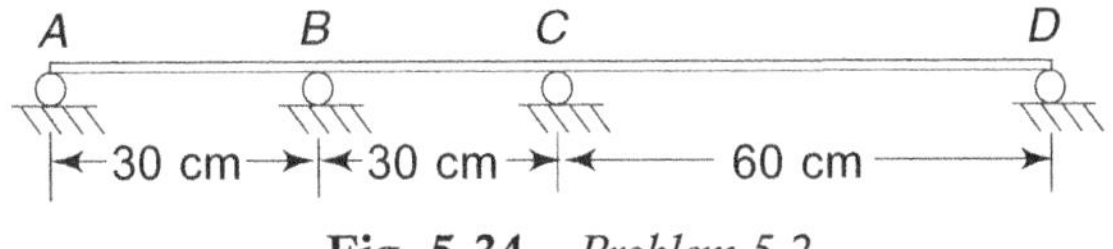

Fig. 5.34 *Problem 5.2*

5.3 A closed circular ring made of inextensible material is subjected to an arbitrary system of forces in its plane. Show that the area enclosed by the frame does not change under this loading. Assume small displacements (Fig. 5.35).

[*Hint:* Subject the ring to uniform internal pressure. Since the material is inextensible, no deformation occurs. Now apply the reciprocal theorem.]

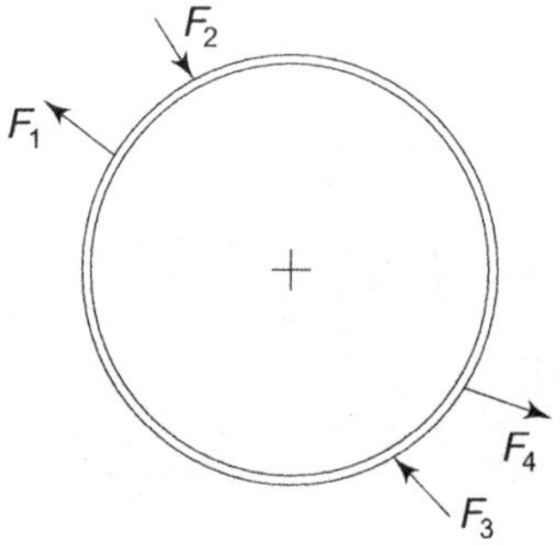

Fig. 5.35 *Problem 5.3*

5.4 Determine the vertical displacement of point A for the structure shown in Fig. 5.36. All members have the same cross-section and the same rigidity EA.

$$\left[Ans.\ \delta_A = \frac{Wl}{EA}\left(7 + 4\sqrt{2}\right) \right]$$

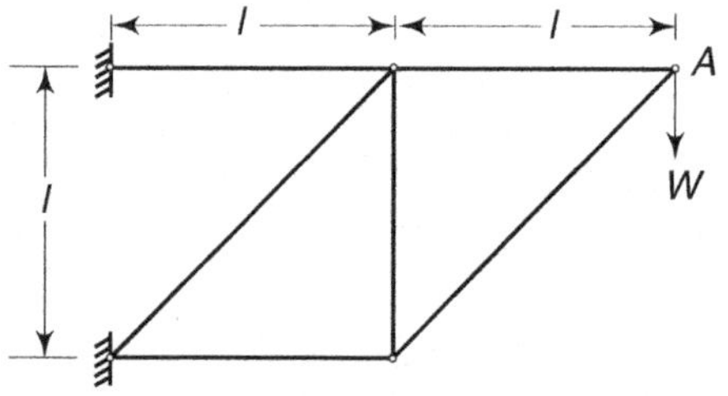

Fig. 5.36 *Problem 5.4*

5.5 Determine the rotation of point C of the beam under the action of a couple M applied at its centre (Fig. 5.37).

$$\left[Ans.\ \theta = \frac{Ml}{12EI} \right]$$

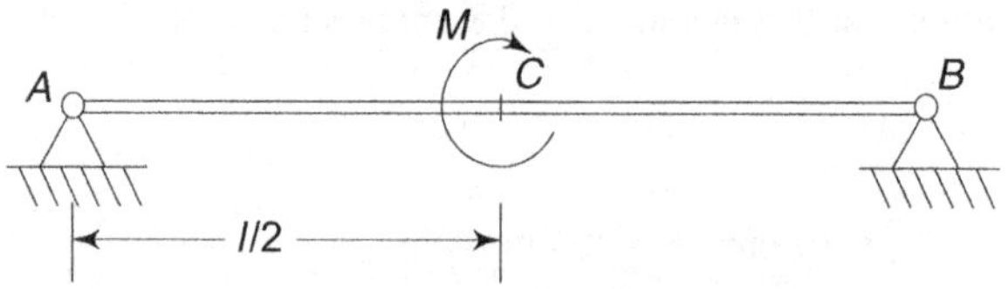

Fig. 5.37 *Problem 5.5*

5.6 What is the relative displacement of points A and B in the framework shown? Consider only bending energy (Fig. 5.38).

$$\left[Ans.\ \delta_{AB} = \frac{Pa^3}{6EI} + \frac{Pa^2 b}{2EI} \right]$$

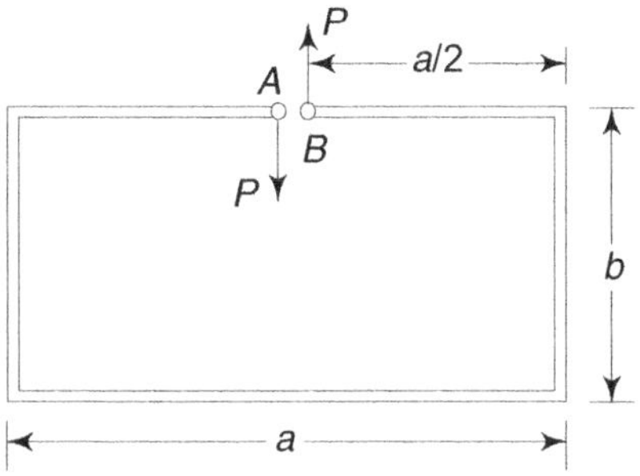

Fig. 5.38 *Problem 5.6*

5.7 What is the relative displacement of points A and B when subjected to forces P. Consider only bending energy (Fig. 5.39).

$$\left[Ans.\ \delta_{AB} = 3\pi \frac{PR^3}{EI} \right]$$

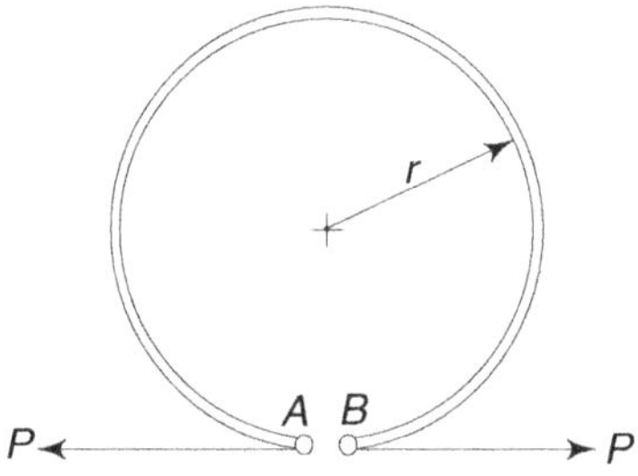

Fig. 5.39 *Problem 5.7*

5.8 Determine the vertical displacement of the point of application of force P. Take all energies into account. The member is of uniform circular cross-section (Fig. 5.40).

$$\left[Ans.\ \ \delta_A = 2P\left(\frac{a^3}{3EI} + \frac{a^2 b}{2EI} + \frac{a^3}{2GI_P} + \frac{ka}{AG} + \frac{b}{2AE} \right) \right]$$

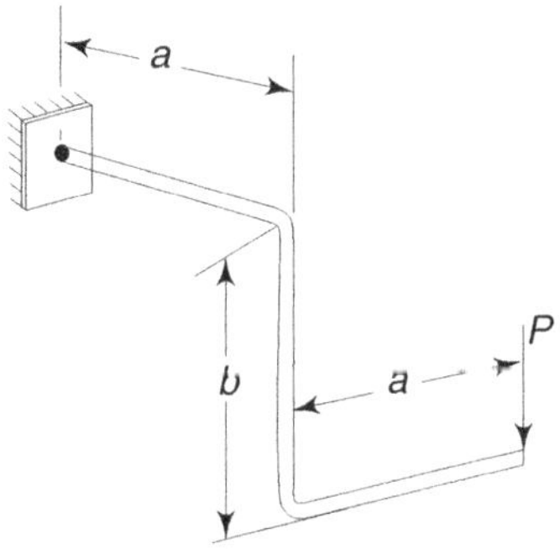

Fig. 5.40 *Problem 5.8*

5.9 What are the horizontal and vertical displacements of point A in Fig. 5.41. Assume AB to be rigid.

$$\left[Ans.\ \ \delta_V = \frac{17Ph}{EA};\ \delta_H = \frac{1.73Ph}{EA}\right]$$

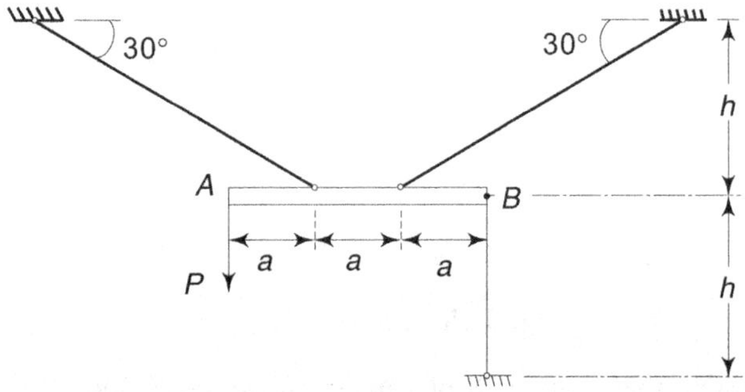

Fig. 5.41 *Problem 5.9*

5.10 Determine the vertical displacement of point B under the action of W. End B is free to rotate but can move only in a vertical direction (Fig. 5.42).

$$\left[Ans.\ \ \delta_B = \frac{Wa^3}{EI}\left(\frac{3\pi}{4} - \frac{1}{9\pi + 8}\right)\right]$$

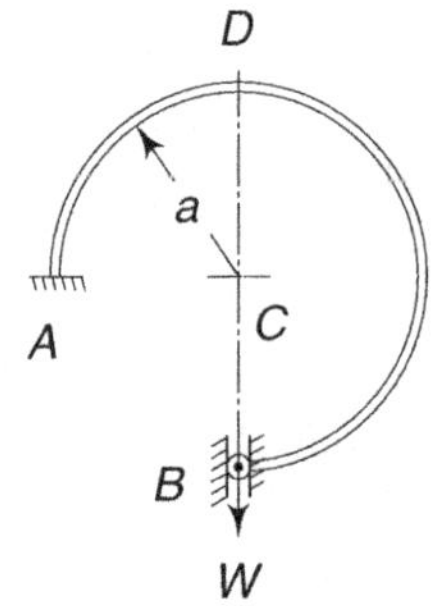

Fig. 5.42 *Problem 5.10*

5.11 Two conditions must be satisfied by an ideal piston ring. (a) It should be truly circular when in the cylinder, and (b) it should exert a uniform pressure all around. Assuming that these conditions are satisfied by specifying the initial shape and the cross-section, show that the initial gap width must be $3\pi pr^4/EI$, if the ring is closed inside the cylinder. p is the uniform pressure per centimetre of circumference. EI is kept constant.

5.12 For the torque measuring device shown in Fig. 5.43 determine the stiffness of the system, i.e. the torque per unit angle of twist of the shaft. Each of the springs has a length l and moment of inertia I for bending in the plane of the moment.

$$\left[Ans.\ \ \frac{M}{\theta} \approx \frac{8EI}{l}\right]$$

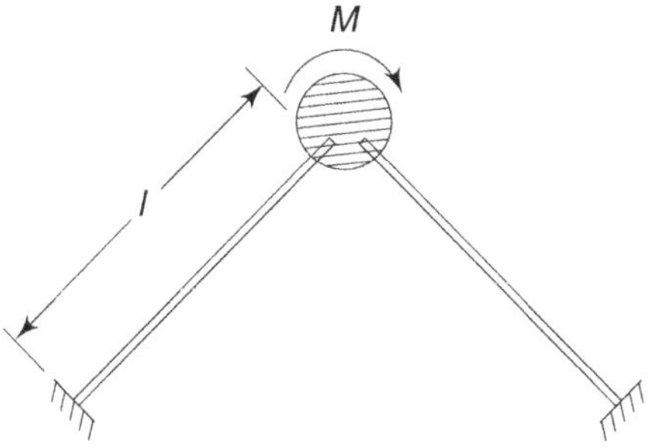

Fig. 5.43 *Problem 5.12*

5.13 A circular steel hoop of square cross-section is used as the controlling element of a high speed governor (Fig. 5.44). Show that the vertical deflection caused by angular velocity ω is given by

$$\delta = \frac{2\rho}{E}\frac{\omega^2 r^5}{t^2}$$

where r is the hoop radius, t the thickness of the section and ρ the weight density of the material.

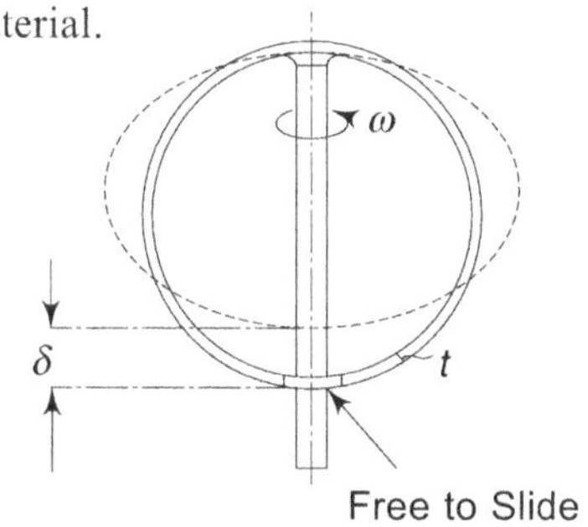

Fig. 5.44 *Problem 5.13*

5.14 A thin circular ring is loaded by three forces P as shown in Fig. 5.45. Determine the changes in the radius of the ring along the line of action of the forces. The included angle between any two forces is 2α and A is the cross-sectional area of the member. Consider both bending and axial energies.

$$\left[Ans. \ \frac{PR^3}{2EI}\left(\cot\frac{\alpha}{2} + \frac{\alpha}{2\sin^2\alpha} - \frac{1}{\alpha}\right) + \frac{PR}{4EA}\left(\cot\alpha + \frac{\alpha}{\sin^2\alpha}\right)\right]$$

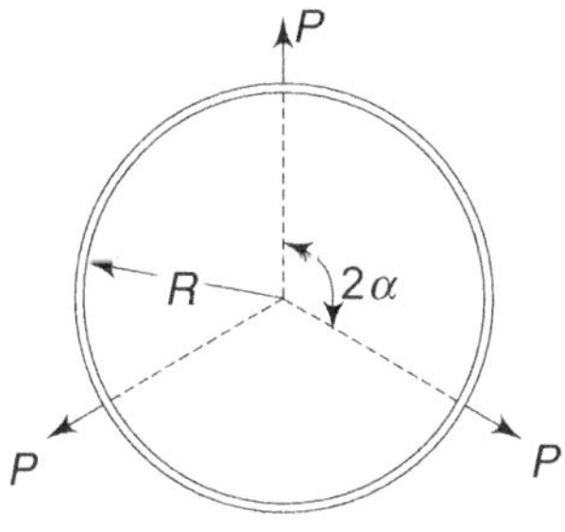

Fig. 5.45 *Problem 5.14*

5.15 For the system shown (Fig. 5.46) determine the load W necessary to cause a displacement δ in the vertical direction of point O. a is the cross-sectional area of each member and l is the length of each member. Use the principle of virtual work.

$$\left[Ans.\ W = \frac{3aE}{2l}\delta \right]$$

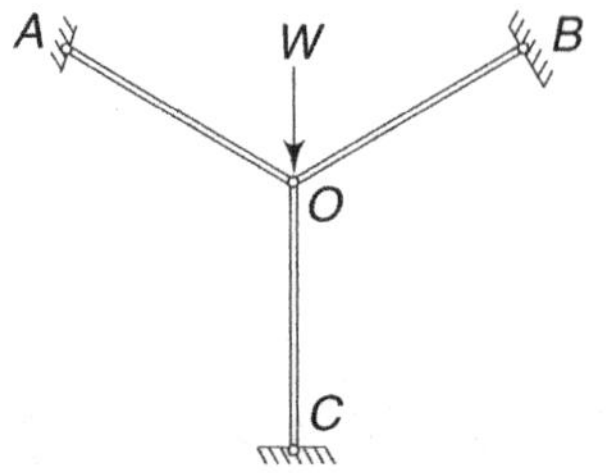

Fig. 5.46 *Problem 5.15*

5.16 In the previous problem determine the force in the member OC by Castigliano's second theorem. [*Ans.* 2W/3]

5.17 Using Castigliano's second theorem, determine the reaction of the vertical support C of the structure shown (Fig. 5.47). Beam ACB has Young's modulus E and member CD has a value E'. The cross-sectional area of CD is a.

$$\left[Ans. \frac{5wl^4aE'}{4\left(6EIh + qE'\,l^3\right)} \right]$$

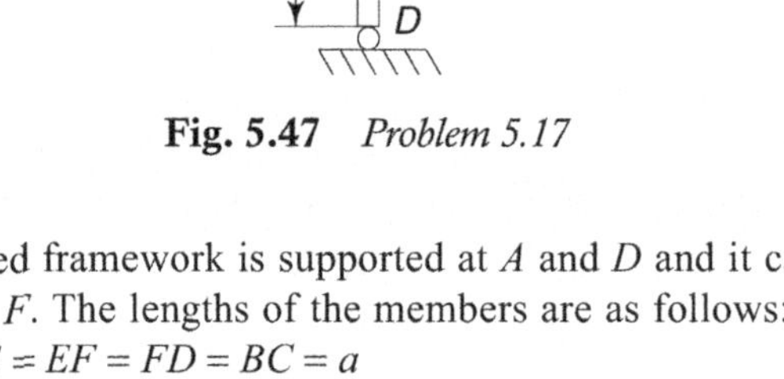

Fig. 5.47 *Problem 5.17*

5.18 A pin jointed framework is supported at A and D and it carries equal loads W at E and F. The lengths of the members are as follows:

$AE = EF = FD = BC = a$

$BE = CF = h$

$BF = CE = AB = CD = l = (a^2 + h^2)^{1/2}$

The cross-sectional areas of BF and CE are A_1 each, and of all the other members are A_2 each. Determine the tensions in BF and CE.

$$\left[Ans. \frac{WA_1\,lh^2}{A_1\left(a^3 + h^3\right) + A_2 l^3} \right]$$

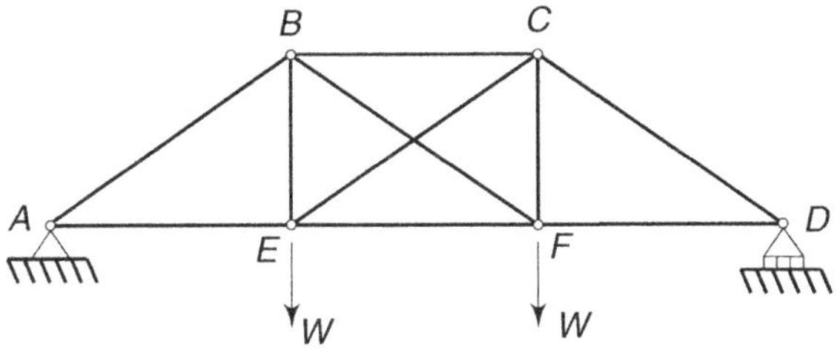

Fig 5.48 *Problem 5.18*

5.19 A ring is made up of two semi-circles of radius a and of two straight lines of length $2a$, as shown in Fig. 5.49. When loaded as shown, determine the change in distance between A and B. Consider only bending energy.

$$\left[Ans. \frac{6-17\pi-6\pi^2}{12(2+\pi)} \cdot \frac{qa^4}{EI} \right]$$

5.20 Determine reaction forces and moments at the fixed ends and also the vertical deflection of the point of loading. Assume $G = 0.4E$ (Fig. 5.50).

$$\left[\begin{array}{l} Ans.\ M = \dfrac{Pa}{2};\ T = 0.387\ Pa \\ \qquad \delta = 0.711 \dfrac{Pa^3}{EI} \end{array} \right]$$

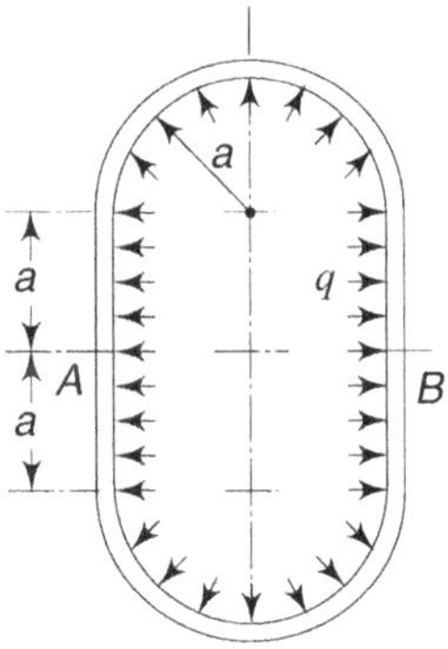

Fig. 5.49 *Problem 5.19*

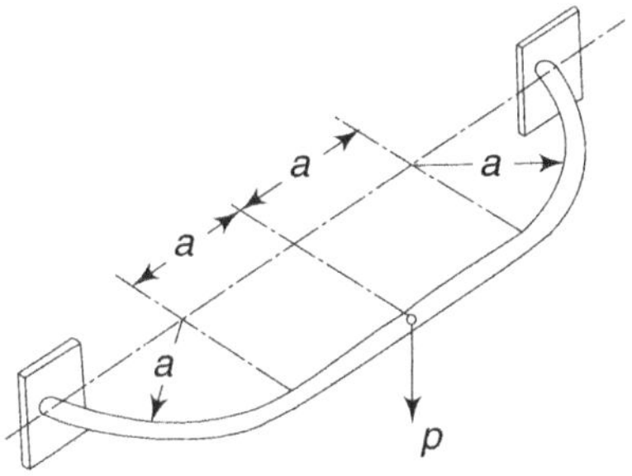

Fig. 5.50 *Problem 5.20*

5.21 A semi-circular member shown in Fig. 5.51 is subjected to a torque T at A. Determine the reactive moments at the built-in ends B and C. Also determine the vertical deflection of A.

$$\left[\begin{array}{l} Ans.\ M = \dfrac{T}{2};\ \text{Torque} = -\dfrac{T}{9\pi} \\ \delta_V = \dfrac{R^2T}{8EI}\left(\dfrac{9\pi}{4} + \dfrac{1}{\pi} - 5\right) \end{array} \right]$$

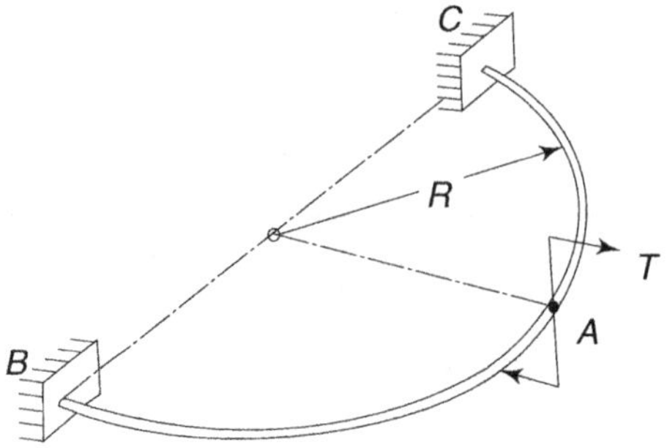

Fig. 5.51 *Problem 5.21*

5.22 In Example 5.12 determine the change in the horizontal diameter

$$\left[Ans.\ \delta_h = -\frac{Pr^3}{EI}\left(\frac{2}{\pi} - \frac{1}{2}\right) \right]$$

CHAPTER

6

Bending of Beams

6.1 INTRODUCTION

In this chapter we shall consider the stresses in and deflections of beams having a general cross-section subjected to bending. In general, the moments causing bending are due to lateral forces acting on the beams. These lateral forces, in addition to causing bending or flexural stresses in transverse sections of the beams, also induce shear stresses.

Flexural stresses are normal to the section. The effects of transverse shear stresses will be discussed in Sec. 6.4–6.6. Because of pure bending moments, only normal stresses are induced. In elementary strength of materials only beams having an axis of symmetry are usually considered. Figure 6.1 shows an initially straight beam having a vertical section of symmetry and subjected to a bending moment acting in this plane of symmetry.

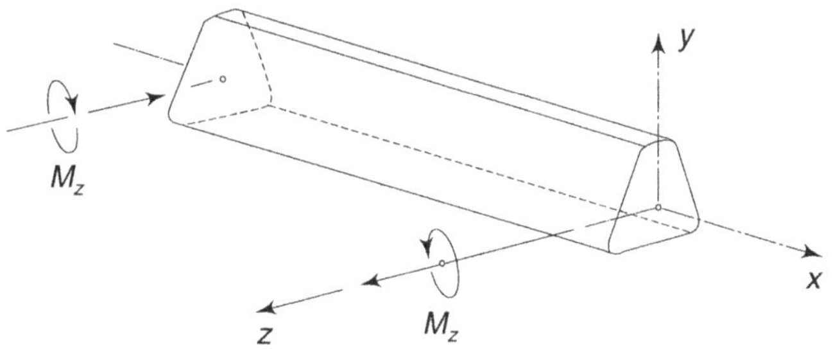

Fig. 6.1 *Beam with a vertical section of symmetry subjected to bending*

The plane of symmetry is the xy plane and the bending moment M_z acts in this plane. Owing to symmetry the beam bends in the xy plane. Assuming that the sections that are plane before bending remain so after bending, the flexural stress σ_x is obtained in elementary strength of materials as

$$\sigma_x = -\frac{M_z y}{I_z} \tag{6.1}$$

The origin of the co-ordinates coincides with the centroid of the cross-section and the z axis coincides with the neutral axis. The minus sign is to take care of the sign of the stress. A positive bending moment M_z, as shown, produces a compressive stress at a point with the positive y co-ordinate. I_z is the area moment of inertia about the neutral axis which passes through the centroid. Further, if E is the Young's modulus of the beam material and R the radius of curvature of the bent beam, the equations from elementary strength of materials give,

$$\frac{M_z}{I_z} = -\frac{\sigma_x}{y} = \frac{E}{R} \tag{6.2}$$

The above set of equation is usually called Euler–Bernoulli equations or Navier–Bernoulli equations.

6.2 STRAIGHT BEAMS AND ASYMMETRICAL BENDING

Now we shall consider the bending of initially straight beams having a uniform cross-section. There are three general methods of solving this problem. We shall consider each one separately. When the bending moment acts in the plane of symmetry, the beam is said to be under symmetrical bending. Otherwise it is said to be under asymmetrical bending.

Method 1 Figure 6.2 shows a beam subjected to a pure bending moment M_z lying in the xy plane. The moment is shown vectorially. The origin O is taken at the centroid of the cross-section. The x axis is along the axis of the beam and the z axis is chosen to coincide with the moment vector. It is once again assumed that sections that are plane before bending remain plane after bending. This is usually known as the Euler–Bernoulli hypothesis. This means that the cross-section will rotate about an axis such that one part of the section will be subjected to tensile stresses and the other part above this axis will be subjected to compression. Points lying on this axis will not experience any stress and consequently this axis is the neutral axis. In Fig. 6.2(b) this is represented by BB and it can be shown that it passes through the centroid O. For this, consider a small area ΔA lying at a distance y' from BB. Since the cross-section rotates about BB during bending, the stretch or contraction of any fibre will be proportional to the perpendicular distance from BB, Hence, the strain in any fibre is

$$\varepsilon_x = k'y'$$

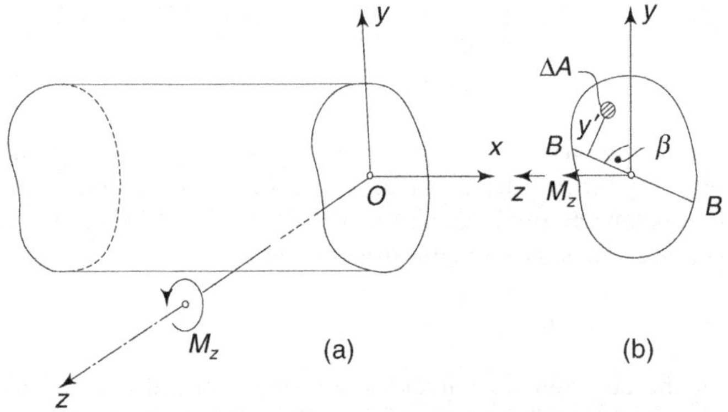

Fig. 6.2 *Beam with a general section subjected to bending*

where k' is some constant. Assuming only σ_x to be acting and $\sigma_y = \sigma_z = 0$, from Hooke's law,

$$\sigma_x = k'Ey' = ky' \tag{6.3}$$

where k is an appropriate constant. The force acting on ΔA is therefore,

$$\Delta F_x = ky' \, \Delta A$$

For equilibrium, the resultant normal force acting over the cross-section must be equal to zero. Hence, integrating the above equation over the area of the section,

$$k\iint y' \, dA = 0 \tag{6.4}$$

The above equation shows that the first moment of the area about BB is zero, which means that BB is a centroidal axis.

It is important to observe that the beam in general will not bend in the plane of the bending moment and the neutral axis BB will not be along the applied moment vector M_z. The neutral axis BB in general will be inclined at an angle β to the y axis. Next, we take moments of the normal stress distribution about the y and z axes. The moment about the y axis must vanish and the moment about the z axis should be equal to $-M_z$. The minus sign is because a positive stress at a positive (y, z) point produces a moment vector in the negative z direction. Hence

$$\iint \sigma_x z \, dA = \iint ky'z \, dA = 0 \tag{6.5a}$$

$$\iint \sigma_x y \, dA = \iint ky'y \, dA = -M_z \tag{6.5b}$$

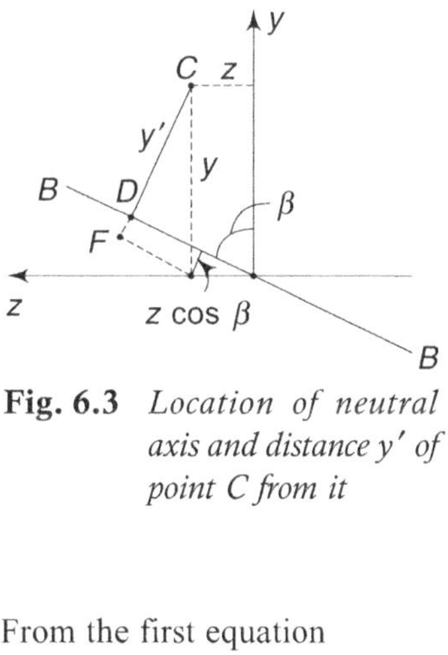

Fig. 6.3 *Location of neutral axis and distance y′ of point C from it*

y' can now be expressed in terms of y and z coordinates (Fig. 6.3) as

$$y' = CF - DF$$
$$= y \sin \beta - z \cos \beta$$

Substituting this in Eqs (6.5)

$$k\iint (yz \sin \beta - z^2 \cos \beta) \, dA = 0$$

and $$k\iint (y^2 \sin \beta - yz \cos \beta) \, dA = -M_z$$

i.e. $$I_{yz} \sin \beta - I_y \cos \beta = 0 \tag{6.6a}$$

and $$k(I_{yz} \cos \beta - I_z \sin \beta) = M_z \tag{6.6b}$$

From the first equation

$$\tan\beta = \frac{I_y}{I_{yz}} \tag{6.7}$$

This gives the location of the neutral axis BB.

Substituting for k from Eq. (6.6b) in Eq. (6.3)

$$\sigma_x = \frac{M_z (y \sin \beta - z \cos \beta)}{I_{yz} \cos \beta - I_z \sin \beta}$$

$$= \frac{y \tan \beta - z}{I_{yz} - I_z \tan \beta} M_z$$

Substituting for tan β from Eq. (6.7),

$$\sigma_x = \frac{yI_y - zI_{yz}}{I_{yz}^2 - I_y I_z} M_z \qquad (6.8)$$

The above equation helps us to calculate the normal stress due to bending. In summary, we conclude that when a beam with a general cross-section is subjected to a pure bending moment M_z, the beam bends in a plane which in general does not coincide with the plane of the moment. The neutral axis is inclined at an angle β to the y axis such that tan $\beta = I_y/I_{yz}$. The stress at any point (y, z) is given by Eq. (6.8).

Method 2 we observe from Eq. (6.7) that $\beta = 90°$ when $I_{yz} = 0$, i.e. if the y and z axes happen to be the principal axes of the cross-section. This means that if the y and z axes are the principal axes and the bending moment acts in the xy plane (i.e. the moment vector M_z is along one of the principal axes), the beam bends in the plane of the moment with the neutral axis coinciding with the z axis. Equation (6.8) then reduces to

$$\sigma_x = -\frac{M_z y}{I_z}$$

This is similar to the elementary flexure formula which is valid for symmetrical bending. This is so because for a symmetrical section, the principal axes coincide with the axes of symmetry. So, an alternative method of solving the problem is to determine the principal axes of the section; next, to resolve the bending moment into components along these axes, and then to apply thc elementary flexure formula. This procedure is shown in Fig. 6.4.

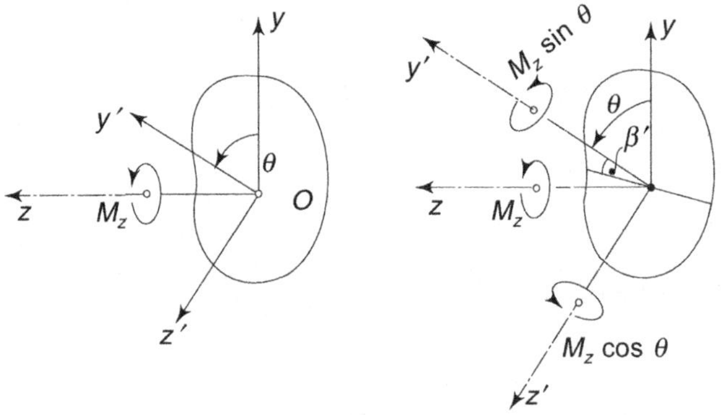

Fig. 6.4 *Resolution of bending moment vector along principal axes*

y and z axes are a set of arbitrary centroidal axes in the section. The bending moment M acts in the xy plane with the moment vector along the z axis. The principal axes Oy' and Oz' are inclined such that

$$\tan 2\theta = \frac{2I_{yz}}{I_z - I_y}$$

The moment resolved along the principal axes Oy' and Oz' are $M_y' = M_z \sin\theta$ and $M_z' = M_z \cos\theta$. For each of these moments, the elementary flexure formula can be used. With the principle of superposition,

$$\sigma_x = \frac{M_{y'} z'}{I_{y'}} - \frac{M_{z'} y'}{I_{z'}} \tag{6.9}$$

It is important to observe that with the positive axes chosen as in Fig. 6.4, a point with a positive y coordinate will be under compressive stress for positive $M_z' = M_z \cos\theta$. Hence, a minus sign is used in the equation.

The neutral axis is determined by equating σ_x to zero, i.e.

$$\frac{M_{y'} z'}{I_{y'}} - \frac{M_{z'} y'}{I_{z'}} = 0$$

or

$$\frac{z'}{y'} = \tan\beta' = \frac{M_{z'} I_{y'}}{M_{y'} I_{z'}} \tag{6.10}$$

The angle β' is with respect to the y' axis.

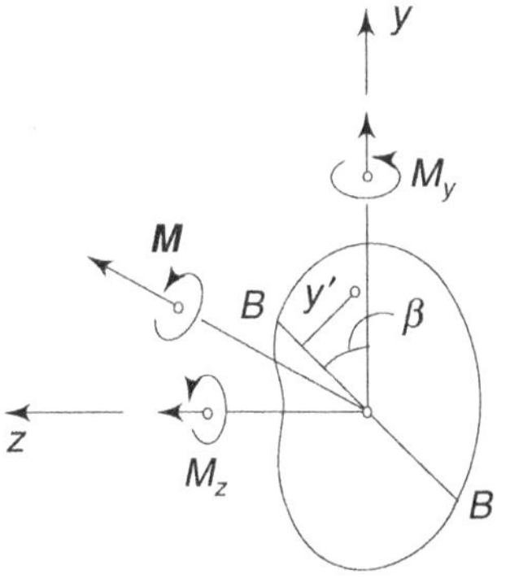

Fig. 6.5 *Resolution of bending moment vector along two arbitrary orthogonal axes*

Method 3 This is the most general method. Choose a convenient set of centroidal axes Oyz about which the moments and product of inertia can be calculated easily. Let $\boldsymbol{M}$ be the applied moment vector (Fig. 6.5).

Resolve the moment vector $\boldsymbol{M}$ into two components M_y and M_z along the y and z axes respectively. We assume the Euler-Bernoulli hypothesis, according to which the sections that were plane before bending remain plane after bending. Hence, the cross-section will rotate about an axis, such as BB. Consequently, the strain at any point in the cross-section will be proportional to the distance from the neutral axis BB.

$$\varepsilon_x = k'y'$$

Assuming that only σ_x is non-zero,

$$\sigma_x = Ek'y' = ky' \tag{a}$$

where k is some constant. For equilibirum, the total force over the cross-section should be equal to zero, since only a moment is acting.

$$\iint \sigma_x \, dA = k \iint y' \, dA = 0$$

As before, this means that the neutral axis passes through the centroid O. Let β be the angle between the neutral axis and the y axis. From geometry (Fig. 6.3).

$$y' = y \sin\beta - z \cos\beta \tag{b}$$

For equilibrium, the moments of the forces about the axes should yield

$$\iint \sigma_x z\, dA = \iint ky' z\, dA = M_y$$

$$\iint \sigma_x y\, dA = \iint ky' y\, dA = -M_z$$

Substituting for y'

$$k \iint \left(yz \sin\beta - z^2 \cos\beta \right) dA = M_y$$

$$k \iint \left(y^2 \sin\beta - yz \cos\beta \right) dA = -M_z$$

i.e.

$$k\left(I_{yz} \sin\beta - I_y \cos\beta\right) = M_y \tag{6.11}$$

and

$$k\left(I_z \sin\beta - I_{yz} \cos\beta\right) = -M_z \tag{6.12}$$

The above two equations can be solved for k and β. Dividing one by the other

$$\frac{I_{yz} \sin\beta - I_y \cos\beta}{I_z \sin\beta - I_{yz} \cos\beta} = -\frac{M_y}{M_z}$$

or

$$\frac{I_{yz} \tan\beta - I_y}{I_z \tan\beta - I_{yz}} = -\frac{M_y}{M_z}$$

i.e.

$$\tan\beta = \frac{I_y M_z + I_{yz} M_y}{I_{yz} M_z + I_z M_y} \tag{6.13}$$

This gives the location of the neutral axis BB. Next, substituting for k from Eq. (6.11) into equations (a) and (b)

$$\sigma_x = \frac{M_y \left(y \sin\beta - z \cos\beta\right)}{I_{yz} \sin\beta - I_y \cos\beta}$$

$$= \frac{M_y \left(y \tan\beta - z\right)}{I_{yz} \tan\beta - I_y}$$

Substituting for $\tan\beta$ from Eq. (6.13)

$$\sigma_x = \frac{M_z \left(yI_y - zI_{yz}\right) + M_y \left(yI_{yz} - zI_z\right)}{I_{yz}^2 - I_y I_z} \tag{6.14}$$

When $M_y = 0$ the above equation for σ_x becomes equivalent to Eq. (6.8).

In recapitulation we have the following three methods to solve unsymmetrical bending.

Method 1 Let $\boldsymbol{M}$ be the applied moment vector.

Choose a centroidal set of axes Oyz such that the z axis is along the $\boldsymbol{M}$ vector. The stress σ_x at any point (y, z) is then given by Eq. (6.8). The neutral axis is given by Eq. (6.7).

Method 2 Let ***M*** be the applied moment vector.

Choose a centroidal set of axes $Oy'z'$, such that the y' and z' axes are the principal axes. Resolve the moment into components M_y' and M_z' along the principal axes. Then the normal stress σ_x at any point (y', z') is given by Eq. (6.9) and the orientation of the neutral axis is given by Eq. (6.10).

Method 3 Choose a convenient set of centroidal axes Oyz about which the product and moments of inertia can easily be calculated. Resolve the applied moment ***M*** into components M_y and M_z. The normal stress σ_x and the orientation of the neutral axis are given by Eqs (6.14) and (6.13) respectively.

Example 6.1 *A cantilever beam of rectangular section is subjected to a load of 1000 N (102 kgf) which is inclined at an angle of 30° to the vertical. What is the stress due to bending at point D (Fig. 6.6) near the built-in-end?*

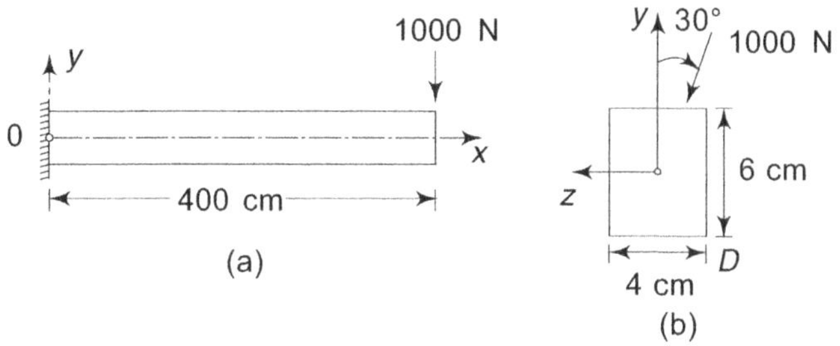

Fig. 6.6 *Example 6.1*

Solution For the section, y and z axes are symmetrical axes and hence these are also the principal axes. The force can be resolved into two components 1000 cos 30° along the vertical axis and 1000 sin 30° along the z axis. The force along the vertical axis produces a negative moment M_z (moment vector in negative z direction).

$$M_z = -(1000 \cos 30°)\, 400 = -400{,}000 \cos 30° \text{ N cm}$$

The horizontal component also produces a negative moment about the y axis, such that

$$M_y = -(1000 \sin 30°)\, 400 = -400{,}000 \sin 30° \text{ N cm}$$

The coordinates of point D are $(y, z) = (-3, -2)$. Hence, the normal stress at D from Eq. (6.9) is

$$\sigma_x = \frac{M_y z}{I_y} - \frac{M_z y}{I_z}$$

$$= \left(-400{,}000 \sin 30^\circ\right)\frac{(-2)}{I_y} - \left(-400{,}000 \cos 30^\circ\right)\frac{(-3)}{I_z}$$

$$= 400{,}000\left(\frac{2 \sin 30°}{I_y} - \frac{3 \cos 30°}{I_z}\right)$$

$$I_y = \frac{6 \times 4^3}{12} = 32 \text{ cm}^4, \qquad I_z = \frac{4 \times 6^3}{12} = 72 \text{ cm}^4$$

$$\therefore \qquad \sigma_x = 400,000\left(\frac{2}{2 \times 32} - \frac{3\sqrt{3}}{2 \times 72}\right)$$

$$= -1934 \text{ N/cm}^2 = -19340 \text{ kPa } (= -197 \text{ kgf/cm}^2)$$

Example 6.2 *A beam of equal-leg angle section, shown in Fig. 6.7, is subjected to its own weight. Determine the stress at point A near the built-in section. It is given that the beam weighs 1.48 N/cm (= 0.151 kgf/cm). The principal moments of inertia are 284 cm⁴ and 74.1 cm⁴.*

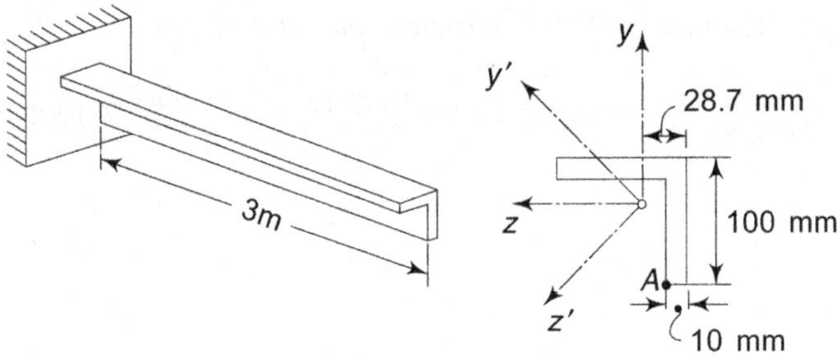

Fig. 6.7 *Example 6.2*

Solution The bending moment at the built-in end is

$$M_z = -\frac{wL^2}{2}$$

$$= \frac{1.48 \times 90,000}{2} = -66,000 \text{ N cm}$$

The centroid of the section is located at

$$\frac{(100 \times 10 \times 50) + (90 \times 10 \times 5)}{(100 \times 10) + (90 \times 10)} = 28.7 \text{ mm}$$

from the outer side of the vertical leg. The principal axes are the y' and z' axes. Since the member has equal legs, the z' axis is at 45° to the z axis. The components of M_z along y' and z' axes are, therefore,

$$M_{y'} = M_z \cos 45° = -47,100 \text{ N cm}$$
$$M_{z'} = M_z \cos 45° = -47,100 \text{ N cm}$$

$$\therefore \qquad \sigma_x = \frac{M_{y'}\, z'}{I_{y'}} - \frac{M_{z'}\, y'}{I_{z'}}$$

For point A

$$y = -(100 - 28.7) = -71.3 \text{ mm} = -7.13 \text{ cm}$$

and

$$z = -(28.7 - 10) = -18.7 \text{ mm} = -1.87 \text{ cm}$$

Hence,

$$y' = y \cos 45^\circ + z \sin 45^\circ$$
$$= -50.42 - 13.22 = -63.6 \text{ mm} = -6.36 \text{ cm}$$

and

$$z' = z \cos 45^\circ - y \sin 45^\circ$$
$$= -13.22 + 50.42 = +37.2 \text{ mm} = 3.72 \text{ cm}$$

$$\therefore \qquad \sigma_x = -\frac{47{,}100 \times 3.72}{74.1} - \frac{47{,}100 \times 6.36}{284}$$
$$= -2364 - 1055 = -3419 \text{ N/cm}^2 = -341{,}900 \text{ kPa}$$

Example 6.3 *Figure 6.8 shows a unsymmetrical one cell box beam with four-corner flange members A, B, C and D. Loads P_x and P_y are acting at a distance of 125 cm from the section ABCD. Determine the stresses in the flange members A and D. Assume that the sheet-metal connecting the flange members does not carry any flexual loads.*

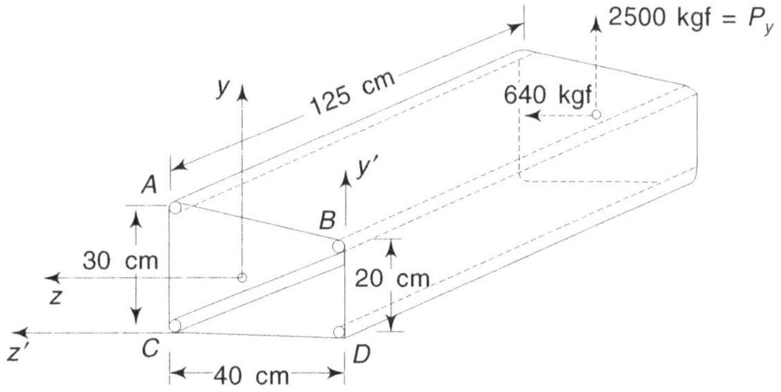

Fig. 6.8 *Example 6.3*

Solution The front face *ABCD* is assumed built-in.

Member	*Area*	y'	z'	Ay'	Az'	y	z
A	6.5	30	40	195	260	14.9	13.7
B	3.5	20	0	70	0	4.9	-26.3
C	5.0	0	40	0	200	−15.1	13.7
D	2.5	0	0	0	0	−15.1	-26.3
Σ =	17.5			265	460		

Therefore, the coordinates of the centroid from *D* are

$$y^* = \frac{\Sigma Ay'}{\Sigma A} = \frac{265}{17.5} = 15.1 \text{ cm}$$

$$z^* = \frac{\Sigma Az'}{\Sigma A} = \frac{460}{17.5} = 26.3 \text{ cm}$$

Member	*Area*	y	z	y^2	z^2	Ay^2	Az^2	Ayz
A	6.5	14.9	13.7	222	187.7	1443	1220.1	1326.8
B	3.5	4.9	–26.3	24	691.7	84	2421	–451
C	5.0	–15.1	13.7	228	187.7	1140	938.5	–1034.4
D	2.5	–15.1	–26.3	228	691.7	570	1729.3	992.8

$$\therefore \quad I_z = \Sigma Ay^2 = 3237 \text{ cm}^4$$
$$I_y = \Sigma Az^2 = 6308.9 \text{ cm}^4$$
$$I_{yz} = \Sigma Ayz = +834.2 \text{ cm}^4$$

One should be careful to observe that the loads P_y and P_z are acting at $x = -125$ cm

$\therefore$ Moment about z axis = $M_z = -312500$ kgf cm = -30646 Nm

Moment about y axis = $M_y = +80000$ kgf cm = $+7845.3$ Nm

From Eq. (6.14)

$$\sigma_x = \frac{-312500\left(6308.9y - 834.2z\right) + 80000\,(834.2y - 3237z)}{\left(834.2\right)^2 - \left(3237 \times 6308.9\right)}$$

$$= -96.57y - 0.09z$$

$$\therefore \quad (\sigma_x)_A = -(96.57 \times 14.9) - (0.09 \times 13.7) = -1440 \text{ kgf.cm}^2$$
$$= -141227 \text{ kPa}$$

$$(\sigma_x)_D = -(-96.57 \times 15.1) - (-0.09 \times 26.3) = +1460 \text{ kgf.cm}^2$$
$$= 143233 \text{ kPa}$$

6.3 REGARDING EULER–BERNOULLI HYPOTHESIS

We were able to solve the flexure problem because of the nature of the cross-section which remained plane after bending. It is natural to question how far this assumption is valid. In order to determine the actual deformation of an intially plane section of a beam subjected to a general loading, we will have to use the methods of the theory of elasticity. Since this is beyond the scope of this book, we shall discuss here the condition necessary for a plane section to remain plane. We have from Hooke's law

$$\varepsilon_x = \frac{1}{E}\left[\sigma_x - \nu\left(\sigma_y + \sigma_z\right)\right]$$

$$\varepsilon_y = \frac{1}{E}\left[\sigma_y - \nu\left(\sigma_z + \sigma_x\right)\right] \qquad \text{(c)}$$

$$\varepsilon_z = \frac{1}{E}\left[\sigma_z - \nu\left(\sigma_x + \sigma_y\right)\right]$$

Solving the above equations for the stress σ_x we get

$$\sigma_x = \frac{\nu E}{\left(1+\nu\right)\left(1-2\nu\right)}(\varepsilon_x + \varepsilon_y + \varepsilon_z) + \frac{E}{1+\nu}\varepsilon_x$$

or from Eq. (3.15)

$$\sigma_x = \lambda J_1 + 2G\varepsilon_x \tag{6.15}$$

where λ is a constant and G is the shear modulus. According to the Euler-Bernoulli hypothesis, we have

$$\sigma_y = \sigma_z = 0$$

Hence,

$$\sigma_x = E\varepsilon_x = E\frac{\partial u_x}{\partial x} \tag{6.16a}$$

Differentiating,

$$\frac{\partial \sigma_x}{\partial x} = E\frac{\partial^2 u_x}{\partial x^2} \tag{6.16b}$$

From equilibrium equation and stress-strain relations

$$\frac{\partial \sigma_x}{\partial x} = -\frac{\partial \tau_{xy}}{\partial y} - \frac{\partial \tau_{xz}}{\partial z}$$

$$= -G\frac{\partial}{\partial y}\left(\frac{\partial u_x}{\partial y} + \frac{\partial u_y}{\partial x}\right) - G\frac{\partial}{\partial z}\left(\frac{\partial u_x}{\partial z} + \frac{\partial u_z}{\partial x}\right)$$

$$= -G\left(\frac{\partial^2 u_x}{\partial y^2} + \frac{\partial^2 u_x}{\partial z^2}\right) - G\frac{\partial}{\partial x}\left(\frac{\partial u_y}{\partial y} + \frac{\partial u_z}{\partial z}\right)$$

$$= -G\left(\frac{\partial^2 u_x}{\partial y^2} + \frac{\partial^2 u_x}{\partial z^2}\right) - G\frac{\partial}{\partial x}\left(\varepsilon_y + \varepsilon_z\right) \tag{6.17a}$$

Since $\sigma_y = \sigma_z = 0$, from Eqs. (a),

$$\varepsilon_y = \varepsilon_z = -\frac{\nu}{E}\,\sigma_x$$

Hence, Eq. (6.17a) becomes

$$\frac{\partial \sigma_x}{\partial x} = -G\left(\frac{\partial^2 u_x}{\partial y^2} + \frac{\partial^2 u_x}{\partial z^2}\right) + \frac{2\nu G}{E}\frac{\partial \sigma_x}{\partial x}$$

i.e.

$$\frac{\partial \sigma_x}{\partial x}\left(1 - \frac{2\nu G}{E}\right) = -G\left(\frac{\partial^2 u_x}{\partial y^2} + \frac{\partial^2 u_x}{\partial z^2}\right)$$

or

$$\frac{\partial \sigma_x}{\partial x} = -\frac{GE}{E - 2\nu G}\left(\frac{\partial^2 u_x}{\partial y^2} + \frac{\partial^2 u_x}{\partial z^2}\right) \tag{6.17b}$$

Substituting in Eq. (6.16b),

$$E\frac{\partial^2 u_x}{\partial x^2} + \frac{GE}{E - 2\nu G}\left(\frac{\partial^2 u_x}{\partial y^2} + \frac{\partial^2 u_x}{\partial z^2}\right) = 0$$

i.e.
$$(E-2\nu G)\frac{\partial^2 u_x}{\partial x^2}+G\left(\frac{\partial^2 u_x}{\partial y^2}+\frac{\partial^2 u_x}{\partial z^2}\right)=0$$

or
$$A\frac{\partial^2 u_x}{\partial x^2}+G\left(\frac{\partial^2 u_x}{\partial y^2}+\frac{\partial^2 u_x}{\partial z^2}\right)=0 \tag{6.18}$$

where A is a constant. From flexure formula and Eq. (6.16a)

$$\sigma_x=\frac{My}{I_z}=E\frac{\partial u_x}{\partial x} \tag{d}$$

In the above equation, M is a function of x only and y is the distance measured from the neutral axis; I_z is the moment of inertia about the neutral axis which is taken as the z axis. Then

$$E\frac{\partial^2 u_x}{\partial z^2}=\frac{y}{I_z}\frac{\partial M}{\partial x}$$

Integrating Eq. (d)

$$Eu_x=\frac{y}{I_z}\int M\,dx+\phi(y,z)$$

where ϕ is a function of y and z only. Differentiating the above expression

$$E\frac{\partial^2 u_x}{\partial y^2}=\frac{\partial^2\phi(y,z)}{\partial y^2}$$

and
$$E\frac{\partial^2 u_x}{\partial z^2}=\frac{\partial^2\phi(y,z)}{\partial z^2}$$

Substituting these in Eq. (6.18),

$$\frac{Ay}{EI_z}\frac{\partial M(x)}{\partial x}+\frac{G}{E}\left[\frac{\partial^2\phi(y,z)}{\partial y^2}+\frac{\partial^2\phi(y,z)}{\partial z^2}\right]=0$$

or
$$K_1\frac{\partial M(x)}{\partial x}=K_2\left[\frac{\partial^2\phi(y,z)}{\partial y^2}+\frac{\partial^2\phi(y,z)}{\partial z^2}\right]$$

The left-hand side quantity is a function of x alone or a constant and the right-hand side quantity is a function of y and z alone or a constant. Hence, both these quantities must be equal to a constant, i.e.

$$\frac{\partial M(x)}{\partial x}=a \text{ constant}$$

or
$$M(x)=K_3x+K_5$$

This means that $M(x)$ can only be due to a concentrated load or a pure moment. In other words, the Euler–Bernoulli hypothesis that $\sigma_x=\frac{My}{I_z}$ (which is equivalent to plane sections remaining plane) will be valid only in those cases where the bending moment is a constant or varies linearly along the axis of the beam.

6.4 SHEAR CENTRE OR CENTRE OF FLEXURE

In the previous sections we considered the bending of beams subjected to pure bending moments. In practice, the beam carries loads which are transverse to the axis of the beam and which cause not only normal stresses due to flexure but also transverse shear stresses in any section. Consider the cantilever beam shown in Fig. 6.9 carrying a load at the free end. In general, this will cause both bending and twisting.

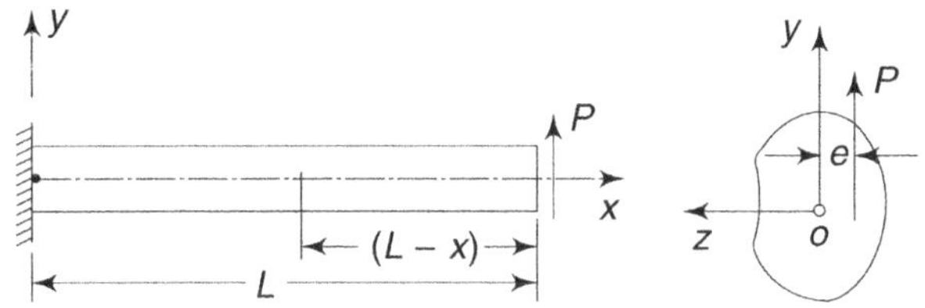

Fig. 6.9 *Cantilever beam loaded by force P*

Let Ox be the centroidal axis and Oy, Oz the principal axes of the section. Let the load be parallel to one of the principal axes (any general load can be resolved into components along the principal axes and each load can be treated separately). This load in general, will at any section, cause

(i) Normal stress σ_x due to flexure;

(ii) Shear stresses τ_{xy} and τ_{xz} due to the transverse nature of the loading and

(iii) Shear stresses τ_{xy} and τ_{xz} due to torsion

In obtaining a solution, we assume that

$$\sigma_x = -\frac{P(L-x)\,y}{I_z}, \qquad \sigma_y = \sigma_z = \tau_{yz} = 0 \tag{6.19}$$

This is known as St. Venant's assumption.

The values of τ_{xy} and τ_{xz} are to be determined with the equations of equilibrium and compatibility conditions. The value of σ_x as given above is derived according to the flexure formula of the previous section. The determination of τ_{xy} and τ_{xz} for a general cross-section can be quite complex. We shall not try to determine these. However, one important point should be noted. As said above, the load P in addition to causing bending will also twist the beam. But P can be applied at such a distance from the centroid that twisting does not occur. For a section with symmetry, the load has to be along the axis of symmetry to avoid twisting. For the same reason, for a beam with a general cross-section, the load P will have to be applied at a distance e from the centroid O. When the force P is parallel to the z-axis, a position can once again be established for which no rotation of the centroidal elements of the cross-sections occur. The point of intersection of these two lines of the bending forces is of significance. If a transverse force is applied at this point, we can resolve it into two components parallel to the y and z-axes and note from the above discussion that these components do not produce rotation of centroidal elements of the cross-sections of the beam. This point is called the shear centre of flexure or flexural centre (Fig. 6.10).

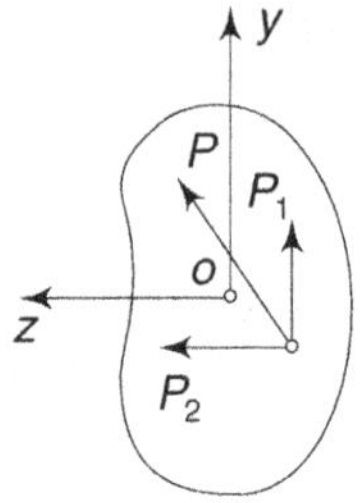

Fig. 6.10 *Load P passing through shear centre*

It is important to observe that the location of the shear centre depends only on the geometry, i.e. the shape of the section. For a section of a general shape, the location of the shear centre depends on the distribution of τ_{xy} and τ_{xz}, which, as mentioned earlier, can be quite complex. However, for thin-walled beams with open sections, approximate locations of the shear-centres can be determined by an elementary analysis, as discussed in the next section.

6.5 SHEAR STRESSES IN THIN-WALLED OPEN SECTIONS: SHEAR CENTRE

Consider a beam having a thin-walled open section subjected to a load V_y, as shown in Fig. 6.11(a). The thickness of the wall is allowed to vary. As mentioned in the previous section, the load V_y produces in general, bending, twisting and shear in the beam. Our object in this section is to locate that point through which the load V_y should act so as to cause no twist, i.e. to locate the shear centre of the section. Let us assume that load V_y is applied at the shear centre. Then there will be normal stress distribution due to bending and shear stress distribution due to vertical load. There will be no shear stress due to torsion.

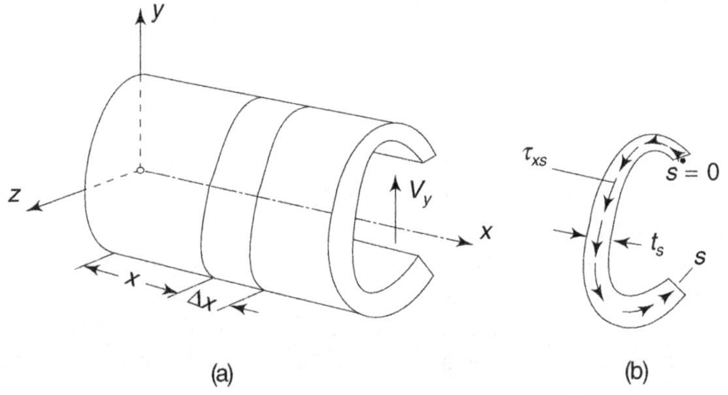

Fig. 6.11 *Thin-walled open section subjected to shear force*

The surface of the beam is not subjected to any tangential stresses and hence, the boundary of the section is an unloaded boundary. Consequently, the shear stresses near the boundary cannot have a component perpendicular to the boundary. In other words, the shear stresses near the boundary lines of the section are parallel to the boundary. Since the section of the beam is thin, the shear stress can be taken to be parallel to the centre line of the section at every point as shown in Fig. 6.11(b).

Consider an element of length Δx of the beam at section x, as shown in Fig. 6.12.

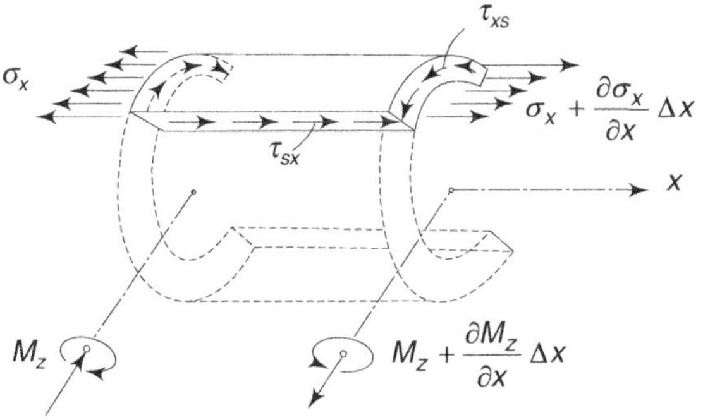

Fig. 6.12 *Free-body diagram of an elementary length of beam*

Let M_z be bending moment at section x and $M_z + \frac{\partial M_z}{\partial x}\Delta x$ the bending moment at section $x + \Delta x$· σ_x and $\sigma_x + \frac{\partial \sigma_x}{\partial x}\Delta x$, are corresponding flexural stresses at these two sections. It is important to observe that for the moments shown the normal stresses should be compressive and not as shown in the figure. However, the sign of the stress will be correctly given by Eq. (6.8). Considering a length s of the section, the unbalanced normal force is balanced by the shear stress τ_{sx} distributed along the length Δx. For equilibrium, therefore,

$$\tau_{sx}\, t_s\, \Delta x - \int_0^s \sigma_x\, t\, ds + \int_0^s \left(\sigma_x + \frac{\partial \sigma_x}{\partial x} \Delta x \right) t\, ds = 0$$

i.e.
$$\tau_{sx} = -\frac{1}{t_s} \int_0^s \frac{\partial \sigma_x}{\partial x} t\, ds \tag{6.20}$$

t_s is the wall thickness at s. Observing that $M_y = 0$, the normal stress σ_x is given by Eq. (6.8) as

$$\sigma_x = \frac{yI_y - zI_{yz}}{I_{yz}^2 - I_y\, I_z} M_z$$

Hence,
$$\frac{\partial \sigma_x}{\partial x} = \frac{yI_y - zI_{yz}}{I_{yz}^2 - I_y\, I_z} \frac{\partial M_z}{\partial x} \tag{6.21}$$

Recalling from elementary strength of materials $\frac{\partial M_z}{\partial x} = -V_y$, and substituting in Eq. (6.20)

$$\tau_{sx} = \frac{V_y}{t_s} \frac{1}{I_{yz}^2 - I_y\, I_z} \int_0^s (I_y y - I_{yz}\, z)\, t\, ds$$

or
$$\tau_{sx} = -\frac{V_y}{t_s \left(I_y\, I_z - I_{yz}^2 \right)} \left[I_y \int_0^s yt\, ds - I_{yz} \int_0^s zt\, ds \right] \tag{6.22}$$

The first integral on the right-hand side represents the first moment of the area between $s = 0$ and s about the z axis. The second integral is the first moment of the same area between $s = 0$ and s about the y axis. Since τ_{xs} is the complementary shear stress, its value at any s is also given by Eq. (6.22).

Let Q_z be the first moment of the area between $s = 0$ and s about the z axis and Q_y the first moment of the same area about the y axis.
Then,

$$\tau_{sx} = \tau_{xs} = -\frac{V_y}{t_s\left(I_y I_z - I_{yz}^2\right)}\left[I_y Q_z - I_{yz} Q_y\right] \tag{6.23}$$

Equation (6.22) gives the shear stress distribution at section x due to the vertical load V_y acting under the explicit assumption that no twisting is caused. Hence, the shear stress distribution τ_{xs} must be statically equivalent to the load V_y. This means the following:

(i) The resultant of τ_{xs} integrated over the section area must be equal to V_y.
(ii) The moment of τ_{xs} about the centroid (or any other convenient point) must be equal to the moment of V_y about the same point. That is,

$V_y\, e_z = moment\ of\ \tau_{xs}$ about O

where e_z is the eccentricity or the distance of V_y from O to avoid twisting (Fig. 6.13).

If a force V_z is acting instead of V_y, we can determine the shear stress τ_{xs} at any s as

$$\tau_{xs} = -\frac{V_z}{t_s\left(I_y I_z - I_{yz}^2\right)}\left[I_z \int_0^s zt\, ds - I_{yz} \int_0^s yt\, ds\right] \tag{6.24}$$

or

$$\tau_{xs} = -\frac{V_z}{t_s\left(I_y I_z - I_{yz}^2\right)}\left[I_z Q_y - I_{yz} Q_z\right] \tag{6.25}$$

If the above shear stress distribution is due to the shear force alone and not due to twisting also, then the moment of V_z about the centroid O must be equal to the moment of τ_{xs} about the same point, i.e.

$$V_y\, e_z = \text{moment of } \tau_{xs} \text{ about } O$$

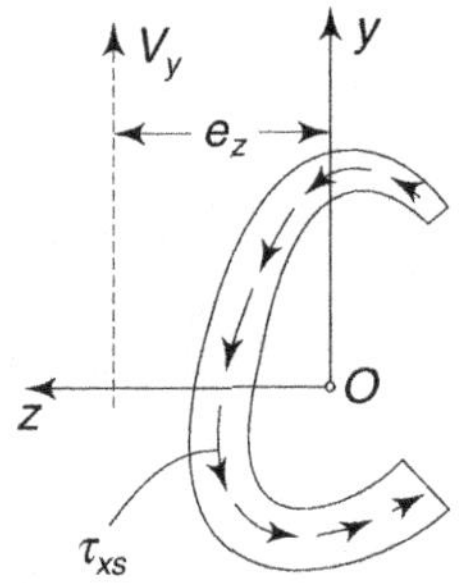

Fig. 6.13 *Location of shear centre and flow of shear stress*

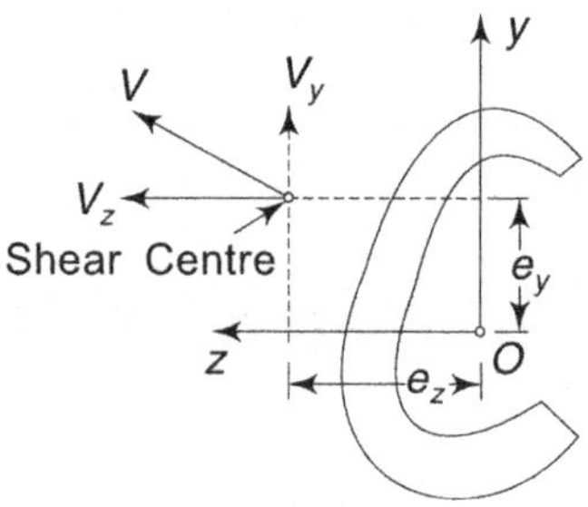

Fig. 6.14 *Location of shear centre for a general shear force*

Any arbitrary load V can be resolved into two components V_y and V_z and the resulting shear stress distribution τ_{xs} at any s is given by superposing Eqs (6.22) and (6.25). The point with coordinates (e_y, e_z), through which V_z and V_y should act to prevent the beam from twisting, is called the shear centre or the centre of flexure, as mentioned in Sec. 6.4. This is shown in Fig. 6.14.

Example 6.4 *Determine the shear stress distribution in a channel section of a cantilever beam subjected to a load F, as shown. Also, locate the shear centre of the section (Fig. 6.15).*

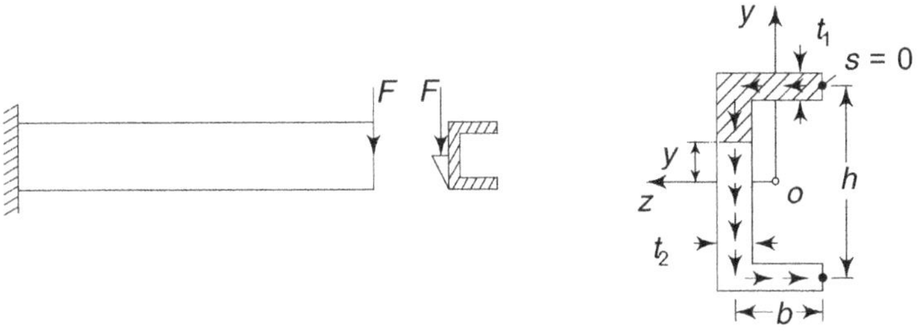

Fig. 6.15 *Example 6.4*

Solution Let Oyz be the principal axes, so that $I_{yz} = 0$. From Eq. (6.23) then, noting that F is negative,

$$\tau_{xs} = \frac{F}{t_s\, I_y\, I_z}\left(I_y\, Q_z\right)$$

or
$$\tau_{xs} = \frac{FQ_z}{t_s\, I_z}$$

where Q_z is the statical moment of the area from $s = 0$ to s about z axis. Considering the top flange, $t_s = t_1$, and the statical moment is

$$Q_z = \frac{t_1\, sh}{2}$$

Hence,
$$\tau_{xs} = \frac{Fsh}{2I_z} \quad \text{for } 0 \le s < b \tag{6.26}$$

i.e. the shear stress increases linearly from $s = 0$ to $s = b$. For s in the vertical web, $t_s = t_2$, and the statical moment is the moment of the shaded area in Fig. (6.15) about the z axis, i.e.

$$Q_z = bt_1\, \frac{h}{2} + \left(\frac{h}{2} - y\right) t_2 \left[y + \frac{1}{2}\left(\frac{h}{2} - y\right)\right]$$

$$= \frac{1}{2}\left[bt_1\, h + \left(\frac{h^2}{4} - y^2\right) t_2\right]$$

Hence,
$$\tau_{xs} = \frac{F}{2t_2\, I_z}\left[bt_1\, h + \left(\frac{h^2}{4} - y^2\right) t_2\right] \quad \text{for } -\frac{h}{2} < y < +\frac{h}{2} \tag{6.27}$$

i.e. the shear varies parabolically from $s = b$ to $s = b + h$. For s in the horizontal flange, $t_s - t_1$ and the statical moment is

$$Q_z = bt_1\, \frac{h}{2} + 0 + (s - b - h)\, t_1 \left(-\frac{h}{2}\right)$$

$$= \left(bh + \frac{h^2}{2} - \frac{h}{2} s\right) t_1$$

Hence, $$\tau_{xs} = \frac{F}{2I_z}\left(bh + \frac{h^2}{2} - \frac{h}{2}s\right) \quad \text{for } 2b + h \geq s > b + h \tag{6.28}$$

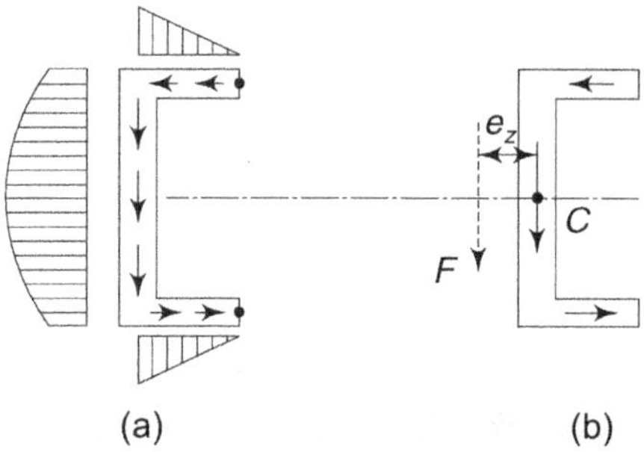

Fig. 6.16 *Example 6.4—Shear stress distribution diagrams*

i.e. the shear varies linearly. When $s = 2b + h$, i.e. the right tip of the bottom flange, the shear is zero. The variation of τ_{xs} is shown in Fig. 6.16.

This shear stress distribution should be statically equivalent to applied shear force F. It is easy to see that this is equal to F in magnitude. On integrating τ_{xs} over the area of the section, the resultant of the stress in the top and bottom flange cancel each other, and therefore, there is no horizontal resultant. Integrating τ_{xs} over the vertical web, we have

$$\int_{-h/2}^{+h/2} \tau_{xs}\, t_2\, dy = \frac{F}{2I_z}\left[\int bt_1 h\, dy + \int\left(\frac{h^2}{4} - y^2\right)t_2\, dy\right]$$

$$= \frac{F}{2I_z}\left[bt_1h^2 + \frac{h^3}{4}t_2 - \frac{h^3}{12}t_2\right]$$

$$= \frac{F}{2I_z}\left[bt_1h^2 + \frac{t_2h^3}{6}\right]$$

Now for the section

$$I_z = bt_1\frac{h^2}{4} + bt_1\frac{h^2}{4} + t_2\frac{h^3}{12}$$

$$= bt_1\frac{h^2}{2} + t_2\frac{h^3}{12} \tag{6.29}$$

Hence, $$\int_{-h/2}^{+h/2} \tau_{xs}\, t_s\, dy = F$$

Hence, the resultant of τ_{xs} over the area is equal to F. In addition, it has a moment. Taking moment about the midpoint of the vertical web [(Fig. 6.15(b)]

$$M = (\text{resultant of } \tau_{xs} \text{ in top flange}) \times \frac{h}{2}$$

$$+ (\text{resultant of } \tau_{xs} \text{ in bottom flange}) \times \frac{h}{2}$$

$$= 2\,(\text{resultant of } \tau_{xs} \text{ in top flange}) \times \frac{h}{2}$$

$$= 2\left(\text{average of } \tau_{xs} \text{ in top flange} \times \text{area} \times \frac{h}{2}\right)$$

$$= 2\left(\frac{Fbh}{4I_z} \times bt_1 \times \frac{h}{2}\right)$$

$$= \frac{Fb^2h^2t_1}{4I_z}$$

This must be equal to the moment of F about the same point. Hence, F must act at a distance e_z from C such that

$$Fe_z = \frac{Fb^2h^2t_1}{4I_z}$$

or

$$e_z = \frac{b^2h^2t_1}{4I_z}$$

Substituting for I_z from Eq. (6.29)

$$e_z = \frac{3b^2h^2t_1}{6bt_1h^2 + t_2h^3}$$

or

$$e_z = \frac{3b^2t_1}{6bt_1 + t_2h}$$

Hence, the shear centre is located at a distance e_z from C [Fig. 6.16(b)].

Example 6.5 *Determine the shear stress distribution for a circular open section under bending caused by a shear force. Locate the shear centre (Fig. 6.17).*

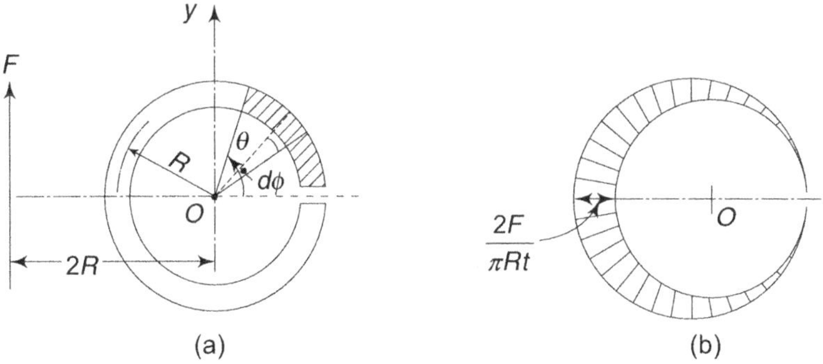

Fig. 6.17 *Example 6.5*

Solution The static moment of the crossed section is

$$Q_z = \int_0^{\theta} (R\, d\phi\, t)\, R \sin\phi$$

$$= R^2 t(1 - \cos\theta)$$

Hence, from Eq. (6.23), noting that $I_{yz} = 0$, and for a vertically upward shear force F,

$$\tau_{xs} = -\frac{FQ_z}{tI_z} = -\frac{F}{tI_z} R^2 t(1 - \cos\theta)$$

But $$I_z = \pi R^3 t$$

Hence, $$\tau_{xs} = -\frac{F}{\pi Rt}(1 - \cos\theta)$$

For $\theta = 180°$ $$\tau_{xs} = -\frac{2F}{\pi Rt}$$

The distribution is shown in Fig. 6.17(b). The moment of this distribution about O is,

$$M = \int_0^{2\pi} \tau_{xs}(R\,d\theta\,t)R$$

$$= -\frac{F}{\pi Rt}\int_0^{2\pi} R^2 t(1 - \cos\theta)d\theta$$

$$= -2FR$$

This should be equal to the moment of the applied transverse force F about O. For F positive, the moment about O is negative since it is directed from $+z$ to $+y$. Hence the, force F must be applied at the shear centre C, which is at a distance of $2R$ from O.

6.6 SHEAR CENTRES FOR A FEW OTHER SECTIONS

In a thin-walled inverted T section, the distribution of shear stress due to transverse shear will be as shown in Fig. 6.18(a). The moment of this distributed stress about C is obviously zero. Hence, the shear centre for this section is C.

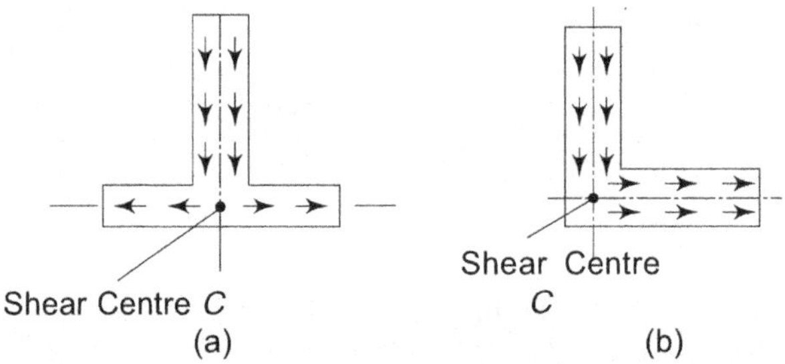

Fig. 6.18 *Location of shear centres for inverted T section and angle section*

For the angle section, the moment of the shear stresses about C is zero and hence, C is the shear centre. Figure 6.19 shows how the beams will twist if the loads are applied through the centroids of the respective sections and not through the shear centres.

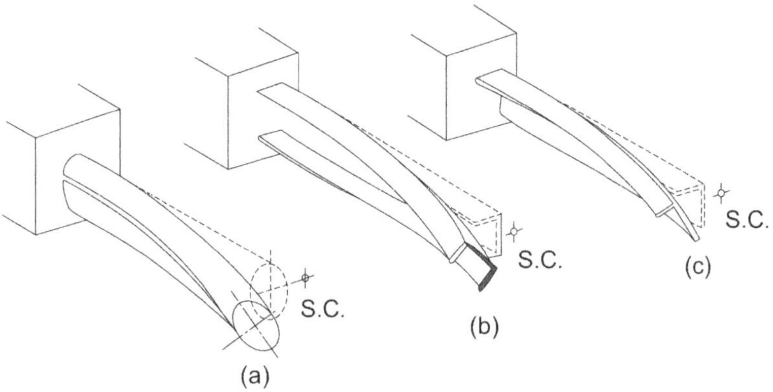

Fig. 6.19 *Twisting effect on some cross-sections if load is not applied through shear centre*

6.7 BENDING OF CURVED BEAMS (WINKLER–BACH FORMULA)

So far we have been discussing the bending of beams which are initially straight. Now we shall study the bending of beams which are initially curved. We consider the case where bending takes place in the plane of curvature. This is possible when the beam section is symmetrical about the plane of curvature and the bending moment M acts in this plane. Let ρ_0 be the initial radius of curvature of the centroidal surface. As in the case of straight beams, it is again assumed that sections which are plane before bending remain plane after bending. Hence, a transverse section rotates about an axis called the neutral axis, as shown in Fig. 6.20.

Consider an elementary length of the curved beam enclosing an angle $\Delta\phi$. Owing to the moment M, let the section AB rotate through $\delta\Delta\phi$ and occupy the position $A'B'$. The section rotates about NN, the neutral axis. SN is the trace of the neutral surface with radius of curvature r_0. Fibres above this surface get compressed and fibres below this surface get stretched. Fibres lying in the neutral surface remain unaltered. Consider a fibre at a distance y from the neutral surface. The unstretched length before bending is $(r_0 - y)\ \Delta\phi$. The change in

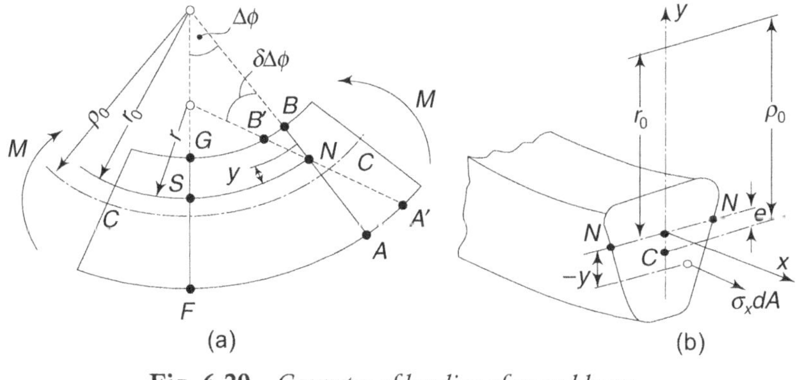

Fig. 6.20 *Geometry of bending of curved beam*

length due to bending is $y(\delta\Delta\phi)$. Noting that for the moment as shown, the strain is negative,

$$\text{strain} \equiv \varepsilon_x = -\frac{y(\delta\,\Delta\phi)}{(r_0 - y)\,\Delta\phi} \tag{6.30}$$

It is assumed here that the quantity y remains unaltered during the process of bending. The value of $(\delta\Delta\phi)/\Delta\phi$ can be obtained from Fig. 6.20(a). It is seen that

$$SN = (\Delta\phi + \delta\,\Delta\phi)\, r$$

where r is the radius of curvature of the neutral surface after bending. Also

$$SN = r_0\,\Delta\phi$$

Hence,

$$\frac{(\Delta\phi + \delta\Delta\phi)r}{\Delta\phi\, r_0} = 1$$

i.e.

$$\frac{\delta\Delta\phi}{\Delta\phi} = \frac{r_0}{r} - 1$$

$$= r_0\left(\frac{1}{r} - \frac{1}{r_0}\right) \tag{6.31}$$

Substituting in Eq. (6.30)

$$\varepsilon_x = -\frac{y}{r_0 - y}\, r_0\left(\frac{1}{r} - \frac{1}{r_0}\right) \tag{6.32a}$$

Now we shall assume that only σ_x is acting and that $\sigma_y = \sigma_z = 0$. This is similar to the Bernoulli–Euler hypothesis for the bending of straight beams. On this assumption,

$$\sigma_x = -\frac{Ey}{r_0 - y}\, r_0\left(\frac{1}{r} - \frac{1}{r_0}\right) \tag{6.32b}$$

The above expression brings out the main distinguishing feature of a curved beam. The value of y must be comparable with that of r_0, i.e. the beam must have a large curvature in which the dimensions of the cross-sections of the beam are comparable with the radius of curvature r_0. On the other hand, if the curvature (i.e. $1/r_0$) is very samll, i.e. r_0 is very large compared to y, then Eq. (6.32b) becomes

$$\sigma_x = -Ey\left(\frac{1}{r} - \frac{1}{r_0}\right)$$

With $r_0 \to \infty$, the above equation reduces to that of the straight beam. For equilibrium, the resultant of σ_x over the area should be equal to zero and the moment about NN should be equal to the applied moment M. It should be observed that the strains in fibres above the neutral axis will be numerically greater than the stains in fibres below the neutral axis. This is evident from Eq. (6.32a), since for positive y, i.e. for a fibre above the neutral axis, the denominator $(r_0 - y)$ will be less than that for a negative y. Since the resultant normal force is zero, the neutral axis gets shifted towards the centre of the curvature. For equilibrium, we have,

$$\int_A \sigma_x \, dA = -Er_0\left(\frac{1}{r} - \frac{1}{r_0}\right)\int_A \frac{y\,dA}{r_0 - y} = 0$$

and
$$-\int_A \sigma_x y \, dA = +\,Er_0\left(\frac{1}{r} - \frac{1}{r_0}\right)\int_A \frac{y^2\,dA}{r_0 - y} = M$$

From the first equation above

$$\int_A \frac{y\,dA}{r_0 - y} = 0 \tag{6.33}$$

The second equation can be written as

$$+\,Er_0\left(\frac{1}{r} - \frac{1}{r_0}\right)\left[-\int_A y\,dA + r_0\int_A \frac{y\,dA}{r_0 - y}\right] = M$$

The first integral represents the static moment of the section with respect to the neutral axis and is equal to $(-Ae)$, where e is the distance of the centroid from the neutral axis NN and this moment is negative. The second integral is zero according to Eq. (6.33). Thus,

$$Er_0\left(\frac{1}{r} - \frac{1}{r_0}\right)Ae = M \tag{6.34}$$

But from Eq. (6.32)

$$Er_0\left(\frac{1}{r} - \frac{1}{r_0}\right) = -\frac{\sigma_x\,(r_0 - y)}{y}$$

Substituting this in Eq. (6.34)

$$-\frac{\sigma_x\,(r_0 - y)}{y}\,Ae = M$$

or
$$\sigma_x = -\frac{M}{Ae}\frac{y}{(r_0 - y)} \tag{6.35}$$

As Eq. (6.35) shows, the normal stress varies non-linearly across the depth. The distribution is hyperbolic and one of its asymptotes coincides with the line passing through the centre of curvature, as shown in Fig. 6.21(a). The maximum stress may occur either at the top or at the bottom of the section, depending on its shapes. Equation (6.35) is often referred to as the Winkler-Bach formula.

In some texts, the origin of the coordinate system is taken at the centroid of the section instead of at the point of intersection of the neutral axis and the y axis. If the origin is taken at the centroid and y' is the distance of any fibre from this origin, then putting $y = y' - e$ and $r_0 = \rho_0 - e$, Eq. (6.35) becomes

$$\sigma_x = -\frac{M}{Ae}\frac{y' - e}{\rho_0 - e - y' + e}$$

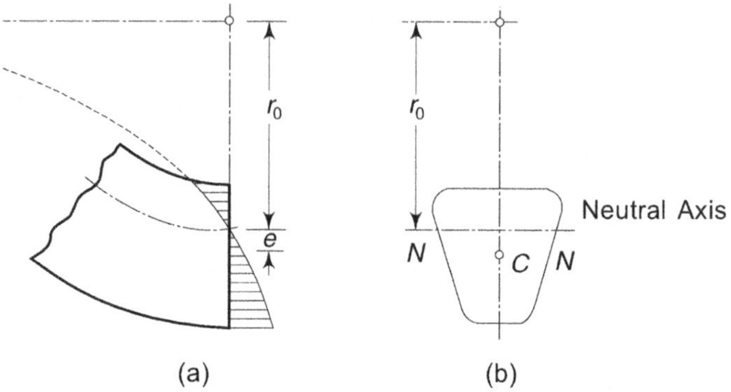

Fig. 6.21 *Distribution of normal stress and location of neutral axis*

or
$$\sigma_x = -\frac{M}{Ae}\frac{y'-e}{\rho_0 - y'} \tag{6.36}$$

To use Eq. (6.35), one requires the value of r_0. For this, consider Eq. (6.33). Introducing the new variable u

$$u = r_0 - y$$

the equation becomes

$$\int_A \frac{r_0 - u}{u}\, dA = 0$$

Hence,
$$r_0 = \frac{A}{\int_A dA/u} \tag{6.37}$$

The integral in the denominator represents a geometrical characteristic of the section. In other words, the values of r_0 and e are independent of the moment within elastic limit. We shall calculate these for a few of the commonly used sections.

Rectangular Section From Fig. 6.22, $dA = b\,du$ and $u = \rho_0 - y'$. Hence,

$$\int_A \frac{dA}{u} = \int_{\rho_0 - h/2}^{\rho_0 + h/2} \frac{b\,du}{u} = b \log_n \frac{\rho_0 + \frac{h}{2}}{\rho_0 - \frac{h}{2}}$$

Hence,
$$r_0 = \frac{h}{\log_n \left(\dfrac{\rho_0 + \frac{h}{2}}{\rho_0 - \frac{h}{2}} \right)} = \frac{h}{\log_n (r_2/r_1)} \tag{6.38}$$

The shift of the neutral axis from the centroid is

$$e = \rho_0 - \frac{h}{\log_n \left(\dfrac{\rho_0 + \frac{h}{2}}{\rho_0 - \frac{h}{2}} \right)} \tag{6.39a}$$

or
$$e = \rho_0 - \frac{h}{\log_n \left(\dfrac{r_2}{r_1} \right)} \tag{6.39b}$$

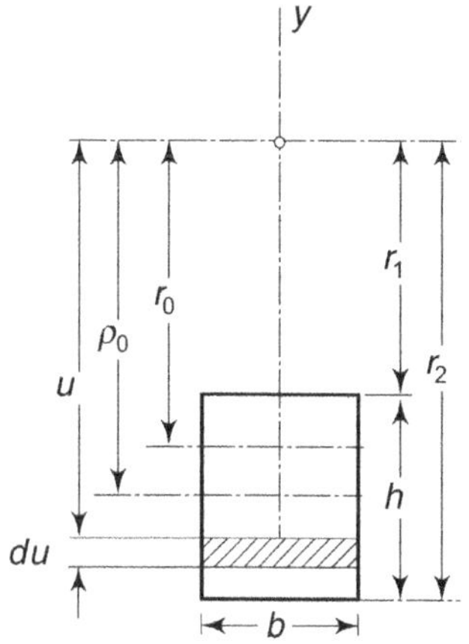

Fig. 6.22 *Parameters for a rectangular section to calculate r_0 according to Eq. (6.31)*

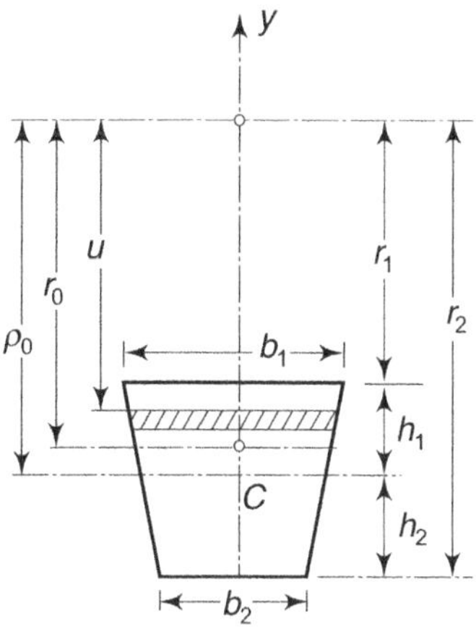

Fig. 6.23 *Parameters for a trapezoidal section to calculate r_0 according to Eq. (6.31)*

Trapezoidal Section (see Fig. 6.23) Let $h_1 + h_2 = h$. The variable width of the section is

$$b = b_2 + \frac{(b_1 - b_2)}{h}(h_2 + e + y)$$

and

$$dA = dy\left[b_2 + (b_1 - b_2)(h_2 + e + y)/h\right]$$

$$u = \rho_0 - e - y$$

$\therefore$

$$\int \frac{dA}{u} = \int_{-h_2 - e}^{h_1 - e} \left[\frac{b_2 + (b_1 - b_2)(h_2 + e + y)/h}{\rho_0 - e - y}\right] dy$$

$$= \left[b_2 + r_2(b_1 - b_2)/h\right] \log \frac{r_2}{r_1} - (b_1 - b_2)$$

When $b_1 = b_2$, the above equation reduces to that of the previous case.

$$r_0 = \frac{(b_1 + b_2)h}{2} \left\{\left[b_2 + r_2(b_1 - b_2)/h\right] \log \frac{r_2}{r_1} - (b_1 - b_2)\right\} \tag{6.40}$$

T-section (see Fig. 6.24) Proceeding as in the previous case, we obtain for the section

$$\int \frac{dA}{u} = b_1 \log \frac{r_3}{r_1} + b_2 \log \frac{r_2}{r_3} \tag{6.41}$$

I-Section For the *I*-section shown in Fig. 6.25, following the same procedure as in the preceding case,

$$\int \frac{dA}{u} = b_1 \log \frac{r_3}{r_1} + b_2 \log \frac{r_4}{r_3} + b_3 \frac{r_2}{r_4} \tag{6.42}$$

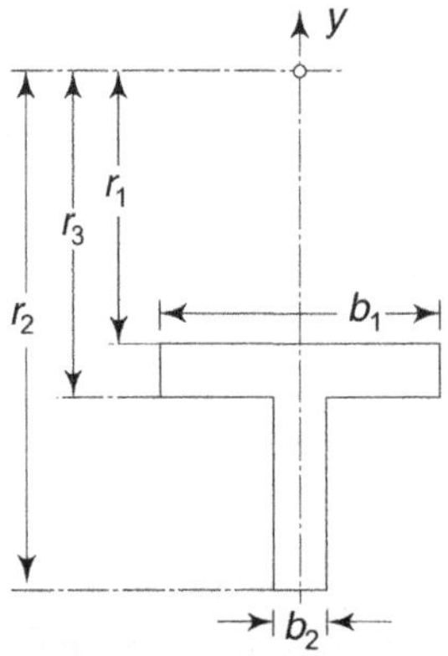

Fig. 6.24 *Parameters for T-section to calculate r_0 according to Eq. (6.31)*

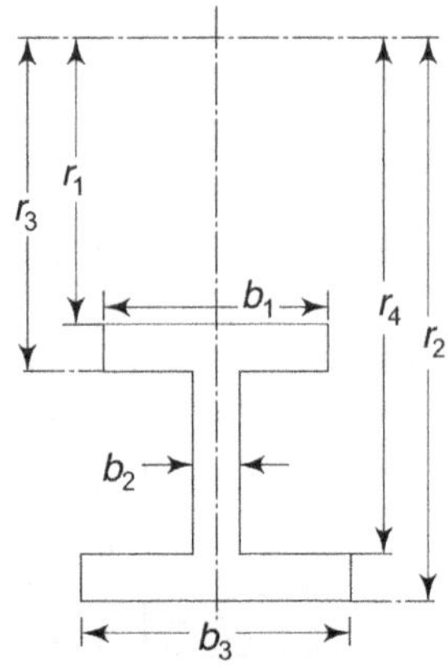

Fig. 6.25 *Parameters for I-section to calculate r_0 according to Eq. (6.31)*

Circular Section (see Fig. 6.26)

$$u = r_0 - y = (\rho_0 - e) - (a \cos \theta - e) = \rho_0 - a \cos \theta$$

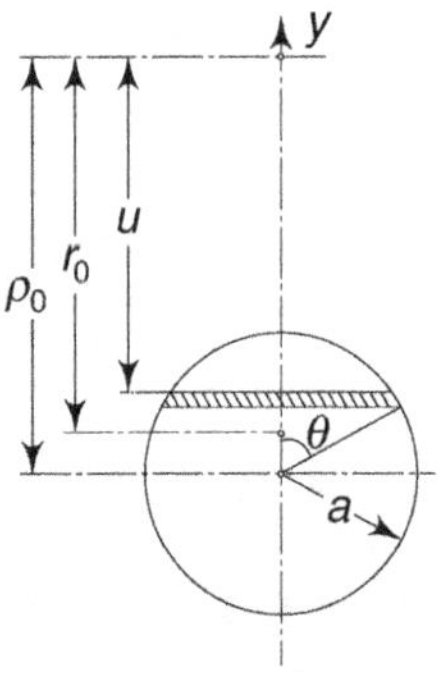

Fig. 6.26 *Parameters for a circular section to calculate r_0 according to Eq. (6.31)*

$$du = a \sin \theta \, d\theta$$

$$dA = 2a \sin \theta \, du = 2a^2 \sin^2 \theta \, d\theta$$

$$\int_A \frac{dA}{u} = \int_0^{\pi} 2a^2 \sin^2 \theta / (\rho_0 - a \cos \theta) \, d\theta$$

$$= 2a \int_0^{\pi} \frac{1 - \cos^2 \theta}{b - \cos \theta} d\theta, \quad \text{where } b = \frac{\rho_0}{a}$$

Adding and substracting ($b \cos \theta + b^2$) to the numerator,

$$\int_A \frac{dA}{u} = 2a\pi \left[b - \left(b^2 - 1 \right)^{1/2} \right]$$

$$= 2\pi \left[\rho_0 - \left(\rho_0^2 - a^2 \right)^{1/2} \right]$$

and

$$r_0 = \frac{a^2}{2\left[\rho_0 - (\rho_0^2 - \alpha^2)^{1/2} \right]}$$

Example 6.6 *Determine the maximum tensile and maximum compressive stresses across the Sec. AA of the member loaded, as shown in Fig. 6. 27. Load P = 2000 kgf (19620 N).*

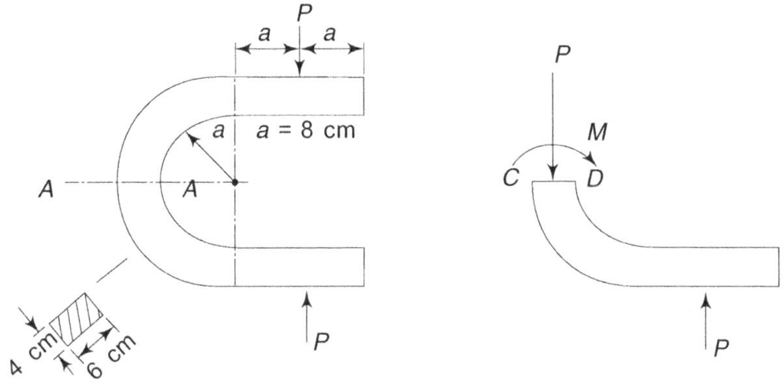

Fig. 6.27 *Example 6.6*

Solution For the section $\rho_0 = 11$ cm, $h = 6$ cm, $b = 4$ cm.

$$\therefore \quad \log \frac{\rho_0 + h/2}{\rho_0 - h/2} = \log \frac{7}{4} = 0.5596$$

From equations (6.38) and (6.39)

$$r_0 = \frac{6}{0.5596} = 10.73, \qquad e = 11 - 10.73 = 0.27$$

From Eq. (6.35), owing to bending moment M

$$\sigma'_x = -\frac{M}{Ae}\frac{y}{(r_0 - y)}$$

$$= -\frac{M}{24 \times 0.27}\frac{y}{(10.73 - y)}$$

For the problem

$$M = P\,(a + a + h/2) = 19P$$

At C, $\quad y = -(e + h/2) = -3.27$

and, at D, $\quad y = \frac{h}{2} - e = 2.73$

Hence, $\quad (\sigma'_x)_C = -\frac{19P}{24 \times 0.27} \times \frac{(-3.27)}{(10.73 + 3.27)} = 0.6848\,P$

and $\quad (\sigma'_x)_D = -\frac{19P}{24 \times 0.27}\frac{2.73}{(10.73 - 2.73)} = -1.001\,P$

The stress due to direct loading is

$$\sigma''_x = -\frac{P}{A} = -\frac{P}{24} = -0.0417\,P$$

Hence the combined stresses are

$$(\sigma_x)_C = (0.6848 - 0.0417)\,P$$
$$= 0.6431P = 129 \text{ kgf/cm}^2 \text{ (12642 kPa)}$$

and

$$(\sigma_x)_D = (-1.001 - 0.0417)\,P$$
$$= -1.0427\,P = -209 \text{ kgf/cm}^2 \text{ (20482 kPa)}$$

Example 6.7 *Determine the stress at point D of a hook (Fig. 6.28) having a trapezoidal section with the following dimensions: b_1 = 4 cm, b_2 = 1 cm, r_1 = 3 cm, r_2 = 10 cm, h = 7 cm, force P = 3000 kgf (29400 N).*

Solution For the section

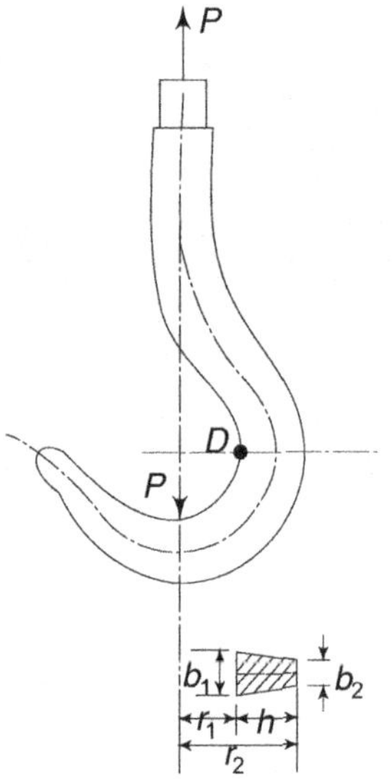

Fig. 6.28 *Example 6.7*

$$\int \frac{dA}{u} = [1 + 10(4-1)/7]\log\frac{10}{3} - (4-1)$$
$$= 3.363 \text{ cm}$$
$$A = \frac{1}{2}(b_1 + b_2)h = \frac{35}{2} = 17.5 \text{ cm}^2$$
$$\therefore \quad r_0 = A/3.363 = 17.5/3.363 = 5.204 \text{ cm}$$
$$\rho_0 = 3 + \frac{(b_1 + 2b_2)h}{3(b_1 + b_2)} = 3 + \frac{14}{5} = 5.80 \text{ cm}$$
$$\therefore \quad e = \rho_0 - r_0 = 0.596$$

The moment across section D is

$$M = -3000\,\rho_0 = -17{,}400 \text{ kgf cm (1705 Nm)}$$

The normal stress due to bending is therefore

$$(\sigma'_x)_D = -\frac{M}{Ae}\frac{y}{r_0 - y}$$
$$= +\frac{17{,}400}{17.5 \times 0.596} \times \frac{2.204}{5.204 - 2.2}$$
$$= 1226 \text{ kgf/cm}^2 \text{ (120,148 kPa)}$$

The normal stress due to axial loading is

$$(\sigma''_x)_D = \frac{3000}{A} = \frac{3000}{17.5} = 171 \text{ kgf/cm}^2$$

The total normal stress is therefore,

$$(\sigma_x)_D = 1397 \text{ kgf/cm}^2, \text{ or } 136{,}907 \text{ kPa}$$

6.8 DEFLECTIONS OF THICK CURVED BARS

In Chapter 5, the problems of thin rings and thin curved members were analyzed using energy methods. In this section, we shall discuss a few problems involving thick rings. The energy method will be used. Consider the member shown in Fig. 6.29(a).

In the straight part of the U-ring, across any section, there is a tangential force P and a moment $(Px - M_0)$. In the curved part of the member, there will

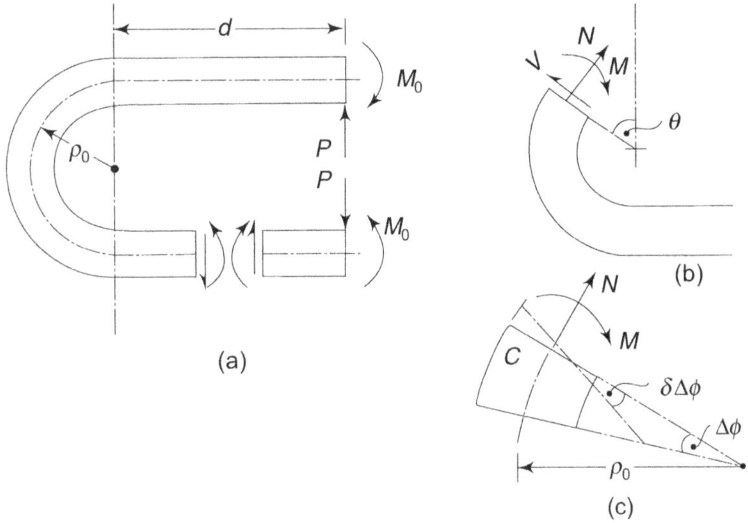

Fig. 6.29 *Geometry of deflection of a curved bar*

be a tangential force V, a normal force N and a bending moment M. Their values are

$$V = P \cos \theta$$
$$N = P \sin \theta$$
$$M = M_0 - (d + \rho_0 \sin \theta)\, P$$

To calculate the strain energy stored we proceed as follows (we make use of the expressions developed in Chapter 5):

(i) In the straight part of the member: Owing to the shear force V, the strain energy stored in a small length Δs is

$$\Delta U_V = \frac{\alpha V^2 \Delta s}{2AG} \tag{6.43}$$

where α is a numerical factor depending on the shape of the cross section, A is the area of the section and G is the shear modulus.

Because of the bending moment M, the energy stored is

$$\Delta U_M = \frac{M^2 \Delta s}{2EI} \tag{6.44}$$

where I is the moment of inertia about the neutral axis, which for a straight beam passes through the centroid of the section.

In general, the strain energy due to V is small as compared to that due to M.

(ii) In the curved part of the member: Owing to the shear force V, the strain energy stored in a small sectoral element, enclosing an angle $\Delta\phi$, is

$$\Delta U_V = \frac{\alpha V^2 \Delta s}{2AG} \tag{6.45}$$

If ρ_0 is the radius of curvature of the centroidal fibre, $\Delta s = \rho_0\, \Delta\phi$.

Because of the normal force N, which is assumed to be acting at the centroid of the cross-section,

$$\Delta U_N = \frac{N^2 \Delta s}{2AE} \tag{6.46}$$

Owing to bending moment M, the energy stored is equal to the work done. If $\delta\,\Delta\phi$ is the change in the angle due to bending [Fig. 6.29 (c)]

$$\Delta U_M = \frac{1}{2} M\,(\delta\,\Delta\phi)$$

From Eq. (6.31),

$$\delta\Delta\phi = \Delta\phi\, r_0 \left(\frac{1}{r} - \frac{1}{r_0}\right)$$

From Eq. (6.34), substituting for the right-hand part in the above equation

$$\delta\Delta\phi = \Delta\phi \frac{M}{AeE}$$

Hence,

$$\Delta U_M = \frac{M^2\,\Delta\phi}{2AeE}$$

Putting

$$\Delta\phi = \frac{\Delta s}{\rho_0}$$

$$\Delta U_M = \frac{M^2\,\Delta s}{2AeE\,\rho_0} \tag{6.47}$$

If N is applied first and then M, owing to the rotation of the section, the centroid C [Fig. 6.29(c)] moves through a distance $\varepsilon_0\,\Delta s$, where ε_0 is the strain at C and consequently, the force N does additional work equal to

$$\Delta U_{MN} = N\,\varepsilon_0\,\Delta s$$

ε_0 from Eq. (6.35) is

$$\varepsilon_0 = \frac{\sigma_x}{E} = -\frac{M}{AeE}\frac{y_0}{(r_0 - y_0)}$$

In the above equation, M is positive, according to the convention followed (Fig. 6.20). y_0 is the distance of the centroidal fibre from the neutral axis and is equal to $-e$. Also, $\rho_0 = r_0 + e$. With these,

$$\varepsilon_0 = +\frac{M}{A\rho_0 E}$$

Hence the work done by N is

$$\Delta_{MN} = \frac{MN\,\Delta s}{A\rho_0 E} \tag{6.48}$$

The same result is obtained if M is applied first and then N. This is according to the principle of superposition, which is valid for small deformations. This can be seen by referring to Fig. 6.30.

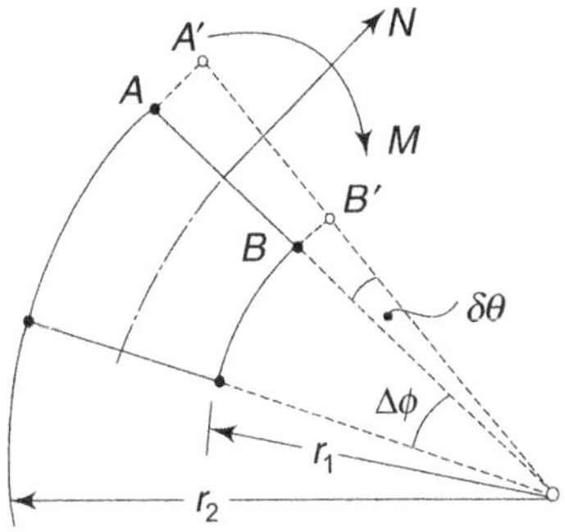

Fig. 6.30 *Deformation of a section of curved bar*

The normal force N acting across the section produces uniform strain ε_n; since the lengths of the fibres are different, face AB will not shift parallel to itself. The extension of the fibre at b will be $\varepsilon_n\, r_1\, \Delta\phi$. The angle enclosed between AB and $A'B'$ is therefore

$$\delta\theta = \frac{\varepsilon_n\, \Delta\phi (r_2 - r_1)}{(r_2 - r_1)} = \varepsilon_n\, \Delta\phi$$

Owing to this rotation of $A'B'$, the moment M does work equal to

$$\Delta U_{NM} = M\varepsilon_n\, \Delta\phi$$

Since

$$\varepsilon_n = \frac{N}{AE}$$

$$\Delta U_{NM} = \frac{MN}{AE}\, \Delta\phi$$

$$= \frac{MN\, \Delta s}{AE\rho_0}$$

For a straight beam, the work done by N when M is applied is zero since the section rotates about the neutral axis which passes through the centroid. This is also confirmed in the above expression where $\rho_0 = \infty$ for a straight beam and therefore $\Delta U_{MN} = 0$. Combining all the energies detailed above, the total strain energy is.

$$U = \int_s (\Delta U_V + \Delta U_N + \Delta U_M + \Delta U_{MN})$$

$$= \int_s \left(\frac{\alpha V^2}{2AG} + \frac{N^2}{2AE} + \frac{M^2}{2AeE\rho_0} + \frac{MN}{AE\rho_0} \right) ds \tag{6.49}$$

For the straight part of the beam, the last expression will be zero and the third expression (which becomes indeterminate since $e = 0$ and $\rho_0 = \infty$) is replaced by $M^2/2EI$. With the strain energy calculated as above and using Castigliano's theorem, one can solve for the unknown—either the deflection or the indeterminate reaction. We shall illustrate this through an example.

Example 6.8 *A ring with a rectangular section is subjected to diametral compression, as shown in Fig. 6.31. Determine the bending moment and stress at point A of the inner radius across a section θ.* r_1 *and* r_2 *are the inner and external radii respectively.*

Solution We observe that the deformation of the ring will be symmetrical about the horizontal and vertical axes. Consequently, there will be no changes in the slopes of the vertical and horizontal faces of the ring [Fig. 6.31(b)]. We can, therefore, consider only a quadrant of the circle for the analysis. This is shown in Fig. 6.31(c). M_0 is the unknown internal moment. Its value can be determined from

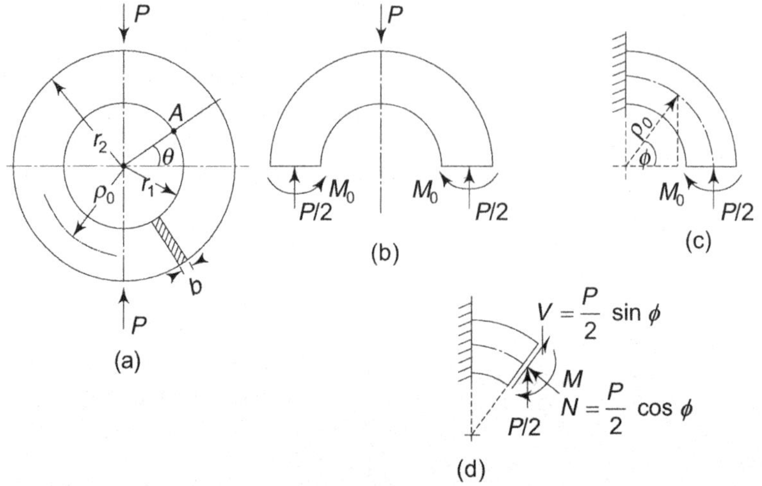

Fig. 6.31 *Example 6.8*

the condition that the change in the slope of this section is zero. We shall use Castigliano's theorem to determine this moment.

Across any section ϕ, the moment is

$$M = M_0 - \frac{P}{2} \rho_0 (1 - \cos \phi)$$

In addition, there is a normal force N and a shear force V, as shown in Fig. 6.31(d). Their values are

$$N = -\frac{P}{2} \rho_0 \cos \phi \quad \text{and} \quad V = -\frac{P}{2} \sin \phi$$

The total strain energy for the quadrant from Eq. (6.49) is

$$U = \int_0^{\pi/2} \frac{\alpha P^2 \sin^2 \phi}{8AG} \rho_0 \, d\phi + \int_0^{\pi/2} \frac{P^2 \cos^2 \phi}{8AE} \rho_0 \, d\phi$$

$$+ \int_0^{\pi/2} \frac{\left[M_0 - \frac{P}{2} \rho_0 \left(1 - \cos \phi\right) \right]^2}{2AeE} d\phi$$

$$- \int_0^{\pi/2} \frac{\left[M_0 - \frac{P}{2} \rho_0 (1 - \cos \phi) \right] P \cos \phi}{2AE} d\phi \tag{6.50a}$$

$$= \left(\frac{\alpha P^2}{8AG} + \frac{P^2}{8AE} \right) \frac{\pi}{4} \rho_0$$

$$+ \frac{1}{2AeE} \left[M_0^2 \frac{\pi}{2} + \frac{P^2}{4} \rho_0^2 \left(\frac{\pi}{2} + \frac{\pi}{4} - 2 \right) - M_0 \, \rho_0 \, P \left(\frac{\pi}{2} - 1 \right) \right]$$

$$- \frac{P}{2AE} \left(M_0 - \frac{P\rho_0}{2} + \frac{P\rho_0}{2} \frac{\pi}{4} \right) \tag{6.50b}$$

In the above expression, M_0 is still an unknown quantity. As the change in slope at the section where M is applied is zero,

$$\frac{\partial U}{\partial M_0} = \frac{1}{2AeE}\left[M_0\,\pi - \rho_0 P\left(\frac{\pi}{2} - 1\right)\right] - \frac{P}{2AE} = 0$$

$$\therefore \qquad M_0 = \frac{P\rho_0}{2}\left(1 - \frac{2}{\pi} + \frac{2e}{\pi\rho_0}\right) \tag{6.51}$$

If we ignore the initial curvature of the member while calculating the strain energy, then

$$U^* = \int_0^{\pi/2} \frac{\alpha P^2 \sin^2\phi}{8AG}\,\rho_0\,d\phi + \int_0^{\pi/2} \frac{P^2 \cos^2\phi}{8AE}\,\rho_0\,d\phi$$

$$+ \int_0^{\pi/2} \frac{\left[M_0 - \frac{P}{2}\rho_0\,(1 - \cos\phi)\right]^2}{2EI}\,d\phi$$

and

$$\frac{\partial U^*}{\partial M_0} = \frac{1}{EI}\int_0^{\pi/2}\left[M_0 - \frac{P}{2}\rho_0\,(1 - \cos\phi)\right]\rho_0\,d\phi = 0$$

i.e.

$$M_0\,\frac{\pi}{2} - \frac{P}{2}\rho_0\,\frac{\pi}{2} + \frac{P}{2}\rho_0 = 0$$

$$\therefore \qquad M_0 = \frac{P\rho_0}{2}\left(1 - \frac{2}{\pi}\right)$$

i.e. same as given in Eq. (6.51) with $e \to 0$ and $\rho_0 \to \infty$. Also, this moment is the same as in Example 5.12, i.e. that of a thin ring.

With the value of M_0 known, the bending moment at any section θ is obtained as

$$M = M_0 - \frac{P}{2}\rho_0\,(1 - \cos\theta)$$

$$= \frac{P\rho_0}{2}\left(\cos\theta + \frac{2e}{\pi\rho_0} - \frac{2}{\pi}\right)$$

The normal stress at A can be calculated using Eq. (6.35) and adding additional stress due to the normal force N.

$$\sigma_A = -\frac{M}{Ae}\cdot\frac{y}{(r_0 - y)} + \frac{N}{A}$$

$$= -\frac{P\rho_0}{2Ae}\left(\cos\theta + \frac{2e}{\pi\rho_0} - \frac{2}{\pi}\right)\frac{y}{r_0 - y} - \frac{P\cos\theta}{2A}$$

For point A, from Eqs (6.38) and (6.39b)

$$y = \frac{h}{2} - e, \qquad r_0 = \frac{r_2 - r_1}{\log(r_2/r_1)}, \qquad e = \rho_0 - \frac{r_2 - r_1}{\log(r_2/r_1)} = \rho_0 - r_0$$

Using these

$$\sigma_A = -\frac{P}{2A}\left\{\frac{\rho_0(\pi\cos\theta - 2) + 2e}{\pi e}\frac{(h-2e)}{(2\rho_0 - h)} + \cos\theta\right\}$$

Example 6.9 *A circular ring of rectangular section, shown in Fig. 6.31, is subjected to diametral compression. Determine the change in the vertical diameter.*

Solution From Eq. (6.50b), the total energy for the complete ring is

$$U = 4\rho_0\left\{\frac{\alpha P^2\pi}{32AG} + \frac{\pi P^2}{32AE} + \frac{1}{2AeE\rho_0}\left[\frac{\pi M_0^2}{2} + \frac{\rho_0^2 P^2}{4}\left(\frac{3\pi}{4} - 2\right)\right.\right.$$

$$\left.\left. - M_0\,\rho_0\,P\left(\frac{\pi}{2} - 1\right)\right] - \frac{P}{2A\rho_0 E}\left[M_0 + \frac{P\rho_0}{2}\left(\frac{\pi}{4} - 1\right)\right]\right\}$$

where $\quad M_0 = \frac{\rho_0 P}{2}\left(1 - \frac{2}{\pi} + \frac{2e}{\pi\rho_0}\right)$

$$\delta_v = \frac{\partial U}{\partial P}$$

Using the above expression for U (remembering that M is also a function of P), and simplifying

$$\delta_v = 4P\rho_0\left\{\frac{\alpha\pi}{16AG} + \frac{1}{2AE}\left(\frac{2}{\pi} - \frac{\pi}{8} - \frac{2e}{\pi\rho_0}\right) + \frac{\rho_0^2}{2AEe\rho_0}\left(\frac{\pi}{8} - \frac{1}{\pi} + \frac{e^2}{\pi\rho_0^2}\right)\right\}$$

If e is small compared to ρ_0, then

$$\delta_v \approx \frac{\alpha\pi\rho_0 P}{4AG} + \frac{2P\rho_0}{AE}\left(\frac{2}{\pi} - \frac{\pi}{8}\right) + \frac{2P\rho_0^3}{AEe\rho_0}\left(\frac{\pi}{8} - \frac{1}{\pi}\right)$$

$$= \frac{\alpha\pi P\rho_0}{4AG} + 0.488\frac{P\rho_0}{AE} + 0.15\frac{P\rho_0^2}{AEe}$$

If we assume that the ring is thin and the effect of the strain energies due to the direct force and shear force are negligible, then the chage in the vertical diameter is obtained as

$$\delta_v = \frac{P\rho_0^3}{EI}\left(\frac{\pi}{4} - \frac{2}{\pi}\right)$$

This can be seen from Eq. (6.35). When ρ_0 is large compared to y and $e \to 0$, $Ae\rho_0$ becomes equal to I according to flexure formula. Also, check with Example 5.13.

Problems

6.1 A rectangular wooden beam (Fig. 6.32) with a 10 cm × 15 cm section is used as a simply supported beam of 3 m span. It carries a uniformly distributed load of 150 kgf (1470 N) per meter. The load acts in a plane making 30° with the vertical. Calculate the maximum flexural stress at midspan and also locate the neutral axis for the same section.

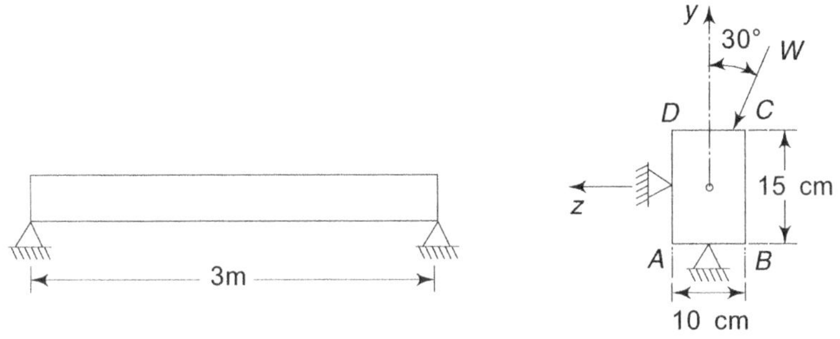

Fig. 6.32 *Problem 6.1*

$$\left[\begin{array}{r} \textit{Ans. } \sigma_A = 73 \text{ kgf/cm}^2 = 7126 \text{ kPa} \\ \text{N.A cuts side } AD \text{ such that } DN = 1.0 \text{ cm} \end{array}\right]$$

6.2 A cantilever beam with a rectangular cross section, 5 cm ×10 cm which is built-in in a tilted position, carries an end load of 45 kgf (441 N), as shown in Fig. 6.33. Calculate the maximum flexural stress at the built-in end and also locate the neutral axis. The length of the cantilever is 1.2 m.

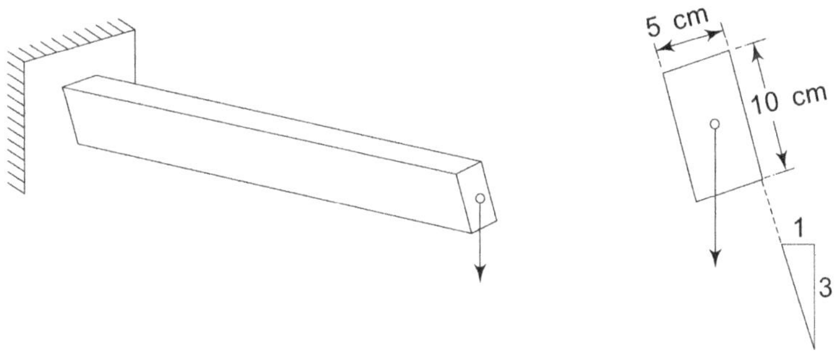

Fig. 6.33 *Problem 6.2*

$$\left[\begin{array}{r} \textit{Ans. } \sigma = \pm 102.5 \text{ kgf/cm}^2 = 10052 \text{ kPa} \\ \text{N.A. is at } 36.8° \text{ to the longerside} \end{array}\right]$$

6.3 A bar of angle section is bent by a couple M acting in the plane of the larger side (Fig. 6.34). Find the centroidal principal axes $Oy'z'$ and the principal moments of inertia. If M = 1.1550 kgf cm (1133 Nm), find the absolute maximum flexural stress in the section.

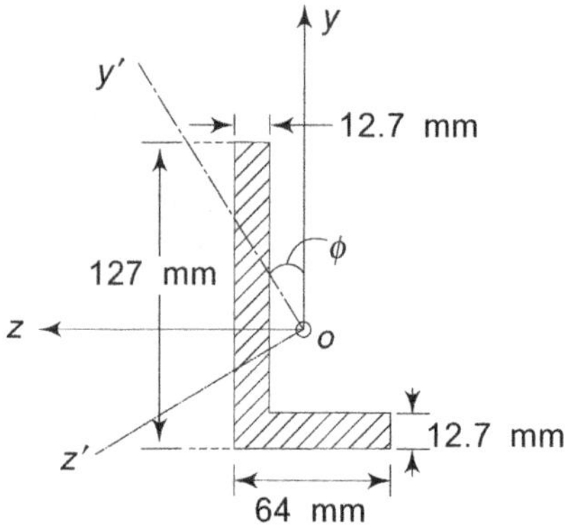

Fig. 6.34 *Problem 6.3*

[*Ans.* $\phi = \pm 14° \, 32'$
$I_{y'} = 41.9 \text{ cm}^4$; $I_{z'} = 391 \text{ cm}^4$
$\sigma_{max} = 33600$ kPa]

6.4 Determine the maximum absolute value of the normal stress due to bending and the position of the neutral axis in the dangerous section of the beam shown in Fig.6.35. Given a = 0.5 m and P = 200 kgf (1960 N). Section properties: equal legs 80 mm; centroid at 2.27 cm from the base; principal moments of inertia 116 cm^4, 30.3 cm^4; $I_z = 73.2$ cm^4.

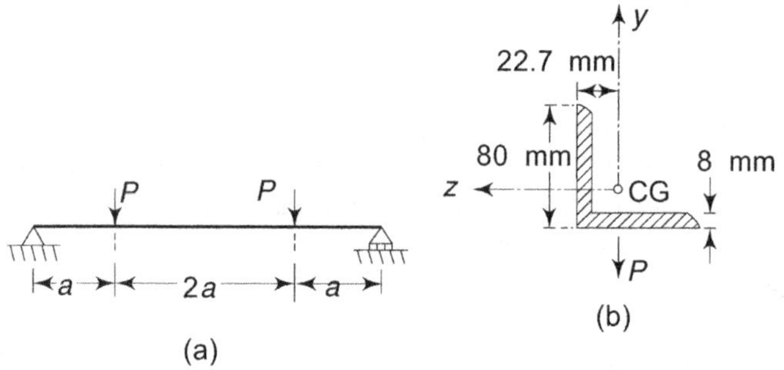

Fig. 6.35 *Problem 6.4*

[*Ans.* $\sigma = 914$ kgf/cm^2 (89640 kPa)
$\phi = 60°$ w.r.t. y axis]

6.5 Determine the maximum absolute value of the normal stress due to bending and the position of the neutral axis in the dangerous section of the beam. (Fig 6.36).

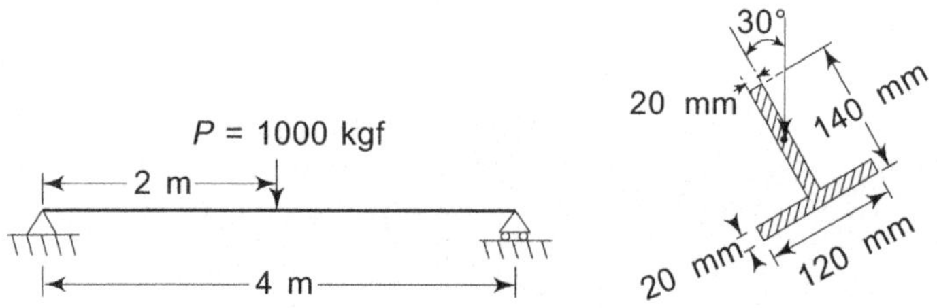

Fig. 6.36 *Problem 6.5*

[*Ans.* 1454 kgf/cm^2 (142588 kPa)
$\phi = 60.1°$ with vertical]

6.6 For the cantilever shown in Fig. 6.37, determine the maximum absolute value of the flexural stress and also locate the neutral axis at the section where this maximum stress occurs. $P = 200$ kgf (1960 N).

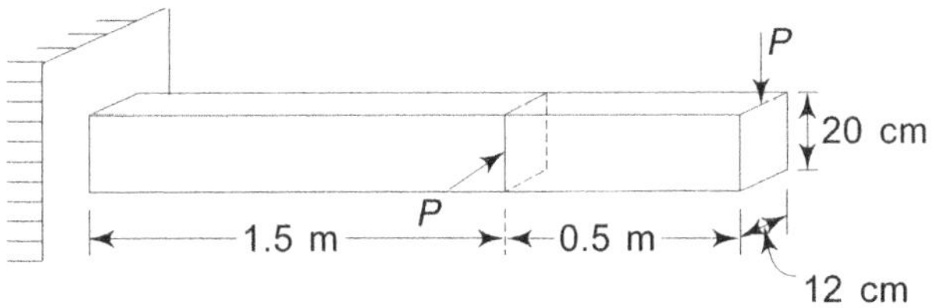

Fig. 6.37 *Problem 6.6*

[*Ans.* 112.5 kgf/cm^2 (11032 kPa)
$\phi = -25°36'$ with vertical]

6.7 A cantilever beam (Fig. 6.38) of length L has right triangular section and is loaded by P at the end. Solve for the stress at A near the built-in end. $P = 500$ kgf (4900 N), $h = 15$ cm, $b = 10$ cm and $L = 150$ cm.

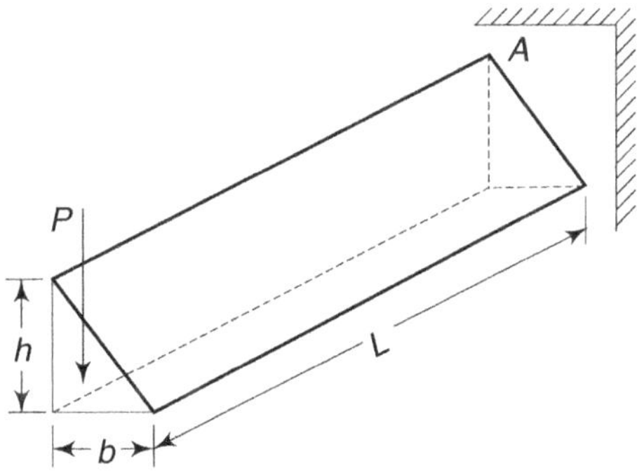

Fig. 6.38 *Problem 6.7*

[*Ans.* 2133 kgf/cm^2 (209175 kPa)]

6.8 Figure 6.39 shows an unsymmetrical beam section composed of four stringers A, B, C and D, each of equal area connected by a thin web. It is assumed that the web will not carry any bending stress. The beam section is subjected to the bending moments M_y and M_z, as indicated. Calculate the stresses in members A and D. The area of each stringer is 0.6 cm^2.

[*Ans.* $(\sigma_x)_A = -464$ kgf/cm^2 (–45503 kPa)
$(\sigma_x)_D = 448$ kgf/cm^2 (43934 kPa)]

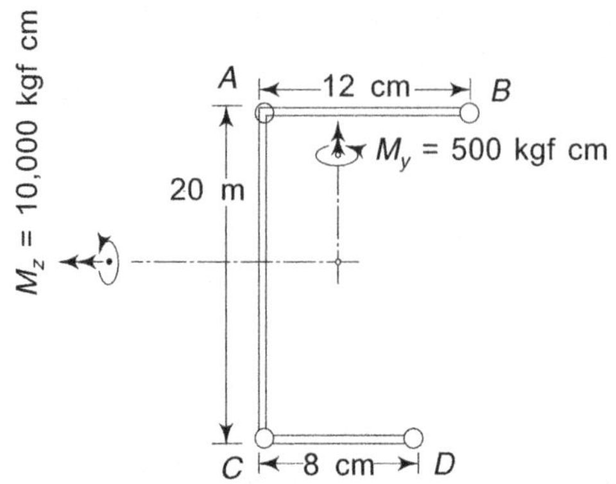

Fig. 6.39 *Problem 6.8*

6.9 In the above problem, if stringers *C* and *D* are made of magnesium alloy and stringers *A* and *B* of stainless steel, what will be the bending stresses in stringers *A* and *D*?

$$E_{\text{st st}} = 2 \times 10^6 \text{ kgf/cm}^2 \ (196 \times 10^6 \text{ kPa})$$
$$E_{\text{mg alloy}} = 0.4 \times 10^6 \text{ kgf/cm}^2 \ (39.2 \times 10^6 \text{ kPa})$$

Hint: Assume once again that sections that are plane before bending remain plane after bending. Hence, to produce the same strain, the stress will be proportional to *E*. Convert all the stringer areas into equivalent areas of one material. For example, the areas of stringers *C* and *D* in equivalent steel will be

$$A'_C = A_C \times \frac{E_{\text{mag}}}{E_{\text{st}}}, \quad \text{and} \quad A'_D = A_D \times \frac{E_{\text{mag}}}{E_{\text{st}}}$$

The areas of *A* and *B* remain unaltered. Solve the problem in the usual manner, using all equivalent steel stringers. Determine the stresses $(\sigma_x)'_A$ and $(\sigma_x)'_D$. Calculate the forces $F_A = (\sigma_x)'_A \, A'_A = (\sigma_x)'_A \, A_A$ and $F_D = (\sigma_x)'_D \, A'_D$. Now, using the original areas calculate the stress as

$$(\sigma_x)_A = (\sigma_x)'_A \, A'_A / A_A = (\sigma_x)'_A$$
$$(\sigma_x)_D = (\sigma_x)'_D \, A'_D / A_D$$

$$\left[\begin{array}{l} Ans. \ (\sigma_x)_A = -480 \text{ kgf/cm}^2 \ (-47072 \text{ kPa}) \\ \qquad (\sigma_x)_D = 425.6 \text{ kgf/cm}^2 \ (41737 \text{ kPa}) \end{array} \right]$$

6.10 Show that the shear centre for the section shown in Fig. 6.40 is at $e = 4R/\pi$ measured from point 0.

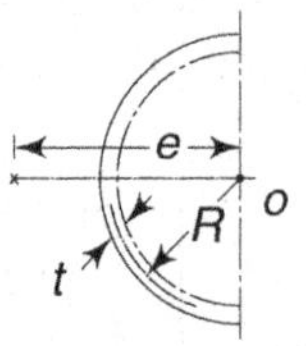

Fig. 6.40 *Problem 6.10*

6.11 For the section shown in Fig. 6.41 show that the shear centre is at a distance

$$e = R\frac{4(\sin\alpha - \alpha\cos\alpha)}{2\alpha - \sin 2\alpha}$$

from the centre of curvature O of the section.

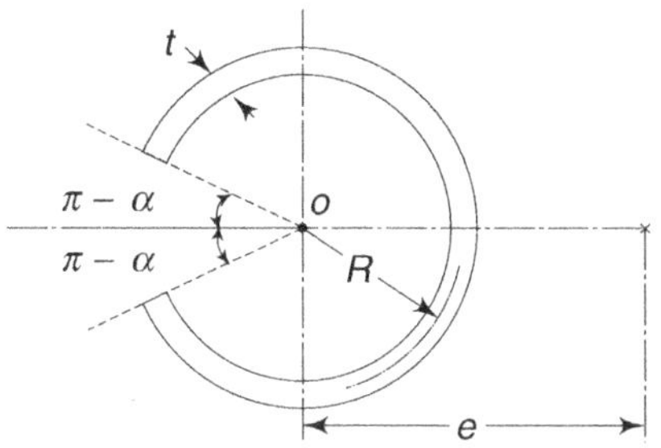

Fig. 6.41 *Problem 6.11*

6.12 Locate the shear centres from $C.Gs$ for the sections shown in Fig. 6.42(a), (b), and (c). In Fig. 6.42(b) the included angle is $\pi/2$.

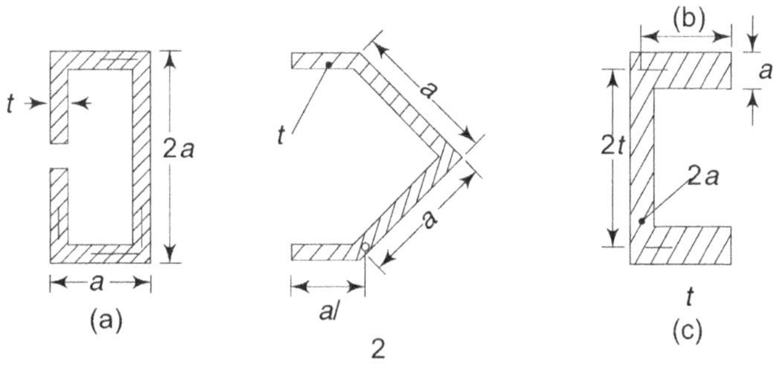

Fig. 6.42 *Problem 6.12*

[*Ans.* (a) 1.2 a, (b) 0.705 a (c) 0.76 a]

6.13 For the section given in Fig. 6.43, show that the shear centre is located at a distance e from O such that

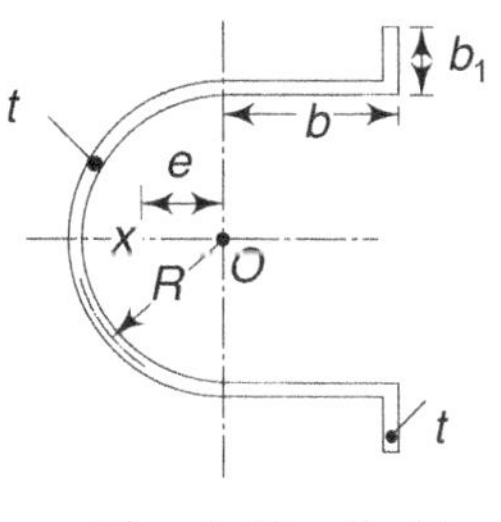

Fig. 6.43 *Problem 6.13*

$$e = \frac{A}{B}$$

where

$$A = 12 + 6\pi \frac{b + b_1}{R} + 6\left(\frac{b}{R}\right)^2 + 12\frac{b}{R}\frac{b_1}{R} + 3\pi\left(\frac{b_1}{R}\right)^2 - 4\left(\frac{b_1}{R}\right)^3 \frac{b}{R}$$

and $$B = 3\pi + 12 \frac{b + b_1}{R} + 3\left(\frac{b_1}{R}\right)^2 \left(4 + \frac{b_1}{R}\right)$$

Note: one can particularise this to the more familiar sections by putting b or b_1 or both equal to zero.

6.14 The open link shown in Fig. 6.44 Is loaded by forces P, each of which is equal to 1500 kgf (14,700 N). Find the maximum tensile and compressive stresses in the curved end at section AB.

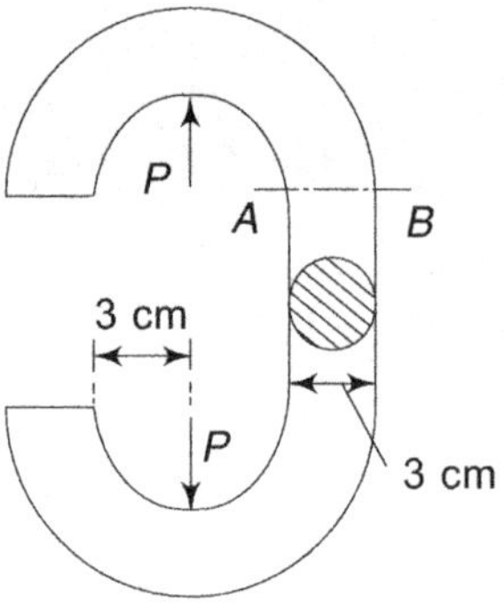

Fig. 6.44 *Problem 6.14*

$$\left[\begin{array}{l} \textit{Ans.}\ (\sigma_x)_A = 3591\,\text{kgf/cm}^2\ (352310\ \text{kPa}) \\ \qquad (\sigma_x)_B = -1796\,\text{kgf/cm}^2\ (-176147\ \text{kPa}) \end{array}\right]$$

6.15 A curved beam has an isosceles triangular section with the base of the triangle in the concave face. Develop the expression for r_0 in terms of the altitude h of the triangle and R the radius of curvature of the centroidal axis.

$$\left[\textit{Ans.}\ r_0 = \frac{3h^2}{2\left[(3R + 2h)\log \dfrac{3R + 2h}{3R - h} - 3h\right]}\right]$$

6.16 Find the maximum tensile stress in the curved part of the hook shown in Fig. 6.45. The web thickness is 1 cm.

[*Ans.* 3299 kgf/cm^2 (328680 kPa)]

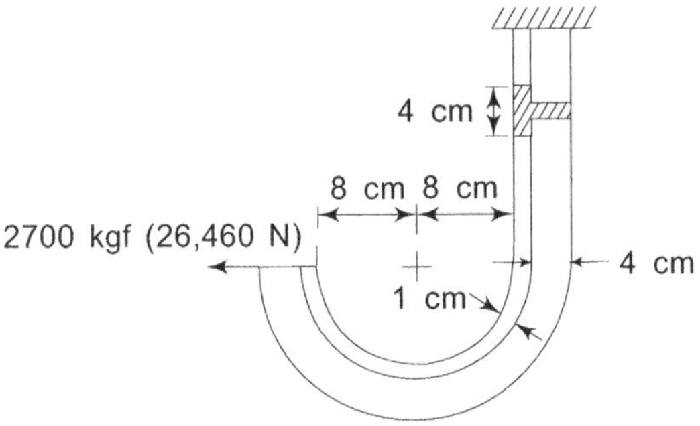

Fig. 6.45 *Problem 6.16*

6.17 Find the maximum tensile stress in the curved part of the hook shown in Fig. 6.46.

[*Ans.* $\sigma_x = 2277$ kgf/cm^2 (223300 kPa)]

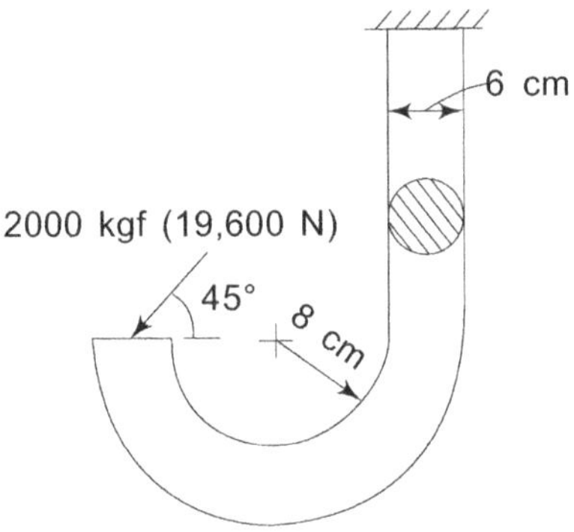

Fig. 6.46 *Problem 6.17*

6.18 Determine the ratio of the numerical value of σ_{max} and σ_{min} for a curved bar of rectangular cross-section in pure bending if $\rho_0 = 5$ cm and $h = r_2 - r_1 = 4$ cm. [*Ans.* 1.76]

6.19 Solve the previous problem if the bar is made of circular cross-section. [*Ans.* 1.89]

6.20 Determine the dimensions b_1 and b_3 of an I-section shown in Fig. 6.25, to make σ_{max} and σ_{min} numerically equal in pure bending. The other dimensions are $r_1 = 3$ cm; $r_3 = 4$ cm; $r_4 = 6$ cm; $r_2 = 7$ cm; $b_2 = 1$ cm; and $b_1 + b_3 = 5$ cm.

[*Ans.* $b_1 = 3.67$ cm, $b_3 = 1.33$ cm]

6.21 For the ring shown in Fig. 6.31 determine the changes in the horizontal diameter.

Hint: Apply two horizontal fictitious forces Q along the diameter. Calculate the total strain energy, Apply Castigliano's theorem.

$$\left[Ans.\ \delta_H = \frac{P\rho_0}{A}\left\{ -\frac{\alpha}{2G} + \frac{1}{E}\left(\frac{4}{\pi} - \frac{1}{2}\right) - \frac{1}{Ee\rho_0}\left[2e^2 - \rho_0^2\left(\frac{2}{\pi} - \frac{1}{2}\right)\right]\right\}\right]$$

CHAPTER

7 Torsion

7.1 INTRODUCTION

The torsion of circular shafts has been discussed in elementary strength of materials. There, we were able to obtain a solution to this problem under the assumption that the cross-sections of the bar under torsion remain plane and rotate without any distortion during twist. To observe this, consider the sheet shown in Fig. 7.1(a), subject to shear stress τ. The sheet deforms through an angle γ, as shown in Fig. 7.1(b).

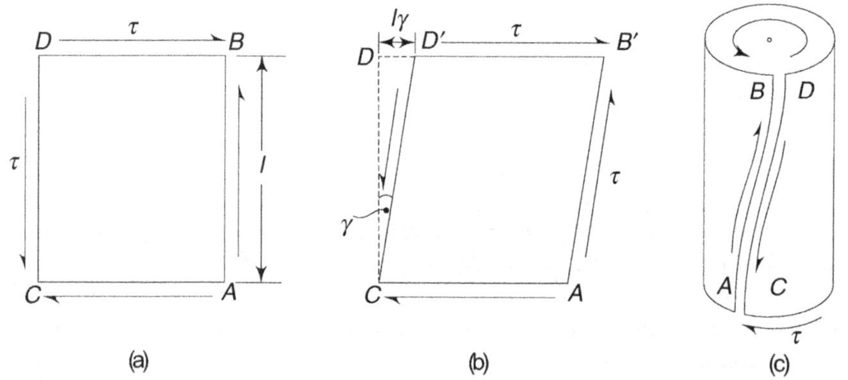

Fig. 7.1 *Deformation of a thin sheet under shear stress and the resulting tube*

If the deformed sheet is now folded to form a tube, the sides AB and CD can be joined without any discontinuity and this joined face will assume the form of a flat helix, as shown in Fig. 7.1(c). If γ is the shear strain, then from Hooke's law

$$\gamma = \frac{\tau}{G} \tag{7.1}$$

where G is the shear modulus. Owing to this strain, point D moves to D' [Fig. 7.1(b)], such that $DD' = l\gamma$. When the sheet is folded into a tube, the top face BD in Fig. 7.1(c), rotates with respect to the bottom face through an angle

$$\theta^* = \frac{l\gamma}{r} \tag{7.2}$$

where r is the radius of the tube. Substituting for γ from Eq. (7.1)

$$\theta^* = \frac{\tau}{G} \cdot \frac{l}{r}$$

or

$$\frac{\theta^*}{l} = \frac{\tau}{Gr} \tag{7.3}$$

Also, the moment about the centre of the tube is

$$T = r \times 2\pi r t \tau$$

or

$$T = \frac{2\pi r^3 t\tau}{r} = \frac{\tau I_p}{r}$$

i.e.

$$\frac{T}{I_p} = \frac{\tau}{r} \tag{7.4}$$

where I_p is the second polar moment of area.

Equations (7.3) and (7.4), therefore, give

$$\frac{T}{I_p} = \frac{\tau}{r} = \frac{G\theta^*}{l} \tag{7.5}$$

the familiar equations from elementary strength of materials. Now one can stack a series of tubes, one inside the other and for each tube, Eq. (7.5) would be valid. These stacked tubes can form the section of a solid (or a hollow) shaft if the top face of each tube has the same rotation $G\theta^*$, i.e. if $\frac{G\theta^*}{l}$ is the same for each tube. Therefore, the ratio $\frac{\tau}{r}$ is the same for each tube, or in other words, τ varies linearly with r. Further, if T_1 is the torque on the first tube with polar moment of inertia I_{p1}, T_2 the torque on the second tube with polar moment of inertia I_{p2}, etc., then

$$\frac{T_1}{I_{p1}} = \frac{T_2}{I_{p2}} = \ldots = \frac{T_n}{I_{pn}} = \frac{T_1 + T_2 + \ldots + T_n}{I_{p1} + I_{p2} + \ldots + I_{pn}} = \frac{T}{I_p}$$

where T is the total torque on the solid (or hollow) shaft and I_p is its polar moment of inertia.

From the above analysis we observe that for circular shafts, the cross-sections remain plane before and after, and there is no distortion of the section. But, for a non-circular section, this will no longer be valid. In the case of circular shafts, the shear stresses are perpendicular to a radial line and vary linearly with the radius. We can see that both these cannot be valid for a non-circular shaft. For, if the shear stress were always perpendicular to the radius *OB* [Fig. 7.2(a)], it would have a component perpendicular to the boundary. This is obviously inadmissible since the surface of the shaft is unloaded and a shear stress cannot cross an unloaded boundary. Hence, at the boundary, the shear stress must be tangential to the boundary. Further, by the same argument, the shear stress at the corner of a rectangular section must be zero, since the shear stresses on both the vertical faces are zero, i.e. both boundaries are unloaded boundaries [Fig. 7.2(b)].

In order to solve the torsion problem in general, we shall adopt St. Venant's semi-inverse method. According to this method, displacements u_x, u_y and u_z are

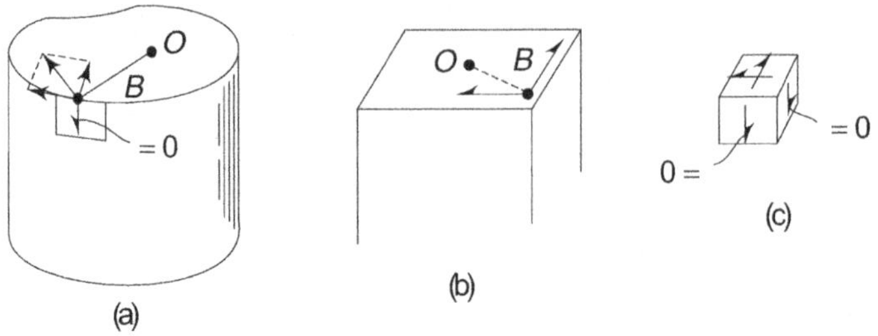

Fig. 7.2 *(a) Figure to show that shear stress must be tangential to boundary; (b) shear stress at the corner of a rectangular section being zero as shown in (c).*

assumed. The strains are then determined from strain-displacement relations [Eqs (2.18) and (2.19)]. Using Hooke's law, the stresses are then determined. Applying the equations of equilibrium and the appropriate boundary conditions, we try to identify the problem for which the assumed displacements and the associated stresses are solutions.

7.2 TORSION OF GENERAL PRISMATIC BARS–SOLID SECTIONS

We shall now consider the torsion of prismatic bars of any cross-section twisted by couples at the ends. It is assumed here that the shaft does not contain any holes parallel to the axis. In Sec. 7.12, multiply-connected sections will be discussed.

On the basis of the solution of circular shafts, we assume that the cross-sections rotate about an axis; the twist per unit length being θ. A section at distance z from the fixed end will, therefore, rotate through θz. A point $P(x, y)$ in this section will undergo a displacement $r\theta z$, as shown in Fig. 7.3. The components of this displacement are

$$u_x = -r\theta z \sin \beta$$
$$u_y = r\theta z \cos \beta$$

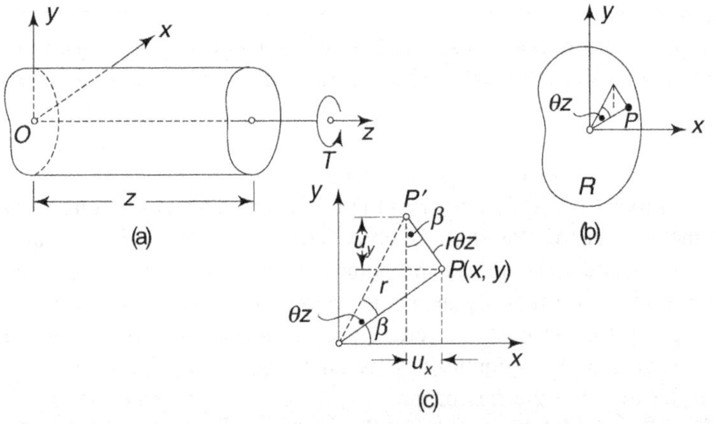

Fig. 7.3 *Prismatic bar under torsion and geometry of deformation*

From Fig. 7.3(c)

$$\sin \beta = \frac{y}{r} \quad \text{and} \quad \cos \beta = \frac{x}{r}$$

In addition to these x and y displacements, the point P may undergo a displacement u_z in z direction. This is called warping; we assume that the z displacement is a function of only (x, y) and is independent of z. This means that warping is the same for all normal cross-sections. Substituting for $\sin \beta$ and $\cos \beta$, St. Venant's displacement components are

$$u_x = -\theta yz \tag{7.6}$$

$$u_y = \theta xz$$
$$u_z = \theta \psi(x, y) \tag{7.7}$$

$\psi(x, y)$ is called the warping function. From these displacement components, we can calculate the associated strain components. We have, from Eqs (2.18) and (2.19),

$$\varepsilon_{xx} = \frac{\partial u_x}{\partial x}, \quad \varepsilon_{yy} = \frac{\partial u_y}{\partial y}, \quad \varepsilon_{zz} = \frac{\partial u_z}{\partial z}$$

$$\gamma_{xy} = \frac{\partial u_x}{\partial y} + \frac{\partial u_y}{\partial x}, \quad \gamma_{yz} = \frac{\partial u_y}{\partial z} + \frac{\partial u_z}{\partial y}, \quad \gamma_{zx} = \frac{\partial u_x}{\partial z} + \frac{\partial u_z}{\partial x}$$

From Eqs (7.6) and (7.7)

$$\varepsilon_{xx} = \varepsilon_{yy} = \varepsilon_{zz} = \gamma_{xy} = 0$$

$$\gamma_{yz} = \theta \left(\frac{\partial \psi}{\partial y} + x \right) \tag{7.8}$$

$$\gamma_{zx} = \theta \left(\frac{\partial \psi}{\partial x} - y \right)$$

From Hooke's law we have

$$\sigma_x = \frac{\nu E}{(1+\nu)(1-2\nu)} \Delta + \frac{E}{1+\nu} \varepsilon_{xx}$$

$$\sigma_y = \frac{\nu E}{(1+\nu)(1-2\nu)} \Delta + \frac{E}{1+\nu} \varepsilon_{yy}$$

$$\sigma_z = \frac{\nu E}{(1+\nu)(1-2\nu)} \Delta + \frac{E}{1+\nu} \varepsilon_{zz}$$

$$\tau_{xy} = G\gamma_{xy}, \quad \tau_{yz} = G\gamma_{yz}, \quad \tau_{zx} = G\gamma_{zx}$$

where $\quad \Delta = \varepsilon_{xx} + \varepsilon_{yy} + \varepsilon_{zz}$

Substituting Eq. (7.8) in the above set

$$\sigma_x = \sigma_y = \sigma_z = \tau_{xy} = 0$$

$$\tau_{yz} = G\theta \left(\frac{\partial \psi}{\partial y} + x \right) \tag{7.9}$$

$$\tau_{zx} = G\theta\left(\frac{\partial \psi}{\partial x} - y\right)$$

The above stress components are the ones corresponding to the assumed displacement components. These stress components should satisfy the equations of equilibrium given by Eq. (1.65), i.e.

$$\frac{\partial \sigma_x}{\partial x} + \frac{\partial \tau_{xy}}{\partial y} + \frac{\partial \tau_{zx}}{\partial z} = 0$$

$$\frac{\partial \tau_{xy}}{\partial x} + \frac{\partial \sigma_y}{\partial y} + \frac{\partial \tau_{yz}}{\partial z} = 0 \tag{7.10}$$

$$\frac{\partial \tau_{zx}}{\partial x} + \frac{\partial \tau_{yz}}{\partial y} + \frac{\partial \sigma_z}{\partial z} = 0$$

Substituting the stress components, the first two equations are satisfied identically. From the third equation, we obtain

$$G\theta\left(\frac{\partial^2 \psi}{\partial x^2} + \frac{\partial^2 \psi}{\partial y^2}\right) = 0$$

i.e.
$$\frac{\partial^2 \psi}{\partial x^2} + \frac{\partial^2 \psi}{\partial y^2} = \nabla^2 \psi = 0 \tag{7.11}$$

Hence, the warping function ψ is harmonic (i.e. it satisfies the Laplace equation) everywhere in region R [Fig. 7.3(b)].

Now let us consider the boundary conditions. If F_x, F_y and F_z are the components of the stress on a plane with outward normal $\boldsymbol{n}$ (n_x, n_y, n_z) at a point on the surface [Fig. 7.4(a)], then from Eq. (1.9)

$$\begin{aligned} n_x\sigma_x + n_y\,\tau_{xy} + n_z\,\tau_{xz} &= F_x \\ n_x\,\tau_{xy} + n_y\,\sigma_y + n_z\,\tau_{yz} &= F_y \\ n_x\,\tau_{xz} + n_y\,\tau_{yz} + n_z\,\sigma_z &= F_z \end{aligned} \tag{7.12}$$

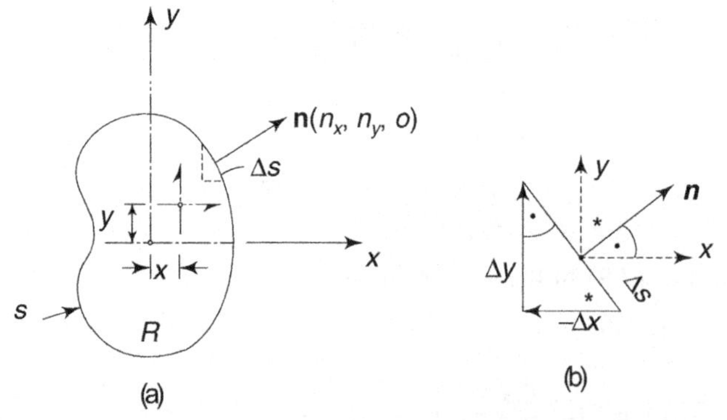

Fig. 7.4 *Cross-section of the bar and the boundary conditions*

In this case, there are no forces acting on the boundary and the normal **n** to the surface is perpendicular to the z-axis, i.e. $n_z \equiv 0$. Using the stress components from Eq. (7.9), we find that the first two equations in the boundary conditions are identically satisfied. The third equation yields

$$G\theta\left(\frac{\partial\psi}{\partial x} - y\right)n_x + G\theta\left(\frac{\partial\psi}{\partial y} + x\right)n_y = 0$$

From Fig. 7.4(b)

$$n_x = \cos(n, x) = \frac{dy}{ds}, \quad n_y = \cos(n, y) = -\frac{dx}{ds} \tag{7.13}$$

Substituting

$$\left(\frac{\partial\psi}{\partial x} - y\right)\frac{dy}{ds} - \left(\frac{\partial\psi}{\partial y} + x\right)\frac{dx}{ds} = 0 \tag{7.14}$$

Therefore, each problem of torsion is reduced to the problem of finding a function ψ which is harmonic, i.e. satisfies Eq. (7.11) in region R, and satisfies Eq. (7.14) on boundary s.

Next, on the two end faces, the stresses as given by Eq. (7.9) must be equivalent to the applied torque. In addition, the resultant forces in x and y directions should vanish. The resultant force in x direction is

$$\iint_R \tau_{zx}\,dx\,dy = G\theta\iint_R\left(\frac{\partial\psi}{\partial x} - y\right)dx\,dy \tag{7.15}$$

The right-hand side integrand can be written by adding $\nabla^2\psi$ as

$$\left(\frac{\partial\psi}{\partial x} - y\right) = \left(\frac{\partial\psi}{\partial x} - y\right) + x\left(\frac{\partial^2\psi}{\partial x^2} + \frac{\partial^2\psi}{\partial y^2}\right)$$

since $\nabla^2\psi = 0$, according to Eq. (7.11). Further,

$$\left(\frac{\partial\psi}{\partial x} - y\right) + x\left(\frac{\partial^2\psi}{\partial x^2} + \frac{\partial^2\psi}{\partial y^2}\right) = \frac{\partial}{\partial x}\left[x\left(\frac{\partial\psi}{\partial x} - y\right)\right] + \frac{\partial}{\partial y}\left[x\left(\frac{\partial\psi}{\partial y} + x\right)\right]$$

Hence, Eq. (7.15) becomes

$$\iint_R \tau_{zx}\,dx\,dy = G\theta\iint_R\left\{\frac{\partial}{\partial x}\left[x\left(\frac{\partial\psi}{\partial x} - y\right)\right] + \frac{\partial}{\partial y}\left[x\left(\frac{\partial\psi}{\partial y} + x\right)\right]\right\}dx\,dy$$

Using Gauss' theorem, the above surface integral can be converted into a line integral. Thus,

$$\begin{aligned}\iint_R \tau_{zx}\,dx\,dy &= G\theta\oint_S\left[x\left(\frac{\partial\psi}{\partial x} - y\right)n_x + x\left(\frac{\partial\psi}{\partial y} + x\right)n_y\right]ds \\ &= G\theta\oint_S x\left[\left(\frac{\partial\psi}{\partial x} - y\right)\frac{dy}{ds} + \left(\frac{\partial\psi}{\partial y} + x\right)\frac{dx}{ds}\right]ds \\ &= 0\end{aligned}$$

according to the boundary condition Eq. (7.14). Similarly, we can show that

$$\iint_R \tau_{yz}\, dx\, dy = 0$$

Now coming to the moment, referring to Fig. 7.4(a) and Eq. (7.9)

$$T = \iint_R (\tau_{yz}\, x - \tau_{zx}\, y)\, dx\, dy$$

$$= G\theta \iint_R \left(x^2 + y^2 + x\frac{\partial \psi}{\partial y} - y\frac{\partial \psi}{\partial x} \right) dx\, dy$$

Writing J for the integral

$$J = \iint_R \left(x^2 + y^2 + x\frac{\partial \psi}{\partial y} - y\frac{\partial \psi}{\partial x} \right) dx\, dy \tag{7.16}$$

we have $$T = GJ\theta \tag{7.17}$$

The above equation shows that the torque T is proportional to the angle of twist per unit length with a proportionality constant GJ, which is usually called the torsional rigidity of the shaft. For a circular cross-section, the quantity J reduces to the familiar polar moment of inertia. For non-circular shafts, the product GJ is retained as the torsional rigidity.

7.3 ALTERNATIVE APPROACH

An alternative approach proposed by Prandtl leads to a simpler boundary condition as compared to Eq. (7.14). In this method, the principal unknowns are the stress components rather than the displacement components as in the previous approach. Based on the result of the torsion of the circular-shaft, let the non-vanishing stress components be τ_{zx} and τ_{yz}. The remaining stress components σ_x, σ_y, σ_z and τ_{xy} are assumed to be zero. In order to satisfy the equations of equilibrium we should have

$$\frac{\partial \tau_{zx}}{\partial z} = 0, \quad \frac{\partial \tau_{yz}}{\partial z} = 0, \quad \frac{\partial \tau_{zx}}{\partial x} + \frac{\partial \tau_{yz}}{\partial y} = 0 \tag{7.18}$$

If it is assumed that in the case of pure torsion, the stresses are the same in every normal cross-section, i.e. independent of z, then the first two conditions above are automatically satisfied. In order to satisfy the third condition, we assume a function $\phi\,(x, y)$ called the stress function, such that

$$\tau_{zx} = \frac{\partial \phi}{\partial y}, \qquad \tau_{yz} = -\frac{\partial \phi}{\partial x} \tag{7.19}$$

With this stress function (called Prandtl's torsion stress function), the third condition is also satisfied. The assumed stress components, if they are to be proper elasticity solutions, have to satisfy the compatibility conditions. We can substitute these directly into the stress equations of compatibility. Alternatively, we can determine the strains corresponding to the assumed stresses and then apply the strain compatibility conditions given by Eq. (2.56). The strain components from Hooke's law are

$$\varepsilon_{xx} = 0, \qquad \varepsilon_{yy} = 0, \qquad \varepsilon_{zz} = 0 \tag{7.20}$$

$$\gamma_{xy} = 0, \qquad \gamma_{yz} = \frac{1}{G}\,\tau_{yz}, \qquad \gamma_{zx} = \frac{1}{G}\,\tau_{zx}$$

Substituting from Eq. (7.19)

$$\gamma_{yz} = -\frac{1}{G}\frac{\partial \phi}{\partial x}, \quad \text{and} \quad \gamma_{zx} = \frac{1}{G}\frac{\partial \phi}{\partial y}$$

From Eq. (2.56), the non-vanishing strain compatibility conditions are (observe that ϕ is independent of z)

$$\frac{\partial}{\partial x}\left(-\frac{\partial \gamma_{yz}}{\partial x} + \frac{\partial \gamma_{zx}}{\partial y}\right) = 0$$

$$\frac{\partial}{\partial y}\left(\frac{\partial \gamma_{yz}}{\partial x} - \frac{\partial \gamma_{zx}}{\partial y}\right) = 0$$

i.e.
$$\frac{\partial}{\partial x}\left(\frac{\partial^2 \phi}{\partial x^2} + \frac{\partial^2 \phi}{\partial y^2}\right) = 0; \quad \frac{\partial}{\partial y}\left(\frac{\partial^2 \phi}{\partial x^2} + \frac{\partial^2 \phi}{\partial y^2}\right) = 0$$

Hence,
$$\frac{\partial^2 \phi}{\partial x^2} + \frac{\partial^2 \phi}{\partial y^2} = \nabla^2 \phi = a \text{ constant } F \tag{7.21}$$

The stress function, therefore, should satisfy Poisson's equation. The constant F is yet unknown. Next, we consider the boundary conditions [Eq. (7.12)]. The first two of these are identically satisfied. The third equation gives

$$n_x \frac{\partial \phi}{\partial y} - n_y \frac{\partial \phi}{\partial x} = 0$$

Substituting for n_x and n_y from Eq. (7.13)

$$\frac{\partial \phi}{\partial y}\frac{dy}{ds} + \frac{\partial \phi}{\partial x}\frac{dx}{ds} = 0$$

i.e.
$$\frac{d\phi}{ds} = 0 \tag{7.22}$$

Therefore, ϕ is constant around the boundary. Since the stress components depend only on the differentials of ϕ, for a simply connected region, no loss of generality is involved in assuming

$$\phi = 0 \text{ on } s \tag{7.23}$$

For a multi-connected region R (i.e. a shaft having holes), certain additional conditions of compatibility are imposed. This will be discussed in Sec. 7.9.

On the two end faces, the resultants in x and y directions should vanish, and the moment about O should be equal to the applied torque T. The resultant in x direction is

$$\iint_R \tau_{zx}\, dx\, dy = \iint_R \frac{\partial \phi}{\partial y}\, dx\, dy$$

$$= \int dx \int \frac{\partial \phi}{\partial y}\, dy$$

$$= 0$$

since ϕ is constant around the boundary. Similarly, the resultant in y direction also vanishes. Regarding the moment, from Fig. 7.4(a)

$$T = \iint_R (x\tau_{zy} - y\tau_{zx})\, dx\, dy$$

$$= -\iint_R \left(x \frac{\partial \phi}{\partial x} + y \frac{\partial \phi}{\partial y} \right) dx\, dy$$

$$= -\iint_R x \frac{\partial \phi}{\partial x}\, dx\, dy - \iint_R y \frac{\partial \phi}{\partial y}\, dx\, dy$$

Integrating by parts and observing that $\phi = 0$ of the boundary, we find that each integral gives

$$\iint \phi\, dx\, dy$$

Thus $$T = 2\iint \phi\, dx\, dy \tag{7.24}$$

Hence, we observe that half the torque is due to τ_{zx} and the other half to $\tau_{yz.}$

Thus, all differential equations and boundary conditions are satisfied if the stress function ϕ obeys Eqs (7.21), (7.23) and (7.24). But there remains an indeterminate constant in Eq. (7.21). To determine this, we observe from Eq. (7.19)

$$\frac{\partial^2 \phi}{\partial x^2} + \frac{\partial^2 \phi}{\partial y^2} = \frac{\partial \tau_{zx}}{\partial y} - \frac{\partial \tau_{yz}}{\partial x}$$

$$= G\left(\frac{\partial \gamma_{zx}}{\partial y} - \frac{\partial \gamma_{yz}}{\partial x} \right)$$

$$= G\left[\frac{\partial}{\partial y}\left(\frac{\partial u_x}{\partial z} + \frac{\partial u_z}{\partial x} \right) - \frac{\partial}{\partial x}\left(\frac{\partial u_y}{\partial z} + \frac{\partial u_z}{\partial y} \right) \right]$$

$$= G\frac{\partial}{\partial z}\left(\frac{\partial u_x}{\partial y} - \frac{\partial u_y}{\partial x} \right)$$

$$= G\frac{\partial}{\partial z}\,(-2\omega_z)$$

where ω_z is the rotation of the element at (x, y) about the z-axis [Eq. (2.25), Sec. 2.8]. $(\partial/\partial z)\,(\omega_z)$ is the rotation per unit length. In this chapter, we have termed it as twist per unit length and denoted it by θ. Hence,

$$\frac{\partial^2 \phi}{\partial x^2} + \frac{\partial^2 \phi}{\partial y^2} = \nabla^2 \phi = -2G\theta \tag{7.25}$$

According to Eq. (7.19),

$$\tau_{zx} = \frac{\partial \phi}{\partial y}, \quad \tau_{yz} = -\frac{\partial \phi}{\partial x}$$

That is, the shear acting in the x direction is equal to the slope of the stress function ϕ (x, y) in the y direction. The shear stress acting in the y direction is equal to the negative of the slope of the stress function in the x direction. This condition may be generalised to determine the shear stress in any direction, as follows. Consider a line of constant ϕ in the cross-section of the bar. Let s be the contour line of ϕ = constant [Fig. 7.5(a)] along this contour

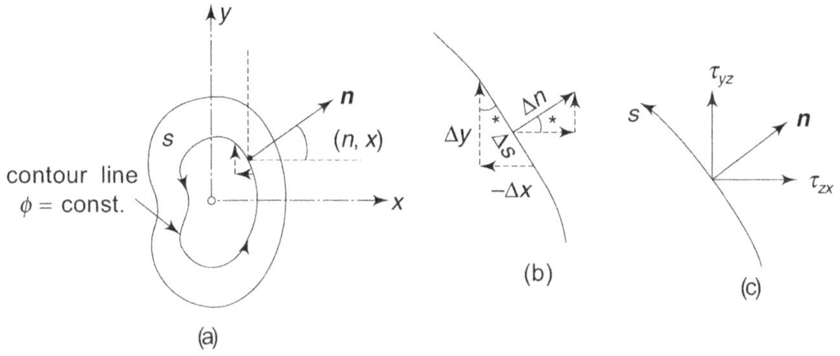

Fig. 7.5 *Cross-section of the bar and contour lines of* ϕ

$$\frac{d\phi}{ds} = 0 \tag{7.26a}$$

i.e.

$$\frac{\partial \phi}{dx}\frac{dx}{ds} + \frac{\partial \phi}{dy}\frac{dy}{ds} = 0 \tag{7.26b}$$

or

$$-\tau_{yz}\frac{dx}{ds} + \tau_{zx}\frac{dy}{ds} = 0 \tag{7.26c}$$

From Fig. 7.5(b)

$$-\frac{dx}{ds} = \cos(\boldsymbol{n}, y) = \frac{dy}{dn}$$

and

$$-\frac{dy}{ds} = \cos(\boldsymbol{n}, x) = \frac{dx}{dn}$$

where $\boldsymbol{n}$ is the outward drawn normal. Therefore, Eq. (7.26c) becomes

$$\tau_{yz}\cos(\boldsymbol{n}, y) + \tau_{zx}\cos(\boldsymbol{n}, x) = 0 \tag{7.27a}$$

From Fig. 7.5(c), the expression on the left-hand side is equal to τ_{zn}, the component of resultant shear in the direction $\boldsymbol{n}$.

Hence,

$$\tau_{zn} = 0 \tag{7.27b}$$

This means that the resultant shear at any point is along the contour line of ϕ = constant at that point. These contour lines are called lines of shearing stress. The resultant shearing stress is therefore

$$\tau_{zs} = \tau_{yz}\sin(\boldsymbol{n}, y) - \tau_{zx}\sin(\boldsymbol{n}, x)$$

$$= \tau_{yz} \cos (\boldsymbol{n}, x) - \tau_{zx} \cos (\boldsymbol{n}, y)$$

$$= \tau_{yz} \frac{dx}{dn} - \tau_{zx} \frac{dy}{dn}$$

$$= -\frac{\partial \phi}{\partial x}\frac{dx}{dn} - \frac{\partial \phi}{\partial y}\frac{dy}{dn} \tag{7.28}$$

or $$\tau_{zs} = -\frac{\partial \phi}{\partial n}$$

Thus, the magnitude of the shearing stress at a point is given by the magnitude of the slope of $\phi (x, y)$ measured normal to the tangent line, i.e. normal to the contour line at the concerned point. The above points are very important in the analysis of a torsion problem by membrane analogy, discussed in Sec. 7.7.

7.4 TORSION OF CIRCULAR AND ELLIPTICAL BARS

(i) The simplest solution to the Laplace equation (Eq. 7.11) is

$$\psi = \text{constant} = c \tag{7.29}$$

With $\psi = c$, the boundary condition given by Eq. (7.14) becomes

$$-y\frac{dy}{ds} - x\frac{dx}{ds} = 0$$

or $$\frac{d}{ds}\frac{x^2 + y^2}{2} = 0$$

i.e. $$x^2 + y^2 = \text{constant}$$

where (x, y) are the coordinates of any point on the boundary. Hence, the boundary is a circle. From Eq. (7.7), $u_z = \theta c$. From Eq. (7.16)

$$J = \iint_R (x^2 + y^2)\, dx\, dy = I_p$$

the polar moment of inertia for the section. Hence, from Eq. (7.17)

$$T = GI_p\theta$$

or $$\theta = \frac{T}{GI_p}$$

Therefore, $$u_z = \theta c = \frac{Tc}{GI_p}$$

which is a constant. Since the fixed end has zero u_z at least at one point, u_z is zero at every cross-section (other than rigid body displacement). Thus, the cross-section does not warp. The shear stresses are given by Eq. (7.9) as

$$\tau_{yz} = G\theta x = \frac{Tx}{I_p}$$

$$\tau_{zx} = -G\theta y = -\frac{Ty}{I_p}$$

Therefore, the direction of the resultant shear τ is such that, from Fig. 7.6

$$\tan \alpha = \frac{\tau_{zy}}{\tau_{zx}} = -\frac{G\theta x}{G\theta y} = -\frac{x}{y}$$

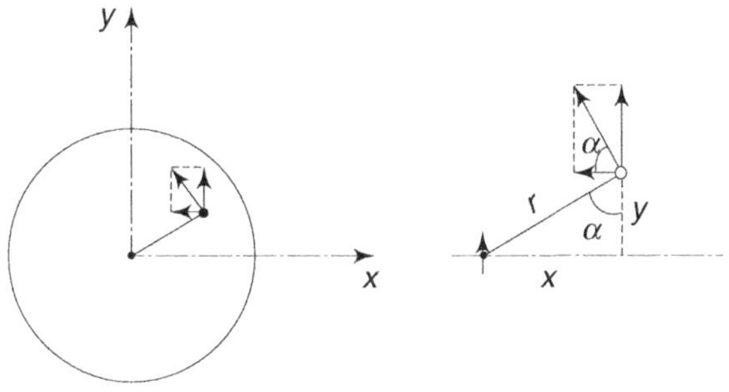

Fig. 7.6 *Torsion of a circular bar*

Hence, the resultant shear is perpendicular to the radius. Further

$$\tau^2 = \tau_{yz}^2 + \tau_{zx}^2 = \frac{T^2 (x^2 + y^2)}{I_p^2}$$

or
$$\tau = \frac{Tr}{I_p}$$

where r is the radial distance of the point (x, y). Thus, all the results of the elementary analysis are justified.

(ii) The next case in the order of simplicity is to assume that

$$\psi = Axy \tag{7.30}$$

where A is a constant. This also satisfies the Laplace equation. The boundary condition, Eq. (7.14) gives,

$$(Ay - y)\frac{dy}{ds} - (Ax + x)\frac{dx}{ds} = 0$$

or
$$y (A - 1)\frac{dy}{ds} - x (A + 1)\frac{dx}{ds} = 0$$

i.e.
$$(A + 1)\, 2x\frac{dx}{ds} - (A - 1)\, 2y\frac{dy}{ds} = 0$$

or
$$\frac{d}{ds}\left[(A + 1)\, x^2 - (A - 1)\, y^2\right] = 0$$

which on integration, yields

$$(1 + A)\, x^2\, (1 - A)\, y^2 - \text{constant} \tag{7.31}$$

This is of the form

$$\frac{x^2}{a^2} + \frac{y^2}{b^2} = 1$$

These two are identical if

$$\frac{a^2}{b^2} = \frac{1-A}{1+A}$$

or
$$A = \frac{b^2 - a^2}{b^2 + a^2}$$

Therefore, the function

$$\psi = \frac{b^2 - a^2}{b^2 + a^2}\, xy$$

represents the warping function for an elliptic cylinder with semi-axes a and b under torsion. The value of J, as given in Eq. (7.16), is

$$J = \iint\limits_R (x^2 + y^2 + Ax^2 - Ay^2)\, dx\, dy$$

$$= (A + 1)\iint x^2\, dx\, dy + (1 - A) \iint y^2\, dx\, dy$$

$$= (A + 1)\, I_y + (1 - A)\, I_x$$

Substituting $I_x = \dfrac{\pi a b^3}{4}$ and $I_y = \dfrac{\pi a^3 b}{4}$, one gets

$$J = \frac{\pi a^3 b^3}{a^2 + b^2}$$

Hence, from Eq. (7.17)

$$T = GJ\theta = G\theta\, \frac{\pi a^3 b^3}{a^2 + b^2}$$

or
$$\theta = \frac{T}{G}\, \frac{a^2 + b^2}{\pi a^3 b^3} \tag{7.32}$$

The shearing stresses are given by Eq. (7.9) as

$$\tau_{yz} = G\theta \left(\frac{\partial \psi}{dy} + x \right)$$

$$= T\frac{a^2 + b^2}{\pi a^3 b^3} \left(\frac{b^2 - a^2}{b^2 + a^2} + 1 \right) x$$

$$= \frac{2Tx}{\pi a^3 b} \tag{7.33a}$$

and similarly,

$$\tau_{zx} = \frac{2Ty}{\pi a b^3} \tag{7.33b}$$

The resultant shearing stress at any point (x, y) is

$$\tau = \left[\tau_{yz}^2 + \tau_{zx}^2\right]^{1/2} = \frac{2T}{\pi a^3 b^3}\left[b^4 x^2 + a^4 y^2\right]^{1/2} \tag{7.33c}$$

To determine where the maximum shear stress occurs, we substitute for x^2 from

$$\frac{x^2}{a^2} + \frac{y^2}{b^2} = 1, \quad \text{or} \quad x^2 = a^2\left(1 - \frac{y^2}{b^2}\right)$$

giving
$$\tau = \frac{2T}{\pi a^3 b^3}\,[a^2 b^4 + a^2(a^2 - b^2)\,y^2]^{1/2}$$

Since all terms under the radical (power 1/2) are positive, the maximum shear stress occurs when y is maximum, i.e. when $y = b$. Thus, τ_{max} occurs at the ends of the minor axis and its value is

$$\tau_{max} = \frac{2T}{\pi a^3 b^3}(a^4 b^2)^{1/2} = \frac{2T}{\pi a b^2} \tag{7.34}$$

With the warping function known, the displacement u_z can easily be determined. We have from Eq. (7.7)

$$u_z = \theta\psi = \frac{T(b^2 - a^2)}{\pi a^3 b^3 G}\,xy$$

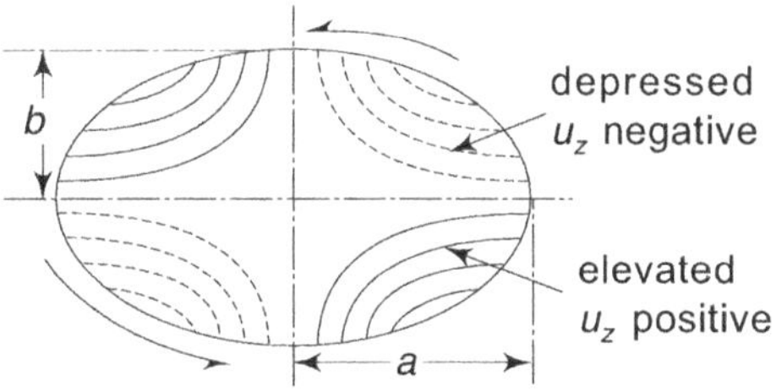

Fig. 7.7 *Cross-section of an elliptical bar and contour lines of* u_z

The contour lines giving u_z = constant are the hyperbolas shown in Fig. 7.7. For a torque T as shown, the convex portions of the cross-section, i.e. where u_z is positive, are indicated by solid lines, and the concave portions or where the surface is depressed, are shown by dotted lines. If the ends are free, there are no normal stresses. However, if one end is built-in, the warping is prevented at that end and consequently, normal stresses are induced which are positive in one quadrant and negative in another. These are similar to bending stresses and are, therefore, called the bending stresses induced because of torsion.

7.5 TORSION OF EQUILATERAL TRIANGULAR BAR

Consider the warping function

$$\psi = A(y^3 - 3x^2 y) \tag{7.35}$$

This satisfies the Laplace equation, which can easily be verified. The boundary condition given by Eq. (7.14) yields

$$(-6Axy - y)\frac{dy}{ds} - (3Ay^2 - 3Ax^2 + x)\frac{dx}{ds} = 0$$

or $$y(6Ax+1)\frac{dy}{ds} + (3Ay^2 - 3Ax^2 + x)\frac{dx}{ds} = 0$$

i.e. $$\frac{d}{ds}\left(3Axy^2 - Ax^3 + \frac{1}{2}x^2 + \frac{1}{2}y^2\right) = 0$$

Therefore,

$$A(3xy^2 - x^3) + \frac{1}{2}x^2 + \frac{1}{2}y^2 = b \tag{7.36}$$

where b is a constant. If we put $A = -\frac{1}{6a}$ and $b = +\frac{2a^2}{3}$, Eq. (7.36) becomes

$$-\frac{1}{6a}(3xy^2 - x^3) + \frac{1}{2}(x^2 + y^2) - \frac{2}{3}a^2 = 0$$

or $$(x - \sqrt{3}\,y + 2a)(x + \sqrt{3}\,y + 2a)(x - a) = 0 \tag{7.37}$$

Equation (7.37) is the product of the three equations of the sides of the triangle shown in Fig. 7.8. The equations of the boundary lines are

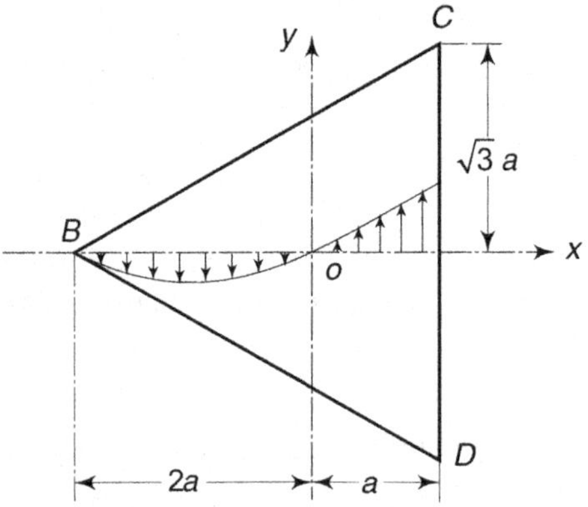

Fig. 7.8 *Cross-section of a triangular bar and plot of τ_{yz} along x-axis*

$$x - a = 0 \quad \text{on } CD$$

$$x - \sqrt{3}\,y + 2a = 0 \quad \text{on } BC$$

$$x + \sqrt{3}\,y + 2a = 0 \quad \text{on } BD$$

From Eq. (7.16)

$$J = \iint_R \left[x^2 + y^2 + Ax\left(3y^2 - 3x^2\right) - Ay(-6xy)\right] dx\, dy$$

$$= \int_0^{\sqrt{3}a} dy \int_{-\sqrt{3}\,y-2a}^{a} \left[x^2 + y^2 + Ax\left(3y^2 - 3x^2\right) - Ay(-6xy)\right] dx$$

$$+ \int_{-\sqrt{3}a}^{a} dy \int_{-\sqrt{3}\,y-2a}^{a} \left[x^2 + y^2 + Ax\left(3y^2 - 3x^2\right) - Ay(-6xy)\right] dx$$

$$= \frac{9\sqrt{3}}{5}a^4 = \frac{3}{5}I_p \tag{7.38}$$

Therefore,

$$\theta = \frac{T}{GJ} = \frac{5}{3}\frac{T}{GI_P} \tag{7.39}$$

I_p is the polar moment of inertia about 0.

The stress components are

$$\tau_{yz} = G\theta\left(\frac{\partial\psi}{\partial y} + x\right)$$

$$= G\theta\left(3Ay^2 - 3Ax^2 + x\right)$$

$$= \frac{G\theta}{2a}\left(x^2 - y^2 + 2ax\right) \tag{7.40}$$

and $$\tau_{zx} = G\theta\left(\frac{\partial\psi}{\partial y} - y\right)$$

$$= \frac{G\theta y}{a}(x-a) \tag{7.41}$$

The largest shear stress occurs at the middle of the sides of the triangle, with a value

$$\tau_{\max} = \frac{3G\theta a}{2} \tag{7.42}$$

At the corners of the triangle, the shear stresses are zero. Along the x-axis, $\tau_{zx} = 0$ and the variation of τ_{yz} is shown in Fig. 7.8. τ_{yz} is also zero at the origin 0.

7.6 TORSION OF RECTANGULAR BARS

The torsion problem of rectangular bars is a bit more involved compared to those of elliptical and triangular bars. We shall indicate only the method of approach without going into the details. Let the sides of the rectangular cross-section be $2a$ and $2b$ with the origin at the centre, as shown in Fig. 7.9(a).

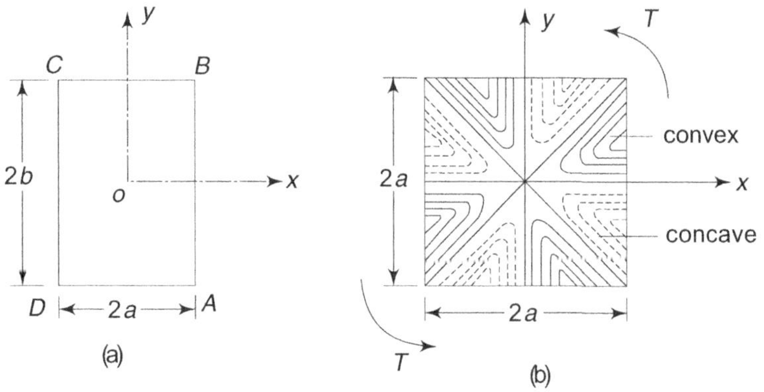

Fig. 7.9 *(a) Cross-section of a rectangular bar (b) Warping of a square section*

Our equations are, as before,

$$\frac{\partial^2 \psi}{\partial x^2} + \frac{\partial^2 \psi}{\partial y^2} = 0$$

over the whole region R of the rectangle, and

$$\left(\frac{\partial \psi}{\partial x} - y\right) n_x + \left(\frac{\partial \psi}{\partial y} + x\right) n_y = 0$$

on the boundary. Now on the boundary lines $x = \pm a$ or AB and CD, we have $n_x = \pm 1$ and $n_y = 0$. On the boundary lines BC and AD, we have $n_x = 0$ and $n_y = \pm 1$. Hence, the boundary conditions become

$$\frac{\partial \psi}{\partial x} = y \qquad \text{on} \quad x = \pm a$$

$$\frac{\partial \psi}{\partial y} = -x \qquad \text{on} \quad y = \pm b$$

These boundary conditions can be transformed into more convenient forms if we introduce a new function ψ_1, such that

$$\psi = xy - \psi_1$$

In terms of ψ_1, the governing equation is

$$\frac{\partial^2 \psi_1}{\partial x^2} + \frac{\partial^2 \psi_1}{\partial y^2} = 0$$

over region R, and the boundary conditions become

$$\frac{\partial \psi_1}{\partial x} = 0 \qquad \text{on} \quad x = \pm a$$

$$\frac{\partial \psi_1}{\partial y} = 2x \qquad \text{on} \quad y = \pm b$$

It is assumed that the solution is expressed in the form of infinite series

$$\psi = \sum_{n=0}^{\infty} X_n(x) Y_n(y)$$

where X_n and Y_n are respectively functions of x alone and y alone. Substitution into the Laplace equation for ψ_1 yields two linear ordinary differential equations with constant coefficients. Further details of the solution can be obtained by referring to books on theory of elasticity. The final results which are important are as follows:

The function J is given by

$$J = Ka^3 b$$

For various b/a ratios, the corresponding values of K are given in Table 7.1. Assuming that $b > a$, it is shown in the detailed analysis that the maximum

Table 7.1

b/a	K	K_1	K_2
1	2.250	1.350	0.600
1.2	2.656	1.518	0.571
1.5	3.136	1.696	0.541
2.0	3.664	1.860	0.508
2.5	3.984	1.936	0.484
3.0	4.208	1.970	0.468
4.0	4.496	1.994	0.443
5.0	4.656	1.998	0.430
10.0	4.992	2.000	0.401
∞	5.328	2.000	0.375

shearing stress is at the mid-points of the long sides $x = \pm a$ of the rectangle. On these sides

$$\tau_{zx} = 0 \qquad \text{and} \qquad \tau_{\max} = K_1 \frac{Ta}{J}$$

The values of K_1 for various values of b/a are given in Table 7.1. Substituting for J, the above expression can be written as

$$\tau_{\max} = K_2 \frac{Ta}{a^2 b}$$

where K_2 is another numerical factor, as given in Table 7.1. For a square section, i.e. $b/a = 1$, the warping is as shown in Fig. 7.9 (b). The zones where u_z is positive are shown by solid lines and the zones where u_z is negative are shown by dotted lines.

Empirical Formula for Squatty Sections

Equation (7.32), which is applicable to an elliptical section, can be written as

$$\frac{T}{\theta} = \frac{\pi a^3 b^3}{a^2 + b^2} G = \frac{1}{4\pi^2} \frac{GA^4}{I_p}$$

where $A = \pi ab$ is the area of the ellipse, and $I_p = \dfrac{(a^2 + b^2)}{4} A$ is the polar moment of inertia. This formula is applicable to a large number of squatty sections with an error not exceeding 10%. If $4\pi^2$ is replaced by 40, the mean error becomes less than 8% for many sections. Hence,

$$\frac{T}{\theta} = \frac{GA^4}{40 I_p}$$

is an approximate formula that can be applied to many sections other than elongated or narrow sections (see Secs 7.10 and 7.11).

7.7 MEMBRANE ANALOGY

From the examples worked out in the previous sections, it becomes evident that for bars with more complicated cross-sectional shapes, analytical solutions tend to become more involved and difficult. In such situations, it is desirable to resort to other techniques—experimental or otherwise. The membrane analogy introduced by Prandtl has proved very valuable in this regard. Let a thin homogeneous membrane like a thin rubber sheet be stretched with uniform tension and fixed at its edge, which is a given curve (the cross-section of the shaft) in the xy-plane (Fig. 7.10).

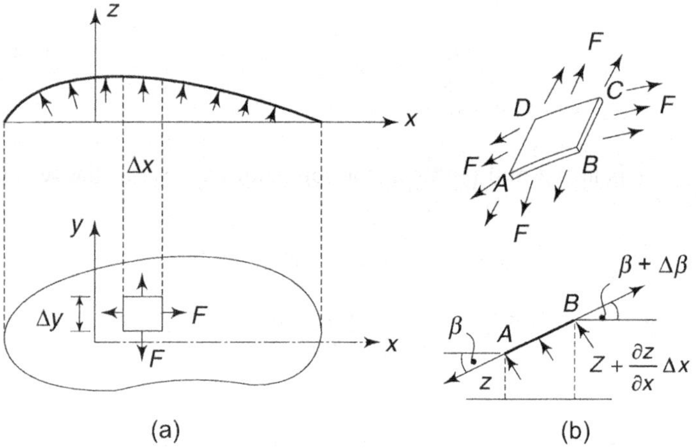

Fig. 7.10 *Stretching of a membrane*

When the membrane is subjected to a uniform lateral pressure p, it undergoes a small displacement z where z is a function of x and y. Consider the equilibrium of an infinitesimal element $ABCD$ of the membrane after deformation. Let F be the uniform tension per unit length of the membrane. The value of the initial tension F is large enough to ignore its change when the membrane is blown up by the small pressure p. On face AD, the force acting is $F\Delta y$. This is inclined at an angle β to the x-axis. $\tan \beta$ is the slope of the face AB and is equal to $\partial z/\partial x$. Hence, the component of $F\Delta y$ in z direction is $\left(-F\Delta y \dfrac{\partial z}{\partial x}\right)$ since $\sin \beta \approx \tan \approx \beta$ for small values of β. The force on face BC is also $F\,\Delta y$ but is inclined at an angle $(\beta + \Delta\beta)$ to the x-axis. Its slope is therefore

$$\frac{\partial z}{\partial x} + \frac{\partial}{\partial x}\left(\frac{\partial z}{\partial x}\right)\Delta x$$

and the component of the force in z direction is

$$F\Delta y\left[\frac{\partial z}{\partial x} + \frac{\partial}{\partial x}\left(\frac{\partial z}{\partial x}\right)\Delta x\right]$$

Similarly, the components of the forces $F\Delta y$ acting on faces AB and CD are

$$-F\Delta x\frac{\partial z}{\partial y} \quad \text{and} \quad F\Delta x\left[\frac{\partial z}{\partial y} + \frac{\partial}{\partial y}\left(\frac{\partial z}{\partial y}\right)\Delta y\right]$$

Therefore, the resultant force in z direction due to tension F is

$$-F\,\Delta y\frac{\partial z}{\partial x}+F\,\Delta y\left[\frac{\partial z}{\partial x}+\frac{\partial^2 z}{\partial x^2}\Delta x\right]-F\Delta x\frac{\partial z}{\partial y}$$

$$+\,F\,\Delta x\left[\frac{\partial z}{\partial y}+\frac{\partial^2 z}{\partial y^2}\Delta y\right]$$

$$=F\left(\frac{\partial^2 z}{\partial x^2}+\frac{\partial^2 z}{\partial y^2}\right)\Delta x\,\Delta y$$

The force p acting upward on the membrane element $ABCD$ is $p\,\Delta x\,\Delta y$, assuming that the membrane deflection is small. For equilibrium, therefore

$$F\left(\frac{\partial^2 z}{\partial x^2}+\frac{\partial^2 z}{\partial y^2}\right)=-p$$

or
$$\frac{\partial^2 z}{\partial x^2}+\frac{\partial^2 z}{\partial y^2}=-\frac{p}{F} \tag{7.43}$$

Now, if we adjust the membrane tension F or the air pressure p such that p/F becomes numerically equal to $2G\theta$, then Eq. (7.43) of the membrane becomes identical to Eq. (7.25) of the torsion stress function ϕ. Further, if the membrane height z remains zero at the boundary contour of the section, then the height z of the membrane becomes numerically equal to the torsion stress function [Eq. (7.23)]. The slopes of the membrane are then equal to the shear stresses and these are in a direction perpendicular to that of the slope. The twisting moment is numerically equivalent to twice the volume under the membrane [Eq. (7.24)].

7.8 TORSION OF THIN-WALLED TUBES

Consider a thin-walled tube subjected to torsion. The thickness of the tube need not be uniform (Fig. 7.11). Since the thickness is small and the boundaries are free, the shear stresses will be essentially parallel to the boundary. Let τ be the magnitude of the shear stress and t the thickness.

Consider the equilibrium of an element of length Δl, as shown. The areas of cut faces AB and CD are respectively $t_1\,\Delta l$ and $t_2\,\Delta l$. The shear stresses (complementary shears) are τ_1 and τ_2. For equilibrium in z direction we should have

$$-\tau_1 t_1\;\Delta l+\tau_2\,t_2\;\Delta l=0$$

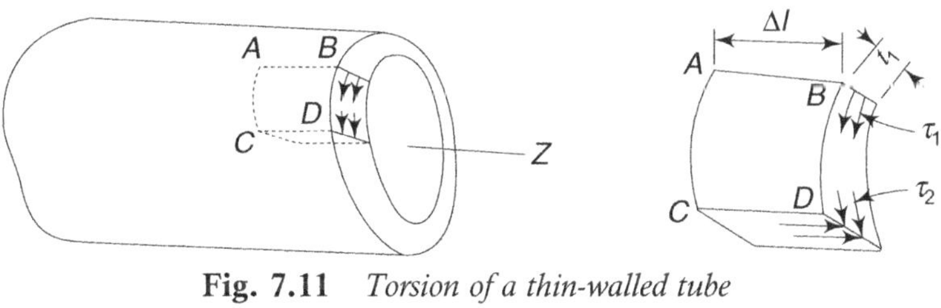

Fig. 7.11 *Torsion of a thin-walled tube*

or $$\tau_1 t_1 = \tau_2 t_2 = q, \quad \text{a constant} \tag{7.44}$$

Hence, the quantity τt is a constant. This is called the shear flow q, since the equation is similar to the flow of an incompressible liquid in a tube of varying area. For continuity, we should have $V_1A_1 = V_2A_2$, where A is the area and V the corresponding velocity of the fluid there.

Consider next the torque of the shear about point O [Fig. 7.12(a)].

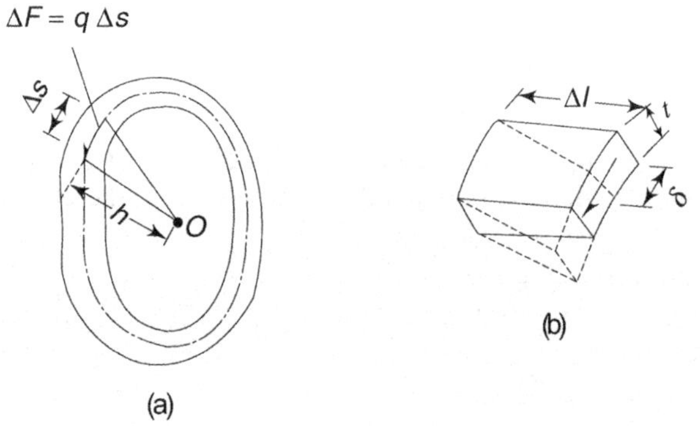

Fig. 7.12 *Cross-section of a thin-walled tube and torque due to shear*

The force acting on an elementary length Δs of the tube is

$$\Delta F = \tau t\, \Delta s = q\, \Delta s$$

The moment arm about O is h and hence, the torque is

$$\Delta T = q\, \Delta s h = 2q\, \Delta A$$

where ΔA is the area of the triangle enclosed at O by the base s. Hence, the total torque is

$$T = \Sigma 2q\, \Delta A = 2qA \tag{7.45}$$

Where A is the area enclosed by the centre line of the tube. Equation (7.45) is generally known as the Bredt–Batho formula.

To determine the twist of the tube, we make use of Castigliano's theorem. Referring to Fig. 7.12(b), the shear force on the element is $\tau t\, \Delta s = q\, \Delta s$. Because of shear strain γ, the force does work equal to

$$\Delta U = \frac{1}{2}(\tau t\, \Delta s)\, \delta$$

$$= \frac{1}{2}(\tau t\, \Delta s)\, \gamma \Delta l$$

$$= \frac{1}{2}(\tau t\, \Delta s)\, \Delta l\, \frac{\tau}{G}$$

$$= \frac{q^2\, \Delta l}{2G}\, \frac{\Delta s}{t} \tag{7.46}$$

$$= \frac{T^2\, \Delta l}{8A^2 G}\, \frac{\Delta s}{t} \tag{7.47}$$

using Eq. (7.45). The total elastic strain energy is therefore

$$U = \frac{T^2 \Delta l}{8A^2G} \oint \frac{ds}{t} \tag{7.48}$$

Hence, the twist or the rotation per unit length ($\Delta l = 1$) is

$$\theta = \frac{\partial U}{\partial T} = \frac{T}{4A^2G} \oint \frac{ds}{t} \tag{7.49}$$

Using once again Eq. (7.45)

$$\theta = \frac{q}{2AG} \oint \frac{ds}{t} \tag{7.50}$$

7.9 TORSION OF THIN-WALLED MULTIPLE-CELL CLOSED SECTIONS

We can extend the analysis of the previous section to torsion of multiple-cell sections. Consider the two-cell section shown in Fig. 7.13.

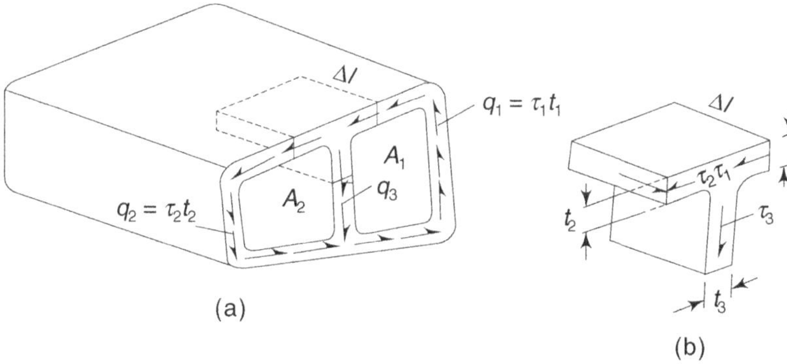

Fig. 7.13 *Torsion of a thin-walled multiple cell closed section*

Consider the equilibrium of an element at the junction, as shown in Fig. 7.13(b). In the direction of the axis of the tube

$$-\tau_1 t_1 \Delta l + \tau_2 t_2 \Delta l + \tau_3 t_3 \Delta l = 0$$

or $$\tau_1 t_1 = \tau_2 t_2 + \tau_3 t_3$$

i.e., $$q_1 = q_2 + q_3 \tag{7.51}$$

This is again equivalent to a fluid flow dividing itself into two streams. Choose any moment axis, such as point O (Fig. 7.14).

The shear flow in the web can be considered to be made up of q_1 and $-q_2$, since $q_3 = q_1 - q_2$. The moment about O due to q_1 flowing in cell 1 (with web included) is [Eq. (7.45)]

$$T_1 = 2q_1 A_1$$

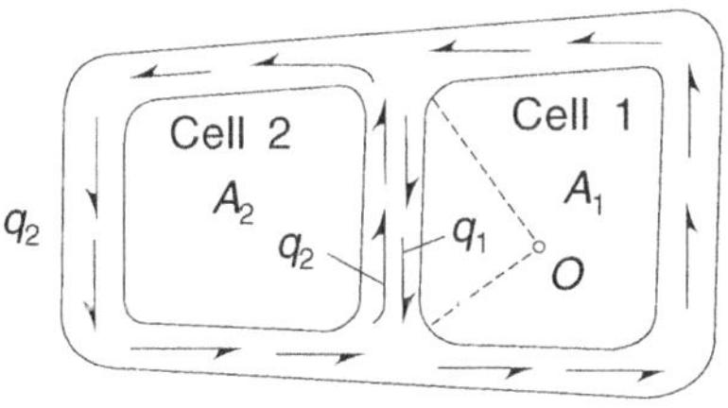

Fig. 7.14 *Section of a thin-walled multiple cell beam and moment axis*

where A_1 is the area of cell 1.
Similarly, the moment about O due to q_2 flowing in cell 2 (with web included), with A_1^* as the area enclosed at O outside cell 2, is

$$T_2 = 2q_2 (A_2 + A_1^*) - 2q_2 A_1^*$$

The second term with the negative sign on the right-hand side is the moment due to the shear flow q_2 in the middle web. Hence, the total torque is

$$T = T_1 + T_2 = 2q_1 A_1 + 2q_2 A_2 \tag{7.52}$$

A_1 and A_2 are the areas of cells 1 and 2 respectively.

Next, we shall consider the twist. For continuity, the twist of each cell should be the same. According to Eq. (7.50), the twist of each cell is given by

$$2G\theta = \frac{1}{A}\oint\frac{q\,ds}{t}$$

Let $$a_1 = \oint\frac{ds}{t} \text{ for cell 1 including the web}$$

$$a_2 = \oint\frac{ds}{t} \text{ for cell 2 including the web}$$

$$a_{12} = \oint\frac{ds}{t} \text{ for the web}$$

Then, for cell 1

$$2G\theta = \frac{1}{A_1}(a_1 q_1 - a_{12} q_2) \tag{7.53}$$

For cell 2

$$2G\theta = \frac{1}{A_2}(a_2 q_2 - a_{12} q_1) \tag{7.54}$$

Equations (7.52)–(7.54) are sufficient to solve for q_1, q_2 and θ.

Example 7.1 *Figure 7.15 shows a two-cell tubular section whose wall thicknesses are as shown. If the member is subjected to a torque T, determine the shear flows and the angle of twist of the member per unit length.*

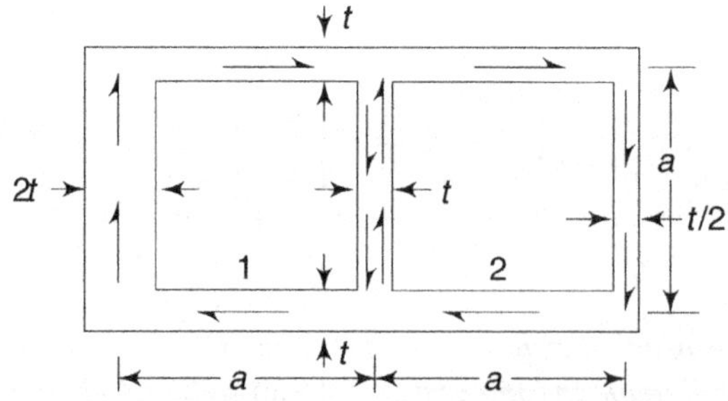

Fig. 7.15 *Example 7.1*

Solution

For cell 1, $\oint \frac{ds}{t} = \frac{a}{t} + \frac{a}{2t} + \frac{a}{t} + \frac{a}{t} = \frac{7a}{2t} = a_1$

For cell 2, $\oint \frac{ds}{t} = \frac{a}{t} + \frac{a}{t} + \frac{a}{t} + \frac{2a}{t} = \frac{5a}{t} = a_2$

For web, $\oint \frac{ds}{t} = \frac{a}{t} = a_{12}$

From Eq. (7.53)

For cell 1,
$$2G\theta = \frac{1}{A_1}(a_1 q_1 - a_{12} q_2)$$
$$= \frac{1}{a^2}\left(\frac{7a}{2t} q_1 - \frac{a}{t} q_2\right) = \frac{1}{at}\left(\frac{7}{2} q_1 - q_2\right)$$

For cell 2,
$$2G\theta = \frac{1}{A_2}(a_2 q_2 - a_{12} q_1)$$
$$= \frac{1}{a^2}\left(\frac{5a}{t} q_2 - \frac{a}{t} q_1\right) = \frac{1}{at}(5q_2 - q_1)$$

Equating, $\frac{7}{2} q_1 - q_2 = 5q_2 - q_1$ or $q_2 = \frac{3}{4} q_1$

From Eq. (7.52)
$$T = 2q_1 A_1 + 2q_2 A_2 = 2a^2\left(q_1 + \frac{3}{4} q_1\right) = \frac{7}{2} a^2 q_1$$

$\therefore$
$$q_1 = \frac{2T}{7a^2} \quad \text{and} \quad q_2 = \frac{3T}{14a^2}$$
$$2G\theta = \frac{1}{at}(5q_2 - q_1)$$
$$= \frac{1}{at}\left(\frac{15}{4} - 1\right) q_1 = \frac{11}{4at} q_1$$

or
$$\theta = \frac{1}{2G}\left(\frac{11}{4at}\right)\left(\frac{2T}{7a^2}\right)$$
$$= \frac{11}{28} \frac{T}{a^3 tG}$$

Example 7.2 *Figure 7.16 shows a two-cell tubular section as formed by a conventional airfoil shape, and having one interior web. An external torque of 10000 Nm (102040 kgf cm) is acting in a clockwise direction. Determine the internal shear flow distribution. The cell areas are as follows:*

$$A_1 = 680 \text{ cm}^2$$
$$A_2 = 2000 \text{ cm}^2$$

The peripheral lengths are indicated in Fig. 7.16.

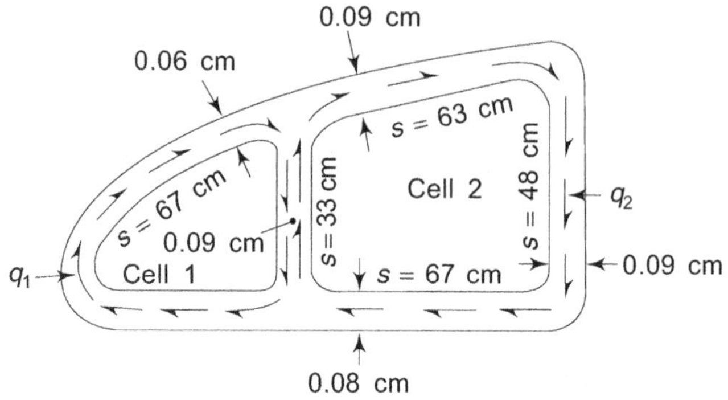

Fig. 7.16 *Example 7.2*

Solution Let us calculate the line integrals $\oint ds/t$.

For cell 1, $$a_1 = \frac{67}{0.06} + \frac{33}{0.09} = 1483$$

For cell 2, $$a_2 = \frac{33}{0.09} + \frac{63}{0.09} + \frac{48}{0.09} + \frac{67}{0.08} = 2409$$

For web, $$a_{12} = \frac{33}{0.09} = 366$$

From Eqs (7.53) and (7.54)

For cell 1, $$2G\theta = \frac{1}{A_1}(a_1 q_1 - a_{12} q_2)$$

$$= \frac{1}{680}(1483 q_1 - 366 q_2)$$

$$= 2.189 q_1 - 0.54 q_2$$

For cell 2, $$2G\theta = \frac{1}{A_2}(a_2 q_2 - a_{12} q_1)$$

$$= \frac{1}{2000}(2409 q_2 - 366 q_1)$$

$$= 1.20\, q_2 - 0.18\, q_1$$

Hence, equating the above two values

$$2.19\, q_1 - 0.54\, q_2 = 1.20\, q_2 - 0.18\, q_1$$

or, $$2.37\, q_1 - 1.74\, q_2 = 0$$

i.e. $$q_2 = 1.36\, q_1$$

The torque due to shear flows should be equal to the applied torque. Hence, from Eq. (7.52)

$$T = 2q_1 A_1 + 2q_2 A_2$$

or $$10000 \times 100 = 2q_1 \times 680 + 2q_2 \times 2000$$

$$= 1360\, q_1 + 4000\, q_2$$

Substituting for q_2

$$10^6 = 1360 q_1 + 5440 q_1 = 6800 q_1$$

$$q_1 = 147\,\frac{\text{N}}{\text{cm}}\left(15.01\,\frac{\text{kgf}}{\text{cm}}\right) \qquad q_2 = 200\,\frac{\text{N}}{\text{cm}}\left(20.4\,\frac{\text{kgf}}{\text{cm}}\right)$$

7.10 TORSION OF BARS WITH THIN RECTANGULAR SECTIONS

Figure 7.17 shows the section of a rectangular bar subjected to a torque T. Let the thickness t be small compared to the width b. The section consists of only one boundary and the value of the stress function ϕ around this boundary is constant. Let $\phi = 0$.

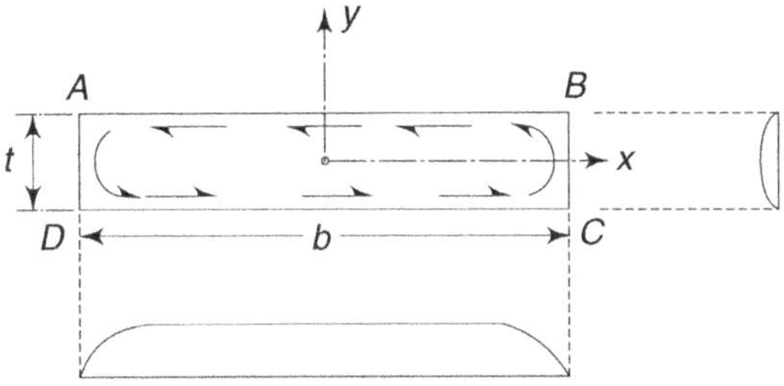

Fig. 7.17 *Torsion of a thin rectangular bar*

From Eq. (7.25)

$$\frac{\partial^2 \phi}{\partial x^2} + \frac{\partial^2 \phi}{\partial y^2} = -2G\theta$$

Excepting at the ends AD and BC, the stress function is fairly uniform and is independent of x. Hence, we can take $\phi(x, y) = \phi(y)$. Therefore, the above equation becomes

$$\frac{\partial^2 \phi}{\partial y^2} = -2G\theta$$

Integrating $\qquad \phi = -G\theta y^2 + a_1 y + a_2$

Since $\phi = 0$ around the boundary, one has $\phi = 0$ at $y = \pm\, t/2$. Substituting these

$$a_1 = 0, \qquad a_2 = \frac{G\theta t^2}{4}$$

and
$$\phi = G\theta\left(\frac{t^2}{4} - y^2\right) \tag{7.55}$$

From Eq. (7.19)

$$\tau_{yz} = -\frac{\partial \phi}{\partial x} = 0$$

and
$$\tau_{zx} = \frac{\partial \phi}{\partial y} = -2G\theta\, y \tag{7.56a}$$

These shears are shown in Fig. 7.17. Obviously, the above equations are not valid near the ends. The maximum shearing stresses are at the surfaces $y = \pm\, t/2$, and

$$(\tau_{zx})_{\max} = \pm\, G\theta\, t \tag{7.56b}$$

From Eq. (7.24),

$$T = 2\iint \phi\, dx\, dy$$

$$= 2G\theta \int_{-b/2}^{b/2} dx \int_{-t/2}^{t/2} \left(\frac{t^2}{4} - y^2\right) dy$$

or
$$T = \frac{1}{3} bt^3 \, G\theta \tag{7.57}$$

The results are

$$\theta = \frac{1}{G}\frac{3T}{bt^3}, \qquad \tau_{zx} = -\frac{6T}{bt^3}y, \qquad (\tau_{zx})_{\max} = \pm\frac{3T}{bt^2} \tag{7.58}$$

7.11 TORSION OF ROLLED SECTIONS

The argument leading to the approximations given by Eqs (7.55) and (7.56) can be applied to any narrow cross-section which has a relatively small curvature, as shown in Figs 7.18(a)–(d). To see this, we imagine a 90° bend in the middle of the rectangle shown in Fig. 7.17, so that the section becomes an angle. This section has only one boundary with ϕ = constant = 0. Excepting for the local effects near the corner, the shape across the thickness will be similar to that shown in Fig. 7.17, for the thin rectangular section. Hence, Eqs (7.55) and (7.57) can be applied, provided b is taken as the total length of both legs of the angle concerned and y is the rectangular coordinate in the direction of the local thickness.

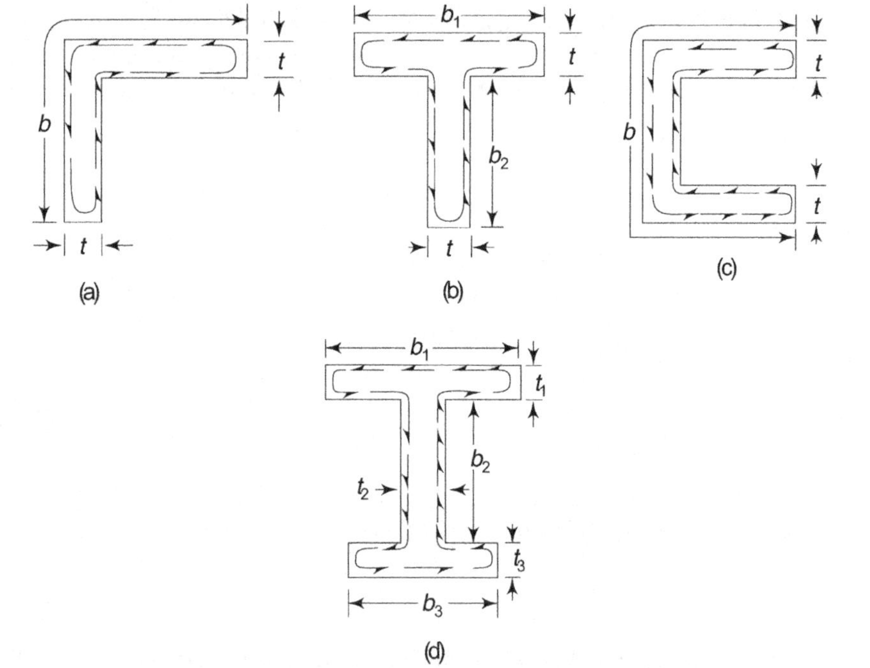

Fig. 7.18 *Torsion of rolled sections*

In the case of a T-section shown in Fig. 7.18(b), the length $b = b_1 + b_2$ if the thickness is uniform. If the thickness changes, as shown in Fig. 7.17(d), Eqs (7.55) and (7.57) become

$$\phi = G\theta\left(\frac{t_i^2}{4} - y^2\right) \qquad (i = 1, 2 \text{ or } 3)$$

and $$T = \frac{1}{3} G\theta (b_1 t_1^3 + b_2 t_2^3 + b_3 t_3^3) \tag{7.59}$$

This is obtained by adding the effect of each rectangular piece.

Example 7.3 *Analyze the torsion of a closed tubular section and the torsion of a tube of the same radius and thickness but with a longitudinal slit, as shown in Figs 7.19(a) and (b).*

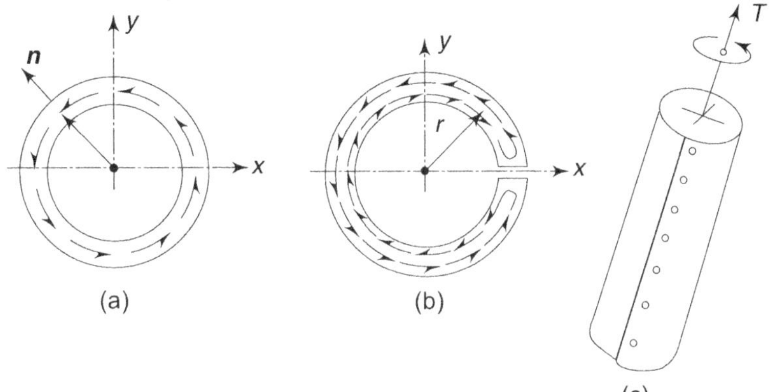

Fig. 7.19 *Example 7.3—Torsion of a closed tubular section and a slit tubular section*

Solution For the closed tube, if τ is the shear stress, we have from elementary analysis

$$T = (2\pi r \tau t).\, r = 2\pi r^2 t\tau \qquad \text{and} \qquad \theta = \frac{\tau}{Gr}$$

Therefore, $T = 2\pi r^3 t G\theta$

For the slit tube, there is only one boundary and on this $\phi = 0$. According to Eq. (7.57)

$$T = \frac{1}{3} bt^3 G\theta = \frac{1}{3} 2\pi r t^3 G\theta$$

Further, following the same analysis as for a thin rectangular section

$$\tau_{max} = \mp G\theta t$$

The shear stress directions in the slit tube are shown in Fig. 7.19(b). The ratio of the torsional rigidities is

$$\frac{T_1}{T_2} = (2\pi r^3 t G\theta) \Big/ \frac{1}{3} (2\pi r t^3 G\theta)$$

$$= 3\left(\frac{r}{t}\right)^2$$

For a thin tube with $r/t = 10$, tube (a) is 300 times as stiff as tube (b).

If the slit tube is riveted along the length to form a closed tube of length l, as shown in Fig. 7.19(c), the force on the rivets will be

$$F = \tau t l = \frac{Tl}{2\pi r^2}$$

where for τ we have put the value

$$\tau = \frac{T}{2\pi r^2 t}$$

as for a non-slit tube. If there are n rivets in a length l, then the average force on each rivet is F/n.

Example 7.4 *(i) A 30- cm I beam (Fig. 7.20), with flanges and with a web 1.25 cm thick, is subjected to a torque T = 50000 kgfcm (4900 Nm). Find the maximum shear stress and the angle of twist per unit length.*

(ii) In order to reduce the stress and the angle of twist, 1.25 cm thick flat plates are welded onto the sides of the section, as shown by dotted lines. Find the maximum shear stress and the angle of twist.

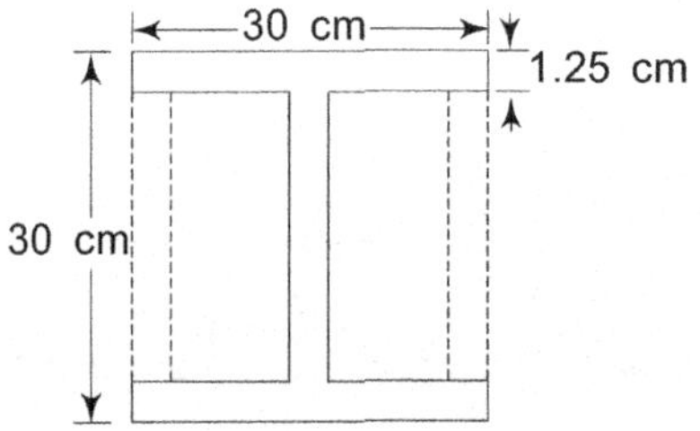

Fig. 7.20 *Example 7.4*

Solution

(i) Using Eq. (7.58)

$$\tau_{max} = 3T/\left(\sum b_i t_i^2\right)$$

$$= \frac{3 \times 50\,000}{30 \times (5/4)^2 + 30 \times (5/4)^2 + (30 - 2.5) \times (5/4)^2}$$

$$= 1097 \text{ kgf/cm}^2 \text{ (107512 kPa)}$$

$$\theta = \frac{1}{G} \frac{3T}{\left(\sum b_i t_i^3\right)}$$

$$= \frac{3 \times 50\,000}{30 \times (5/4)^3 + 30 \times (5/4)^3 + (30 - 2.5) \times (5/4)^3} \times \frac{1}{G}$$

$$= 878/G \text{ radians per cm length}$$

(ii) When the two side plates are welded, the section becomes a two-cell structure for which we can apply Eqs (7.52)–(7.54). For each cell

$$a_1 = a_2 = \oint \frac{ds}{t} = \frac{1}{1.25}\left(\frac{28.75}{2} + \frac{28.75}{2} + 28.75 + 28.75\right)$$

$$= 69.00$$

$$T = 2q_1 A_1 + 2q_2 A_2$$

$$= 4q_1 A_1 = 4\tau \times \frac{1}{1.25} \times \frac{28.75}{2} \times 28.75$$

or $\quad T = 1322.5\tau$

Therefore,

$$\tau = 50000/1322.5 = 37.81\ \text{kgf/cm}^2\ (3705\ \text{kPa})$$

$$2G\theta = \frac{1}{A_1}(a_1 q_1 - a_{12} q_2)$$

$$= \frac{1}{A_1} q_1 (a_1 - a_{12})$$

Therefore

$$\theta = \frac{1}{2G} \times \frac{2}{28.75} \times \frac{1}{28.75}\left(69 - \frac{28.75}{1.25}\right)$$

$$= 0.06/G \text{ radians per cm length}$$

7.12 MULTIPLY CONNECTED SECTIONS

In Sec. 7.2 and 7.3, we considered the torsion of shafts with sections which do not have holes. It is easy to extend the same analysis for the solution of shafts, the cross-sections of which contain one or more holes. Figure 7.21 shows the section of a shaft subjected to a torque T. The holes have boundaries C_1 and C_2.

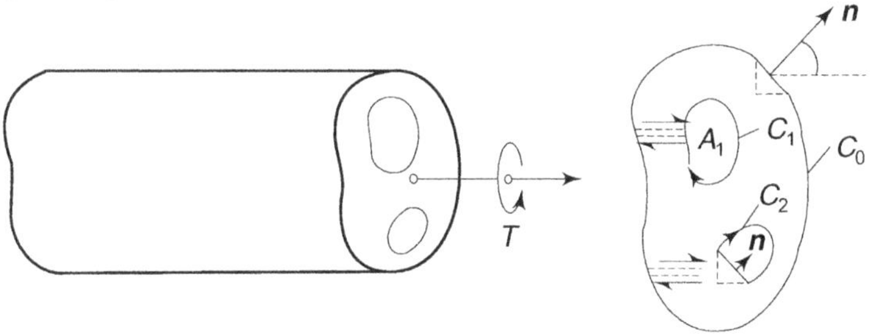

Fig. 7.21 *Torsion of multiply-connected sections*

Once again, as in Sec. 7.3, we assume that τ_{yz} and τ_{zx} are the only non-vanishing stress components. The equations of equilibrium yield

$$\frac{\partial \tau_{yz}}{\partial z} = 0, \qquad \frac{\partial \tau_{zx}}{\partial z} = 0, \qquad \frac{\partial \tau_{zx}}{\partial x} + \frac{\partial \tau_{yz}}{\partial y} = 0$$

Let $\phi\ (x, y)$ be a stress function, such that

$$\tau_{zx} = \frac{\partial \phi}{\partial y}, \qquad \tau_{yz} = -\frac{\partial \phi}{\partial x}$$

The non-vanishing strain components are

$$\gamma_{zx} = \frac{1}{G}\tau_{zx} = \frac{1}{G}\frac{\partial \phi}{\partial y}$$

and $$\gamma_{yz} = \frac{1}{G}\tau_{yz} = -\frac{1}{G}\frac{\partial \phi}{\partial x}$$

The compatibility conditions given by Eq. (2.56) yield

$$\frac{\partial^2 \phi}{\partial x^2} + \frac{\partial^2 \phi}{\partial y^2} = \nabla^2 \phi = \text{a constant } F \tag{7.60}$$

So far, the analysis is identical to that given in Sec. 7.3. Considering the boundary conditions, we observe that there are several boundaries and on each boundary the conditions given by Eq. (7.12), Sec. 7.2 should be satisfied. Since each boundary is a free boundary, we should have

$$n_x \frac{\partial \phi}{\partial y} - n_y \frac{\partial \phi}{\partial x} = 0$$

Substituting for n_x and n_y from Eq. (7.13)

$$\frac{\partial \phi}{\partial y}\frac{dy}{ds} + \frac{\partial \phi}{\partial x}\frac{dx}{ds} = 0$$

or

$$\frac{d\phi}{ds} = 0$$

i.e.

$$\phi \text{ for } C_i = K_i \tag{7.61}$$

i.e. on each boundary ϕ is a constant. Unlike the case where the section did not contain holes, we cannot assume that $\phi = 0$ on each boundary. We can assume that $\phi = 0$ on one boundary, say on C_0, and then determine the corresponding values of K_i on each of the remaining boundaries C_i. To do this, we observe that the displacement of the section in z direction, i.e. $u_z = \theta\psi(x, y)$, from Eq. (7.7), must be single valued. Consequently, the value of $d\psi$ integrated around any closed contour C_i should be equal to zero, i.e.

$$\oint_{C_i} d\psi = \oint_{C_i}\left(\frac{\partial \psi}{\partial x}dx + \frac{\partial \psi}{\partial y}dy\right) = 0 \tag{7.62}$$

From Eq. (7.9), and using the stress function

$$\frac{\partial \psi}{\partial x} = \frac{1}{G\theta}\tau_{zx} + y = \frac{1}{G\theta}\frac{\partial \phi}{\partial y} + y$$

and

$$\frac{\partial \psi}{\partial y} = \frac{1}{G\theta}\tau_{yz} - x = -\frac{1}{G\theta}\frac{\partial \phi}{\partial x} - x \tag{7.63}$$

Hence, for the single valuedness of u_z

$$\frac{1}{G\theta}\oint_{C_i}\left(\frac{\partial \phi}{\partial y}dx - \frac{\partial \phi}{\partial x}dy\right) + \oint_{C_i}(y\,dx - x\,dy) = 0 \tag{7.64}$$

The second integral on the left-hand side is equal to twice the area enclosed by the contour C_i. This can be seen from Fig. 7.22(a).

$$\begin{aligned}\oint_{C_i} y\,dx &= \int_{GKH} y\,dx + \int_{HLG} y\,dx \\ &= \text{area } G'GKHH' - \text{area } H'\,HLGG' \\ &= \text{area enclosed by } C_i = A_i\end{aligned}$$

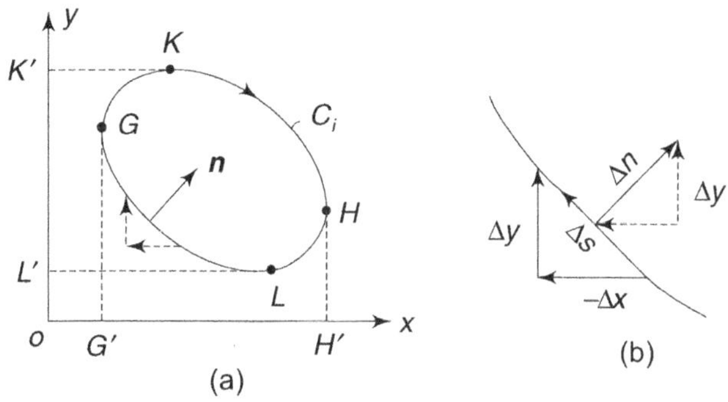

Fig. 7.22 *Evaluation of the integral around contour* C_i

Similarly, $\oint x\,dy = \int_{LGK} x\,dy + \int_{KHL} x\,dy$

$= \text{area } L'LGKK' - \text{area } K'KHLL'$

$= -A_i$

Therefore, $\oint_{C_i} (y\,dx - x\,dy) = 2A_i$ (7.65)

The first integral in Eq. (7.64) can be written as

$$\oint_{C_i}\left(\frac{\partial \phi}{\partial y}dx - \frac{\partial \phi}{\partial x}dy\right) = \oint_{C_i}\left(\frac{\partial \phi}{\partial y}\frac{dx}{ds} - \frac{\partial \phi}{\partial x}\frac{dy}{ds}\right)ds \tag{7.66a}$$

and from Fig. 7.22(b)

$$= -\oint_{C_i}\left(\frac{\partial \phi}{\partial y}\frac{dy}{dn} + \frac{\partial \phi}{\partial x}\frac{dx}{dn}\right)ds$$

$$= -\oint_{C_i}\frac{\partial \phi}{\partial n}ds \tag{7.66b}$$

where $\boldsymbol{n}$ is the outward drawn normal to the boundary C_i. Therefore, Eq. (7.64) becomes

$$\oint_{C_i}\frac{\partial \phi}{\partial n}ds = 2G\theta A_i \tag{7.67}$$

on each boundary C_i. A_i is the area enclosed by C_i.

The remaining equations of Sec. 7.3 remain unaltered, i.e. Eqs (7.24) and (7.25) are

$$\text{Torque } T = 2\iint_R \phi\,dx\,dy$$

and
$$\frac{\partial^2 \phi}{\partial x^2} + \frac{\partial^2 \phi}{\partial y^2} = \nabla^2 \phi = -2G\theta$$

The value of J defined in Eq. (7.16) can be obtained for a multiply-connected body in terms of the stress function ϕ, as follows. Using Eq. (7.63)

$$J = \iint_R \left(x^2 + y^2 + x\frac{\partial \psi}{\partial y} - y\frac{\partial \psi}{\partial x}\right)dx\,dy$$

$$= \iint_R \left(x^2 + y^2 - \frac{1}{G\theta}\frac{\partial \phi}{\partial x}x - x^2 - \frac{1}{G\theta}\frac{\partial \phi}{\partial y}y - y^2\right)dx\,dy$$

$$= -\frac{1}{G\theta}\iint_R \left(\frac{\partial\phi}{\partial x}x + \frac{\partial\phi}{\partial y}y\right) dx\, dy$$

$$= \frac{1}{G\theta}\iint_R \left[2\phi - \frac{\partial}{\partial x}(x\phi) - \frac{\partial}{\partial y}(y\phi)\right] dx\, dy$$

$$= \frac{2}{G\theta}\iint_R \phi\, dx\, dy + \frac{1}{G\theta}\oint_C \phi(y\, dx - x\, dy)$$

where we have made use of Gauss' theorem and the subscript C on the line integral means that the integration is to be performed in appropriate directions over all the contours C_i ($i = 0, 1, 2, \ldots$,) shown in Fig. 7.21. Since we have chosen ϕ to be zero over the boundary C_0

$$J = \frac{2}{G\theta}\iint_R \phi\, dx\, dy + \frac{1}{G\theta}\oint_{C_1}(y\, dx - x\, dy)K_1 + \frac{1}{G\theta}\oint_{C_2}(y\, dx - x\, dy)K_2 + \ldots$$

where $K_1, K_2, \ldots$, are the values of ϕ on $C_1, C_2 \ldots$,
And from Eq. (7.9)

$$J = \frac{2}{G\theta}\iint_R \phi\, dx\, dy + \frac{1}{G\theta}(2K_1A_1 + 2K_2A_2 + \ldots)$$

where A_i is the area enclosed by curve C_i. Hence,

$$J = \frac{2}{G\theta}\left[\iint \phi\, dx\, dy + \Sigma K_i A_i\right] \tag{7.68}$$

Equation (7.17), therefore, assumes the form

$$T = GJ\theta = 2\left(\iint \phi\, dx\, dy + \Sigma K_i A_i\right) \tag{7.69}$$

For a solid shaft with no holes, the above equation reduces to Eq. (7.24).

Example 7.5 *Analyse the torsion problem of a thin-walled, multiple-cell closed section, using equations (7.28), (7.56) and (7.69). Assume uniform thickness t.*

Solution Consider the two-celled section shown in Fig. 7.23. According to Eq. (7.61), the stress function ϕ is constant around each boundary. Put

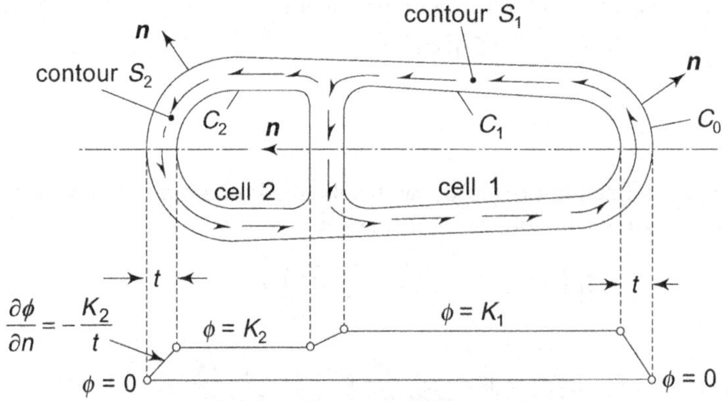

Fig. 7.23 *Example 7.5*

$\phi = 0$ on boundary C_0, $\phi = K_1$ on C_1 and $\phi = K_2$ on C_2. From Eq. (7.28), the resultant shear stress is given by $\tau_{zs} = -\frac{\partial \phi}{\partial n}$ where $\boldsymbol{n}$ is the normal to the contour of ϕ, i.e. the line of shear stress. Since the thickness t is small, the lines of shear stress follow the contours of the cells. Further, since $\phi = 0$ on C_0 and $\phi = K_1$ on C_1, we have for contour S_1 of cell 1,

$$-\frac{\partial \phi}{\partial n} = -\frac{0 - K_1}{t} = \frac{K_1}{t} = \tau_1 \tag{7.70}$$

For the web contour S_{12}

$$-\frac{\partial \phi}{\partial n} = -\frac{K_2 - K_1}{t} = \tau_{12} \tag{7.71}$$

and for contour S_2 of cell 2

$$-\frac{\partial \phi}{\partial n} = -\frac{0 - K_2}{t} = \frac{K_2}{t} = \tau_2 \tag{7.72}$$

From Eq. (7.67), for cell 1

$$\tau_1(S_1 - S_{12}) + \tau_{12}S_{12} = 2G\theta A_1$$

where S_1 is the peripheral length of cell 1 including the web, and S_{12} the length of the web. Substituting for τ_1 and τ_{12}

$$\frac{K_1}{t}\left(S_1 - S_{12}\right) - \frac{K_2 - K_1}{t} S_{12} = 2G\theta A_1$$

or

$$K_1 \frac{S_1}{t} - K_2 \frac{S_{12}}{t} = 2G\theta A_1 \tag{7.73}$$

Similarly, for cell 2

$$\tau_2(S_2 - S_{12}) - \tau_{12}S_{12} = 2G\theta A_2$$

or

$$\frac{K_2}{t}\left(S_2 - S_{12}\right) + \frac{K_2 - K_1}{t} S_{12} = 2G\theta A_2$$

i.e.

$$K_2 \frac{S_2}{t} - K_1 \frac{S_{12}}{t} = 2G\theta A_2 \tag{7.74}$$

From Eq. (7.69)

$$T = 2\left(\iint \phi \, dx \, dy + \Sigma_i K_i A_i\right)$$

Compared to A_i, the area of the solid part of the tube section is very small and hence, the integral on the right-hand side can be omitted. With this

$$T = 2\left(K_1 A_1 + K_2 A_2\right) \tag{7.75}$$

Equations (7.73)–(7.75) will enable us to solve for K_1, K_2 and θ.

Example 7.6 *Using equations (7.28) and (7.61), prove that the shear flow is constant for a thin-walled tube (shown in Fig. 7.24) subjected to torsion.*

Solution Let S be the contour of the centre line and t_s the thickness at any section. According to Eq. (7.61), the stress function ϕ is constant around

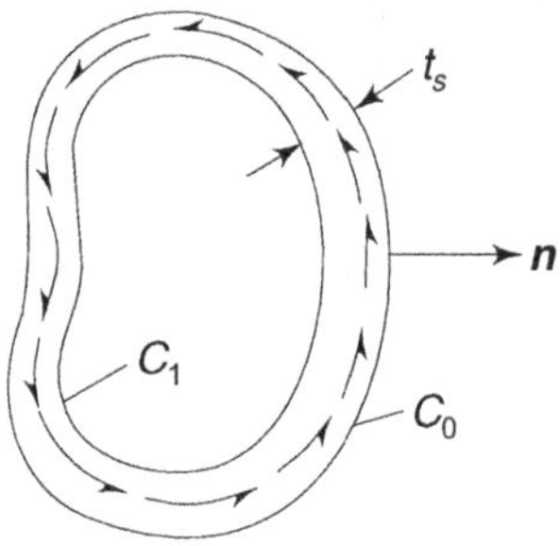

Fig 7.24 *Example 7.6*

each boundary. Let $\phi = 0$ on C_0 and $\phi = K_1$ on C_1. Then, from Eq. (7.28), at any section

$$-\frac{\partial \phi}{\partial n} = -\frac{0 - K_1}{t_s} = \frac{K_1}{t_s} = \tau_s$$

Therefore,

$$\tau_s t_s = K_1 \text{ a constant}$$

i.e. q is constant.

7.13 CENTRE OF TWIST AND FLEXURAL CENTRE

We have assumed in all the previous analyses in this chapter that when a twisting moment or a torque is applied to the end of a shaft, the section as a whole will rotate and only one point will remain at rest. This point is termed the centre of twist. Similarly, it was stated in Sec. 6.5 that there exists a point in the cross-section, such that when a transverse force is applied passing through this point, the beam bends without the section rotating. This point is called shear centre or flexural centre. Consider a cylindrical rod with one end firmly fixed so that no deformation occurs at the built-in section (Fig. 7.25).

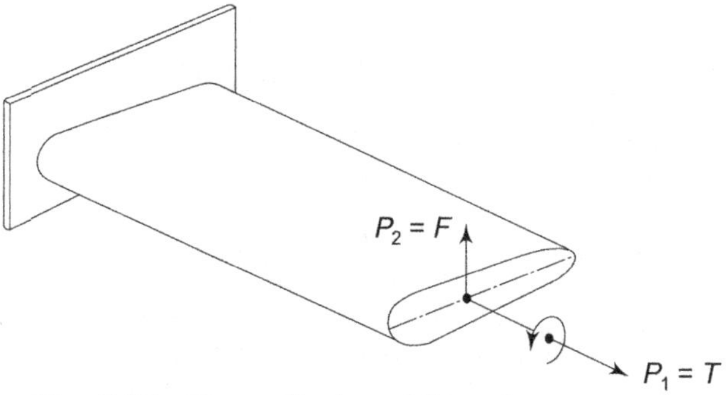

Fig. 7.25 *Centre of twist and flexural centre*

For such a built-in cylinder, it can be shown that the centre of twist and the flexural centre coincide. To see this, let the twisting couple be $T = P_1$ and the bending force be $F = P_2$. It is assumed that P_1 is applied at point 1, which is the centre of twist and P_2, through point 2, the flexural centre. Let δ_1 be the rotation caused at point 1 due to force P_2 (= F) and let δ_2 be the deflection (i.e. displacement) of point 2 due to force P_1 (= T). But δ_1, the rotation, is zero since the force P_2 is acting through the flexural centre. That is, $a_{12} = 0$. Consequently, from the reciprocal theorem, $a_{21} = 0$. But a_{21} is the deflection (i.e. displacement) of the flexural centre due to torque. Since this is equal to zero, and since during twisting, the only point which does not undergo rotation, i.e. deflection, is the centre of twist, the flexural centre and the centre of twist coincide. It is important to note that for this analysis to be valid it is necessary for the end to be firmly built-in.

Problems

7.1 (a) Verify that

$$\psi = -\frac{G\theta}{2}\left(x^2 + y^2 - 2ax + \frac{2b^2 ax}{x^2 + y^2} - b^2\right)$$

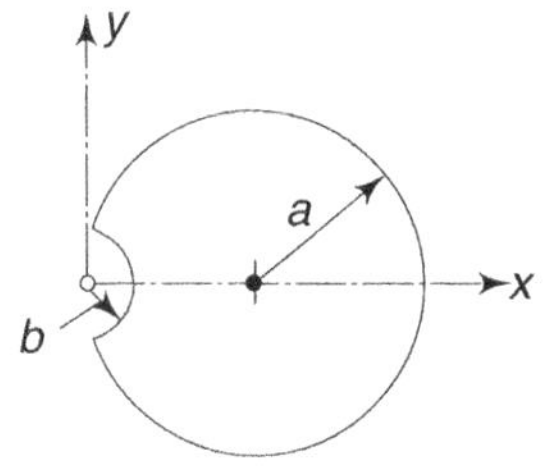

Fig. 7.26 *Problem 7.1*

where a and b are as shown in Fig. 7.26 and C is a constant, is the Saint-Venant warping function (also Prandtl stress function) for the torsion of a round shaft with a semi-circular keyway.

(b) Obtain an expression for the maximum stress in the section.

(c) What is the ratio of the maximum stress in a shaft without a groove to the maximum stress in a shaft with a groove where b tends to be very small.

[*Ans.* (b) $\tau_{max} = G\theta\,(2a - b)$
(c) Ratio $\to$ 2 as $b \to 0$]

7.2 The two tubular sections shown in Fig. 7.27 have the same wall thickness t and same circumference. Neglecting stress concentration, find the ratio of the shear stresses for

(a) equal twisting moments in the two cases and
(b) equal angles of twist in the two cases.

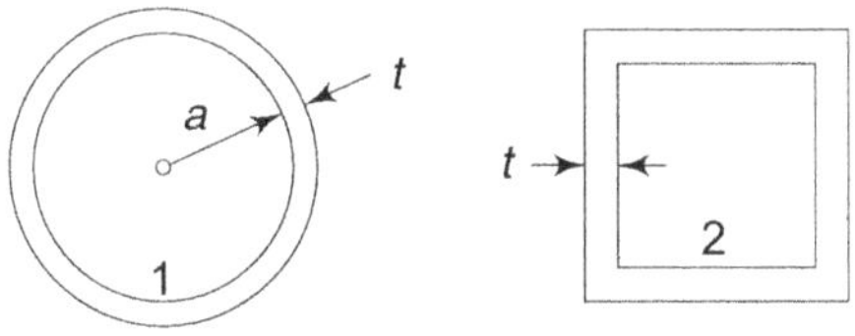

Fig. 7.27 *Problem 7.2*

[*Ans.* (a) 1 : $4/\pi$
(b) 1 : $\pi/4$]

7.3 A thin-walled box section of dimensions $2a \times a \times t$ is to be compared with a solid section of diameter a (Fig. 7.28). Find the thickness t so that the two sections have

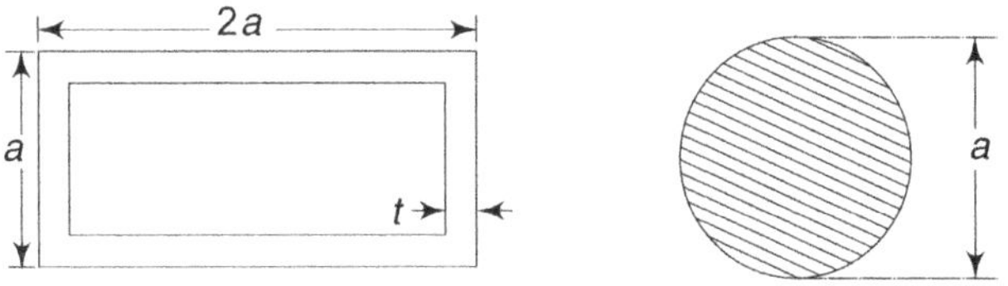

Fig. 7.28 *Problem 7.3*

(a) the same maximum stress for the same torque and
(b) the same stiffness.

[*Ans.* (a) $t = \pi a/64$
(b) $t = \frac{3}{4}\frac{\pi a}{64}$]

7.4 A hollow aluminium section is designed, as shown in Fig. 7.29(a), for a maximum shear stress of 35000 kPa (357 kgf/cm^2), neglecting stress concentrations. Find the twisting moment that can be taken up by the section and the angle of twist if the length of the member is 3 m. If the member is redesigned, as shown in Fig. 7.29(b), find the allowable twisting moment and the angle of twist. Take $G = 157.5 \times 10^6$ kPa.

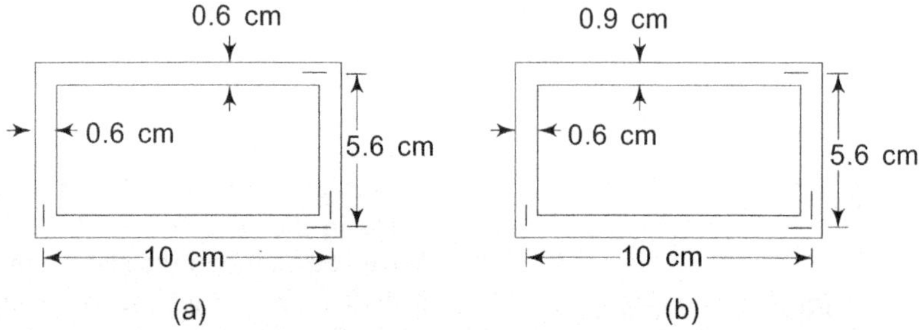

Fig 7.29 *Problem 7.4*

[*Ans.* (a) 2352 Nm; 1.06°
(b) 2352 Nm; 0.837°]

7.5 A steel girder has the cross-section shown in Fig. 7.30. The wall thickness is uniformly 1.25 cm. The stress due to twisting should not exceed 350000 kPa (3570 kgf/cm^2). Neglect stress concentrations.

(a) What is the maximum allowable torque?

(b) What is the twist per metre length under that torque?

(c) What is the shear stress in the middle web?

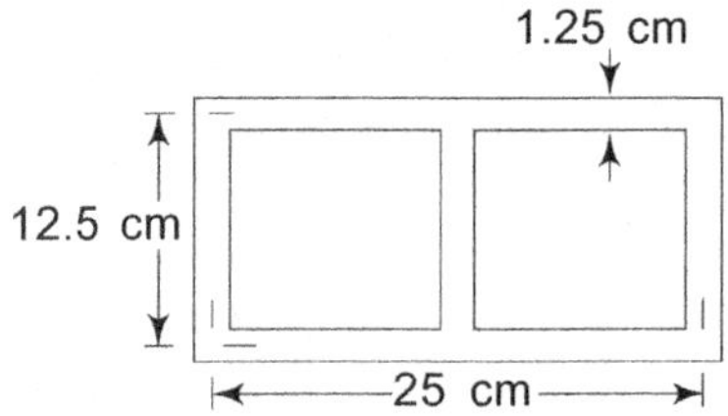

Fig. 7.30 *Problem 7.5*

[*Ans.* (a) 273.44 kNm
(b) $\frac{1}{G}(4.2 \times 10^8)$ radians
(c) zero]

7.6 A thin-walled box shown in Fig. 7.31 is subjected to a torque T. Determine the shear stresses in the walls and the angle of twist per unit length of the box.

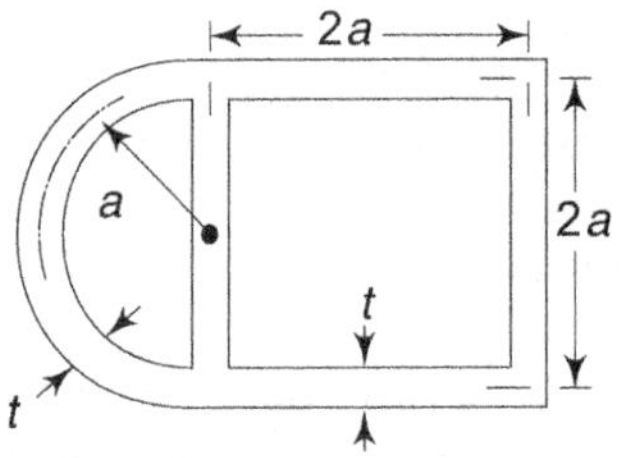

Fig. 7.31 *Problem 7.6*

[*Ans.* $q_1 = \dfrac{(\pi + 2)T}{a^2(\pi^2 + 12\pi + 16)}$; $q_2 = \dfrac{5\pi + 8}{4\pi + 18} q_1$

$\theta = \dfrac{(2\pi + 3)T}{2Ga^3 t(\pi^2 + 12\pi + 16)}$]

7.7 Figure 7.32 shows a tubular section with three cells. The thin-walled tube is subjected to a torque $T = 113000$ Nm (115455 kgf cm). Determine the shear stresses in the walls of the section.

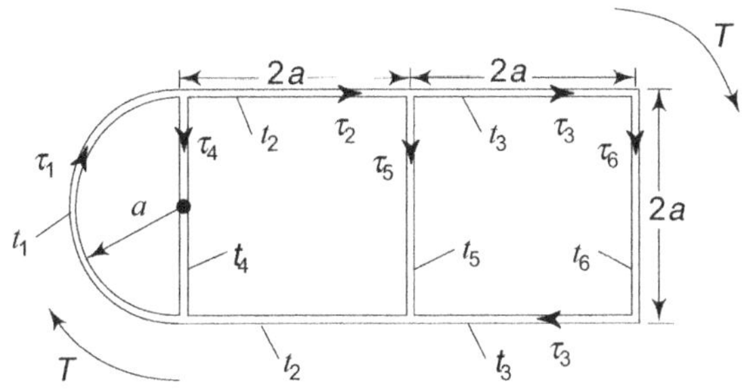

Fig. 7.32 *Problem 7.7*

$a = 12.7$ cm, $t_1 = 0.06$ cm, $t_2 = 0.08$ cm, $t_3 = 0.08$ cm, $t_4 = 0.13$ cm, $t_5 = 0.08$ cm, $t_6 = 0.10$ cm

Ans. $\tau_1 = 394649$ kPa
$\tau_2 = 518388$ kPa
$\tau_3 = 460472$ kPa
$\tau_4 = -136862$ kPa
$\tau_5 = 57920$ kPa
$\tau_6 = 368377$ kPa

7.8 A thin tubular bar shown in Fig. 7.33 is subjected to a torque $T = 113000$ Nm (115455 kgf cm). The dimensions are as indicated. Determine the shear stresses in the walls.
Given $a = 12.7$ cm, $t_1 = 0.06$ cm, $t_2 = 0.08$ cm,
$t_3 = 0.06$ cm, $t_4 = 0.10$ cm, $t_5 = 0.13$ cm

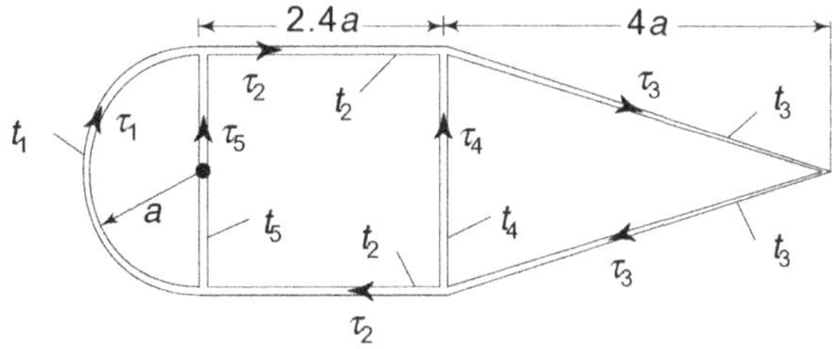

Fig. 7.33 *Problem 7.8*

Ans. $\tau_1 = 441.2$ MPa
$\tau_2 = 558.6$ MPa
$\tau_3 = 393.1$ MPa
$\tau_4 = -211$ MPa
$\tau_5 = 140$ MPa

7.9 A thin-walled box section has two compartments, as shown in Fig. 7.34. It has a constant wall thickness t. What is the shear stress for a given torque and what is the stiffness, i.e. the torque per unit radian of twist? [*Hint:* Treat cell 1 as a closed box and cell 2 as made of two narrow rectangular members. The shear flow near the junction is shown in Fig.7.33(b).]

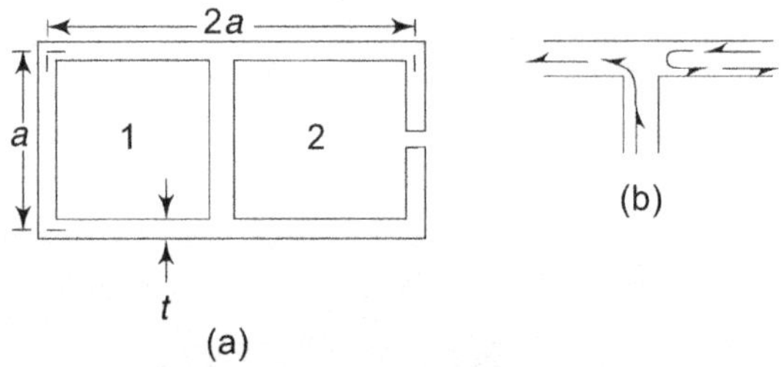

Fig. 7.34 *Problem 7.9*

$$\left[Ans.\ \tau_1 = \frac{T}{2t\,(a^2+t^2)} \approx \frac{T}{2a^2 t} \qquad \frac{T}{\theta_1} = Gat\,(a^2+t^2) \approx Ga^3 t \right]$$

7.10 A section which is subjected to twisting is as shown in Fig. 7.35. Determine the allowable twisting moment for a maximum shear stress of 68950 kPa (703.6 kgf/cm^2). Calculate the shear stresses in the different parts of the section, neglecting stress concentrations.

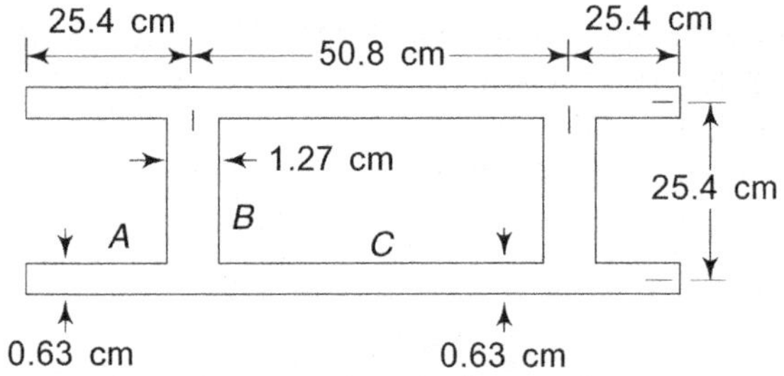

Fig. 7.35 *Problem 7.10*

$$\left[\begin{array}{ll} Ans. & T = 112{,}126 \text{ Nm } (1.13 \times 10^6 \text{ kgf cm}) \\ & \tau_A = 2151 \text{ kPa } (21.95 \text{ kgf/cm}^2) \\ & \tau_B = 34475 \text{ kPa } (351.8 \text{ kgf/cm}^2) \\ & \tau_C = 68950 \text{ kPa } (703.6 \text{ kgf/cm}^2) \end{array} \right]$$

CHAPTER

8

Axisymmetric Problems

8.1 INTRODUCTION

Many problems of practical importance are concerned with solids of revolution which are deformed symmetrically with respect to the axis of revolution. Examples of such solids are circular cylinders subjected to uniform internal and external pressures, rotating circular disks, spherical shells subjected to uniform internal and external pressures, etc. In this chapter, a few of these problems will be investigated. Let the axis of revolution be the z-axis. The deformation being symmetrical with respect to the z-axis, it is convenient to use cylindrical coordinates. Since the deformation is symmetrical about the axis, the stress components do not depend on θ. Further, $\tau_{r\theta}$ and $\tau_{\theta z}$ do not exist. Consequently, the differential equations of equilibrium [Eqs (1.67)–(1.69)] can be reduced to our special case. However, it is instructive to derive the relevant equations applicable to axisymmetric problems from first principles. Consider an axisymmetric body shown in Fig. 8.1. Let an elementary radial element be isolated. The stress vectors acting on its faces are as shown.

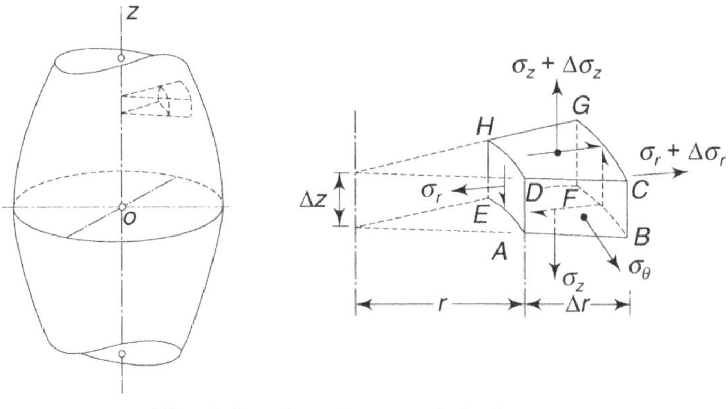

Fig. 8.1 *An axisymmetric body*

On faces $ABCD$ and $EFGH$, the normal stresses are σ_θ and there are no shear stresses. On face $ABFE$, the stresses are σ_z and τ_{zr}. On face $CDHG$, the normal and shear stresses are

$$\sigma_z + \Delta\sigma_z = \sigma_z + \frac{\partial \sigma_z}{\partial z}\,\Delta z$$

$$\tau_{rz} + \Delta\tau_{rz} = \tau_{rz} + \frac{\partial \tau_{rz}}{\partial z}\,\Delta z$$

On face $AEHD$, the normal and shear stresses are σ_r and τ_{rz}. On face $BCGF$, the stresses are $\sigma_r + \dfrac{\partial \sigma_r}{\partial r}\,\Delta r$ and $\tau_{rz} + \dfrac{\partial \tau_{rz}}{\partial r}\,\Delta r$

For equilibrium in z direction

$$\left(\sigma_z + \frac{\partial \sigma_z}{\partial z}\Delta z\right)\left(r + \frac{\Delta r}{2}\right)\Delta\theta\,\Delta r + \left(\tau_{rz} + \frac{\partial \tau_{rz}}{\partial r}\Delta r\right)(r + \Delta r)\,\Delta\theta\,\Delta z$$

$$- \tau_{rz} r\,\Delta\theta\,\Delta z - \sigma_z\left(r + \frac{\Delta r}{2}\right)\Delta\theta\,\Delta r + \gamma_z\left(r + \frac{\Delta r}{2}\right)\Delta\theta\,\Delta r\,\Delta z = 0$$

where γ_z is the body force per unit volume in z direction. Hence,

$$\frac{\partial \sigma_z}{\partial z}\left(r + \frac{\Delta r}{2}\right)\Delta r\,\Delta\theta\,\Delta z + \frac{\partial \tau_{rz}}{\partial r}(r + \Delta r)\,\Delta r\,\Delta\theta\,\Delta z$$

$$+ \tau_{rz}\,\Delta r\,\Delta\theta\,\Delta z + \gamma_z\left(r + \frac{\Delta r}{2}\right)\Delta r\,\Delta\theta\,\Delta z = 0$$

Cancelling $\Delta r\,\Delta\theta\,\Delta z$ and going to limits

$$\frac{\partial \sigma_z}{\partial z} + \frac{\partial \tau_{rz}}{\partial r} + \frac{\tau_{rz}}{r} + \gamma_z = 0 \tag{8.1}$$

Similarly, for equilibrium in r direction we get

$$\frac{\partial \sigma_r}{\partial r} + \frac{\partial \tau_{rz}}{\partial z} + \frac{\sigma_r - \sigma_\theta}{r} + \gamma_r = 0 \tag{8.2}$$

where γ_r is the body force per unit volume in r direction. Since the stress components are independent of θ, the equilibrium equation for θ direction is identically satisfied.

For the problems that we are going to discuss in this chapter, we need expressions for the circumferential strain ε_θ and the radial strain ε_r.

Referring to Fig. 8.2(a), consider an arc AE at distance r, subtending an angle $\Delta\theta$ at the centre. The arc length is $r\Delta\theta$. The radial displacement is u_r. Consequently, the length of the arc becomes $(r + u_r)\,\Delta\theta$. Hence, the circumferential strain is

$$\varepsilon_\theta = \frac{(r + u_r)\Delta\theta - r\Delta\theta}{r\Delta\theta} = \frac{u_r}{r} \tag{8.3}$$

The radial strain is, from Fig. 8.2(b),

$$\varepsilon_r = \frac{\partial u_r}{\partial r} \tag{8.4}$$

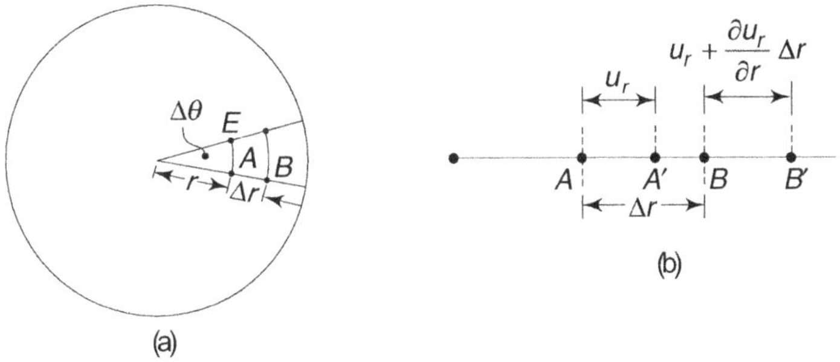

Fig. 8.2 *Displacements along a radius*

The axial strain is

$$\varepsilon_z = \frac{\partial u_z}{\partial z} \tag{8.5a}$$

where u_z is the axial displacement. In subsequent sections we shall consider the following problems:

Circular cylinder subjected to internal or external pressure
Sphere subjected to internal or external pressure
Sphere subjected to mutual gravitational attraction
Rotating disk of uniform thickness
Rotating disk of variable thickness
Rotating shaft and cylinder

8.2 THICK-WALLED CYLINDER SUBJECTED TO INTERNAL AND EXTERNAL PRESSURES—LAME'S PROBLEM

Consider a cylinder of inner radius a and outer radius b (Fig. 8.3). Let the cylinder be subjected to an internal pressure p_a and an external pressure p_b. It is possible to treat this problem either as a plane stress case ($\sigma_z = 0$) or as a plane strain case ($\varepsilon_z = 0$). Appropriate solutions will be obtained for each case.

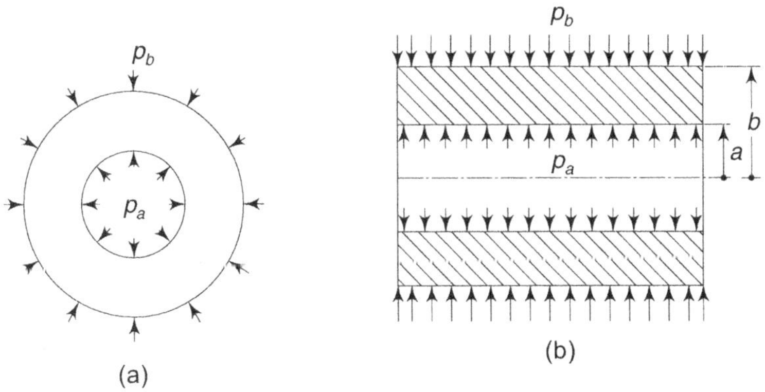

Fig. 8.3 *Thick-walled cylinder under internal and external pressures*

Case (a) Plane Stress

Let the ends of the cylinder be free to expand. We shall assume that $\sigma_z = 0$ and our results will justify this assumption. Owing to uniform radial deformation, $\tau_{rz} = 0$. Neglecting body forces, Eq. (8.2) reduces to

$$\frac{\partial \sigma_r}{\partial r} + \frac{\sigma_r - \sigma_\theta}{r} = 0 \tag{8.5b}$$

Since r is the only independent variable, the above equation can be written as

$$\frac{d}{dr}(r\sigma_r) - \sigma_\theta = 0 \tag{8.5c}$$

Equation (8.1) is identically satisfied. From Hooke's law

$$\varepsilon_r = \frac{1}{E}(\sigma_r - \nu\sigma_\theta), \qquad \varepsilon_\theta = \frac{1}{E}(\sigma_\theta - \nu\sigma_r)$$

or the stresses in terms of strains are

$$\sigma_r = \frac{E}{1-\nu^2}(\varepsilon_r + \nu\varepsilon_\theta) \qquad \sigma_\theta = \frac{E}{1-\nu^2}(\varepsilon_\theta + \nu\varepsilon_r)$$

Substituting for ε_r and ε_θ from Eqs (8.3) and (8.4)

$$\sigma_r = \frac{E}{1-\nu^2}\left(\frac{du_r}{dr} + \nu\frac{u_r}{r}\right) \tag{8.6a}$$

$$\sigma_\theta = \frac{E}{1-\nu^2}\left(\frac{u_r}{r} + \nu\frac{du_r}{dr}\right) \tag{8.6b}$$

Substituting these in the equation of equilibrium given by Eq. (8.5c)

$$\frac{d}{dr}\left(r\frac{du_r}{dr} + \nu u_r\right) - \left(\frac{u_r}{r} + \nu\frac{du_r}{dr}\right) = 0$$

or $$\frac{du_r}{dr} + r\frac{d^2u_r}{dr^2} + \nu\frac{du_r}{dr} - \frac{u_r}{r} - \nu\frac{du_r}{dr} = 0$$

i.e. $$\frac{d^2u_r}{dr^2} + \frac{1}{r}\frac{du_r}{dr} - \frac{u_r}{r^2} = 0$$

This can be reduced to

$$\frac{d}{dr}\left(\frac{du_r}{dr} + \frac{u_r}{r}\right) = 0$$

or $$\frac{d}{dr}\left[\frac{1}{r}\frac{d}{dr}(u_r r)\right] = 0 \tag{8.7}$$

If the function u_r is found from this equation, the stresses are then determined from Eqs. (8.6a) and (8.6b).

The solution to Eq. (8.7) is

$$u_r = C_1 r + \frac{C_2}{r} \tag{8.8}$$

where C_1 and C_2 are constants of integration. Substituting this function in Eqs. (8.6a) and (8.6b)

$$\sigma_r = \frac{E}{1-\nu^2}\left[C_1(1+\nu) - C_2(1-\nu)\frac{1}{r^2}\right] \tag{8.9a}$$

$$\sigma_\theta = \frac{E}{1-\nu^2}\left[C_1(1+\nu) + C_2(1-\nu)\frac{1}{r^2}\right] \tag{8.9b}$$

The constants C_1 and C_2 are determined from the boundary conditions.

When $r = a$, $\sigma_r = -p_a$

when $r = b$, $\sigma_r = -p_b$

Hence,

$$\frac{E}{1-\nu^2}\left[C_1(1+\nu) - C_2(1-\nu)\frac{1}{a^2}\right] = -p_a$$

$$\frac{E}{1-\nu^2}\left[C_1(1+\nu) - C_2(1-\nu)\frac{1}{b^2}\right] = -p_b$$

whence,

$$C_1 = \frac{1-\nu}{E}\,\frac{p_a a^2 - p_b b^2}{b^2 - a^2}$$

$$C_2 = \frac{1+\nu}{E}\,\frac{a^2 b^2}{b^2 - a^2}(p_a - p_b)$$

Substituting these in Eqs (8.8) and (8.9) we get

$$u_r = \frac{1-\nu}{E}\,\frac{p_a a^2 - p_b b^2}{b^2 - a^2}\,r + \frac{1+\nu}{E}\,\frac{a^2 b^2}{r}\,\frac{p_a - p_b}{b^2 - a^2} \tag{8.10}$$

$$\sigma_r = \frac{p_a a^2 - p_b b^2}{b^2 - a^2} - \frac{a^2 b^2}{r^2}\,\frac{p_a - p_b}{b^2 - a^2} \tag{8.11}$$

$$\sigma_\theta = \frac{p_a a^2 - p_b b^2}{b^2 - a^2} + \frac{a^2 b^2}{r^2}\,\frac{p_a - p_b}{b^2 - a^2} \tag{8.12}$$

It is interesting to observe that the sum $\sigma_r + \sigma_\theta$ is constant through the thickness of the wall of the cylinder, i.e. independent of r. Hence, according to Hooke's Law, the stresses σ_r and σ_θ produce a uniform extension or contraction in z direction, and cross-sections perpendicular to the axis of the cylinder remain plane. If we consider two adjacent cross-sections, the deformation undergone by the element does not interfere with the deformation of the neighbouring element. Hence, the elements can be considered to be in a state of plane stress, i.e. $\sigma_z = 0$, as we assumed at the beginning of the discussion. It is important to note that in Eqs (8.10)–(8.12), p_a and p_b are the numerical values of the compressive pressures applied.

Cylinder Subjected to Internal Pressure In this case $p_b = 0$ and $p_a = p$. Then Eqs (8.11) and (8.12) become

$$\sigma_r = -\frac{pa^2}{b^2 - a^2}\left(1 - \frac{b^2}{r^2}\right) \tag{8.13}$$

$$\sigma_\theta = \frac{pa^2}{b^2 - a^2}\left(1 + \frac{b^2}{r^2}\right) \tag{8.14}$$

These equations show that σ_r is always a compressive stress and σ_θ a tensile stress. Figure 8.4 shows the variation of radial and circumferential stresses across the thickness of the cylinder under internal pressure. The circumferential stress is greatest at the inner surface of the cylinder, where

$$(\sigma_\theta)_{\max} = \frac{p\left(a^2 + b^2\right)}{b^2 - a^2} \tag{8.15}$$

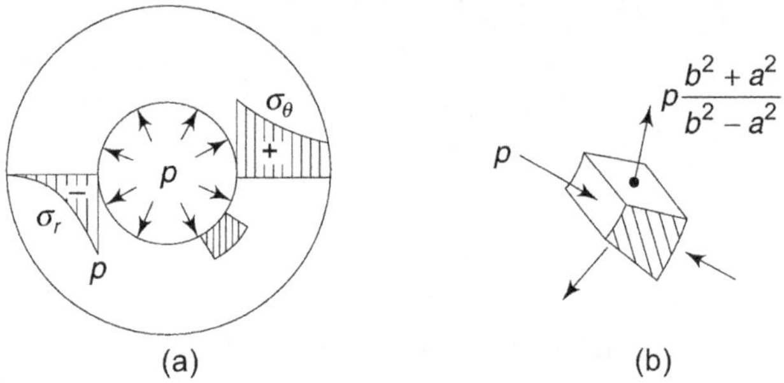

Fig. 8.4 *Cylinder subjected to internal pressure*

Hence, $(\sigma_\theta)_{\max}$ is always greater than the internal pressure and approaches this value as b increases so that it can never be reduced below p_a irrespective of the amount of material added on the outside.

Cylinder Subjected to External Pressure In this case, $p_a = 0$ and $p_b = p$. Equations (8.11) and (8.12) reduce to

$$\sigma_r = -\frac{pb^2}{b^2 - a^2}\left(1 - \frac{a^2}{r^2}\right) \tag{8.16}$$

$$\sigma_\theta = -\frac{pb^2}{b^2 - a^2}\left(1 + \frac{a^2}{r^2}\right) \tag{8.17}$$

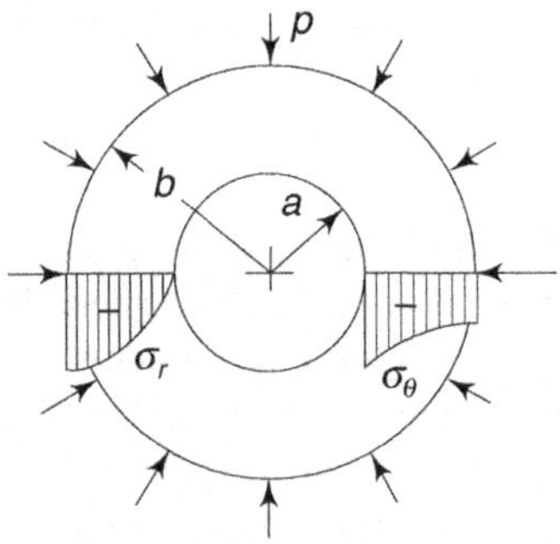

Fig. 8.5 *Cylinder subjected to external pressure*

The variations of these stresses across the thickness are shown in Fig. 8.5. If there is no inner hole, i.e. if $a = 0$, the stresses are uniformly distributed in the cylinder with $\sigma_r = \sigma_\theta = -p$.

Example 8.1 *Select the outer radius b for a cylinder subjected to an internal pressure p = 500 atm with a factor of safety 2. The yield point for the material (in tension as well as in compression) is σ_{yp} = 5000 kgf/cm² (490000 kPa). The inner radius is 5 cm. Assume that the ends of the cylinder are closed.*

Solution The critical point lies on the inner surface of the cylinder, where

$$\sigma_r = -p, \qquad \sigma_\theta = p\frac{b^2+a^2}{b^2-a^2}, \qquad \sigma_z = p\frac{a^2}{b^2-a^2} \qquad \text{(assumed)}$$

In the above expressions, it is assumed that away from the ends, σ_z caused by p is uniformly distributed across the thickness. The maximum and minimum principal stresses are $\sigma_1 = \sigma_\theta$ and $\sigma_3 = \sigma_r$. Hence,

$$\tau_{\max} = \frac{1}{2}(\sigma_1 - \sigma_3) = p\frac{b^2}{b^2-a^2}$$

Substituting the numerical values (1 atm = 98.07 kPa),

$$b = \sqrt{\frac{5}{3}}\,a = 6.45 \text{ cm}$$

Example 8.2 *A thick-walled steel cylinder with radii a = 5 cm and b = 10 cm is subjected to an internal pressure p. The yield stress in tension for the material is 350 MPa. Using a factor of safety of 1.5, determine the maximum working pressure p according to the major theories of failure. E = 207 × 10⁶ kPa, v = 0.25.*

Solution

(i) *Maximum normal stress theory*

Maximum normal stress = σ_θ at $r = a$

$$= p\frac{(b^2+a^2)}{(b^2-a^2)}$$

$$\therefore \qquad p\frac{(b^2+a^2)}{(b^2-a^2)} = \frac{\sigma_y}{N}$$

or

$$p = \frac{350\times 10^6}{1.5}\times\frac{100-25}{100+25} = 140\times 10^3 \text{ kPa}$$

(ii) *Maximum shear stress theory*

Maximum shear stress = $\frac{1}{2}(\sigma_\theta - \sigma_r)$ at $r = a$

$$= \frac{1}{2}p\left(\frac{2b^2}{b^2-a^2}\right)$$

$$\therefore \qquad p\frac{b^2}{b^2-a^2} = \frac{1}{2}\frac{\sigma_y}{N}$$

or $$p = \frac{350 \times 10^6}{3} \times \frac{100 - 25}{100} = 87.5 \times 10^3 \text{ kPa}$$

(iii) *Maximum strain theory*

Maximum strain $= \varepsilon_\theta = \frac{1}{E}(\sigma_\theta - \nu\,\sigma_r)$ at $r = a$

$$= \frac{p}{E}\frac{a^2}{(b^2 - a^2)}\left[\left(1 + \frac{b^2}{a^2}\right) - \nu\left(1 - \frac{b^2}{a^2}\right)\right]$$

$\therefore$ $$\frac{p}{E(b^2 - a^2)}\left[(a^2 + b^2) - \nu(a^2 - b^2)\right] = \frac{\sigma_y}{NE}$$

or $$\frac{p}{(100 - 25)}\left[125 + (0.25 \times 75)\right] = \frac{350 \times 10^6}{1.5}$$

$\therefore$ $$p = \frac{350 \times 10^6 \times 75}{1.5 \times 143.75} = 121.7 \times 10^3 \text{ kPa}$$

(iv) *Octahedral shear stress theory*

$$\tau_{\text{oct}} = \frac{1}{3}\left[\sigma_\theta^2 + \sigma_r^2 + (\sigma_r - \sigma_\theta)^2\right]^{1/2} \quad \text{at} \quad r = a$$

$$= \frac{1}{3}\left[2(\sigma_r - \sigma_\theta)^2 + 2\sigma_r\,\sigma_\theta\right]^{1/2}$$

$$= \frac{\sqrt{2}}{3}\left\{\left[-p - \frac{p(b^2 + a^2)}{(b^2 - a^2)}\right]^2 - p^2\frac{(b^2 + a^2)}{(b^2 - a^2)}\right\}^{1/2}$$

$$= \frac{\sqrt{2}}{3}\frac{\sigma_y}{N}$$

$\therefore$ $$\frac{\sqrt{2}}{3}\,p\left[\frac{4b^4}{(b^2 - a^2)^2} - \frac{(b^2 + a^2)}{(b^2 - a^2)}\right]^{1/2} = \frac{\sqrt{2}}{3}\frac{\sigma_y}{N}$$

or $$p\left(\frac{40000}{5625} - \frac{125}{75}\right)^{1/2} = \frac{350}{1.5}$$

$\therefore$ $$p = 100 \times 10^3 \text{ kPa}$$

(v) *Energy of distortion theory*

This will give a value identical to that obtained based on octahedral shear stress theory, i.e. $p = 100 \times 10^3$ kPa.

Example 8.3 *A pipe made of steel has a tensile elastic limit $\sigma_y = 275$ MPa and $E = 207 \times 10^6$ kPa. If the pipe has an internal radius $a = 5$ cm and is subjected to an internal pressure $p = 70 \times 10^3$ kPa, determine the proper thickness for the pipe wall according to the major theories of failure. Use a factor of safety $N = \frac{4}{3}$.*

Solution

(i) *Maximum principal stress theory*

Maximum principal stress = σ_θ at $r = a$

$$= p\frac{(b^2 + a^2)}{(b^2 - a^2)} = \frac{\sigma_y}{N}$$

$$\therefore \quad \frac{70 \times 10^6\left[(25 \times 10^{-4}) + b^2\right]}{\left[b^2 - (25 \times 10^{-4})\right]} = \frac{275 \times 10^6 \times 3}{4}$$

or $\quad 1750 \times 10^{-4} + 70b^2 = 825b^2 - \dfrac{20625}{4} \times 10^{-4}$

or $\quad 136.25b^2 = 6906.25 \times 10^{-4}$

$\therefore \quad b = 7.12 \times 10^{-2}$ m = 7.12 cm

$\therefore$ Wall thickness $t = 2.12$ cm

(ii) *Maximum shear stress theory*

$$\tau_{max} = \frac{1}{2}(\sigma_\theta - \sigma_r) \quad \text{at} \quad r = a$$

$$= \frac{pb^2}{(b^2 - a^2)} = \frac{\sigma_y}{N}$$

$$\therefore \quad \frac{70 \times 10^6\, b^2}{[b^2 - (25 \times 10^{-4})]} = \frac{3}{8} \times 275 \times 10^6$$

or $\quad 70b^2 = 103.13\, b^2 - 2578.1 \times 10^{-4}$

$\therefore \quad b = 8.82 \times 10^{-2}$ m = 8.82 cm

$\therefore$ Wall thickness $t = 3.82$ cm

(iii) *Maximum strain theory (with $\nu = 0.25$)*

$$\varepsilon_{max} = \frac{1}{E}(\sigma_\theta - \nu\sigma_r) \quad \text{at} \quad r = a$$

$$= \frac{p}{E\,(b^2 - a^2)}\left[(a^2 + b^2) - \nu\,(a^2 - b^2)\right] = \frac{\sigma_y}{NE}$$

$$\therefore \quad \frac{70 \times 10^6}{[b^2 - (25 \times 10^{-4})]}\Big[(0.75 \times 25 \times 10^{-4})$$

$$+ (1.25 \times b^2)\Big] = \frac{3}{4} \times 275 \times 10^6$$

or $\quad 1312.5 \times 10^{-4} + 87.5b^2 = 206.25b^2 - 5156.25 \times 10^{-4}$

$\therefore \quad b = 7.38 \times 10^{-2}$ m = 7.38 cm

$\therefore$ Wall thickness $t = 2.38$ cm

(iv) *Maximum distortion energy theory*

From Eq. (4.12) with

$$\sigma_1 = \sigma_\theta, \quad \sigma_2 = 0, \quad \sigma_3 = \sigma_r = -p$$

$$U^* = \frac{1}{12G}[\sigma_\theta^2 + \sigma_r^2 + (\sigma_r - \sigma_\theta)^2]$$

$$= \frac{(1+\nu)}{6E}(2\sigma_\theta^2 + 2\sigma_r^2 - 2\sigma_r\,\sigma_\theta)$$

$$= \frac{(1+\nu)}{3E}(\sigma_\theta^2 + \sigma_r^2 - \sigma_\theta\,\sigma_r) = \frac{1+\nu}{E}\frac{\sigma_y^2}{N^2}$$

$$\therefore \quad \sigma_\theta^2 + \sigma_r^2 - \sigma_\theta\,\sigma_r = \frac{\sigma_y^2}{N^2}$$

i.e.
$$p^2\left[\frac{(b^2+a^2)^2}{(b^2-a^2)^2} + 1 + \frac{(b^2-a^2)}{(b^2+a^2)}\right] = \frac{\sigma_y^2}{N^2}$$

Putting $\left(\frac{\sigma_y}{pN}\right) = f_y$ and simplifying one gets

$$(3 - f_y^2)\,b^4 + 2a^2 f_y^2\, b^2 + (1 - f_y^2)\,a^4 = 0$$

$$\therefore \quad b^2 = \frac{-2a^2 f_y^2 \pm \sqrt{\left[4a^4 f_y^4 - 4a^4(1-f_y^2)(3-f_y^2)\right]}}{2(3-f_y^2)}$$

$$= \frac{a^2\left[-f_y^2 \pm \sqrt{(4f_y^4 - 3)}\right]}{2(3-f_y^2)}$$

With $a = 5 \times 10^{-2}$

$$fy = \frac{275 \times 10^6 \times 3}{70 \times 10^6 \times 4} = 2.946$$

$\therefore \quad b^2 = (63 \text{ or } 13.4)\,10^{-4}$ or $b = 7.9 \times 10^{-2}\text{ m} = 7.9\text{ cm}$

Wall thickness $t = 2.9$ cm

Case (b) Plane Strain

When the cylinder is fairly long, sections that are far from the ends can be considered to be in a state of plane strain and we can assume that σ_z does not vary along the z-axis. As in the case of plane stress, the equation is

$$\frac{d}{dr}(r\sigma_r) - \sigma_\theta = 0$$

From Hooke's law

$$\varepsilon_r = \frac{1}{E}[\sigma_r - \nu(\sigma_\theta + \sigma_z)]$$

$$\varepsilon_\theta = \frac{1}{E}[\sigma_\theta - \nu(\sigma_r + \sigma_z)]$$

$$\varepsilon_z = \frac{1}{E}[\sigma_z - \nu(\sigma_r + \sigma_\theta)]$$

Since $\varepsilon_z = 0$ in this case, one has from the last equation

$$\sigma_z = \nu(\sigma_r + \sigma_\theta)$$

$$\varepsilon_r = \frac{1+\nu}{E}[(1-\nu)\,\sigma_r - \nu\sigma_\theta] \tag{8.18}$$

$$\varepsilon_\theta = \frac{1+\nu}{E}\left[(1-\nu)\,\sigma_\theta - \nu\sigma_r\right]$$

Solving for σ_θ and σ_r

$$\sigma_\theta = \frac{E}{(1-2\nu)(1+\nu)}\left[\nu\varepsilon_r + (1-\nu)\,\varepsilon_\theta\right] \tag{8.19a}$$

$$\sigma_r = \frac{E}{(1-2\nu)(1+\nu)}\left[(1-\nu)\,\varepsilon_r + \nu\varepsilon_\theta\right] \tag{8.19b}$$

On substituting for ε_r and ε_θ from Eqs (8.3) and (8.4), the above equations become

$$\sigma_\theta = \frac{E}{(1-2\nu)(1+\nu)}\left[\nu\frac{du_r}{dr} + (1-\nu)\frac{u_r}{r}\right] \tag{8.20}$$

$$\sigma_r = \frac{E}{(1-2\nu)(1+\nu)}\left[(1-\nu)\frac{du_r}{dr} + \nu\frac{u_r}{r}\right] \tag{8.21}$$

Substituting these in the equation of equilibrium, Eq. (8.5c)

$$\frac{d}{dr}\left[(1-\nu)\,r\frac{du_r}{dr} + \nu u_r\right] - \nu\frac{du_r}{dr} - (1-\nu)\frac{u_r}{r} = 0$$

or

$$\frac{du_r}{dr} + r\frac{d^2u_r}{dr^2} - \frac{u_r}{r} = 0$$

i.e.

$$\frac{d}{dr}\left(\frac{du_r}{dr} + \frac{u_r}{r}\right) = 0$$

This is the same as Eq. (8.7) for the plane stress case. The solution is the same as in Eq. (8.8).

$$u_r = C_1 r + \frac{C_2}{r}$$

where C_1 and C_2 are constants of integration. From Eqs (8.20) and (8.21)

$$\sigma_\theta = \frac{E}{(1-2\nu)(1+\nu)}\left[C_1 + (1-2\nu)\frac{C_2}{r^2}\right] \tag{8.22a}$$

$$\sigma_r = \frac{E}{(1-2\nu)(1+\nu)}\left[C_1 - (1-2\nu)\frac{C_2}{r^2}\right] \tag{8.22b}$$

Once again, we observe that $\sigma_r + \sigma_\theta$ is a constant independent of r. Further, the axial stress from Eq. (8.18) is

$$\sigma_z = -\frac{2\nu E}{(1-2\nu)(1+\nu)}C_1 \tag{8.22c}$$

Applying the boundary conditions

$$\sigma_r = -p_a \quad \text{when } r = a, \qquad \sigma_r = -p_b \quad \text{when } r = b$$

$$\frac{E}{(1-2\nu)(1+\nu)}\left[C_1 - (1-2\nu)\frac{C_2}{a^2}\right] = -p_a$$

$$\frac{E}{(1-2\nu)(1+\nu)}\left[C_1 - (1-2\nu)\frac{C_2}{b^2}\right] = -p_b$$

Solving, $$C_1 = \frac{(1-2\nu)(1+\nu)}{E}\,\frac{p_b b^2 - p_a a^2}{a^2 - b^2}$$

and $$C_2 = \frac{1+\nu}{E}\,\frac{(p_b - p_a)\,a^2 b^2}{a^2 - b^2}$$

Substituting these, the stress components become

$$\sigma_r = \frac{p_a a^2 - p_b b^2}{b^2 - a^2} - \frac{p_a - p_b}{b^2 - a^2}\,\frac{a^2 b^2}{r^2} \tag{8.23}$$

$$\sigma_\theta = \frac{p_a a^2 - p_b b^2}{b^2 - a^2} + \frac{p_a - p_b}{b^2 - a^2}\,\frac{a^2 b^2}{r^2} \tag{8.24}$$

$$\sigma_z = 2\nu\,\frac{p_b a^2 - p_a b^2}{b^2 - a^2} \tag{8.25}$$

It is observed that the values of σ_r and σ_θ are identical to those of the plane stress case. But in the plane stress case, $\sigma_z = 0$, whereas in the plane strain case, σ_z has a constant value given by Eq. (8.25).

8.3 STRESSES IN COMPOSITE TUBES—SHRINK FITS

The problem which will be considered now, involves two cylinders made of two different materials and fitted one inside the other. Before assembling, the inner cylinder has an internal radius a and an external radius c. The internal radius of the outer cylinder is less than c by Δ, i.e. its internal radius is $c - \Delta$. Its external radius is b. If the inner cylinder is cooled and the outer cylinder is heated, then the two cylinders can be assembled, one fitting inside the other. When the cylinders come to room temperature, a shrink fit is obtained. The problem lies in determining the contact pressure p_c between the two cylinders.

The above construction is often used to obtain thick-walled vessels to withstand high pressures. For example, if we need a vessel to withstand a pressure of say 15000 atm, the yield point of the material must be at least 30000 kgf/cm^2 (2940000 kPa). Since no such high-strength material exists, shrink-fitted composite tubes are designed.

The contact pressure p_c acting on the outer surface of the inner cylinder reduces its outer radius by u_1. On the other hand, the same contact pressure increases the inner radius of the outer cylinder by u_2. The sum of these two quantities, i.e. $(-u_1 + u_2)$ must be equal to Δ, the difference in the radii of the cylinders. To determine u_1 and u_2, we make use of Eq. (8.10), assuming a plane stress case.

For the inner tube

$$u_1 = \frac{1-\nu_1}{E_1}\left(-p_c\frac{c^2}{c^2-a^2}\right)c + \frac{1+\nu_1}{E_1}\frac{a^2c^2}{c}\left(-\frac{p_c}{c^2-a^2}\right)$$

or

$$u_1 = -\frac{cp_c}{E_1(c^2-a^2)}\left[(1-\nu_1)\,c^2 + (1+\nu_1)\,a^2\right]$$

For the outer tube

$$u_2 = -\frac{1-\nu_2}{E_2}\left(p_c\frac{c^2}{b^2-c^2}\right)c + \frac{1+\nu_2}{E_2}\frac{c^2b^2}{c}\left(\frac{p_c}{b^2-c^2}\right)$$

or

$$u_2 = -\frac{cp_c}{E_2(b^2-c^2)}\left[(1-\nu_2)\,c^2 + (1+\nu_2)b^2\right]$$

In calculating u_2, we have neglected Δ since it is very small as compared to c. Noting that u_1 is negative and u_2 is positive, we should have

$$-u_1 + u_2 = \Delta$$

i.e.

$$\frac{cp_c}{E_1(c^2-a^2)}\left[(1-\nu_1)c^2 + (1+\nu_1)a^2\right]$$

$$+\frac{cp_c}{E_2(b^2-c^2)}\left[(1-\nu_2)c^2 + (1+\nu_2)b^2\right] = \Delta \tag{8.26a}$$

Regrouping, the contact pressure p_c is given by

$$p_c = \frac{\Delta/c}{\frac{1}{E_1}\left[\frac{c^2+a^2}{c^2-a^2} - \nu_1\right] + \frac{1}{E_2}\left[\frac{b^2+c^2}{b^2-c^2} + \nu_2\right]} \tag{8.26b}$$

If the two cylinders are made of the same material, then $E_1 = E_2$ and $\nu_1 = \nu_2$. Equation (8.26) will then reduce to

$$p_c = \frac{E\Delta}{2c^3}\frac{(c^2-a^2)(b^2-c^2)}{(b^2-a^2)} \tag{8.27}$$

It is important to note that in Eqs (8.26) and (8.27), Δ is the difference in radii between the inner cylinder and the outer jacket. Because of shrink fitting, therefore, the inner cylinder is under external pressure p_c. The stress distribution in the assembled cylinders is shown in Fig. 8.6.

If the composite cylinder made up of the same material is now subjected to an internal pressure p, then the two parts will act as a single unit and the additional stresses induced in the composite can be determined from Eqs (8.13) and (8.14). At the inner surface of the inner cylinder, the internal pressure p causes a tensile tangential stress σ_θ, Eq. (8.14), but, the contact pressure p_c causes at the same points a compressive tangential stress, Eq (8.17). Hence, a composite cylinder can

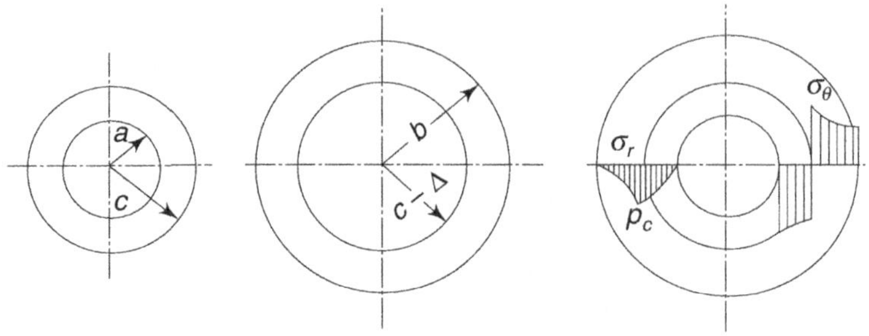

Fig. 8.6 *Streses in composite tubes*

support greater internal pressure than an ordinary one. However, at the inner points of the jacket or the outer cylinder, the internal pressure p and the contact pressure p_c both will induce tensile tangential (i.e. circumferential) stresses σ_θ. For design purposes, one can choose the shrink-fit allowance Δ such that the strengths of the two cylinders are equal. To determine this value of Δ, one can proceed as follows.

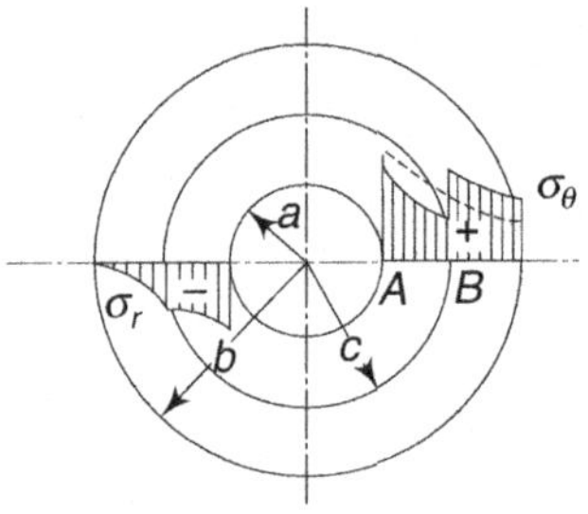

Fig. 8.7 *Equal strength composite tube*

Let a and c be the radii of the inner cylinder, and c and b the radii of the jacket (see Fig. 8.7) c is the common radius of the two cylinders at the contact surface when the composite cylinder is experiencing an internal pressure p and the shrink-fit pressure p_c. If the strengths of the two cylinders are the same, then according to the maximum shear stress theory, $(\sigma_1 - \sigma_3)$ at point A of the inner cylinder should be equal to $(\sigma_1 - \sigma_3)$ at point B of the outer cylinder. σ_1 and σ_3 are the maximum and minimum normal stresses, which are respectively equal to σ_θ and σ_r.

At point A, due to internal pressure p, from Eqs (8.13) and (8.14),

$$(\sigma_\theta - \sigma_r)_A = p\frac{b^2 + a^2}{b^2 - a^2} - (-p)$$

$$= 2p\frac{b^2}{b^2 - a^2}$$

Because of shrink-fitting pressure p_c, at the same point, from Eqs (8.16) and (8.17),

$$(\sigma_\theta - \sigma_r)_A = -2p_c\frac{c^2}{c^2 - a^2}$$

Hence, the resultant value of $(\sigma_\theta - \sigma_r)$ at A is

$$(\sigma_\theta - \sigma_r)_A = 2p\frac{b^2}{b^2 - a^2} - 2p_c\frac{c^2}{c^2 - a^2} \tag{8.28}$$

At point B of the outer cylinder, since the composite involves the same material, due to the pressure p, from Eqs (8.13) and (8.14), and observing that $r = c$ in these equations,

$$(\sigma_\theta - \sigma_r)_B = p\left[\frac{a^2(c^2+b^2)}{c^2(b^2-a^2)} - \frac{a^2(c^2-b^2)}{c^2(b^2-a^2)}\right]$$

$$= 2p\frac{a^2b^2}{c^2(b^2-a^2)}$$

At the same point B, due to the contact pressure p_c, from Eqs (8.13) and (8.14), with internal radius equal to c and external radius b,

$$(\sigma_\theta - \sigma_r)_B = p_c\left[\frac{c^2+b^2}{b^2-c^2} - \frac{c^2-b^2}{b^2-c^2}\right]$$

$$= 2p_c\frac{b^2}{b^2-c^2}$$

The resultant value of $(\sigma_\theta - \sigma_r)$ at B is therefore

$$(\sigma_\theta - \sigma_r)_B = 2p\frac{a^2b^2}{c^2(b^2-a^2)} + 2p_c\frac{b^2}{(b^2-c^2)} \tag{8.29}$$

For equal strength, equating Eqs (8.28) and (8.29)

$$2p\frac{b^2}{(b^2-a^2)} - 2p_c\frac{c^2}{(c^2-a^2)} = 2p\frac{a^2b^2}{c^2(b^2-a^2)} + 2p_c\frac{b^2}{(b^2-c^2)}$$

or $$p_c\left[\frac{b^2}{b^2-c^2} + \frac{c^2}{(c^2-a^2)}\right] = p\left[\frac{b^2}{(b^2-a^2)} - \frac{a^2b^2}{c^2(b^2-a^2)}\right] \tag{8.30}$$

The shrink-fitting pressure p_c is related to the negative allowance Δ through Eq. (8.27) and it is this value of Δ that is required now for equal strength. Hence, substituting for p_c from Eq. (8.27), Eq. (8.30) becomes

$$\frac{\Delta E}{2c^3}\frac{(c^2-a^2)(b^2-c^2)}{(b^2-a^2)}\left[\frac{b^2}{(b^2-c^2)} + \frac{c^2}{(c^2-a^2)}\right] = p\left[\frac{b^2}{(b^2-a^2)} - \frac{a^2b^2}{c^2(b^2-a^2)}\right]$$

or $$\frac{\Delta E}{2c^3}\frac{(2b^2c^2-b^2a^2-c^4)}{(b^2-a^2)} = \frac{pb^2(c^2-a^2)}{c^2(b^2-a^2)}$$

or $$\Delta = \frac{2p}{E}\frac{b^2c(c^2-a^2)}{b^2(c^2-a^2)-c^2(b^2-c^2)} \tag{8.31a}$$

Also, from Eq. (8.30),

$$p_c = p\frac{b^2(c^2-a^2)^2(b^2-c^2)}{c^2(b^2-a^2)\left[b^2(c^2-a^2)+c^2(b^2-c^2)\right]} \tag{8.31b}$$

The value of $(\sigma_\theta - \sigma_r)$ either at A or at B, from Eqs (8.28) and (8.31), is

$$\sigma_\theta - \sigma_r = p\frac{2b^2}{(b^2-a^2)}\left[1 - \frac{(c^2-a^2)(b^2-c^2)}{b^2(c^2-a^2)+c^2(b^2-c^2)}\right]$$

or
$$\sigma_\theta - \sigma_r = p\frac{2b^2}{(b^2-a^2)}\left[1 - \frac{1}{\dfrac{b^2}{b^2-c^2}+\dfrac{c^2}{c^2-a^2}}\right] \tag{8.32}$$

Therefore, for composites made of the same material, in order to have equal strength according to the shear stress theory, the shrink-fit allowance Δ that is necessary is given by Eq. (8.31a), and this depends on the internal pressure p. Further, Δ depends upon the difference between the external radius of the inner cylinder and the internal radius of the jacket. In other words, this depends on $c(+)$ and $c(-)$. With a, b and p fixed, one can determine the optimum value of c for minimum $(\sigma_\theta - \sigma_r)$ at A and B. From Eq. (8.32), the minimum value of $(\sigma_\theta - \sigma_r)$ is obtained when the denominator of the second expression within the square brackets is a maximum, i.e. when D is a maximum, where

$$D = \frac{b^2}{b^2-c^2}+\frac{c^2}{c^2-a^2}$$

Differentiating with respect to c and equating the differential to zero,

$$\frac{dD}{dc} = \frac{2cb^2}{(b^2-c^2)^2}+\frac{2c(c^2-a^2)-2c^3}{(c^2-a^2)^2}=0$$

Simplifying, one gets

$$c = \sqrt{ab}$$

The corresponding value of $(\sigma_\theta - \sigma_r)$, from Eq. (8.32), is

$$(\sigma_\theta - \sigma_r)_{\min} = p\frac{2b^2}{(b^2-a^2)}\left[1 - \frac{1}{\dfrac{b^2}{b(b-a)}+\dfrac{ab}{a(b-a)}}\right]$$

$$= p\frac{2b^2}{(b^2-a^2)}\left[1 - \frac{(b-a)}{2b}\right]$$

or
$$(\sigma_\theta - \sigma_r)_{\min} = p\frac{b}{(b-a)} \tag{8.33}$$

Also, the optimum value of Δ is from Eq. (8.31a),

$$\Delta_{\text{opt}} = \frac{1}{E}pc = \frac{p}{E}\sqrt{ab} \tag{8.34}$$

Example 8.4 *Determine the diameters 2c and 2b and the negative allowance Δ for a two-layer barrel of inner diameter 2a = 100 mm. The maximum pressure the barrel is to withstand is p_{max} = 2000 kgf/cm^2 (196000 kPa). The material is steel with E = 2(10)^6 kgf/cm^2 (196 $\times$ 10^5 kPa); σ_{yp} in tension or compression is 6000 kgf/cm^2 (588 $\times$ 10^3 kPa). The factor of safety is 2.*

Solution From Eq. (8.33),

$$\frac{6000}{2} = 2000\frac{b}{b-a}$$

Therefore, $b = 3a$

Since $c = \sqrt{ab}, c = \sqrt{3}\,a$. The numerical values are therefore, $2a$ = 100 mm, $2b$ = 300 mm, $2c$ = 173 mm. With $c = ab$, the value of Δ is, from Eq. (8.34),

$$\Delta = \frac{p}{E}\sqrt{ab} = \frac{2000}{2 \times 10^6}\sqrt{(50 \times 150)} = 0.866 \text{ mm}$$

Example 8.5 *A steel shaft of 10 cm diameter is shrunk inside a bronze cylinder of 25 cm outer diameter. The shrink allowance is 1 part per 1000 (i.e. 0.005 cm difference between the radii). Find the circumferential stresses in the bronze cylinder at the inside and outer radii and the stress in the shaft.*

$$E_{\text{steel}} = 2.18 \times 10^6 \text{ kgf/cm}^2\ (214 \times 10^6 \text{ kPa})$$
$$E_{\text{bronze}} = 1.09 \times 10^6 \text{ kgf/cm}^2\ (107 \times 10^6 \text{ kPa})$$

and $\nu = 0.3$ for both metals.

Solution In Eq. (8.26),

$$a = 0, \quad c = 5, \quad b = 12.5, \quad \Delta = 0.005, \quad \nu_1 = \nu_2 = 0.3$$

Substituting in Eq. (8.26a),

$$\frac{5p_c}{2.18 \times 10^6 \times 25}(0.7 \times 25) + \frac{5p_c}{1.09 \times 10^6 \times (156.25 - 25)} \times \left(0.7 \times 25 + 1.3 \times 156.25\right) = 0.005$$

or, $p_c = 610 \text{ kgf/cm}^2$ (59780 kPa)

For the bronze tube, the circumferential stress is, from Eq. (8.14),

$$\sigma_\theta = \frac{610 \times 25}{(156.25 - 25)}\left(1 + \frac{156.25}{r^2}\right)$$

When r = 5 cm and r = 12.5 cm

$$\sigma_\theta = 842.4 \text{ kgf/cm}^2\ (82555 \text{ kPa})$$
$$\sigma_\theta = 232.4 \text{ kgf/cm}^2\ (22775 \text{ kPa})$$

The shaft experiences equal σ_r and σ_θ at every point, from Eqs (8.16) and (8.17). Hence,

$$\sigma_r = \sigma_\theta = -610 \text{ kgf/cm}^2\ (59780 \text{ kPa})$$

Example 8.6 *A compound cylinder made of copper inner tube of radii a = 10 cm and c = 20 cm is snug fitted (Δ = 0) inside a steel jacket of external radius b = 40 cm. If the compound cylinder is subjected to an internal pressure p = 1500 kgf/cm^2 (147009 kPa), determine the contact pressure p_c and the values of σ_r and σ_θ at the inner and external points of the inner cylinder and of the jacket. Use the following data:*

$$E_{st} = 2 \times 10^6 \text{ kgf/cm}^2\ (196 \times 10^6 \text{ kPa}),$$
$$E_{cu} = 1 \times 10^6 \text{ kgf/cm}^2\ (98 \times 10^6 \text{ kPa}), \qquad \nu_{st} = 0.3, \qquad \nu_{cu} = 0.34$$

Solution Since the initial shrink-fit allowance Δ is zero, the initial contact pressure is zero. When the compound cylinder is subjected to an internal pressure p, the increase in the external radius of the copper cylinder under p and contact pressure p_c should be equal to the increase in the internal radius of the jacket under the contact pressure p_c,

i.e. $\left[(u_r)_p + (u_r)_{p_c} \text{ at } r = c\right]_{cu} = \left[(u_r)_{p_c} \text{ at } r = c\right]_{st}$

For copper cylinder, from Eq. (8.10),

$$(u_r)_p = \frac{1-\nu_{cu}}{E_{cu}}\frac{pa^2c}{(c^2-a^2)} + \frac{1+\nu_{cu}}{E_{cu}}\frac{a^2c^2}{c}\frac{p}{(c^2-a^2)}$$

$$(u)_{p_c} = -\frac{1-\nu_{cu}}{E_{cu}}\frac{p_c c^3}{(c^2-a^2)} - \frac{1+\nu_{cu}}{E_{cu}}\frac{a^2c^2}{c}\frac{p_c}{(c^2-a^2)}$$

$$(u_r)_{\text{total}} = \frac{2pa^2c}{E_{cu}(c^2-a^2)} - \frac{p_c c}{E_{cu}(c^2-a^2)}\left[c^2(1-\nu_{cu}) + a^2(1+\nu_{cu})\right]$$

For steel jacket, from Eq. (8.10),

$$(u)_{p_c} = \frac{1-\nu_{st}}{E_{st}}\frac{p_c c^3}{(b^2-c^2)} + \frac{1+\nu_{st}}{E_{st}}\frac{c^2b^2}{c}\frac{p_c}{(b^2-c^2)}$$

$$= \frac{p_c c}{E_{st}(b^2-c^2)}\left[c^2(1-\nu_{st}) + b^2(1+\nu_{st})\right]$$

Equating the (u_r)s

$$\frac{2pa^2c}{E_{cu}(c^2-a^2)} - \frac{p_c c}{E_{cu}(c^2-a^2)}\left[c^2(1-\nu_{cu}) + a^2(1+\nu_{cu})\right]$$

$$= \frac{p_c c}{E_{st}(b^2-c^2)}\left[c^2(1-\nu_{st}) + b^2(1+\nu_{st})\right]$$

or $$p_c\left[\frac{(c^2+a^2) - \nu_{cu}(c^2-a^2)}{E_{cu}(c^2-a^2)} + \frac{(b^2+c^2) + \nu_{st}(b^2-c^2)}{E_{st}(b^2-c^2)}\right]$$

$$= p\frac{2a^2}{E_{cu}(c^2-a^2)}$$

With $p = 1500$ kgf/cm^2, $a = 10$, $c = 20$, $b = 40$, $\nu_{st} = 0.3$, $\nu_{cu} = 0.34$,

$$p_c\left[\frac{500 - 300 \times 0.34}{300 \times 10^6} + \frac{2000 + 1200 \times 0.3}{2 \times 1200 \times 10^6}\right] = \frac{3000 \times 100}{300 \times 10^6}$$

$\therefore$ $p_c = 433$ kgf/cm^2 (42453 kPa)

Now, p_c will act as an external pressure on the copper tube and as an internal pressure on the steel jacket. For copper tube, from Eqs (8.11) and (8.12),

(i) Inner surface:,

σ_r at $r = a$ is -1500 kgf/cm^2 and,

σ_θ at $r = a$ is

$$= \frac{(1500 \times 100) - (433 \times 400)}{(400 - 100)} + \frac{100 \times 400}{100} \times \frac{(1500 - 433)}{(400 - 100)}$$

$= 1357$ kgf/cm^2 (compressive)

(ii) Outer surface:

σ_r at $r = c$ is -433 kgf/cm^2 and,

σ_θ at $r = c$ is

$$= \frac{(1500 \times 100) - (433 \times 400)}{(400 - 100)} + \frac{100 \times 400}{400} \frac{(1500 - 433)}{(400 - 100)}$$

$= 279$ kgf/cm^2

For steel jacket, from Eqs (8.11) and (8.12),

(i) Inner surface:

σ_r at $r = c$ is -433 kgf/cm^2 and,

σ_θ at $r = c$ is

$$= \frac{(433 \times 400)}{(1600 - 400)} = \frac{400 \times 1600}{400} \times \frac{433}{(1600 - 400)}$$

$= 722$ kgf/cm^2

(ii) Outer surface:

σ_r at $r = b$ is zero and,

σ_θ at $r = b$ is

$$= \frac{(433 \times 400)}{(1600 - 400)} + \frac{400 \times 1600}{1600} \times \frac{433}{(1600 - 400)}$$

$= 289$ kgf/cm^2

8.4 SPHERE WITH PURELY RADIAL DISPLACEMENTS

Consider a uniform sphere or spherical shell subjected to radial forces only, such as internal or external pressures. The sphere or the spherical shell will then undergo radial displacements only. Consider a particle situated at radius r before deformation. After deformation, the spherical surface of radius r becomes a surface of radius $(r + u_r)$ and the particle undergoes a displacement u_r. Similarly, another particle at distance $(r + \Delta r)$ along the same radial line will undergo a displacement $\left(u_r + \frac{\partial u_r}{\partial r}\Delta r\right)$.

Hence, the radial strain is

$$e_r = \frac{\partial u_r}{\partial r}$$

Before deformation, the circumference of any great circle on the surface of radius r is $2\pi r$. After deformation, the radius becomes $(r + u_r)$ and the circumference of the great circle is $2\pi(r + u_r)$ Hence, the circumferential strain is

$$\varepsilon_\phi = \frac{2\pi(r + u_r) - 2\pi r}{2\pi r} = \frac{u_r}{r}$$

This is the strain in every direction perpendicular to the radius r. Because of complete symmetry, we can choose a frame of reference, as shown in Fig. 8.8.

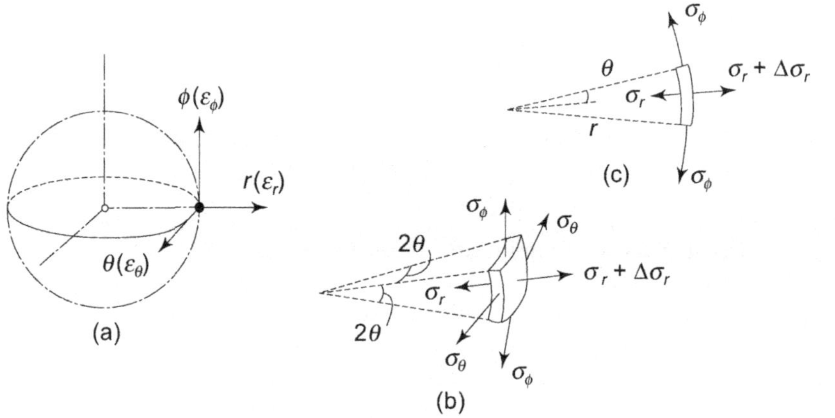

Fig. 8.8 *Sphere with purely radial displacement*

Thus, the three extensional strains along the three axes are

$$\varepsilon_r = \frac{\partial u_r}{\partial r}, \qquad \varepsilon_\theta = \frac{u_r}{r}, \qquad \varepsilon_\phi = \frac{u_r}{r} \tag{8.35}$$

Because of symmetry, there are no shear stresses and shear strains. Let γ_r be the body force per unit volume in the radial direction.

The stress equations of equilibrium can also be derived easily. Consider a spherical element of thickness Δr at distance r, subtending a small angle 2θ at the centre. Because of spherical symmetry, $\sigma_\theta = \sigma_\phi$. For equilibrium in the radial direction,

$$-\sigma_r(2\theta r)(2\theta r) + (\sigma_r + \Delta\sigma_r)(r + \Delta r)2\theta(r + \Delta r)2\theta$$

$$-2\left(r + \frac{\Delta r}{2}\right)2\theta\Delta r\sigma_\phi \sin\theta - 2\left(r + \frac{\Delta r}{2}\right)2\theta\,\Delta r\sigma_\theta \sin\theta + \gamma_r 4\theta^2 r^2 \Delta r = 0$$

Putting $\sigma_\theta = \sigma_\phi$ and $\Delta\sigma_r = \frac{\sigma_r}{r}\Delta r$, the above equation reduces in the limit to

$$r^2 \frac{\partial\sigma_r}{\partial r} + 2r\sigma_r - 2r\sigma_\phi + r^2\gamma_r = 0$$

Since r is the only independent variable, the above equation can be rewritten as

$$\frac{d}{dr}(r^2\sigma_r) - 2r\sigma_\phi + r^2\gamma_r = 0 \tag{8.36}$$

If body force is ignored,

$$\frac{1}{r^2}\frac{d}{dr}(r^2\sigma_r) = \frac{2}{r}\sigma_\phi \tag{8.37}$$

From Hooke's law

$$\varepsilon_r = \frac{1}{E}\left[\sigma_r - \nu(\sigma_\theta + \sigma_\phi)\right]$$

or
$$\frac{du_r}{dr} = \frac{1}{E}(\sigma_r - 2\nu\sigma_\phi) \tag{8.38}$$

and

$$\varepsilon_\phi = \frac{1}{E}\left[\sigma_\phi - \nu(\sigma_\phi + \sigma_r)\right]$$

or

$$\frac{u_r}{r} = \frac{1}{E}\left[(1-\nu)\,\sigma_\phi - \nu\sigma_r\right] \tag{8.39}$$

Equations (8.37)–(8.39) can be solved. From Eq. (8.39)

$$u_r = \frac{1}{E}\left[(1-\nu)r\,\sigma_\phi - \nu r\sigma_r\right]$$

Differentiating with respect to r

$$\frac{du_r}{dr} = \frac{1}{E}\left[(1-\nu)\frac{d(r\sigma_\phi)}{dr} - \nu\frac{d(r\sigma_r)}{dr}\right]$$

Subtracting the above equation from Eq. (8.38)

$$0 = -(1-\nu)\frac{d(r\sigma_\phi)}{dr} + \nu\frac{d(r\sigma_r)}{dr} + \sigma_r - 2\nu\sigma_\phi$$

Substituting for σ_ϕ from Eq. (8.36)

$$\frac{1}{2}(1-\nu)\frac{d^2(r^2\sigma_r)}{dr^2} - \nu\frac{d(r\sigma_r)}{dr} - \sigma_r + \frac{\nu}{r}\frac{d(r^2\sigma_r)}{dr} = 0 \tag{8.40}$$

If $r^2\,\sigma_r = y$,

$$\frac{d}{dr}(r\sigma_r) = \frac{d}{dr}\left(\frac{y}{r}\right) = \frac{1}{r}\frac{dy}{dr} - \frac{1}{r^2}y$$

Therefore, Eq. (8.40) becomes

$$\frac{1}{2}(1-\nu)\frac{d^2y}{dr^2} - \frac{\nu}{r}\frac{dy}{dr} + \frac{\nu y}{r^2} - \frac{y}{r^2} + \frac{\nu}{r}\frac{dy}{dr} = 0$$

or
$$\frac{d^2y}{dr^2} - 2\frac{y}{r^2} = 0 \tag{8.41}$$

This is a homogeneous linear equation with the solution

$$y = Ar^2 + \frac{B}{r}$$

where A and B are constants. Hence,

$$\sigma_r = A + \frac{B}{r^3} \tag{8.42}$$

And from Eq. (8.37)

$$\sigma_\phi = \frac{1}{2r}\frac{d}{dr}\left(Ar^2 + \frac{B}{r}\right) = A - \frac{B}{2r^3} \tag{8.43}$$

The constants A and B are determined from the boundary conditions.

Problem of Thick Hollow Sphere

Consider a spherical body formed by the boundaries of two spherical surfaces of radii a and b respectively. Let the hollow sphere be subjected to an internal pressure p_a and an external pressure p_b. The boundary conditions are therefore

$$\sigma_r = -p_a \quad \text{when } r = a, \quad \text{and} \quad \sigma_r = -p_b \quad \text{when } r = b$$

From Eq. (8.42)

$$-p_a = A + \frac{B}{a^3} \quad \text{and} \quad -p_b = A + \frac{B}{b^3}$$

Solving,

$$A = -\frac{b^3 p_b - a^3 p_a}{b^3 - a^3}, \qquad B = \frac{a^3 b^3}{b^3 - a^3}(p_b - p_a)$$

Thus, the general expressions for σ_r and σ_ϕ are

$$\sigma_r = \frac{1}{b^3 - a^3}\left[-b^3 p_b + a^3 p_a + \frac{a^3 b^3}{r^3}(p_b - p_a)\right] \tag{8.44}$$

$$\sigma_\phi = \sigma_\theta = \frac{1}{b^3 - a^3}\left[-b^3 p_b + a^3 p_a - \frac{a^3 b^3}{2r^3}(p_b - p_a)\right] \tag{8.45}$$

If the sphere is subjected to internal pressure only, $p_b = 0$, and

$$\sigma_r = p_a \frac{a^3}{b^3 - a^3}\left(1 - \frac{b^3}{r^3}\right) \tag{8.46}$$

$$\sigma_\phi = \sigma_\theta = p_a \frac{a^3}{b^3 - a^3}\left(1 + \frac{b^3}{2r^3}\right) \tag{8.47}$$

The above two equations can also be written as

$$\sigma_r = p_a \frac{a^3}{1 - (a^3/b^3)}\left(\frac{1}{b^3} - \frac{1}{r^3}\right)$$

$$\sigma_\phi = \sigma_\theta = p_a \frac{a^3}{1 - (a^3/b^3)}\left(\frac{1}{b^3} + \frac{1}{2r^3}\right)$$

In the case of a cavity inside an infinite or a large medium, $b \to \infty$ and the above equations reduce to

$$\sigma_r = -p_a \frac{a^3}{r^3} \tag{8.48}$$

$$\sigma_\phi = \sigma_\theta = + p_a \frac{a^3}{2r^3} \tag{8.49}$$

The above equations can also be used to calculate stresses in a body of any shape with a spherical hole under an internal pressure p_a, provided the outer surface of the body is free from pressure and provided that every point of this outer surface is at a distance greater than four or five times the diameter of the hole from its centre.

Example 8.7 *Calculate the thickness of the shell of a bomb calorimeter of spherical form of 10 cm inside diameter if the working stress is σ kgf/cm² (98 σ kPa) and the internal pressure is σ/2 kgf/cm² (49 σ kPa).*

Solution From equations (8.46) and (8.47), the maximum tensile stress is due to σ_ϕ, which occurs at $r = a$. Hence,

$$\sigma_\phi = \frac{\sigma}{2} \frac{5^3}{b^3 - 5^3} \left(1 + \frac{b^3}{2 \times 5^3}\right)$$

Equating this to the working stress σ

$$\frac{5^3}{b^3 - 5^3} \left(1 + \frac{b^3}{2 \times 5^3}\right) = 2$$

$\therefore$ $\quad b \approx 6.3$ cm

Hence, the thickness of the shell is 1.3 cm.

Example 8.8 *Express the stress equation of equilibrium, i.e.*

$$\frac{1}{r^2} \frac{d}{dr} (r^2 \sigma_r) - \frac{2}{r} \sigma_\phi = 0$$

given by Eq. (8.37), in terms of the displacement component u_r, using Hooke's law and strain–displacement relations.

Solution We have $\varepsilon_r = \frac{1}{E}\left[(\sigma_r - 2\nu\sigma_\phi)\right]$

$$\varepsilon_\phi = \frac{1}{E}\left[(1 - \nu)\, \sigma_\phi - \nu\sigma_r\right]$$

Solving for σ_r and σ_ϕ,

$$\sigma_r = \frac{E}{(1 + \nu)(1 - 2\nu)} \left[\varepsilon_r\, (1 - \nu) + 2\nu\varepsilon_\phi\right] \tag{8.50}$$

$$\sigma_\phi = \frac{E}{(1 + \nu)(1 - 2\nu)} (\nu\varepsilon_r + \varepsilon_\phi) \tag{8.51}$$

Using the strain–displacement relations

$$\varepsilon_r = \frac{du_r}{dr} \quad \text{and} \quad \varepsilon_\phi = \frac{u_r}{r}$$

and substituting in the equilibrium equation, we get

$$\frac{E(1-\nu)}{(1+\nu)(1-2\nu)}\left[\frac{d^2u_r}{dr^2}+\frac{2}{r}\frac{du_r}{dr}-\frac{2}{r^2}u_r\right]=0$$

or

$$\frac{E(1-\nu)}{(1+\nu)(1-2\nu)}\frac{d}{dr}\left(\frac{du_r}{dr}+2\frac{u_r}{r}\right)=0 \tag{8.52}$$

8.5 STRESSES DUE TO GRAVITATION

When body forces are operative, the stress equation of equilibrium is, from Eq. (8.36),

$$\frac{1}{r^2}\frac{d}{dr}\left(r^2\sigma_r\right)-\frac{2}{r}\sigma_\phi+\gamma_r=0 \tag{8.53}$$

where γ_r is the body force per unit volume. The problem of a sphere strained by the mutual gravitation of its parts will now be considered . It is known from the theory of attractions

$$\gamma_r=-\rho g\frac{r}{a}$$

where a is the radius of the sphere, ρ is the mass density, r is the radius of any point from the centre and g is the acceleration due to gravity. Expressing the equations of equilibrium in terms of displacement u_r [Eq. (8.52)], we have

$$\frac{E(1-\nu)}{(1+\nu)(1-2\nu)}\frac{d}{dr}\left(\frac{du_r}{dr}+2\frac{u_r}{r}\right)-\rho g\frac{r}{a}=0 \tag{8.54}$$

The complementary solution is

$$u_r=Cr+\frac{C_1}{r^2}$$

and the particular solution is

$$u_r=\frac{1}{10}\frac{(1+\nu)(1-2\nu)}{E(1-\nu)a}\rho g r^3$$

Hence, the complete solution is

$$u_r=Cr+\frac{C_1}{r^2}+\frac{1}{10}\frac{(1+\nu)(1-2\nu)}{E(1-\nu)a}\rho g r^3$$

For a solid sphere, C_1 should be equal to zero as otherwise the displacement will become infinite at $r = 0$. The remaining constant is determined from the boundary condition $\sigma_r = 0$ at $r = a$. From the general solution

$$\frac{du_r}{dr}=C+\frac{3(1+\nu)(1-2\nu)}{10E(1-\nu)a}\rho g r^2,\qquad \frac{u_r}{r}=C+\frac{(1+\nu)(1-2\nu)}{10E(1-\nu)a}\rho g r^2$$

and from Eq. (8.50)

$$\sigma_r=\frac{E}{(1+\nu)(1-2\nu)}\left[\varepsilon_r(1-\nu)+2\nu\varepsilon_\phi\right]$$

$$= \frac{E}{(1+\nu)(1-2\nu)}\left[\frac{du_r}{dr}(1-\nu) + 2\nu\frac{u_r}{r}\right]$$

$$= \frac{E}{(1+\nu)(1-2\nu)}\left[C(1-\nu) + \frac{3(1+\nu)(1-2\nu)}{10Ea}\rho g r^2 + 2\nu C\right.$$

$$\left. + \frac{\nu(1+\nu)(1-2\nu)}{5E(1-\nu)a}\rho g r^2\right]$$

$$= \frac{E}{(1+\nu)(1-2\nu)}\left[C(1+\nu) + \frac{(3-\nu)(1+\nu)(1-2\nu)}{10Ea(1-\nu)}\rho g r^2\right]$$

From the boundary condition $\sigma_r = 0$ at $r = a$,

$$C = -\frac{(3-\nu)(1-2\nu)}{10E(1-\nu)}\rho g a$$

Hence,
$$\sigma_r = -\frac{1}{10}\frac{(3-\nu)}{(1-\nu)}(a^2 - r^2)\frac{\rho g}{a} \tag{8.55}$$

and from Eq. (8.53)

$$\sigma_\phi = \sigma_\theta = -\frac{1}{10}\frac{(3-\nu)a^2 - (1+3\nu)r^2}{(1-\nu)}\frac{\rho g}{a} \tag{8.56}$$

It will be observed that both stress components σ_r and σ_θ are compressive at every point. At the centre ($r = 0$), they are equal and have a magnitude

$$\sigma_r = \sigma_\phi = \sigma_\theta = \frac{1}{10}\frac{3-\nu}{1-\nu}\rho g a \text{ (compressive)}$$

Further,

$$\frac{du_r}{dr} = C + \frac{1}{10}\frac{3(1+\nu)(1-2\nu)}{E(1-\nu)a}\rho g r^2$$

$$= -\frac{1}{10}\frac{(3-\nu)(1-2\nu)}{E(1-\nu)}\rho g a + \frac{1}{10}\frac{3(1+\nu)(1-2\nu)}{E(1-\nu)a}\rho g r^2$$

$$= \frac{1}{10}\frac{(1+\nu)(1-2\nu)}{E(1-\nu)}\frac{\rho g}{a}\left[3r^2 - \frac{(3-\nu)}{(1+\nu)}a^2\right]$$

The above value is zero when

$$r^2 = \frac{(3-\nu)}{3(1+\nu)}a^2$$

Hence, if ν is positive (which is true for all known materials), there is a definite surface outside which the radial strain is an extension. In other words, for

$$r > a\left[\frac{(1+\nu)}{3(1+\nu)}\right]^{1/2}$$

the radial strain ε_r is positive though the radial stress σ_r is compressive everywhere. This result is due, of course, to the 'Poisson effect' of the large circumferential stress, i.e. hoop stress, which is compressive.

8.6 ROTATING DISKS OF UNIFORM THICKNESS

We shall now consider the stress distribution in rotating circular disks which are thin. We assume that over the thickness, the radial and circumferential stresses do not vary and that the stress σ_z in the axial direction is zero. The equation of equilibrium given by Eq. (8.5b) can be used, provided we add the inertia force term $\rho\omega^2 r$, i.e. in the general equation of equilibrium [Eq. (8.2)] we put the body force term equal to the inertia term $\rho\omega^2 r$, where ω is the angular velocity of the rotating disk and ρ is the density of the disk material. The z-axis is the axis of rotation. Then;

$$\frac{\partial \sigma_r}{\partial r} + \frac{\sigma_r - \sigma_\theta}{r} + \rho\omega^2 r = 0 \tag{8.57a}$$

or

$$\frac{d}{dr}(r\sigma_r) - \sigma_\theta + \rho\omega^2 r^2 = 0 \tag{8.57b}$$

The strain components are, as before,

$$\varepsilon_r = \frac{du_r}{dr} \quad \text{and} \quad \varepsilon_\theta = \frac{u_r}{r} \tag{8.58}$$

From Hooke's law, with $\sigma_z = 0$,

$$\varepsilon_r = \frac{1}{E}(\sigma_r - \nu\sigma_\theta)$$

$$\varepsilon_\theta = \frac{1}{E}(\sigma_\theta - \nu\sigma_r)$$

From Eq. (8.58)

$$\varepsilon_r = \frac{d}{dr}(r\varepsilon_\theta)$$

From Hooke's law

$$\frac{1}{E}(\sigma_r - \nu\sigma_\theta) = \varepsilon_r = \frac{d}{dr}(r\varepsilon_\theta) = \frac{1}{E}\frac{d}{dr}(r\sigma_\theta - \nu r\sigma_r) \tag{8.59}$$

Let

$$r\sigma_r = y \tag{8.60a}$$

Then, from Eq. (8.57b)

$$\sigma_\theta = \frac{dy}{dr} + \rho\omega^2 r^2 \tag{8.60b}$$

Substituting these in Eq. (8.59) and rearranging

$$r^2\frac{d^2y}{dr^2} + r\frac{dy}{dr} - y + (3+\nu)\,\rho\omega^2 r^3 = 0 \tag{8.61}$$

The solution of the above differential equation is

$$y = Cr + C_1\frac{1}{r} - \frac{(3+\nu)}{8}\rho\omega^2 r^3 \tag{8.62}$$

From Eq. (8.60)

$$\sigma_r = C + C_1\frac{1}{r^2} - \frac{(3+\nu)}{8}\rho\omega^2 r^3 \tag{8.63}$$

$$\sigma_\theta = C - C_1 \frac{1}{r^2} - \frac{(1+3\nu)}{8} \rho\omega^2 r^3 \tag{8.64}$$

The integration constants are determined from boundary conditions.

Solid Disk

For a solid disk, we must take $C_1 = 0$, since otherwise the stresses σ_r and σ_θ become infinite at the centre. The constant C is determined from the condition at the periphery ($r = b$) of the disk. If there are no forces applied there, then,

$$(\sigma_r)_{r=b} = C - \frac{3+\nu}{8} \rho\omega^2 b^2 = 0$$

Hence,

$$C = \frac{3+\nu}{8} \rho\omega^2 b^2$$

and the stress components become

$$\sigma_r = \frac{3+\nu}{8} \rho\omega^2 (b^2 - r^2) \tag{8.65a}$$

$$\sigma_\theta = \frac{3+\nu}{8} \rho\omega^2 b^2 - \frac{1+3\nu}{8} \rho\omega^2 r^2 \tag{8.65b}$$

These stresses attain their maximum values at the centre of the disk, where

$$\sigma_r = \sigma_\theta = \frac{3+\nu}{8} \rho\omega^2 b^2 \tag{8.66}$$

Circular Disk with a Hole of Radius *a*

If there are no forces applied at the boundaries a and b, then

$$(\sigma_r)_{r=a} = 0, \qquad (\sigma_r)_{r=b} = 0$$

from which we find that

$$C = \frac{3+\nu}{8} \rho\omega^2 \left(b^2 + a^2\right), \qquad C_1 = -\frac{3+\nu}{8} \rho\omega^2 a^2 b^2$$

Substituting these in Eqs (8.63) and (8.64)

$$\sigma_r = \frac{3+\nu}{8} \rho\omega^2 \left(b^2 + a^2 - \frac{a^2 b^2}{r^2} - r^2\right) \tag{8.67}$$

$$\sigma_\theta = \frac{3+\nu}{8} \rho\omega^2 \left(b^2 + a^2 + \frac{a^2 b^2}{r^2} - \frac{1+3\nu}{3+\nu} r^2\right) \tag{8.68}$$

The radial stress σ_r reaches its maximum at r = $\sqrt{ab}$ where

$$(\sigma_r)_{\max} = \frac{3+\nu}{8} \rho\omega^2 (b-a)^2 \tag{8.69}$$

The maximum circumferential stress is at the inner boundary, where

$$(\sigma_\theta)_{\max} = \frac{3+\nu}{4} \rho\omega^2 \left(b^2 + \frac{1-\nu}{3+\nu} a^2\right) \tag{8.70}$$

It can be seen that $(\sigma_\theta)_{\max}$ is greater than $(\sigma_r)_{\max}$.

When the radius a of the hole approaches zero, the maximum circumferential stress approaches a value twice as great as that for a solid disk [Eq. (8.66)]. In other words, by making a small circular hole at the centre of a solid rotating disk, we double the maximum stress.

The displacement u_r for all the cases considered above can be calculated from Eq. (8.58), i.e.

$$u_r = r\varepsilon_\theta = \frac{r}{E}(\sigma_\theta - \nu\sigma_r) \tag{8.71}$$

Example 8.9 *A flat steel disk of 75 cm outside diameter with a 15 cm diameter hole is shrunk around a solid steel shaft. The shrink-fit allowance is 1 part in 1000 (i.e. an allowance of 0.0075 cm in radius). $E = 2.18 \times 10^6$ kgf/cm² (214×10^6 kPa).*

(i) What are the stresses due to shrink-fit?
(ii) At what rpm will the shrink-fit loosen up as a result of rotation?
(iii) What is the circumferential stress in the disk when spinning at the above speed?

Assume that the same equations as for the disk are applicable to the solid rotating shaft also.

Solution

(i) To calculate the shrink-fit pressure, we have from Eq. (8.27)

$$p_c = \frac{2.18 \times 10^6 \times 0.0075}{2 \times 7.5^3} \times \frac{(7.5^2 - 0)(37.5^2 - 7.5^2)}{(37.5^2 - 0)}$$

or $\quad p_c = 1044$ kgf/cm² (102312 kPa)

The tangential stress at the hole will be the largest stress in the system and from Eq. (8.24)

$$\sigma_\theta = \frac{1044 \times 7.5^2}{(37.5^2 - 7.5^2)}\left(1 + \frac{37.5^2}{7.5^2}\right)$$

$$= 1131 \text{ kgf/cm}^2 \text{ (110838 kPa)}$$

(ii) When the shrink-fit loosens up as a result of rotation, there will be no radial pressure on any boundary. When the shaft and the disk are rotating, the radial displacement of the disk at the hole will be greater than the radial displacement of the shaft at its boundary. The difference between these two radial displacements should equal $\Delta = 0.0075$ cm at 7.5 cm radius. From Eqs (8.71), (8.67) and (8.68)

$$u_{\text{disk}} = \frac{r}{E}(\sigma_\theta - \nu\sigma_r)$$

$$= \frac{r}{E}\,\frac{3+\nu}{8}\rho\omega^2\left[b^2 + a^2 + \frac{a^2b^2}{r^2} - \frac{1+3\nu}{3+\nu}r^2 - \nu\left(b^2 + a^2 - \frac{a^2b^2}{r^2} - r^2\right)\right]$$

$$= \frac{r}{E}\frac{(3+\nu)(1-\nu)}{8}\rho\omega^2\left(b^2 + a^2 + \frac{1+\nu}{1-\nu}\frac{b^2a^2}{r^2} - \frac{1+\nu}{3+\nu}r^2\right)$$

$$= \frac{7.5}{2.18\times10^6}\times\frac{3.3\times0.7}{8}\rho\omega^2$$

$$\times\left(37.5^2 + 7.5^2 + \frac{1.3}{0.7}\times\frac{37.5^2\times7.5^2}{7.5^2} - \frac{1.3}{3.3}\times7.5^2\right)$$

$$= 4052\times10^{-6}\rho\omega^2$$

From equations (8.71), (8.65a) and (8.65b)

$$u_{\text{shaft}} = \frac{r}{E}(\sigma_\theta - \nu\sigma_r)$$

$$= \frac{1-\nu}{8E}\rho\omega^2 r\left[(3+\nu)b^2 - (1+\nu)r^2\right]$$

$$= \frac{0.7}{8\times2.18\times10^6}\rho\omega^2\times7.5\,(3.3\times7.5^2 - 1.3\times7.5^2)$$

$$= 34\times10^6\rho\omega^2$$

Therefore,

$$(4052 - 34)\times10^{-6}\rho\omega^2 = 0.0075$$

or

$$\omega^2 = 0.0075\times10^6\times\frac{1}{4018}\times\frac{981}{0.0081}$$

$$= 226066\ (\text{rad/s})^2$$

Therefore,

$$\omega = 475 \text{ rad/s or } 4536 \text{ rpm}$$

(iii) The stresses in the disk can be calculated from Eq. (8.68)

$$\sigma_r = \frac{3.3}{8}\rho\omega^2\left(37.5^2 + 7.5^2 + 37.5^2 - \frac{1.9}{3.3}\times7.5^2\right)$$

$$= 1170\rho\omega^2$$

$$= 1170\times\frac{0.0081}{981}\times226066$$

$$= 2184 \text{ kgf/cm}^2\ (214024 \text{ kPa})$$

Example 8.10 *A flat steel turbine disk of 75 cm outside diameter and 15 cm inside diameter rotates at 3000 rpm, at which speed the blades and shrouding cause a tensile rim loading of 44kgf/cm² (4312 kPa). The maximum stress at this speed is to be 1164 kgf/cm² (114072 kPa). Find the maximum shrinkage allowance on the diameter when the disk and the shift are rotating.*

Solution Let c be the radius of the shaft and b that of the disk. From Eq. (8.70), the maximum circumferential stress due to rotation alone is

$$(\sigma_\theta)_1 = \frac{3+\nu}{4}\rho\omega^2\left(b^2 + \frac{1-\nu}{3+\nu}c^2\right)$$

$$= \frac{3.3}{4}\rho\omega^2\left(37.5^2 + \frac{0.7}{3.3}\times 7.5^2\right)$$

$$= 1170\,\rho\omega^2$$

Owing to shrinkage pressure p_c, and the tensile rim loading p_b, from Eq. (8.12)

$$(\sigma_\theta)_2 = \frac{p_c c^2}{b^2 - c^2}\left(1 + \frac{b^2}{c^2}\right) + 2\,\frac{p_b b^2}{b^2 - c^2}$$

$$= p_c\,\frac{7.5^2}{37.5^2 - 7.5^2}\left(1 + \frac{37.5^2}{7.5^2}\right) + 2\,\frac{44\times 37.5^2}{37.5^2 - 7.5^2}$$

$$= 1.08p_c + 91.7$$

Hence, the combined stress at 7.5 cm radius is

$$\sigma_\theta = 1170\rho\omega^2 + 1.08p_c + 91.7$$

This should be equal to 1164 kgf/cm^2. Hence,

$$1.08\,p_c = 1164 - 1170\rho\omega^2 - 91.7$$

$$= 1164 - 1170\times(100\pi)^2\times\frac{0.0081}{981} - 91.7$$

$$= 1164 - 953.5 - 91.7$$

$$= 118.8$$

Hence,

$$p_c = 110\ \text{kgf/cm}^2$$

The corresponding shrink-fit allowance is obtained from Eq. (8.27), i.e.

$$110 = \frac{E\Delta}{2\times 7.5^3}\times\frac{7.5^2\left(37.5^2 - 7.5^2\right)}{37.5^2}$$

$$= 0.064\,E\Delta$$

or

$$\Delta = \frac{110\times 10^{-6}}{0.064\times 2.18} = 0.0008\ \text{cm}$$

8.7 DISKS OF VARIABLE THICKNESS

Assuming that the stresses do not vary over the thickness of the disk, the method of analysis developed in the previous section for thin disks of constant thickness can be extended also to disks of variable thickness. Let h be the thickness of the disk, varying with radius r. The equation of equilibrium can be obtained by referring to Fig. 8.9.

For equilibrium in the radial direction

$$\left(h\sigma_r + \frac{\partial(h\sigma_r)}{\partial r}\Delta r\right)(r + \Delta r)\,\Delta\theta + \rho\left(r + \frac{\Delta r}{2}\right)\Delta\theta\left(h + \frac{\Delta h}{2}\right)\omega^2\left(r + \frac{\Delta r}{2}\right)$$

$$-h\sigma_r\,r\Delta\theta - 2\sigma_\theta\left(h + \frac{\Delta h}{2}\right)\Delta r\sin\frac{\Delta\theta}{2} = 0$$

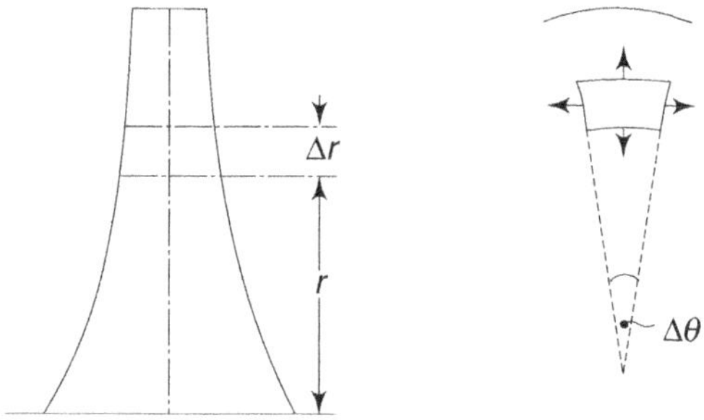

Fig 8.9 *Rotating disk of variable thickness*

Simplifying and going to the limit

$$r\frac{d}{dr}(h\sigma_r) + h\sigma_r + \rho\omega^2 r^2 h - \sigma_\theta h = 0$$

or $$\frac{d}{dr}(r\,h\sigma_r) - \sigma_\theta h + \rho\omega^2 r^2 h = 0 \tag{8.72}$$

Putting $$y = rh\sigma_r \tag{8.73a}$$

$$h\sigma_\theta = \frac{dy}{dr} + h\rho\omega^2 r^2 \tag{8.73b}$$

The strain components remain as in Eq. (8.58), i.e.

$$\varepsilon_r = \frac{du_r}{dr} \quad \text{and} \quad \varepsilon_\theta = \frac{u_r}{r}$$

Hence, $$\varepsilon_r = \frac{d}{dr}(r\varepsilon_\theta)$$

From Hooke's law and Eq. (8.59)

$$\frac{1}{E}(\sigma_r - \nu\sigma_\theta) = \frac{1}{E}\frac{d}{dr}(r\sigma_\theta - \nu r\sigma_r)$$

Substituting for σ_r and σ_θ from Eqs (8.73a) and (8.73b)

$$r^2\frac{d^2y}{dr^2} + r\frac{dy}{dr} - y + (3+\nu)\,\rho\omega^2 hr^3 - \frac{r}{h}\frac{dh}{dr}\left(r\frac{dy}{dr} - \nu y\right) = 0 \tag{8.74}$$

In the particular case where the thickness varies according to the equation

$$y = Cr^n \tag{8.75}$$

in which C is a constant and n any number, Eq. (8.74) can easily be integrated. The general solution has the form

$$y = mr^{n+2} + Ar^\alpha + Br^\beta$$

in which $$m = -\frac{(3+\nu)\rho\omega^2 c}{(\nu n + 3n + 8)}$$

and α and β are the roots of the quadratic equation

$$x^2 - nx + rn - 1 = 0$$

A and B are constants which are determined from the boundary conditions.

Example 8.11 *Determine the shape for a disk with uniform stress, i.e.* $\sigma_r = \sigma_\theta$.

Solution From Hooke's law

$$\varepsilon_r = \frac{1}{E}(\sigma_r - \nu\sigma_\theta), \qquad \varepsilon_\theta = \frac{1}{E}(\sigma_\theta - \nu\sigma_r)$$

if $\sigma_r = \sigma_\theta$ then $\varepsilon_r = \varepsilon_\theta$. From strain–displacement relations

$$\varepsilon_r = \frac{du_r}{dr} \quad \text{and} \quad \varepsilon_\theta = \frac{u_r}{r}$$

we get $$\varepsilon_r = \frac{d}{dr}(r\varepsilon_\theta) = \frac{d}{dr}(r\varepsilon_r)$$

Since $\varepsilon_r = \varepsilon_\theta$, the above equation gives

$$\frac{d\varepsilon_r}{dr} = 0$$

i.e. $$\varepsilon_r = \text{constant}$$

Hence, from Hooke's law, σ_r and σ_θ are not only equal but also constant throughout the disk. Let $\sigma_r = \sigma_\theta = \sigma$. Equilibrium Eq. (8.73) gives

$$h\sigma = \frac{d}{dr}(rh\sigma) + h\rho\omega^2 r^2$$

$$= h\sigma + r\sigma\frac{dh}{dr} + \rho\omega^2 hr^2$$

or $$\frac{1}{h}\frac{dh}{dr} = -\frac{1}{\sigma}\rho\omega^2 r$$

which upon integration gives

$$\log h = -\frac{\rho\omega^2}{2\sigma}r^2 + C_1$$

or $$h = \exp\left[-\frac{\rho\omega^2}{2\sigma}r^2 + C_1\right] = C\exp\left(-\rho\omega^2\frac{r^2}{2\sigma}\right)$$

8.8 ROTATING SHAFTS AND CYLINDERS

In Sec. 8.5 and 8.6, we assumed that the disk was thin and that it was in a state of plane stress with $\sigma_z = 0$. It is also possible to treat the problem as a plane strain problem as in the case of a uniformly rotating long circular shaft or a cylinder. Let the z-axis be the axis of rotation. The equation of equilibrium is the same as in Eq. (8.57):

$$\frac{d}{dr}(r\sigma_r) - \sigma_\theta + \rho\omega^2 r^2 = 0 \tag{8.76}$$

The strain components are, as before,

$$\varepsilon_r = \frac{du_r}{dr}, \qquad \varepsilon_\theta = \frac{u_r}{r}, \qquad \varepsilon_z = \frac{\partial u_z}{\partial z} = 0 \tag{8.77}$$

From Hooke's law

$$\varepsilon_r = \frac{1}{E}\left[\sigma_r - \nu(\sigma_\theta + \sigma_z)\right]$$

$$\varepsilon_\theta = \frac{1}{E}\left[\sigma_\theta - \nu(\sigma_r + \sigma_z)\right]$$

$$\varepsilon_z = \frac{1}{E}\left[\sigma_z - \nu(\sigma_r + \sigma_\theta)\right]$$

Since $\varepsilon_z = 0$ (plane strain),

$$\sigma_z = \nu(\sigma_r + \sigma_\theta)$$

and hence, substituting in equations for ε_r and ε_θ

$$\varepsilon_r = \frac{1+\nu}{E}\left[(1-\nu)\,\sigma_r - \nu\sigma_\theta\right]$$

$$\varepsilon_\theta = \frac{1+\nu}{E}\left[(1-\nu)\,\sigma_\theta - \nu\sigma_r\right]$$

From strain–displacement relations given in Eq. (8.77)

$$\varepsilon_r = \frac{d}{dr}(r\varepsilon_\theta)$$

and using the above expressions for ε_r and ε_θ, we get

$$(1-\nu)\,\sigma_r - \nu\sigma_\theta = \frac{d}{dr}\left[(1-\nu)r\,\sigma_\theta - \nu\, r\, \sigma_r\right] \tag{8.78}$$

With $r\sigma_r = y$, Eq. (8.76) gives for σ_θ

$$\sigma_\theta = \frac{dy}{dr} + \rho\omega^2 r^2$$

Substituting for σ_r and σ_θ in Eq. (8.78)

$$(1-\nu)\frac{y}{r} - \nu\frac{dy}{dr} - \nu\rho\omega^2 r^2 = \frac{d}{dr}\left[(1-\nu)\left(r\frac{dy}{dr} + \rho\omega^2 r^3\right) - \nu y\right]$$

or
$$r^2\frac{d^2y}{dr^2} + r\frac{dy}{dr} - y + \frac{3-2\nu}{1-\nu}\rho\omega^2 r^3 = 0$$

The solution for this differential equation is

$$y = Cr + C_1\frac{1}{r} - \frac{(3-2\nu)}{8(1-\nu)}\rho\omega^2 r^3$$

Hence,
$$\sigma_r = C + C_1\frac{1}{r^2} - \frac{(3-2\nu)}{8(1-\nu)}\rho\omega^2 r^2 \tag{8.79a}$$

$$\sigma_\theta = C - C_1\frac{1}{r^2} - \frac{(1+2\nu)}{8(1-\nu)}\rho\omega^2 r^2 \tag{8.79b}$$

and
$$\sigma_z = \nu\left[2C - \frac{1}{2(1-\nu)}\rho\omega^2 r^2\right] \tag{8.79c}$$

(i) For a hollow shaft or a long cylinder, $\sigma_r = 0$ at $r = a$ and $r = b$ which are the inner and outer radii. From these

$$C = K(a^2 + b^2) \quad \text{and} \quad C_1 = -Ka^2b^2$$

where $$K = \frac{(3-2\nu)}{8(1-\nu)}\rho\omega^2$$

Hence, from Eqs (8.79a–c)

$$\sigma_r = \frac{(3-2\nu)}{8(1-\nu)}\left[(a^2+b^2) - \frac{a^2b^2}{r^2} - r^2\right]\rho\omega^2 \tag{8.80}$$

$$\sigma_\theta = \frac{(3-2\nu)}{8(1-\nu)}\left[(a^2+b^2) + \frac{a^2b^2}{r^2} - \frac{1+2\nu}{3-2\nu}r^2\right]\rho\omega^2 \tag{8.81}$$

$$\sigma_z = \frac{\nu}{4(1-\nu)}\left[(a^2+b^2)(3-2\nu) - 2r^2\right]\rho\omega^2 \tag{8.82}$$

σ_θ assumes a maximum value at $r = a$ and its value is

$$(\sigma_\theta)_{\max} = \frac{(3-2\nu)}{8(1-\nu)}\left(2b^2 + a^2 - \frac{1+2\nu}{3-2\nu}a^2\right)\rho\omega^2$$

If a^2/b^2 is very small, we find that

$$(\sigma_\theta)_{\max} \approx \frac{(3-2\nu)}{4(1-\nu)}b^2\rho\omega^2 \tag{8.83}$$

(ii) For a long solid shaft, the constant C_1 must be equal to zero, since otherwise the stresses would become infinite at $r = 0$. Using the other boundary condition that $\sigma_r = 0$ when $r = b$, the radius of the shaft, we find that

$$C = \frac{(3-2\nu)}{8(1-\nu)}\rho\omega^2 b^2$$

Hence, the stresses are

$$\sigma_r = \frac{(3-2\nu)}{8(1-\nu)}(b^2 - r^2)\rho\omega^2 \tag{8.84}$$

$$\sigma_\theta = \frac{(3-2\nu)}{8(1-\nu)}\left(b^2 - \frac{1+2\nu}{3-2\nu}r^2\right)\rho\omega^2 \tag{8.85}$$

$$\sigma_z = \frac{\nu}{4(1-\nu)}\left[b^2(3-2\nu) - 2r^2\right]\rho\omega^2 \tag{8.86}$$

The value of σ_θ at $r = 0$ is

$$(\sigma_\theta)_{\max} = \frac{(3-2\nu)}{8(1-\nu)}b^2\rho\omega^2 \tag{8.87}$$

Comparing Eq. (8.87) with Eq. (8.83), we find that by drilling a small hole along the axis in a solid shaft, the maximum circumferential stress is doubled in its magnitude.

Example 8.12 *A solid steel propeller shaft, 60 cm in diameter, is rotating at a speed of 300 rpm. If the shaft is constrained at its ends so that it cannot expand or contract longitudinally, calculate the total longitudinal thrust over a cross-section due to rotational stresses. Poisson's ratio may be taken as 0.3. The weight of steel may be taken as 0.0081 kgf/cm³ (0.07938 N/cm³).*

Solution The total axial force is

$$F_z = \int_0^b \sigma_z 2\pi r\, dr$$

and from Eq. (8.86), substituting for σ_z,

$$F_z = \frac{\nu}{4(1-\nu)}\left[b^2 (3-2\nu)\,\pi b^2 - \pi b^4\right]\rho\omega^2$$

$$= \frac{\nu}{2}\pi b^4 \rho\omega^2$$

Substituting the numerical values

$$F_z = \frac{0.3}{2} \times \pi \times 30^4 \times \frac{0.0081}{981} \times \frac{300^2}{60^2} \times \pi 4^2$$

$$= 3120 \text{ kgf } (31576 \text{ N}) \text{ Tensile force}$$

8.9 SUMMARY OF RESULTS FOR USE IN PROBLEMS

(i) For a tube of internal radius a and external radius b subjected to an internal pressure p_a and an external pressure p_b, the radial and circumferential stresses are given by (according to plane stress theory)

$$\sigma_r = \frac{p_a a^2 - p_b b^2}{b^2 - a^2} - \frac{a^2 b^2}{r^2}\frac{p_a - p_b}{b^2 - a^2}$$

$$\sigma_\theta = \frac{p_a a^2 - p_b b^2}{b^2 - a^2} + \frac{a^2 b^2}{r^2}\frac{p_a - p_b}{b^2 - a^2}$$

$$\sigma_z = 0$$

The stress $\sigma_r < 0$ for all values of p_a and p_b, whereas σ_θ can be greater or less than zero depending on the values of p_a and p_b. σ_θ is greater than zero if

$$p_a > \frac{p_b}{2}\left(\frac{b^2}{a^2} + 1\right)$$

The maximum and minimum stresses are

$$(\sigma_r)_{max} = \sigma_r (\text{at } r = b) = -p_b$$

$$(\sigma_r)_{min} = \sigma_r (\text{at } r = a) = -p_a$$

$$(\sigma_\theta)_{max} = \sigma_\theta(\text{at } r = a) = \frac{p_a(a^2 + b^2) - 2p_b b^2}{b^2 - a^2}$$

$$(\sigma_\theta)_{\min} = \sigma_\theta(\text{at } r = b) = \frac{2p_a a^2 - p_b(a^2 + b^2)}{b^2 - a^2}$$

If $$p_a = \frac{1}{2} p_b \left(\frac{b^2}{a^2} + 1 \right)$$

then, $$(\sigma_\theta)_{\min} = 0$$

$$(u_r)_{r=a} = \frac{a}{E}\left[p_a \left(\frac{b^2 + a^2}{b^2 - a^2} + \nu \right) - 2p_b \frac{b^2}{b^2 - a^2} \right]$$

$$(u_r)_{r=b} = \frac{b}{E}\left[2p_a \frac{a^2}{b^2 - a^2} - p_b \left(\frac{b^2 + a^2}{b^2 - a^2} - \nu \right) \right]$$

(ii) Built-up cylinders: When the cylinders are of equal length, the contact pressure p_c due to difference Δ between the outer radius of the inner tube and the inner radius of the outer tube is given by

$$p_c = \frac{\Delta / c}{\left[\frac{1}{E_1}\left(\frac{c^2 + a^2}{c^2 - a^2} - \nu_1 \right) + \frac{1}{E_2}\left(\frac{b^2 + c^2}{b^2 - c^2} - \nu_2 \right) \right]}$$

where E_1, ν_1, a and c refer to the inner tube's modulus, Poisson's ratio, inner radius and outer radius respectively. E_2, ν_2, c and b are the corresponding values for the outer tube.

If $E_1 = E_2$ and $\nu_1 = \nu_2$, then

$$p_c = \frac{\Delta E}{2c^3} \frac{(b^2 - c^2)(c^2 - a^2)}{b^2 - a^2}$$

(iii) For a sphere subjected to an internal pressure p_a and an external pressure p_b, the radial and circumferential stresses are given by

$$\sigma_r = \frac{1}{b^3 - a^3}\left[-b^3 p_b + a^3 p_a + \frac{a^3 b^3}{r^3} - (p_b - p_a) \right]$$

$$\sigma_\theta = \sigma_\phi = \frac{1}{b^3 - a^3}\left[-b^3 p_b + a^3 p_a - \frac{a^3 b^3}{2r^3} - (p_b - p_a) \right]$$

(iv) For a thin solid disk of radius b rotating with an angular velocity ω, the stresses are given by

$$\sigma_r = \frac{3 + \nu}{8} \rho\omega^2 (b^2 - r^2)$$

$$\sigma_\theta = \frac{3 + \nu}{8} \rho\omega^2 b^2 - \frac{1 + 3\nu}{8} \rho\omega^2 r^2$$

These stresses attain their maximum values at the centre $r = 0$, where

$$\sigma_\theta = \sigma_r = \frac{3 + \nu}{8} \rho\omega^2 b^2$$

The radial outward displacement at $r = b$ is

$$(u_r)_{r=b} = \frac{1-\nu}{4E}\rho\omega^2 b^3$$

(v) For a thin disk with a hole of radius a, rotating with an angular velocity ω, the stresses are

$$\sigma_r = \frac{3+\nu}{8}\rho\omega^2\left(b^2 + a^2 - \frac{a^2b^2}{r^2} - r^2\right)$$

$$\sigma_\theta = \frac{3+\nu}{8}\rho\omega^2\left(b^2 + a^2 + \frac{a^2b^2}{r^2} - \frac{1+3\nu}{3+\nu}r^2\right)$$

$$(\sigma_r)_{\max} = \sigma_r\left(\text{at } r = \sqrt{ab}\right) = \frac{3+\nu}{8}\rho\omega^2\,(b-a)^2$$

$$(\sigma_\theta)_{\max} = \sigma_\theta\,(\text{at } r = a) = \frac{3+\nu}{4}\rho\omega^2\left(b^2 + \frac{1+\nu}{3+\nu}a^2\right)$$

and $\quad (\sigma_\theta)_{\max} > (\sigma_r)_{\max}$

The radial displacements are

$$(u_r)_{r=a} = \frac{3+\nu}{4E}\rho\omega^2 a\left(b^2 + \frac{1-\nu}{3+\nu}a^2\right)$$

$$(u_r)_{r=b} = \frac{3+\nu}{4E}\rho\omega^2 b\left(a^2 + \frac{1-\nu}{3+\nu}b^2\right)$$

Problems

8.1 A thick-walled tube has an internal radius of 4 cm and an external radius of 8 cm. It is subjected to an external pressure of 1000 kPa (10.24 kgf/cm^2). If $E = 1.2 \times 10^8$ kPa (1.23×10^6 kgf/cm^2) and $\nu = 0.24$, determine the internal pressure according to Mohr's theory of failure, which says that

$$(\sigma)_{\max} - n\,(\sigma)_{\min} \le \sigma_{\text{tenslie strength}}$$

where n is the ratio of σ-tensile strength to σ-compressive strength. For the present problem, assume σ-tensile strength = 30000 kPa (307.2 kgf/cm^2) and σ-compressive strength = 120000 kPa (1228.8 kgf/cm^2).

[*Ans.* p = 17000 kPa (174 kgf/cm^2)]

8.2 In the above problem, determine the changes in the radii.

$$\left[\textit{Ans.}\quad \begin{aligned}\Delta r_1 &= 0.01 \text{ mm}\\ \Delta r_2 &= 0.007 \text{ mm}\end{aligned}\right]$$

8.3 In Example 8.1, if one uses the energy of distortion theory, what will be the external radius of the cylinder? The rest of the data remain the same.

[*Ans.* = 6.05 cm]

8.4 A thick-walled tube with an internal radius of 10 cm is subjected to an internal pressure of 2000 kgf/cm^2 (196000 kPa). $E = 2 \times 10^6$ kgf/cm^2

(196 × 10^6 kPa) and $\nu = 0.3$. Determine the value of the external radius if the maximum shear stress developed is limited to 3000 kgf/cm^2 (294 × 10^6 kPa). Calculate the change in the internal radius due to the pressure.

$$\left[\begin{array}{l} Ans. \quad r_2 = 17.3 \text{ cm} \\ \quad\quad \Delta r_1 = 0.023 \text{ cm} \end{array}\right]$$

8.5 A thick-walled tube is subjected to an external pressure p_2. Its internal and external radii are 10 cm and 15 cm respectively, $\nu = 0.3$ and $E = 200000$ MPa (2041 × 10^3 kgf/cm^2). If the maximum shear stress is limited to 200000 kPa (2041 kgf/cm^2), determine the value of p_2 and also the change in the external radius.

$$\left[\begin{array}{l} Ans. \quad p_2 = 111 \text{ MPa } (1133 \text{ kgf/cm}^2) \\ \quad\quad \Delta r_2 = -0.19 \text{ mm} \end{array}\right]$$

8.6 Determine the pressure p_0 between the concrete tube and the perfectly rigid core. Assume $E_c = 2 \times 10^6$ kgf/cm^2, $r_c = 0.16$. Take $r_1/r_2 = 0.5$ (Fig. 8.10).

[*Ans.* $p_0 = 17.4$ kgf/cm^2]

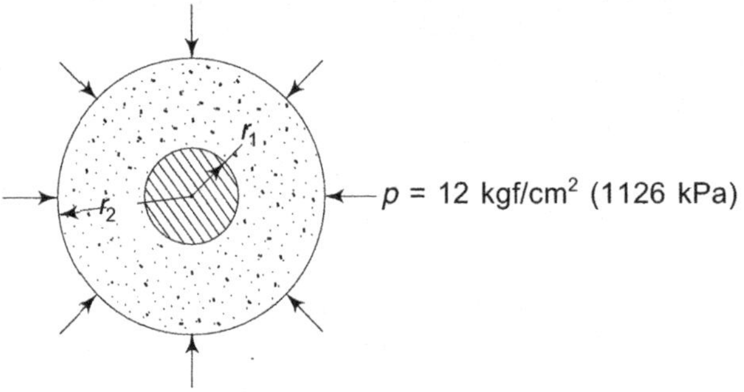

Fig. 8.10 *Problem 8.6*

8.7 Determine the dimensions of a two-piece composite tube of optimum dimensions if the internal pressure is 2000 kgf/cm^2 (196000 kPa), external pressure $p_2 = 0$, internal radius $r_1 = 8$ cm and $E = 2 \times 10^6$ kgf/cm^2 (196 × 10^6 kPa). The maximum shear stress is to be limited to 1500 kgf/cm^2 (147 × 10^6 kPa). Check the strength according to the maximum shear theory.

$$\left[\begin{array}{l} Ans. \quad r_2 \approx 14 \text{ cm}; r_3 = 24 \text{ cm} \\ \quad\quad \Delta = 0.014 \text{ cm} \\ \quad\quad p_c = 500 \text{ kgf/cm}^2 \\ \quad\quad\quad (49030 \text{ kPa}) \end{array}\right]$$

8.8 Determine the radial and circumferential stresses due to the internal pressure p = 2000 kgf/cm^2 (196,000 kPa) in a composite tube consisting of an inner copper tube of radii 10 cm and 20 cm and an outer steel tube of external radius 40 cm. $\nu_{st} = 0.3$, $\nu_{cu} = 0.34$, $E_{st} = 2 \times 10^6$ kgf/cm^2 (196 ×10^6 kPa) and $E_{cu} = 10^6$ kgf/cm^2 (98 × 10^6 kPa). Calculate the stresses at the inner and outer radius points of each tube. Determine the contact pressure also.

Ans. For inner tube:

$\sigma_r = -2000$ kgf/cm^2 (-196000 kPa)
$\sigma_r = -577$ kgf/cm^2 (-56546 kPa)
$\sigma_t = 1800$ kgf/cm^2 (176400 kPa)
$\sigma_t = 371$ kgf/cm^2 (36358 kPa)

For outer tube:

$\sigma_r = -577$ kgf/cm^2 (-56546 kPa)
$\sigma_r = 0$
$\sigma_t = 962$ kgf/cm^2 (94276 kPa)
$\sigma_t = 385$ kgf /cm^2 (37730 kPa)
$p_c = 577$ kgf/cm^2 (56546)

8.9 In problem 8.7, if the inner tube is made of steel (radii 10 cm and 20 cm) and the outer tube is of copper (outer radius 40 cm), determine the circumferential and radial stresses at the inner and outer radii points of each tube.

Ans. For inner tube:

$\sigma_r = -2000$ kgf/cm^2 (-196000 kPa)
$\sigma_r = -248$ kgf/cm^2 (-24304 kPa)
$\sigma_t = 2672$ kgf/cm^2 (262032 kPa)
$\sigma_t = 920$ kgf/cm^2 (90221 kPa)

For outer tube:

$\sigma_r = -248$ kgf/cm^2 (-24304 kPa)
$\sigma_r = 0$
$\sigma_t = 413$ kgf/cm^2 (40474 kPa)
$\sigma_t = 165$ kgf/cm^2 (16170 kPa)
$p_c = 248$ kgf/cm^2 (24304 kPa)

8.10 A composite tube is made of an inner copper tube of radii 10 cm and 20 cm and an outer steel tube of external radius 40 cm. If the temperature of the assembly is raised by 100°C, determine the radial and tangential stresses at the inner and outer radius points of each tube. $\alpha_{cu} = 16.5 \times 10^{-6}$; $\alpha_{st} = 12.5 \times 10^{-6}$; E_{st}, ν_{st}, E_{cu} and ν_{cu} are as in Problem 8.

Ans. For inner tube:

$\sigma_r = 0$
$\sigma_r = -173$ kgf/cm^2 (-16954 kPa)
$\sigma_t = -461$ kgf/cm^2 (-45080 kPa)
$\sigma = -288$ kgf/cm^2 (-28243 kPa)

For outer tube:

$\sigma_r = -173$ kgf/cm^2 (-16954 kPa)
$\sigma_r = 0$
$\sigma_t = 288$ kgf/cm^2 (28243 kPa)
$\sigma_t = 115$ kgf/cm^2 (11270 kPa)

8.11 Determine for the composite three-piece tube (Fig. 8.11):

(a) Stresses due to the heavy-force fits with interferences of $\Delta_1 = 0.06$ mm and $\Delta_2 = 0.12$ mm in diameters

(b) Stresses due to the internal pressure $p = 2400$ kgf/cm^2
$r_1 = 80$ mm, $r_2 = 100$ mm, $r_3 = 140$ mm, $r_4 = 200$ mm,
$E = 2.2 \times 10^6$ kgf/cm^2.

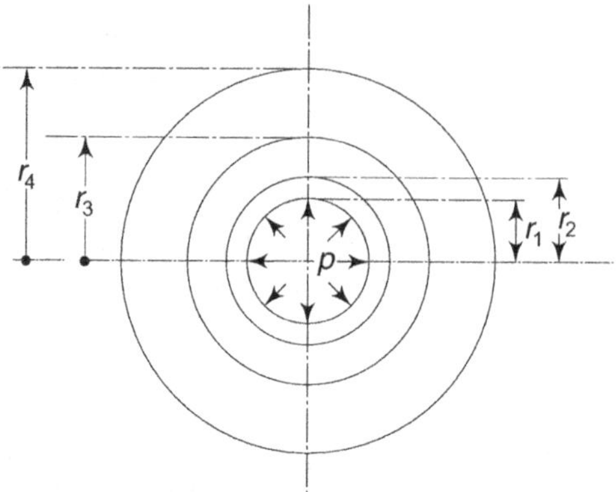

Fig 8.11 *Problem 8.11*

8.12 The radial displacement at the outside of a thick cylinder subjected to an internal pressure p_a is

$$\frac{\left(2p_a r_b r_a^2\right)}{E\left(r_b^2 - r_a^2\right)}$$

By Maxwell's reciprocal theorem, find the inward radial displacement at the inside of a thick cylinder subjected to external pressure.

$$\left[Ans. \quad u_r = \frac{2p_b ab^2}{E(b^2 - a^2)} \right]$$

8.13 A thin spherical shell of thickness h and radius R is subjected to an internal pressure p. Determine the mean radial stress, the circumferential stress and the radial displacement.

$$\left[\begin{array}{l} Ans. \quad u_r = pR^2(1-\nu)/(2Eh) \\ \sigma_\theta = \sigma_\phi = \dfrac{pR}{2h} \\ (\sigma_r) \text{ average} = \dfrac{1}{2}p \end{array} \right]$$

8.14 An infinite elastic medium with a spherical cavity of radius R is subjected to hydrostatic compression p at the outside. Determine the radial and circumferential stresses at point r. Show that the circumferential stress at the surface of the cavity exceeds the pressure at infinity.

$$\left[\begin{array}{l} Ans. \quad \sigma_r = -p\left(1 - \dfrac{R^2}{r^3}\right) \\ \sigma_\theta = \sigma_\phi = -p\left[1 + \dfrac{R^3}{2r^3}\right] \\ (\sigma_\theta) \text{ at cavity} = -\dfrac{3}{2}p \end{array} \right]$$

8.15 A perfectly rigid spherical body of radius a is surrounded by a thick spherical shell of thickness h. If the shell is subjected to an external pressure p, determine the radial and circumferential stresses at the inner surface of the shell ($b = a + h$).

$$\left[Ans. \quad \sigma_r = \frac{3(1-\nu)b^3 p}{2(1-2\nu)a^3 + (1+\nu)b^3} \right]$$

8.16 A steel disk of 50 cm outside diameter and 10 cm inside diameter is shrunk on a steel shaft so that the pressure between the shaft and disk at standstill is 364 kgf/cm^2 (3562 kPa). Take $\rho = 0.0081/g$ kgm/cm^3.

(a) Assuming that the shaft does not change its dimensions because of its own centrifugal force, find the speed at which the disk is just free on the shaft.

(b) Solve the problem without making assumption (a).

[*Ans.* (a) 4013 rpm
(b) 4028 rpm.]

8.17 A steel disk of 75 cm diameter is shrunk on a steel shaft of 7.5 cm diameter. The interference on the diameter is 0.0045 cm

(a) Find the maximum tangential stress in the disk when it is at a standstill.

(b) Find the rotation speed at which the contact pressure is zero.

(c) What is the maximum tangential stress at the above speed.

[*Ans.* (a) 647 kgf/cm^2 (6349 kPa)
(b) 4990 rpm
(c) 2622 kgf/cm^2 (257129 kPa)]

8.18 A disk of thickness t and outside diameter $2b$ is shrunk on to a shaft of diameter $2a$, producing a radial interface pressure p in the non-rotating condition. It is then rotated with an angular velocity ω rad/s. If f is the coefficient of friction between disk and shaft and ω_0 is the value of the angular velocity for which the interface pressure falls to zero, show that

(a) the maximum horsepower is transmitted when $\omega = \omega_0/\sqrt{3}$ and

(b) this maximum horsepower is equal to $0.000366\, a^2 t f p \omega_0$, where dimensions are in inches and pounds.

8.19 A steel shaft of 7.5 cm diameter has an aluminium disk of 25 cm outside diameter shrunk on it. The shrink allowance is 0.001 cm/cm. Calculate the rpm of rotation at which the shrink-fit loosens up. Neglect the expansion of the shaft caused by rotation. $\nu_{al} = 0.3$, $E_{st} = 7.3 \times 10^5$ kgf/cm^2 (175×10^5 kPa); $\gamma = 2.76\ 10^{-3}$ kgf/cm^3. [*Ans.* 13420 rpm]

CHAPTER

9 Thermal Stresses

9.1 INTRODUCTION

It is well known that changes in temperature cause bodies to expand or contract. The increase in the length of a uniform bar of length L, when its temperature is raised from T_0 to T, is

$$\Delta L = \alpha L\,(T - T_0)$$

where α is the coefficient of thermal expansion. If the bar is prevented from completely expanding in the axial direction, then the average compressive stress induced is

$$\sigma = E\frac{\Delta L}{L}$$

where E is the modulus of elasticity. Thus, for complete restraint, the thermal stress needed is

$$\sigma = -\alpha E\,(T - T_0)$$

where the negative sign indicates the compressive nature of the stress. If the expansion is prevented only partially, then the stress induced is

$$\sigma = -k\alpha E\,(T - T_0)$$

where k represents a restraint coefficient. It is assumed in the above analysis that E and α are independent of temperature. In general, in an elastic continuum, the temperature change is not uniform throughout. It is a function of time and the space coordinates (x, y, z), i.e.

$$T = T(t, x, y, z)$$

The body under consideration may be restrained from expansion or movement in some regions and external tractions may be applied to other regions. The determination of stresses under such situations may be quite complex. In this chapter, we shall restrict ourselves to the analysis of the following problems:

(i) Thin circular disks with symmetrical temperature variation;
(ii) Long circular cylinders—hollow and solid;
(iii) Spheres with purely radial temperature variation—hollow and solid;
(iv) Straight beams of arbitrary cross-section;
(v) Curved beams.

Before these specific problems are analysed, we shall develop the general thermoelastic stress–strain relations and discuss two important general results.

9.2 THERMOELASTIC STRESS–STRAIN RELATIONS

Consider a body to be made up of a large number of small cubical elements. If the temperatures of all these elements are uniformly raised and if the boundary of the body is unconstrained, then all the cubical elements will expand uniformly and all will fit together to form a continuous body. If, however, the temperature rise is not uniform, each element will tend to expand by a different amount and if these elements have to fit together to form a continuous body, then distortions of the elements and consequently stresses should occur in the body.

The total strains at each point of a body are thus made up of two parts. The first part is a uniform expansion proportional to the temperature rise T. For any elementary cubical element of an isotropic body, this expansion is the same in all directions and in this manner only normal strains and no shearing strains occur. If the coefficient of linear thermal expansion is α, this normal strain in any direction is equal to αT. The second part of the strains at each point is due to the stress components. The total strains at each point can, therefore, be written as

$$\varepsilon_x = \frac{1}{E}\left[\sigma_x - \nu\left(\sigma_y + \sigma_z\right)\right] + \alpha T$$

$$\varepsilon_y = \frac{1}{E}\left[\sigma_y - \nu\left(\sigma_x + \sigma_z\right)\right] + \alpha T \qquad (9.1a)$$

$$\varepsilon_z = \frac{1}{E}\left[\sigma_z - \nu\left(\sigma_x + \sigma_y\right)\right] + \alpha T$$

$$\gamma_{xy} = \frac{1}{G}\tau_{xy}, \qquad \gamma_{yz} = \frac{1}{G}\tau_{yz}, \qquad \gamma_{zx} = \frac{1}{G}\tau_{zx} \qquad (9.1b)$$

The stresses can be expressed explicitly in terms of strains by solving Eq. (9.1a). These are

$$\sigma_x = \lambda e + 2\mu\varepsilon_x - (3\lambda + 2\mu)\,\alpha T$$

$$\sigma_y = \lambda e + 2\mu\varepsilon_y - (3\lambda + 2\mu)\,\alpha T \qquad (9.2a)$$

$$\sigma_z = \lambda e + 2\mu\varepsilon_z - (3\lambda + 2\mu)\,\alpha T$$

$$\tau_{xy} = \mu\gamma_{xy}, \qquad \tau_{yz} = \mu\gamma_{yz}, \qquad \tau_{zx} = \mu\gamma_{zx} \qquad (9.2b)$$

The Lame constants λ and μ (= G) are given by

$$\lambda = \frac{\nu E}{(1+\nu)(1-2\nu)}, \qquad \mu = G = \frac{E}{2(1+\nu)} \qquad (9.3)$$

9.3 EQUATIONS OF EQUILIBRIUM

The equations of equilibrium are the same as those of isothermal elasticity since they are based on purely mechanical considerations. In rectangular coordinates these are given by Eq. (1.65). These are repeated for convenience.

$$\frac{\partial \sigma_x}{\partial x} + \frac{\partial \tau_{xy}}{\partial y} + \frac{\partial \tau_{xz}}{\partial z} + \gamma_x = 0$$

$$\frac{\partial \tau_{xy}}{\partial x} + \frac{\partial \sigma_y}{\partial y} + \frac{\partial \tau_{yz}}{\partial z} + \gamma_y = 0 \tag{9.4}$$

$$\frac{\partial \tau_{xz}}{\partial x} + \frac{\partial \tau_{yz}}{\partial y} + \frac{\partial \sigma_z}{\partial z} + \gamma_z = 0$$

where γ_x, γ_y and γ_z are body force components.

9.4 STRAIN–DISPLACEMENT RELATIONS

Only geometrical considerations are involved in deriving strain–displacement relations. Hence, the equations are the same as in isothermal elasticity. In rectangular coordinates, these are given by Eqs (2.18) and (2.19). To repeat, these are

$$\varepsilon_x = \frac{\partial u_x}{\partial x}, \qquad \varepsilon_y = \frac{\partial u_y}{\partial y}, \qquad \varepsilon_z = \frac{\partial u_z}{\partial z} \tag{9.5a}$$

$$\gamma_{xy} = \frac{\partial u_x}{\partial y} + \frac{\partial u_y}{\partial x}, \qquad \gamma_{yz} = \frac{\partial u_y}{\partial z} + \frac{\partial u_z}{\partial y}, \qquad \gamma_{zx} = \frac{\partial u_z}{\partial x} + \frac{\partial u_x}{\partial z} \tag{9.5b}$$

9.5 SOME GENERAL RESULTS

When the temperature distribution is known, the problem of thermoelasticity consists in determining the following 15 functions:

6 stress components σ_x, σ_y, σ_z, τ_{xy}, τ_{yz}, τ_{zx}
6 strain components ε_x, ε_y, ε_z, γ_{xy}, γ_{yz}, γ_{zx}
3 displacement components u_x, u_y, u_z

so as to satisfy the following 15 equations throughout the body

3 equilibrium equations, Eq. (9.4)
3 stress–strain relations, Eq. (9.1)
6 strain–displacement relations, Eq. (9.5)

and the prescribed boundary conditions. In most problems, the boundary conditions belong to one of the following two cases:

Traction Boundary Conditions In this case, the stress components determined must agree with the prescribed surface traction at the boundary.

Displacement Boundary Conditions Here, the displacement components determined should agree with the prescribed displacements at the boundary.

In some cases, the prescribed boundary conditions may be a combination of the above two, i.e. on a part of the boundary, the surface tractions are prescribed and on the remaining part, displacements are prescribed.

(i) The method of arriving at a solution depends in general on the specific nature of the problem. It is shown in books on thermoelasticity that if the temperature distribution in a body is a linear function of the rectangular Cartesian space coordinates, i.e. if

$$T(x, y, z, t) = a(t) + b(t)x + c(t)y + d(t)z \tag{9.6}$$

where t represents time, then all the stress components are identically zero throughout the body, provided that all external restraints, body forces and displacement discontinuities are absent. Conversely, under those provisions, this is the only temperature distribution for which all stress components are identically zero. These results are obtained immediately by considering the stress compatibility relations.

It, therefore, follows from the above statement and from the linearity of the boundary-value problem as formulated through Eqs (9.1), (9.4) and (9.5) that a linear function may be added to or subtracted from a given temperature distribution without affecting the resulting stress distribution. However, the strains and displacements are altered, as is obvious.

(ii) We shall now show that if a body is subjected to a uniform temperature rise $T = T_0(t)$ and if the boundary of the body is prevented from having any displacements, then the solution of the corresponding thermoelastic problem is

$$u_x = 0, \quad u_y = 0, \quad u_z = 0$$

$$\varepsilon_x = \varepsilon_y = \varepsilon_z = 0, \quad \gamma_{xy} = \gamma_{yz} = \gamma_{zx} = 0$$

$$\tau_{xy} = \tau_{yz} = \tau_{zx} = 0, \quad \sigma_x = \sigma_y = \sigma_z = -\frac{E\alpha}{1-2\nu} T_0$$

To show this, we shall apply the principle of superposition. We shall first allow free expansion of the body due to temperature rise T_0 with no restraint whatsoever. Since all cubical elements of the body expand freely, no stresses develop and all elements expand in an identical manner. This has been discussed in Sec. 9.2. Consequently,

$$\sigma_x = \sigma_y = \sigma_z = 0, \quad \tau_{xy} = \tau_{yz} = \tau_{zx} = 0$$

$$\gamma_{xy} = \gamma_{yz} = \gamma_{zx} = 0, \quad \varepsilon_x = \varepsilon_y = \varepsilon_z = \alpha T_0$$

Therefore,
$$u_x = \alpha T_0 x, \quad u_y = \alpha T_0 y, \quad u_z = \alpha T_0 z \tag{9.7a}$$

Now we apply boundary tractions to prevent this displacement. If the body is subject to a hydrostatic state of stress, then all elements of the body will experience the same state of stress ($-p$). With this state of stress, i.e. $\sigma_x = \sigma_y = \sigma_z = -p$ and $\tau_{xy} = \tau_{yz} = \tau_{zx} = 0$, the equations of equilibrium are identically satisfied.

Corresponding to this state of stress, the strain components are

$$\varepsilon_\xi = \varepsilon_y = \varepsilon_z = \frac{1}{E}\left[-p - \nu(-p - p)\right]$$

$$= -\frac{1}{E}(1-2\nu)p$$

$$\gamma_{xy} = \gamma_{yz} = \gamma_{zx} = 0$$

Therefore,
$$u_x = -\frac{1}{E}(1-2\nu)px\,; \quad u_y = -\frac{1}{E}(1-2\nu)py,$$

$$u_z = -\frac{1}{E}(1-2\nu)pz \tag{9.7b}$$

To get the original problem, the above values of u_x, u_y and u_z together with the values of u_x, u_y and u_z [Eq. (9.7a)] corresponding to free thermal expansion, should give zero displacements. Hence,

$$\alpha T_0 x - \frac{1}{E}(1-2\nu)\,px = 0$$

or,
$$p = \frac{E\alpha}{1-2\nu} T_0$$

Hence,
$$\sigma_x = \sigma_y = \sigma_z = -p = -\frac{E\alpha}{1-2\nu} T \tag{9.8}$$

as stated earlier.

9.6 THIN CIRCULAR DISK: TEMPERATURE SYMMETRICAL ABOUT CENTRE

Consider a thin disk subjected to a temperature distribution which varies only with r and is independent of θ. It is assumed further that it does not vary over the thickness and consequently, it is taken that the stresses and displacements also do not vary over the thickness. The stresses σ_r and σ_θ, therefore, satisfy the equilibrium equation

$$\frac{d\sigma_r}{dr} + \frac{\sigma_r - \sigma_\theta}{r} = 0 \tag{9.9}$$

Body forces are ignored. Also, because of symmetry, $\tau_{r\theta} = 0$. With $\sigma_z = 0$, the stress–strain relations given by Eq. (9.1a) take the form, in polar coordinates,

$$\varepsilon_r = \frac{1}{E}(\sigma_r - \nu\sigma_\theta) + \alpha T \tag{9.10}$$

$$\varepsilon_\theta = \frac{1}{E}(\sigma_\theta - \nu\sigma_r) + \alpha T$$

Solving the above equations for σ_r and σ_θ, we find

$$\sigma_r = \frac{E}{1-\nu^2}\left[\varepsilon_r + \nu\varepsilon_\theta - (1+\nu)\,\alpha T\right] \tag{9.11}$$

$$\sigma_\theta = \frac{E}{1-\nu^2}\left[\varepsilon_\theta + \nu\varepsilon_r - (1+\nu)\,\alpha T\right]$$

Substituting these in the equation of equilibrium

$$r\frac{d}{dr}(\varepsilon_r + \nu\varepsilon_\theta) + (1-\nu)(\varepsilon_r - \varepsilon_\theta) = (1+\nu)\alpha r\frac{dT}{dr} \tag{9.12}$$

The strain-displacement relation for a symmetrically strained body, from Eqs (8.3) and (8.4), are

$$\varepsilon_r = \frac{du_r}{dr}, \qquad \varepsilon_\theta = \frac{u_r}{r} \tag{9.13}$$

Substituting in Eq. (9.12)

$$\frac{d^2u_r}{dr^2} + \frac{1}{r}\frac{du_r}{dr} - \frac{u_r}{r^2} = (1+\nu)\alpha\frac{dT}{dr}$$

This may be written as

$$\frac{d}{dr}\left[\frac{1}{r}\frac{d}{dr}(ru_r)\right] = (1+\nu)\alpha\frac{dT}{dr}$$

Integration of the above equation yields,

$$u_r = (1+\nu)\alpha\frac{1}{r}\int_a^r Tr\,dr + C_1 r + \frac{C_2}{r} \tag{9.14}$$

It can be observed that the above expression becomes identical to Eq. (8.8) if T is put equal to zero.

The lower limit a in the integral above depends on the disk. For a disk with a hole, a is the inner radius and for a solid disk, a is zero.

The stress components are determined by substituting the value of u_r in Eq. (9.13) and using the results in Eq. (9.11). The results are

$$\sigma_r = -\alpha E\frac{1}{r^2}\int_a^r Tr\,dr + \frac{E}{1-\nu^2}\left[C_1\,(1+\nu) - C_2\,(1-\nu)\frac{1}{r^2}\right] \tag{9.15}$$

$$\sigma_\theta = \alpha E\frac{1}{r^2}\int_a^r Tr\,dr - \alpha ET + \frac{E}{1-\nu^2}\left[C_1\,(1+\nu) + C_2\,(1-\nu)\frac{1}{r^2}\right] \tag{9.16}$$

The constants C_1 and C_2 are determined by the boundary conditions. We shall now consider two specific cases. It should be observed that a linear variation of temperature with r will also induce stresses. This does not contradict the statement made in Sec. 9.5 that the stresses in a body are zero if the temperature distribution is linear with respect to a Cartesian frame of reference and if the body is free from external restraints and body forces. A linear radial variation will not give a linear variation with respect to the x, y and z-axes. In fact

$$T = kr = k\sqrt{x^2 + y^2 + z^2}$$

Solid Disk of Radius b In the case of a solid circular disk, $a = 0$ and in Eq. (9.14) it is observed from L′ Hospital's rule that

$$\lim_{r\to 0}\frac{1}{r}\int_0^r Tr\,dr = 0$$

Hence, the constant C_2 should be equal to zero, as otherwise u_r would become infinite at $r = 0$. The remaining constant C_1 is determined from the condition that $\sigma_r = 0$ at $r = b$, the outer radius of the disk. From Eq. (9.15), therefore, we get

$$C_1 = (1-\nu)\frac{\alpha}{b^2}\int_0^b Tr\,dr$$

Substituting this, the stresses are

$$\sigma_r = \alpha E\left(\frac{1}{b^2}\int_0^b Tr\,dr - \frac{1}{r^2}\int_0^r Tr\,dr\right) \tag{9.17}$$

$$\sigma_\theta = \alpha E\left(-T + \frac{1}{b^2}\int_0^b Tr\,dr + \frac{1}{r^2}\int_0^r Tr\,dr\right) \tag{9.18}$$

From L′Hospital's rule

$$\lim_{r\to 0}\frac{1}{r^2}\int_0^r Tr\,dr = \frac{1}{2}T_0$$

where T_0 is the temperature at the centre of disk. Hence, at $r = 0$

$$\sigma_r(0) = \sigma_\theta(0) = \alpha E\left(\frac{1}{b^2}\int_0^r Tr\,dr - \frac{1}{2}T_0\right) \tag{9.19}$$

Disk with a Hole of Radius a For a disk with a hole and traction free surfaces, $\sigma_r = 0$ at $r = a$ and $r = b$. Substituting these in Eq. (9.15)

$$C_1(1+\nu) - C_2(1-\nu)\frac{1}{a^2} = 0$$

$$-\alpha E\frac{1}{b^2}\int_a^b Tr\,dr + \frac{E}{1-\nu^2}\left[C_1(1+\nu) - C_2(1-\nu)\frac{1}{b^2}\right] = 0$$

Solving, we get

$$C_1 = \alpha(1-\nu)\frac{1}{b^2-a^2}\int_a^b Tr\,dr;\quad C_2 = \alpha(1-\nu)\frac{a^2}{b^2-a^2}\int_a^b Tr\,dr$$

Substituting in Eqs (9.15) and (9.16)

$$\sigma_r = \frac{\alpha E}{r^2}\left[\frac{r^2-a^2}{b^2-a^2}\int_a^b Tr\,dr - \int_a^r Tr\,dr\right] \tag{9.20}$$

$$\sigma_\theta = \frac{\alpha E}{r^2}\left[\frac{r^2+a^2}{b^2-a^2}\int_a^b Tr\,dr + \int_a^r Tr\,dr - Tr^2\right] \tag{9.21}$$

and from Eq. (9.14)

$$u_r = \frac{\alpha}{r}\left[(1+\nu)\int_a^r Tr\,dr + \frac{(1-\nu)r^2 + (1+\nu)a^2}{b^2-a^2}\int_a^b Tr\,dr\right] \tag{9.22}$$

If the temperature T is constant, then all the stress components are zero and the radial displacement is $u_r = \alpha rT$.

9.7 LONG CIRCULAR CYLINDER

We shall now consider the nature of the thermal stresses induced in a long circular cylinder when the temperature is symmetrical about the axis and does not vary along the axis. If the z-axis is the axis of the cylinder and r the radius, then T is a function of r alone and is independent of z. Since the cylinder is long, sections far from the ends can be considered to be in a state of plane strain and we can analyse this problem with u_z, the axial displacement, assumed to be zero.

Once again, owing to symmetry, all the shear stress components are zero and there are now three normal stress components σ_r, σ_θ and σ_z. The stress-strain relations are

$$\varepsilon_r = \frac{1}{E}\left[\sigma_r - \nu(\sigma_\theta + \sigma_z)\right] + \alpha T$$

$$\varepsilon_\theta = \frac{1}{E}\left[\sigma_\theta - \nu(\sigma_r + \sigma_z)\right] + \alpha T$$

$$\varepsilon_z = \frac{1}{E}\left[\sigma_z - \nu(\sigma_r + \sigma_\theta)\right] + \alpha T$$

Since $u_z = 0$, we have $\varepsilon_z = 0$. Hence, from the last equation we get

$$\sigma_z = \nu(\sigma_r + \sigma_\theta) - ET\alpha \tag{9.23}$$

Substituting this in the expressions for ε_z and ε_θ

$$\varepsilon_r = \frac{1-\nu^2}{E}\left(\sigma_r - \frac{\nu}{1-\nu}\sigma_\theta\right) + (1+\nu)\,\alpha T$$

$$\varepsilon_\theta = \frac{1-\nu^2}{E}\left(\sigma_\theta - \frac{\nu}{1-\nu}\sigma_r\right) + (1+\nu)\,\alpha T \tag{9.24}$$

Let $$\frac{E}{1-\nu^2} = E_1, \qquad \frac{\nu}{1-\nu} = \nu_1, \qquad (1+\nu)\,\alpha = \alpha_1 \tag{9.25}$$

Then, Eqs (9.24) can be written as

$$\varepsilon_r = \frac{1}{E_1}(\sigma_r - \nu_1\sigma_\theta) + \alpha_1 T$$

$$\varepsilon_\theta = \frac{1}{E_1}(\sigma_\theta - \nu_1\sigma_r) + \alpha_1 T \tag{9.26}$$

Comparing the above expressions with Eq. (9.10), it is immediately observed that the expressions for ε_r and ε_θ in the plane strain case is similar to those in the plane stress case if we use E_1, ν_1 and α_1, given by Eq. (9.25), in place of E, ν and α respectively. Since the equation of equilibrium is the same as in the plane stress case, further analysis is identical to that in the plane stress case. The expressions for u_r, σ_r and σ_z can, therefore, be written from equations (9.14) and (9.16) as

$$u_r = \frac{1+\nu}{1-\nu}\alpha\frac{1}{r}\int_a^r Tr\,dr + C_1 r + \frac{C_2}{r} \tag{9.27}$$

$$\sigma_r = -\frac{\alpha E}{1-\nu}\frac{1}{r^2}\int_a^r Tr\,dr + \frac{E}{1+\nu}\left(\frac{C_1}{1-2\nu} - \frac{C_2}{r^2}\right) \tag{9.28}$$

$$\sigma_\theta = \frac{E}{1-\nu}\frac{1}{r^2}\int_a^r Tr\,dr - \frac{\alpha ET}{1-\nu} + \frac{E}{1+\nu}\left(\frac{C_1}{1-2\nu} + \frac{C_2}{r^2}\right) \tag{9.29}$$

and from Eqs. (9.23), (9.28), and (9.29)

$$\sigma_z = -\frac{\alpha ET}{1-\nu} + \frac{2\nu EC_1}{(1+\nu)(1-2\nu)} \tag{9.30}$$

When $T = 0$, the equations become identical to Eqs (8.22a–c). Normal force given by Eq. (9.30) is necessary to keep $u_z = 0$ throughout. The constants C_1 and C_2 are determined from the boundary conditions. We shall now consider two particular cases.

Solid Cylinder of Radius b As before, from L′Hospital's rule

$$\lim_{r\to 0}\frac{1}{r}\int_0^r Tr\,dr = 0$$

Hence, from Eq. (9.27) we observe that C_2 should be equal to zero, as otherwise u_r would be infinite at $r = 0$. Since $\sigma_r = 0$ at $r = b$, we get from Eq. (9.28) with $C_2 = 0$ and $a = 0$ in the lower limit of the integration

$$\frac{C_1}{(1+\nu)(1-2\nu)} = \frac{\alpha}{1-\nu}\frac{1}{b^2}\int_0^b Tr\,dr$$

or

$$C_1 = \frac{(1+\nu)(1-2\nu)}{(1-\nu)}\frac{\alpha}{b^2}\int_0^b Tr\,dr$$

Comparing this with the value of C_1 obtained for the plane stress case, we observe that C_1 for the plane strain case can be obtained from C_1 for the plane stress case, merely by changing E, ν and α to E_1, ν_1 and α_1, and then converting these according to Eq. (9.25). The values of σ_r and σ_θ are, accordingly,

$$\sigma_r = \frac{\alpha E}{(1-\nu)}\left(\frac{1}{b^2}\int_0^b Tr\,dr - \frac{1}{r^2}\int_0^r Tr\,dr\right) \tag{9.31}$$

$$\sigma_\theta = \frac{\alpha E}{(1-\nu)}\left(-T + \frac{1}{b^2}\int_0^b Tr\,dr + \frac{1}{r^2}\int_0^r Tr\,dr\right) \tag{9.32}$$

and at $r = 0$

$$\sigma_r(0) = \sigma_\theta(0) = \frac{\alpha E}{(1-\nu)}\left(\frac{1}{b^2}\int_0^b Tr\,dr - \frac{1}{2}T_0\right) \tag{9.33}$$

where T_0 is the temperature at $r = 0$. Further, from Eq. (9.23)

$$\sigma_z = \frac{\alpha E}{(1-\nu)}\left(\frac{2\nu}{b^2}\int_0^b Tr\,dr - T\right) \tag{9.34}$$

The radial displacement is given by

$$u_r = \frac{1+\nu}{1-\nu}\alpha\left[(1-2\nu)\frac{r}{b^2}\int_0^b Tr\,dr + \frac{1}{r}\int_0^b Tr\,dr\right] \tag{9.35}$$

Note: In obtaining Eq. (9.34), we have assumed a plane strain condition with $\varepsilon_z = 0$. Consequently, a stress distribution σ_z as given by Eq. (9.30) was necessary to maintain $u_z = 0$. If, however, the ends of the cylinder are free, then the resultant force in z direction should be equal to zero. This condition can be achieved by superposing a uniform stress distribution $\sigma_z' = C_3$ so that the resultant force is zero.

For the solid cylinder, the resultant of σ_z from Eq. (9.30) is

$$\int_0^b 2\pi r\sigma_z\,dr = -\frac{2\pi\alpha E}{(1-\nu)}\int_0^b Tr\,dr + \frac{2\nu EC_1}{(1+\nu)(1-2\nu)}\pi b^2$$

The resultant of the superimposed uniform stress $\sigma_z' = C_3$ is $\pi b^2 C_3$. The value of C_3 to make the total force zero in z direction is, therefore, given by

$$C_3\pi b^2 = \frac{2\pi\alpha E}{(1-\nu)}\int_0^b Tr\,dr - \frac{2\nu EC_1}{(1+\nu)(1-2\nu)}\pi b^2$$

The resultant σ_z distribution is given by

$$\sigma_z = \frac{\alpha E}{(1-\nu)}\left(\frac{2}{b^2}\int_0^b Tr\,dr - T\right)$$

The u_r displacement is then given by Eq. (9.35) plus $(-\nu C_3 r/E)$.

Hollow Cylinder with Inner Radius a We shall write the solutions for this from the plane stress case results, Eqs (9.20)–(9.22), by putting E_1, ν_1, σ_1 in place of E, ν, σ and then converting these according to Eq. (9.25). Accordingly, we get

$$\sigma_r = \frac{\alpha E}{(1-\nu)}\frac{1}{r^2}\left[\frac{r^2-a^2}{b^2-a^2}\int_a^b Tr\,dr - \int_a^r Tr\,dr\right] \tag{9.36}$$

$$\sigma_\theta = \frac{\alpha E}{(1-\nu)}\frac{1}{r^2}\left[\frac{r^2+a^2}{b^2-a^2}\int_a^b Tr\,dr + \int_a^r Tr\,dr - Tr^2\right] \tag{9.37}$$

and

$$u_r = \frac{(1+\nu)\alpha}{(1-\nu)r}\left[\int_a^r Tr\,dr + \frac{(1-2\nu)\,r^2+a^2}{(b^2-a^2)}\int_a^b Tr\,dr\right] \tag{9.38}$$

Also

$$\sigma_z = \frac{\alpha E}{(1-\nu)}\left(\frac{2}{(b^2-a^2)}\int_a^b Tr\,dr - T\right) \tag{9.39}$$

Example 9.1 *The inner surface of a hollow tube is at temperature T_i and the outer surface at zero temperature.*

Assuming steady-state conditions, calculate the stresses. What are the values of σ_θ and σ_z near the inner and outer surfaces?

Solution Under steady heat flow conditions, the temperature at any distance r from the centre is given by the expression

$$T = \frac{T_i}{\log(b/a)}\log(b/r)$$

Substituting this in Eqs (9.36)–(9.39)

$$\sigma_r = \frac{\alpha E T_i}{2(1-\nu)\log(b/a)}\left[-\log\frac{b}{r} - \frac{a^2}{b^2-a^2}\left(1-\frac{b^2}{r^2}\right)\log\frac{b}{a}\right]$$

$$\sigma_\theta = \frac{\alpha E T_i}{2(1-\nu)\log(b/a)}\left[1-\log\frac{b}{r} - \frac{a^2}{b^2-a^2}\left(1+\frac{b^2}{r^2}\right)\log\frac{b}{a}\right]$$

$$\sigma_z = \frac{\alpha E T_i}{2(1-\nu)\log(b/a)}\left[1-2\log\frac{b}{r} - \frac{2a^2}{b^2-a^2}\log\frac{b}{a}\right]$$

$\sigma_r = 0$ at $r = 0$ and $r = b$. The stress components σ_θ and σ_z attain their maximum positive and negative values at $r = a$ and $r = b$. These values are

$$(\sigma_\theta)_{r=a} = (\sigma_z)_{r=a} = \frac{\alpha E T_i}{2(1-\nu)\log\frac{b}{a}}\left(1 - \frac{2b^2}{b^2 - a^2}\log\frac{b}{a}\right)$$

$$(\sigma_\theta)_{r=b} = (\sigma_z)_{r=b} = \frac{\alpha E T_i}{2(1-\nu)\log\frac{b}{a}}\left(1 - \frac{2a^2}{b^2 - a^2}\log\frac{b}{a}\right)$$

If T_i is positive, the radial stress is compressive at all points, whereas σ_θ and σ_z are compressive at the inner surface and tensile at the outer surface. These tensile stresses cause cracks in brittle materials such as stone, brick and concrete.

9.8 THE PROBLEM OF A SPHERE

We shall now consider the problem of a sphere subjected to purely radial temperature variation, i.e. T is a function of r alone. Because of symmetry, the shear stresses are all zero and the normal stresses are such that $\sigma_\theta = \sigma_\phi$. The equation of equilibrium in the radial direction is, from Eq. (8.37),

$$\frac{1}{r^2}\frac{d}{dr}(r^2\sigma_r) - \frac{2}{r}\sigma_\phi = 0$$

or

$$\frac{d}{dr}(r^2\sigma_r) - 2r\sigma_\phi = 0$$

The stress–strain relations are

$$\varepsilon_r = \frac{1}{E}(\sigma_r - 2\nu\sigma_\phi) + \alpha T$$

$$\varepsilon_\phi = \varepsilon_\theta = \frac{1}{E}\left[\sigma_\phi - \nu(\sigma_r + \sigma_\phi)\right] + \alpha T$$

Solving the above equations for σ_r and σ_ϕ

$$\sigma_r = \frac{E}{(1+\nu)(1-2\nu)}\left[(1-\nu)\,\varepsilon_r + 2\nu\varepsilon_\phi - (1+\nu)\,\alpha T\right] \quad (9.40)$$

$$\sigma_\phi = \frac{E}{(1+\nu)(1-2\nu)}\left[\varepsilon_\phi + \nu\varepsilon_r - (1+\nu)\,\alpha T\right] \quad (9.41)$$

From strain-displacement relations, we have

$$\varepsilon_r = \frac{du_r}{dr} \quad \text{and} \quad \varepsilon_\phi = \varepsilon_\theta = \frac{u_r}{r} \quad (9.42)$$

Substituting these in the expressions for σ_r and σ_θ and then substituting these in the equilibrium equation, we get

$$\frac{d^2u_r}{dr^2} + \frac{2}{r}\frac{du_r}{dr} - \frac{2u_r}{r^2} = \frac{1+\nu}{1-\nu}\,\alpha\,\frac{dT}{dr}$$

This can also be written as

$$\frac{d}{dr}\left[\frac{1}{r^2}\frac{d}{dr}(r^2u_r)\right] = \frac{1+\nu}{1-\nu}\alpha\frac{dT}{dr}$$

The solution is

$$u_r = \frac{1+\nu}{1-\nu}\alpha\frac{1}{r^2}\int_a^r Tr^2 dr + C_1 r + \frac{C_2}{r^2} \tag{9.43}$$

where C_1 and C_2 are constants to be determined from boundary conditions. The lower limit of the integral in the above equation is zero if the sphere is solid or is equal to a, the inner radius, if the sphere is hollow. From the expression for u_r, the strain components ε_r and ε_ϕ can be determined from Eq. (9.42) and substituted in Eq. (9.40). The results are

$$\sigma_r = -\frac{2\alpha E}{1-\nu}\frac{1}{r^3}\int_a^r Tr^2\,dr + \frac{EC_1}{1-2\nu} - \frac{2EC_2}{(1+\nu)r^3} \tag{9.44}$$

$$\sigma_\theta = \sigma_\phi = \frac{\alpha E}{1-\nu}\frac{1}{r^3}\int_a^r Tr^2\,dr + \frac{EC_1}{1-2\nu} + \frac{EC_2}{(1+\nu)r^3} - \frac{ET\alpha}{1-\nu} \tag{9.45}$$

we shall consider two specific cases.

Solid Sphere In this case, the lower limit a in the integrals may be taken as zero. In Eq. (9.43), the limit

$$\lim_{r\to 0}\left(-\frac{1}{r^2}\int_0^r Tr^2\,dr\right) = 0$$

according to L′ Hospital's rule. Consequently, the constant C_2 should be equal to zero, as otherwise, the displacement u_r would become infinite at $r = 0$. The remaining constant C_1 is determined from the condition that $\sigma_r = 0$ at $r = b$. Hence, from Eq. (9.44),

$$-\frac{2\alpha E}{(1-\nu)}\frac{1}{b^3}\int_0^b Tr^2\,dr + \frac{EC_1}{1-2\nu} = 0$$

or

$$C_1 = \frac{2\alpha(1-2\nu)}{(1-\nu)}\frac{1}{b^3}\int_0^b Tr^2\,dr$$

Substituting this in Eqs (9.44) and (9.45)

$$\sigma_r = \frac{2\alpha E}{(1-\nu)}\left(\frac{1}{b^3}\int_0^b Tr^2\,dr - \frac{1}{r^3}\int_0^r Tr^2\,dr\right) \tag{9.46}$$

$$\sigma_\theta = \sigma_\phi = \frac{\alpha E}{(1-\nu)}\left(\frac{2}{b^3}\int_0^b Tr^2\,dr + \frac{1}{r^3}\int_0^r Tr^2\,dr - T\right) \tag{9.47}$$

Hollow Sphere Let a be the radius of the inner cavity and b the outer radius of the sphere. The boundary conditions are $\sigma_r = 0$ at $r = a$ and $r = b$. Hence, from Eq. (9.44),

$$\frac{EC_1}{1-2\nu} - \frac{2EC_2}{1+\nu}\cdot\frac{1}{a^3} = 0$$

$$-\frac{2\alpha E}{1-\nu}\frac{1}{b^3}\int_a^b Tr^2\,dr + \frac{EC_1}{1-2\nu} - \frac{2EC_2}{1+\nu}\cdot\frac{1}{b^3} = 0$$

The above equations can be solved for C_1 and C_2 and substituted in Eqs (9.44) and (9.45). The result is

$$\sigma_r = \frac{2\alpha E}{1-\nu}\left[\frac{r^3 - a^3}{(b^3 - a^3)r^3}\int_a^b Tr^2\,dr - \frac{1}{r^3}\int_a^r Tr^2\,dr\right] \tag{9.48}$$

$$\sigma_\phi = \frac{2\alpha E}{1-\nu}\left[\frac{2r^3 + a^3}{2(b^3 - a^3)r^3}\int_a^b Tr^2\,dr + \frac{1}{2r^2}\int_a^r Tr^2\,dr - \frac{1}{2}T\right] \tag{9.49}$$

Therefore, the stress components can be calculated if the distribution of temperature is known.

Example 9.2 *Let the inner surface of a hollow sphere be at temperature T_i and the outer surface at temperature zero. Let the system be in a steady heat flow condition. The temperature distribution is then given by*

$$T = \frac{T_i a}{b-a}\left(\frac{b}{r} - 1\right)$$

Determine the stress distribution.

Solution Substituting the above expression for T in Eqs (9.48) and (9.49), we get

$$\sigma_r = \frac{\alpha E T_i}{1-\nu}\frac{ab}{b^3 - a^3}\left[a + b - \frac{1}{r}(b^2 + ab + a^2) + \frac{a^2b^2}{r^3}\right]$$

$$\sigma_\theta = \sigma_\phi = \frac{\alpha E T_i}{1-\nu}\frac{ab}{b^3 - a^3}\left[a + b - \frac{1}{2r}(b^2 + ab + a^2) - \frac{a^2b^2}{2r^3}\right]$$

As can be seen, $\sigma_r = 0$ at $r = a$ and $r = b$, according to the boundary conditions. Differentiating the expression for σ_r with respect to r and equating the resulting expression to zero, it is observed that σ_r is a maximum or a minimum when

$$r^2 = \frac{3a^2b^2}{b^2 + ab + a^2}$$

The expression for σ_ϕ shows that its value increases with r for T_i positive, and

$$(\sigma_\phi)_{r=a} = -\frac{\alpha E T_i}{2(1-\nu)}\frac{b(b-a)(a+2b)}{b^3 - a^3}$$

$$(\sigma_\phi)_{r=b} = -\frac{\alpha E T_i}{2(1-\nu)}\frac{a(b-a)(2a+b)}{b^3 - a^3}$$

9.9 NORMAL STRESSES IN STRAIGHT BEAMS DUE TO THERMAL LOADING

In this section, we shall develop an elementary formula for normal stresses in free beams subjected to thermal loadings. We shall make use of the Bernoulli–Euler assumption mentioned in Chapter 6. According to this assumption, sections which are plane and perpendicular to the axis before loading remain so after loading and the effect of lateral contraction (due to Poisson effect) may be neglected. The beam is assumed to be statically determinate and free of external loads. The temperature variation is arbitrary and the cross-section of the beam is also arbitrary.

Let the y and z-axes lie in the plane of the section and let the x-axis be the axis of the beam (Fig. 9.1). x, y and z-axes form a set of centroidal axes.

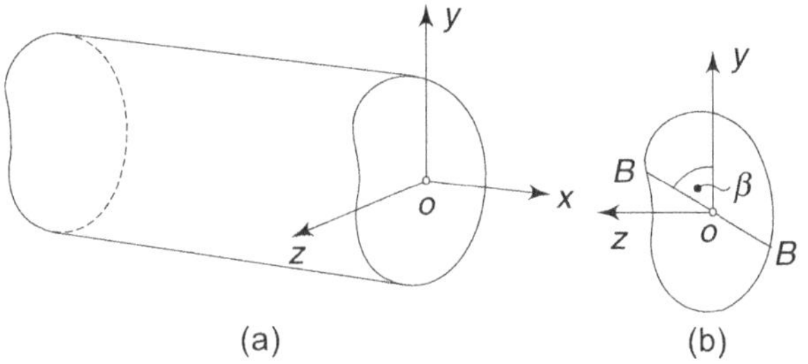

Fig. 9.1 *Beam subjected to thermal loading*

The analysis is similar to the one used in Chapter 6 for the bending of beams. If the beam is prevented from bending and if warping is not allowed, then the displacement of any section in the axial direction due to temperature rise will be a function of the axial coordinate x. Let this be $f_0(x)$. If now, the beam is allowed to undergo bending with the plane section remaining plane, then the displacement in x direction of any point (y, z) in a plane will be a linear function of the coordinates y and z. This is equivalent to saying that the cross-section rotates about an axis. The section that was plane before bending will, therefore, remain plane after bending and axial displacement. Hence, the total axial displacement, according to the Euler–Bernoulli hypothesis, will be

$$u_x = f_0(x) + yf_1(x) + zf_2(x)$$

where f_1 and f_2 are functions of x alone. The axial strain ε_x is, therefore,

$$\varepsilon_x = \frac{\partial u_x}{\partial x} = f_0'(x) + yf_1'(x) + zf_2'(x) \tag{9.50}$$

The strain represented by the last two terms on the right-hand side is similar to the one expressed in Chapter 6. We can also assume that the section rotates about an axis, such as BB in Fig. 9.1(b), and write the strain as $\varepsilon_x = f_0'(x) + ky'$, where y' is the perpendicular distance of a point from BB, which is inclined at β to the y-axis. This is what was done in Chapter 6. The unknowns k and β are now replaced by $f_1'(x)$ and $f_2'(x)$. From Hooke's law, since σ_y and σ_z are assumed to be zero,

$$\sigma_x = E(\varepsilon_x - \alpha T)$$

Substituting for ε_x

$$\sigma_x = E[f_0'(x) + yf_1'(x) + zf_2'(x) - \alpha T] \tag{9.51}$$

Since a free beam without external loading is considered, the conditions to be satisfied at any section are

$$\iint \sigma_x \, dA = 0; \qquad \iint \sigma_x y \, dA = M_z = 0, \qquad \iint \sigma_x z \, dA = M_y = 0$$

i.e. the resultant force over the section is zero and the moments about the y and z axes should individually vanish. Substituting the expression for σ_x, the above conditions become

$$\begin{aligned} f_0' \iint dA + f_1' \iint y \, dA + f_2' \iint z \, dA &= \iint \alpha T \, dA \\ f_0' \iint y \, dA + f_1' \iint y^2 \, dA + f_2' \iint yz \, dA &= \iint \alpha Ty \, dA \\ f_0' \iint z \, dA + f_1' \iint yz \, dA + f_2' \iint z^2 \, dA &= \iint \alpha Tz \, dA \end{aligned} \tag{9.52}$$

The integrations extend over the entire cross-section. The expressions

$$\iint y \, dA = \iint z \, dA = 0$$

because of the selection of the centroidal axes. Further,

$$\iint dA = A, \qquad \iint y^2 \, dA = I_z, \qquad \iint z^2 \, dA = I_y, \qquad \iint yz \, dA = I_{yz}$$

Substituting these, Eq. (9.52) can be written as

$$\begin{aligned} Af_0'(x) &= \iint \alpha T \, dA \\ f_1' I_z + f_2' I_{yz} &= \iint \alpha Ty \, dA \\ f_1' I_{yz} + f_2' I_y &= \iint \alpha Tz \, dA \end{aligned} \tag{9.53}$$

Let $E\iint \alpha T \, dA = p_t, \qquad E\iint \alpha Ty \, dA = -M_{zt}, \qquad E\iint \alpha Tz \, dA = M_{yt}$

A minus sign is used in the second expression in order to make the final result similar to the result of Chapter 6. The solutions for f_0', f_1' and f_2' are then given by

$$f_0' = \frac{p_t}{EA}, \qquad f_1' = \frac{-I_y M_{zt} - I_{yz} M_{yt}}{E(I_y I_z - I^2_{yz})}, \qquad f_2' = \frac{I_z M_{yt} + I_{yz} M_{zt}}{E(I_y I_z - I^2_{yz})} \tag{9.54}$$

Substituting these, the axial stress σ_x is, from Eq. (9.51),

$$\sigma_x = -\alpha ET + \frac{p_t}{A} - \frac{(I_y M_{zt} + I_{yz} M_{yt})}{(I_y I_z - I^2_{yz})} y + \frac{(I_z M_{yt} + I_{yz} M_{zt})}{(I_y I_z - I^2_{yz})} z$$

or

$$\sigma_x = -\alpha ET + \frac{p_t}{A} + \frac{M_{zt}(yI_y - zI_{yz}) + M_{yt}(yI_{yz} - zI_z)}{I^2_{yz} - I_y I_z} \tag{9.55}$$

Equation (9.55) bears a very close resemblance to Eq. (6.14) since the analyses in both cases have proceeded on similar lines.

If the axes chosen happen to be the principal axes of the section, then $I_{yz} = 0$ and Eq. (9.55) reduces to

$$\sigma_x = -\alpha ET + \frac{P_t}{A} - \frac{M_{zt}}{I_z} y + \frac{M_{yt}}{I_y} z \tag{9.56}$$

9.7 STRESSES IN CURVED BEAMS DUE TO THERMAL LOADING

An elementary analysis of the stresses developed in curved beams may be developed on the same basic assumptions as in the case of straight beams. Consider a free curved beam of arbitrary constant cross-section, the centre line of which is an arc of a circle (Fig. 9.2). It is assumed that this arc lies in one of the principal planes of the beam. Let the temperature vary as a function of r and θ, i.e. $T(r, \theta)$. We shall follow the notations used in Chapter 6.

In the isothermal case, the radius of curvature of the neutral surface is given by r_0 [Eq. (6.33)] such that

$$\iint \frac{y\,dA}{r_0 - y} = 0 \tag{9.57}$$

As in Sec. 6.7, the origin 0 lies on the neutral axis and y is measured towards the centre of curvature. A view of the deformed element is given in Fig. 9.3. Let the elementary length of an undeformed element enclose an angle $\Delta\theta$.

Because of thermal loading, the element deforms and it is assumed that sections which were plane before, remain plane after deformation. A fibre at a distance y from the chosen origin has a length $(r_0 - y)\,\Delta\theta$ before deformation. After deformation, the length of the same fibre becomes

$$\left[r_0' - y - \int_0^y \alpha T\,dy\right](\Delta\theta + \delta\Delta\theta) \tag{9.58}$$

The third term in the first bracket above represents the thermal expansion in y direction. The change in the length of the fibre is therefore

$$\left[r_0' - y - \int_0^y \alpha T\,dy\right](\Delta\theta + \delta\Delta\theta) - (r_0 - y)\,\Delta\theta$$

$$= \left[r_0' - r_0 - \int_0^y \alpha T\,dy\right]\Delta\theta + \left[r_0' - y - \int_0^y \alpha T\,dy\right](\delta\,\Delta\theta)$$

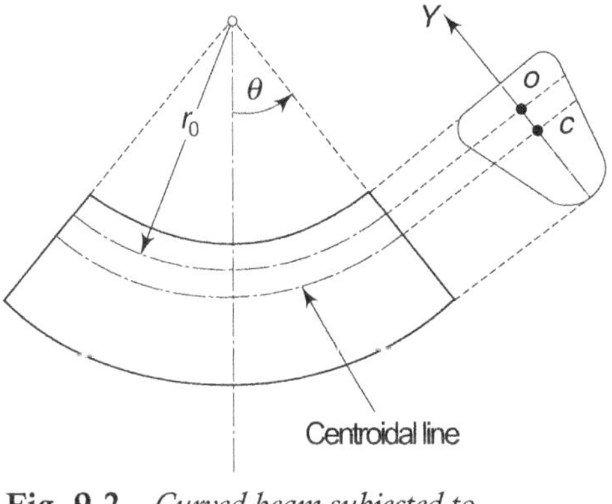

Fig. 9.2 *Curved beam subjected to thermal loading*

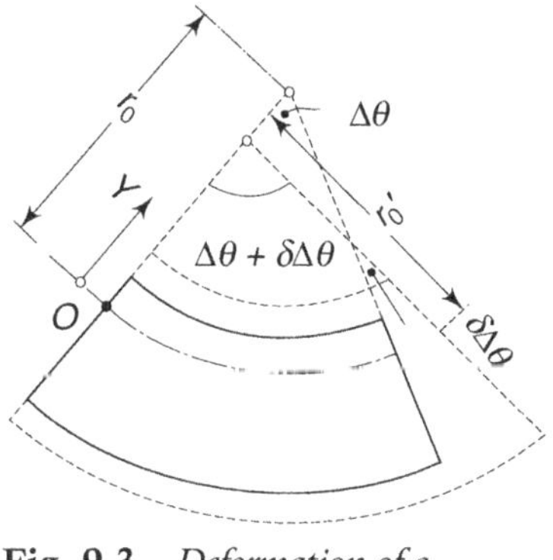

Fig. 9.3 *Deformation of a curved beam*

Hence, the strain is

$$\varepsilon_\theta = \frac{1}{(r_0 - y)}\left[\left(r_0' - r_0 - \int_0^y \alpha T\, dy\right) + \frac{\delta\Delta\theta}{\Delta\theta}\left(r_0' - y - \int_0^y \alpha T\, dy\right)\right]$$

We observe that

$$\int_0^y \alpha T\, dy \ll y$$

Hence,

$$\varepsilon_\theta = \frac{1}{(r_0 - y)}\left[r_0' - r_0 - \int_0^y \alpha T\, dy\right] + \frac{\delta\Delta\theta}{\Delta\theta}(r_0' - y) \tag{9.59}$$

From Hooke's law, taking only σ_θ into account

$$\sigma_\theta = E(\varepsilon_\theta - \alpha T)$$

Therefore,

$$\sigma_\theta = E\left\{\frac{1}{r_0 - y}\left[\left(r_0' - r_0 - \int_0^y \alpha T\, dy\right) + \frac{\delta\,\Delta\theta}{\Delta\theta}(r_0' - y)\right] - \alpha T\right\}$$

The two unknowns r_0' and $\dfrac{\delta\,\Delta\theta}{\Delta\theta}$ are determined from the boundary conditions of the beam. Since the beam is free of external loadings, we should have

$$\iint \sigma_\theta\, dA = 0; \qquad \iint \sigma_\theta\, y\, dA = 0$$

Beam with Rectangular Section

A somewhat more accurate result can be obtained for a curved beam with a rectangular cross-section and temperature independent of θ. This is obtained by superposing the result for a thin circular disk subjected to radial thermal loading with the result for the bending of a curved beam subjected to pure bending moment. If a sectoral element is isolated from a disk, as shown in Fig. 9.4(b), the ends of the element will be found (Examples 9.3 and 9.4) to be subjected to zero resultant circumferential force and some moment F_m, i.e.

$$\int_A \sigma_\theta\, dA = F_\theta = 0$$

$$\int_A \sigma_\theta y\, dA = F_m$$

where F_m is the moment about the median line.

If, on this curved beam, we apply an equal and opposite moment F_m, as shown in Fig. 9.4(c), then we get a free curved beam subjected to thermal loading only.

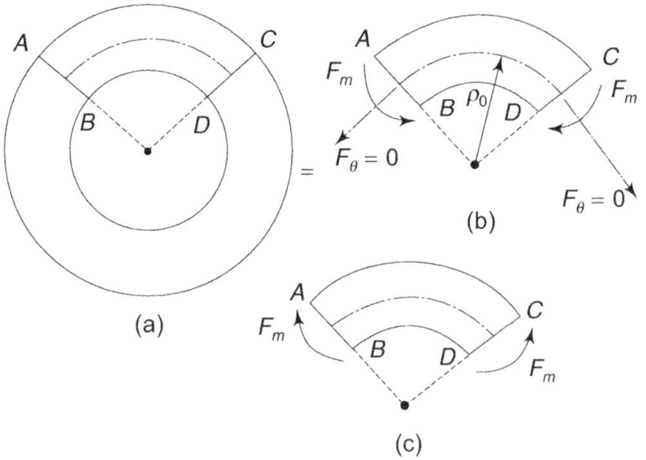

Fig. 9.4 *Curved beam with rectangular section*

Example 9.3 *Show that the resultant circumferential force across any radial section of a hollow disk subjected to thermal loading is zero.*

Solution From Eq. (9.21), the value of the circumferential stress σ_θ is

$$\sigma_\theta = \frac{\alpha E}{r^2}\left[\frac{r^2 + a^2}{b^2 - a^2}\int_a^b Tr\, dr + \int_a^r Tr'\, dr' - Tr^2\right]$$

Let the disk be of unit thickness perpendicular to the plane of the paper. The resultant circumferential force across any section is

$$F_\theta = \int_a^b \sigma_\theta\, dr = \frac{\alpha E}{b^2 - a^2}\left[\int_a^b dr \int_a^b Tr\, dr + \int_a^b \frac{a^2}{r^2}\, dr \int_a^b Tr\, dr\right]$$

$$+ \alpha E\left[\int_a^b \frac{1}{r^2} dr \int_a^r Tr'\, dr' - \int_a^b T\, dr\right]$$

Let $\int_a^b Tr\, dr = \beta$

Then, $$F_\theta = \frac{\alpha E}{b^2 - a^2}\left[\beta(b - a) - \beta a^2\left(\frac{1}{b} - \frac{1}{a}\right)\right]$$

$$+ \alpha E\left[\left(-\frac{1}{r}\int_a^r Tr'\, dr'\right)\Bigg|_a^b + \int_a^b T\, dr - \int_a^b T\, dr\right]$$

In the above expression, we have made use of the formula

$$\frac{d}{d\alpha}\int_{U(\alpha)}^{V(\alpha)} F(\alpha, x)\,dx = \int_{U(\alpha)}^{V(\alpha)} \frac{dF(\alpha, x)}{d\alpha}\,dx + F(V, \alpha)\frac{dV}{d\alpha} - F(U, \alpha)\frac{dU}{d\alpha}$$

Substituting the limits, it is observed that

$$F_\theta = 0$$

There is no resultant circumferential force across any section.

Example 9.4 *Determine the bending moment due to the circumferential stress across a section of a thin hollow disk subjected to radial thermal variation.*

Solution If ρ_0 is the radius of the median line and σ_θ the circumferential stress on a fibre at r from the centre of curvature (Fig. 9.4b), then the moment about the median line is

$$F_m = \int_a^b \sigma_\theta\,(r - \rho_0)\,dr$$

$$= \int_a^b \sigma_\theta r\,dr - \rho_0 \int_a^b \sigma_\theta\,dr$$

The second integral on the right-hand side is zero, from Example 9.3. Using Eq. (9.21), the moment becomes

$$F_m = \frac{\alpha E}{b^2 - a^2}\left[\int_a^b r\,dr \int_a^b Tr\,dr + \int_a^b \frac{a^2}{r}\,dr \int_a^b Tr\,dr\right]$$

$$+ \alpha E\left[\int_a^b \frac{1}{r}\,dr \int_a^r Tr'\,dr' - \int_a^b Tr\,dr\right]$$

Putting $\int_a^b Tr\,dr = \beta$, the above expression becomes

$$F_m = \frac{\alpha E}{b^2 - a^2}\left[\frac{\beta}{2}(b^2 - a^2) + a^2\beta \log\frac{b}{a}\right]$$

$$+ \alpha E\left[\log r \int_a^r Tr'\,dr'\right]\Bigg|_a^b - \alpha E\int_a^b (\log r)\,Tr\,dr - \alpha E\beta$$

$$= \frac{\alpha E\beta}{2(b^2 - a^2)}\left[b^2 + a^2\left(2\log\frac{b}{a} - 1\right)\right] + \alpha E\left[\beta(\log b - 1) - \int_a^b (\log r)\,Tr\,dr\right]$$

Problems

9.1 A thin hollow tube has its inner surface at temperature T_i and its outer surface at zero temperature. Assuming steady-state conditions, calculate the stresses. The inner radius is a and the thickness of the tube is t.

$$\left[\begin{aligned} Ans. (\sigma_\theta)_{r=a} = (\sigma_z)_{r=a} &= -\frac{\alpha E T_i}{2(1-\nu)}\left(1+\frac{t}{3a}\right) \\ (\sigma_\theta)_{r=b} = (\sigma_z)_{r=b} &= \frac{\alpha E T_i}{2(1-\nu)}\left(1-\frac{t}{3a}\right) \end{aligned}\right]$$

9.2 A solid sphere of radius b is subjected to thermal loading $T = T(r)$. Show that the radial stress σ_r at any radius r is proportional to the difference between the mean temperature of the whole sphere and the mean temperature of a sphere of radius r. Also, show that the circumferential stress at any point is equal to $\dfrac{2\alpha E}{3(1-\nu)}$ multiplied by the following expression:

[(mean temperature of the whole sphere)
+ (1/2 the mean temperature within the sphere of radius r) $- \dfrac{3}{2}T$]

9.3 A thin disk of inner radius a and outer radius b is subjected to a temperature variation which is symmetrical about the axis, i.e. $T = T(r)$. Consider a sectoral element, as shown in Fig. 9.4. Calculate the resultant moment due to σ_θ about the median line of the section across any radial section.

$$\left[\begin{aligned} Ans.\ F_m = \frac{\alpha E \beta}{2(b^2-a^2)}\left[b^2 + a^2\left(2\log\frac{b}{a} - 1\right)\right] + \alpha E\Big[\beta(\log b - 1) \\ -\int_a^b \log r\, Tr\, dr\Big] \\ \text{where} \quad \beta = \int_a^b Tr\, dr \end{aligned}\right]$$

9.4 A thin, uniform disk of radius b is surrounded by a heavy ring of the same material. The assembly just fits when the disk and the ring are at a uniform temperature. The faces of the disk are kept at temperature T_i and the circumference is kept at temperature T_0. The temperature variation along r from the centre is given by

$$T = T_i - (T_i - T_0)\frac{r^2}{b^2}$$

The heavy ring is at temperature T_0 and its strain is assumed to be negligible. Show that the radial compressive stress in the disk at radius r is

$$\sigma_r = \frac{1}{4}E\alpha(T_i - T_0)\left(\frac{3-\nu}{1-\nu} - \frac{r^2}{b^2}\right)$$

9.5 The temperature distribution in a long cylindrical conductor due to the passage of current is given by

$$T = \lambda(b^2 - r^2)$$

where λ is a constant. Determine the stresses due to thermal loading only.

$$\left[\begin{aligned} \textit{Ans. } \sigma_r &= -\frac{E\alpha\lambda}{4(1-\nu)}(b^2 - r^2) \\ \sigma_\theta &= -\frac{E\alpha\lambda}{4(1-\nu)}(3r^2 - b^2) \\ \sigma_z &= \frac{E\alpha\lambda}{2(1-\nu)}(2r^2 - b^2) \end{aligned}\right]$$

9.6 A beam of rectangular section (see Fig. 9.5) is subjected to a temperature distribution of the form

$$T = T_0\left(1 - \frac{y^2}{c^2}\right)$$

Show that the normal stress induced is given by

$$\sigma_x = \frac{2}{3}\alpha T_0 E - \alpha T_0 E\left(1 - \frac{y^2}{c^2}\right)$$

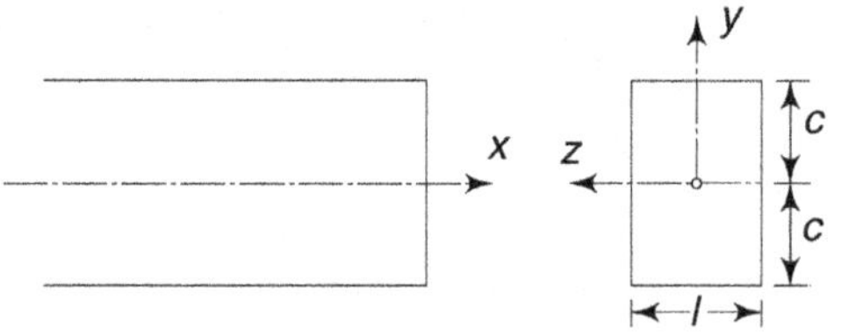

Fig. 9.5 *Problem 9.6*

CHAPTER

10

Elastic Stability

10.1 EULER'S BUCKLING LOAD

Consider a long slender column subjected to an axial force P. If the column is perfectly straight and is ideal in every respect, then it will remain straight and will be in equilibrium. If now a small lateral force Q is applied in addition to the axial force P (Fig. 10.1), the member will act as a beam and will assume a deflected form and will remain deflected as long as the lateral force Q is acting. When Q is removed, the member will return to its straight equilibrium position. However, there exists a critical axial load P_{cr}, such that under the action of P_{cr}, if the column is given a small lateral deflection by a force Q and the lateral force is removed, the column will continue to remain in the slightly buckled form and will be in equilibrium.

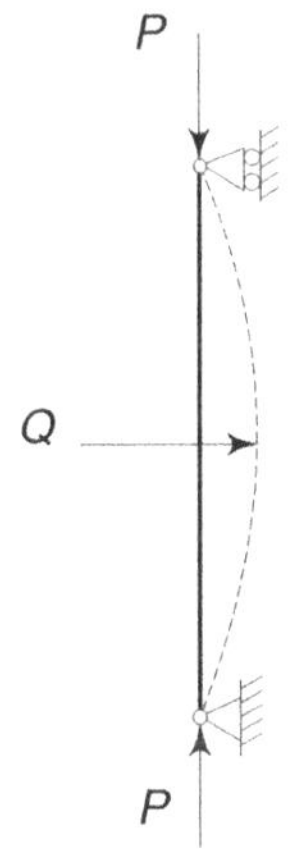

Fig. 10.1 *Column with lateral load*

The value of P_{cr}, known as the Euler's critical load or the buckling load can, therefore, be obtained by considering the equilibrium of a slightly buckled column. In elementary strength of materials, following this approach, Euler's critical loads for the columns shown in Fig. 10.2 have been obtained.

The critical loads for the first modes shown in Fig. 10.2 are as follows:

$$\frac{P^1 EI}{4L^2} \text{ for case (a),} \qquad \frac{P^2 EI}{L^2} \text{ for case (b), and} \qquad \frac{4P^2 EI}{L^2} \text{ for case (c)}$$

The method followed in elementary strength of materials to derive the above formulas will be applied to the following problem, which is slightly more complicated than the above cases.

Consider a centrally loaded column with the lower end built-in and the upper end hinged (Fig. 10.3). The critical value of the compressive load is that value of P_{cr} which can keep the strut in a slightly buckled shape. It may be observed that in order to keep point A in line with B, a lateral reaction R will be necessary.

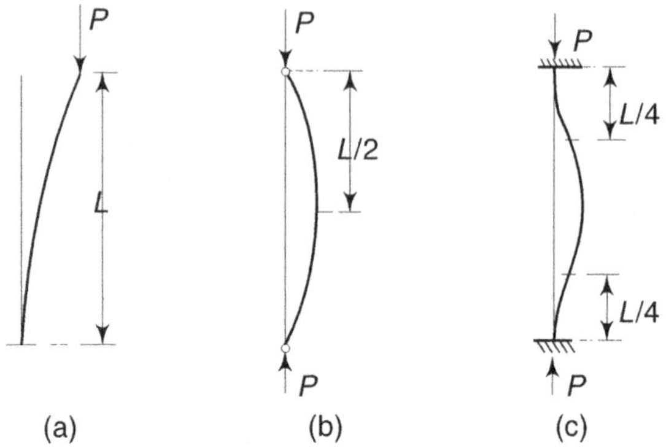

Fig. 10.2 *(a) Column with one end fixed and the other end free, (b) Column with both ends hinged; (c) Column with both ends fixed*

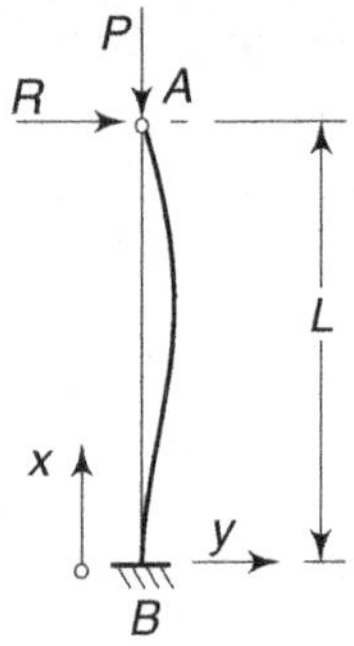

Fig. 10.3 *Column with one end fixed and the other end hinged*

The bending moment at any section x is

$$M = Py - R(L - x)$$

Using the expression

$$M = -EI \frac{d^2y}{dx^2}$$

$$EI \frac{d^2y}{dx^2} = -Py + R(L - x)$$

Let $$k^2 = \frac{P}{EI}$$

The differential equation then becomes

$$\frac{d^2 y}{dx^2} = -k^2y + \frac{R}{EI} (L - x) \qquad (10.1)$$

The general solution of this equation is

$$y = C_1 \cos kx + C_2 \sin kx + \frac{R}{P}(L - x)$$

The constants C_1 and C_2 and the reaction R will have to be determined from the boundary conditions. These are

$$y = 0 \quad \text{at } x = 0 \quad \text{and} \quad \text{at } x = L, \qquad \frac{dy}{dx} = 0 \quad \text{at } x = 0$$

Substituting these, we obtain the following equations:

$$C_1 + \frac{R}{P} L = 0$$

$$C_1 \cos kL + C_2 \sin kL = 0$$

$$kC_2 - \frac{R}{P} = 0$$

The trivial solution is $C_1 = C_2 = R = 0$, which means that the column remains straight. The non-trivial solution is

$$C_1 = -\frac{R}{P}L, \qquad C_2 = \frac{1}{k}\frac{R}{P}$$

Substituting into the second equation, we obtain the transcendental equation

$$\tan kL = kL \tag{10.2}$$

A solution to this can be obtained from a graphical plot. The smallest value of kL satisfying this equation is $kL = 4.493$, which means

$$P_{cr} = k^2 EI = \frac{20.19\, EI}{L^2} \approx \frac{\pi^2 EI}{(0.699L)^2}$$

As another example, we shall analyse by the elementary method, a fairly general problem of a column with a varying cross-section and with end as well as intermediate loading.

Example 10.1 *A column AB with hinged ends (Fig. 10.4) is compressed by two forces P_1 and P_2. The moment of inertia for the length L_1 of the column is I_1 and for the remaining length L_2, it is I_2. Determine the critical value of the force $P_1 + P_2$.*

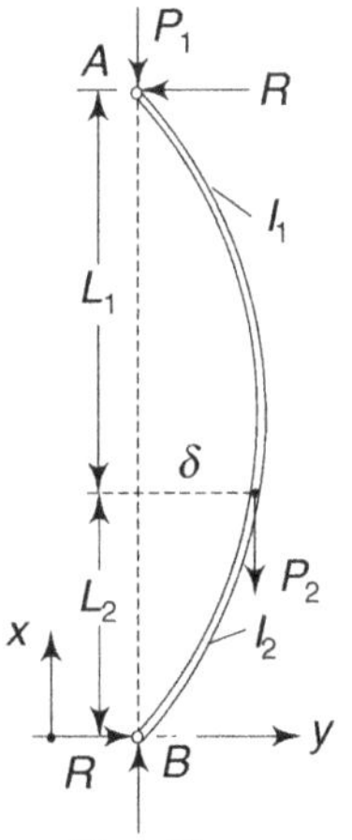

Fig. 10.4 *Example 10.1*

Solution If the equilibrium position of the buckled column is as shown in Fig. 10.4, then to have zero moments at the hinged ends A and B, it is necessary to have a horizontal reaction R such that

$$RL = P_2\delta \quad \text{or,} \quad R = P_2\delta / L \tag{a}$$

Let y_1 be the deflection at any section of the L_1 portion and y_2 the deflection at any section of the L_2 portion. Let $L = L_1 + L_2$

For the L_1 portion, the moment is

$$M = P_1 y_1 + R(L - x)$$

Using Eq. (a),

$$-EI_1 \frac{d^2 y_1}{dx^2} = P_1 y_1 + \frac{\delta P_2}{L}(L - x)$$

and for the L_2 portion, the moment is

$$M = P_1 y_2 + R(L - x) - P_2(\delta - y_2)$$

or

$$EI_2 \frac{d^2 y_2}{dx^2} = P_1 y_2 + \frac{\delta P_2}{L}(L - x) - P_2(\delta - y_2)$$

Using the notations

$$\frac{P_1}{EI_1} = k_1^2, \qquad \frac{P_2}{EI_2} = k_2^2, \qquad \frac{P_1 + P_2}{EI_2} = k_3^2, \qquad \frac{P_2}{EI_1} = k_4^2$$

the differential equations become

$$\frac{d^2 y_1}{dx^2} = -k_1^2 y_1 - \frac{\delta}{L} k_4^2 (L - x);$$

$$\frac{d^2 y_2}{dx^2} = -k_3^2 y_2 - \frac{\delta}{L} k_2^2 x$$

The solutions of the above equations are

$$y_1 = C_1 \sin k_1 x + C_2 \cos k_2 x - \frac{\delta}{L} \frac{k_4^2}{k_1^2} (L - x)$$

$$y_2 = C_3 \sin k_3 x + C_4 \cos k_4 x + \frac{\delta}{L} \frac{k_2^2}{k_3^2} x$$

The boundary conditions are

$$y_1 = 0 \quad \text{at } x = L, \qquad y_1 = \delta \quad \text{at } x = L_2, \qquad y_2 = \delta \quad \text{at } x = L_2,$$

$$y_2 = 0 \text{ at } x = 0, \qquad \left(\frac{dy_1}{dx}\right) = \left(\frac{dy_2}{dx}\right) \quad \text{at } x = L_2$$

The first four conditions yield

$$C_1 = \frac{\delta (k_1^2 L + k_4^2 L_1)}{k_1^2 L (\sin k_1 L_2 - \tan k_1 L \cos k_1 L_2)}$$

$$C_2 = -C_1 \tan k_1 L, \qquad C_3 = \frac{\delta (k_3^2 L - k_2^2 L_2)}{k_3^2 L \sin k_3 L_2}, \qquad C_4 = 0$$

Substituting the values of these constants into the continuity condition, i.e.

$$\left(\frac{dy_1}{dx}\right) = \left(\frac{dy_2}{dx}\right) \quad \text{at } x = L_2$$

the following transcendental equation is obtained:

$$\frac{k_4^2}{k_1^2} - \frac{k_1^2 L + k_4^2 L_1}{k_1 \tan k_1 L_1} = \frac{k_2^2}{k_3^2} + \frac{k_3^2 L - k_2^2 L_2}{k_3 \tan k_3 L_2}$$

For any particular case, the above equation can be solved to give the critical load. If, as an example, $L_1 = L_2$, $I_1 = I_2 = I$ and $P_1 = P_2$ are taken, we get

$$(P_1 + P_2)_{cr} = \frac{\pi^2 EI}{(0.87L)^2}$$

In this chapter, we shall discuss three specific topics:

(i) Beam columns;

(ii) Stability problem as an eigenvalue problem and

(iii) Energy methods to obtain approximate solutions to buckling problems.

I. BEAM COLUMNS

10.2 BEAM COLUMN

In the theory of bending discussed in Chapter 6, it was found that stresses were directly proportional to the applied loads. Similarly, in determining the deflections, we could apply the principle of superposition because of the linear relationship between the load acting and the deflection produced. In these cases, it is assumed that the deformations produced by one load do not affect the action of the other loads. Figure 10.5(a) shows a cantilever loaded by forces Q_1 and Q_2. If δ_1 is the deflection caused at point S due to Q_1 alone, and δ_2 the deflection at the same point S due to Q_2 alone, then the deflection δ due to the combined action of Q_1 and Q_2 is $\delta_1 + \delta_2$.

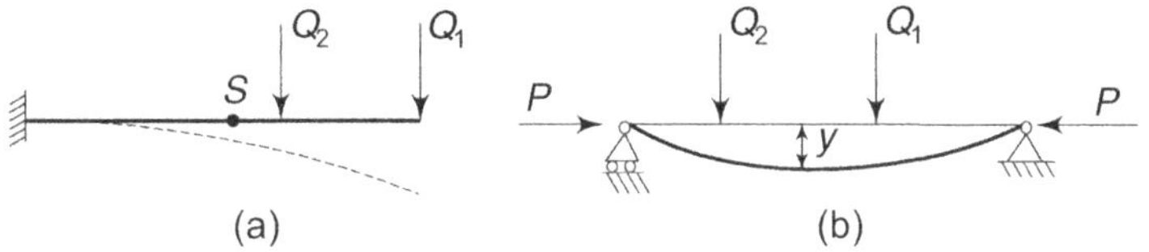

Fig. 10.5 *(a) Cantilever with loads Q_1 and Q_2, (b) Beam column with axial and lateral loads*

In arriving at this result it is assumed that the deflection δ_1 caused by Q_1 does not affect the action of Q_2. However, in the case of a beam which is subjected to lateral forces Q_1 and Q_2 as well as to axial forces P as shown in Fig.10.5(b), it can be seen that the bending moment caused by P depends on the deflection y produced by the lateral forces Q_1 and Q_2. In such cases, the principle of superposition cannot be applied without certain modifications. The beams that are subjected to axial loads in addition to lateral loads are known as beam columns. We shall restrict our analysis to beam columns having symmetrical cross-sections.

10.3 BEAM COLUMN EQUATIONS

Consider the beam shown in Fig. 10.6(a). The beam carries a distributed lateral load of intensity q, which is a function of x. In addition, the beam is subjected to an axial compressive force P. An elementary length Δx of the beam before deflection is considered. The lateral load q will be assumed to be positive when it is acting downward. The free body diagram of length Δx is shown in Fig.10.6(b).

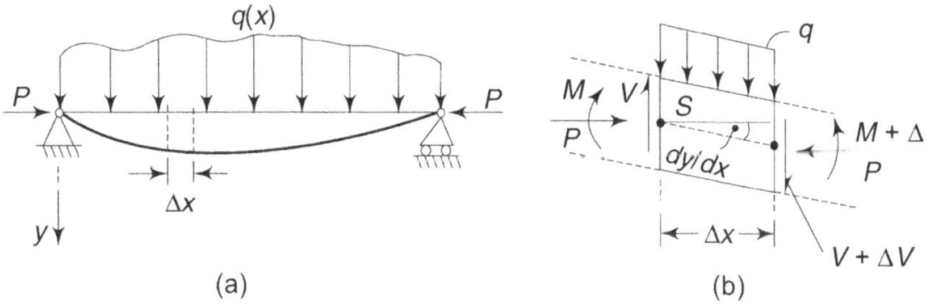

Fig. 10.6 *Beam column with varying lateral load*

The shearing force V and bending moment M acting on the sides of the element are assumed to be positive in the directions shown.

The relations among the load, shearing force V and bending moment M are obtained from the equilibrium considerations of the element. Summing forces in Y directions

$$-V + q\,\Delta x + (V + \Delta V) = 0$$

or, in the limit as $\Delta x \to 0$

$$q = -\frac{dV}{dx} \tag{10.3}$$

Taking moments about point S and assuming that the angle between the axis of the deformed beam and the horizontal is small, we get

$$M + q\,\Delta x\frac{\Delta x}{2} + (V + \Delta V)\,\Delta x - (M + \Delta M) + P\frac{dy}{dx}\,\Delta x = 0$$

or,
$$q\frac{\Delta x}{2} + V + \Delta V - \frac{\Delta M}{\Delta x} + P\frac{dy}{dx} = 0$$

In the limit, as $\Delta x \to 0$

$$V = \frac{dM}{dx} - P\frac{dy}{dx} \tag{10.4}$$

As in the case of the bending of beams, we ignore the effects of shearing deformation and assume that the curvature of the beam axis is given by

$$EI\,\frac{d^2y}{dx^2} = -M \tag{10.5}$$

where E is the Young's modulus of the beam material and I is the moment of inertia about the neutral axis. Using Eq. (10.5), Eqs (10.4) and (10.3) can be written as

$$EI\,\frac{d^3y}{dx^3} + P\,\frac{dy}{dx} = -V \tag{10.6}$$

and
$$EI\,\frac{d^4y}{dx^4} + P\,\frac{d^2y}{dx^2} = q \tag{10.7}$$

Equations (10.3)–(10.7) are the basic differential equations for the bending of beam columns. These equations reduce to the familiar beam bending equations when P is equal to zero.

10.4 BEAM COLUMN WITH A CONCENTRATED LOAD

Consider a uniform beam of span L (Fig. 10.7) simply supported and carrying a load Q at distance a from the right hand support. The beam is subjected to an axial force P. The bending moment at any section x is due to Q as well as P. However, the bending moment due to P cannot be calculated until the deflection is determined. The beam column is therefore statically indeterminate.

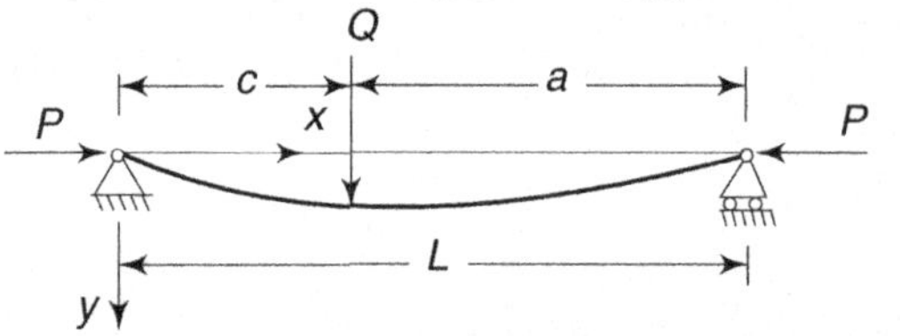

Fig. 10.7 *Beam column with concentrated load*

The bending moment at any x is

$$M = \frac{Qa}{L}\,x + Py \quad \text{for} \quad x \le (L-a)$$

$$M = \frac{Q(L-a)}{L}\,(L-x) + Py \quad \text{for} \quad x \ge (L-a)$$

Hence, from Eq. (10.5)

$$EI\,\frac{d^2y}{dx^2} = -\frac{Qa}{L}\,x - Py \quad \text{for} \quad x \le (L-a) \tag{10.8}$$

$$EI\,\frac{d^2y}{dx^2} = -\frac{Q\,(L-a)}{L}\,(L-x) - Py \quad \text{for } x \ge (L-a) \tag{10.9}$$

Putting $\quad k^2 = \dfrac{P}{EI}$

the above equations become

$$\frac{d^2y}{dx^2} + k^2 y = -\frac{Qa}{EIL}\,x$$

$$\frac{d^2y}{dx^2} + k^2 y = -\frac{Q(L-a)\,(L-x)}{EIL}$$

The general solutions of these equations are

$$y = A\cos kx + B\sin kx - \frac{Qa}{PL}\,x \quad \text{for} \quad x \le (L-a)$$

$$y = C\cos kx + D\sin kx - \frac{Q(L-a)(L-x)}{PL} \quad \text{for} \quad x \ge (L-a)$$

The constants A, B, C and D are to be determined from the conditions of the beam. The conditions are

(i) $y = 0$ at $x = 0$ and at $x = L$

(ii) y at $x = (L-a)$ should be the same according to both solutions.

(iii) The tangent at $x = (L-a)$ would be the same according to both solutions.

From condition (i)

$A = 0 \quad$ and $\quad C = -D\tan kL$

Conditions (ii) and (iii) give

$$B\sin k(L-a) - \frac{Qa}{PL}\,(L-a)$$

$$= D\left[\sin k(L-a) - \tan kL\cos k\,(L-a)\right] - \frac{Qa}{PL}\,(L-a)$$

$$Bk\cos k(L-a) - \frac{Qa}{PL}$$

$$= Dk\left[\cos k(L-a) + \tan kL \sin k(L-a)\right] + \frac{Q(L-a)}{PL}$$

From the above two equations we get

$$B = \frac{Q \sin ka}{Pk \sin kL}, \qquad D = -\frac{Q \sin k(L-a)}{Pk \tan kL}$$

Substituting these constants, the solutions are

$$y = \frac{Q \sin ka}{Pk \sin kL} \sin kx - \frac{Qa}{PL} x \quad \text{for } x \le (L-a) \tag{10.10a}$$

$$y = \frac{Q \sin k(L-a)}{Pk \sin kL} \sin k\,(L-x) - \frac{Q\,(L-a)}{PL}(L-x) \quad \text{for } x \ge (L-a) \tag{10.10b}$$

By the differentiation of Eqs (10.10a) and (10.10b), we obtain the following formulae, which are useful.

$$\left.\begin{aligned} \frac{dy}{dx} &= \frac{Q \sin ka}{P \sin kL} \cos kx - \frac{Qa}{PL}; \quad 0 \le x \le L-a \\ \frac{dy}{dx} &= -\frac{Q \sin k(L-a)}{P \sin kL} \cos k\,(L-x) + \frac{Q(L-a)}{PL}; \quad x \ge L-a \le L \end{aligned}\right\} \tag{10.11}$$

$$\left.\begin{aligned} \frac{d^2y}{dx^2} &= -\frac{Qk \sin ka}{P \sin kL} \sin kx; \quad 0 \le x \le L-a \\ \frac{d^2y}{dx^2} &= \frac{Qk \sin k(L-a)}{P \sin kL} \sin k(L-a); \quad x \ge L-a \le L \end{aligned}\right\} \tag{10.12}$$

As a particular case, if $a = L/2$, i.e. the load acts at midspan, then

$$\delta = y \quad \text{at} \quad \frac{L}{2} = \frac{Q}{2Pk}\left[\tan \frac{kL}{2} - \frac{kL}{2}\right]$$

Putting

$$u = \frac{kL}{2} = \frac{L}{2}\left(\frac{P}{EI}\right)^{1/2} \tag{10.13}$$

$$\delta = \frac{QL^3}{48EI} \frac{3(\tan u - u)}{u^3} \tag{10.14}$$

It is observed from the above equation that δ becomes infinite when $u = \pi/2$, i.e. when

$$\pi/2 = \frac{L}{2}\sqrt{\frac{P}{EI}}$$

or

$$P = \frac{\pi^2 EI}{L^2} = P_{cr}$$

So, however small Q is, when P becomes equal to P_{cr}, the lateral deflections becomes very large. We should recall that P_{cr} given above is the Euler buckling load for a slender column with hinged ends.

10.5 BEAM COLUMN WITH SEVERAL CONCENTRATED LOADS

Equations (10.10a and b) show that deflection y is proportional to lateral load Q, whereas the relation between the deflection and axial force P is more complicated. Because of the linear relationship between deflection y and load Q, if Q is doubled (with P remaining unaltered), then the deflection also is doubled. Hence, the principle of superposition in a modified form can be used for the effect of the lateral load, provided the same axial force acts on the bar.

Consider the beam shown in Fig. 10.8, which is subjected to an axial force P and three lateral loads Q_1, Q_2 and Q_3 acting at distances a_1, a_2 and a_3 respectively from the right hand side support B.

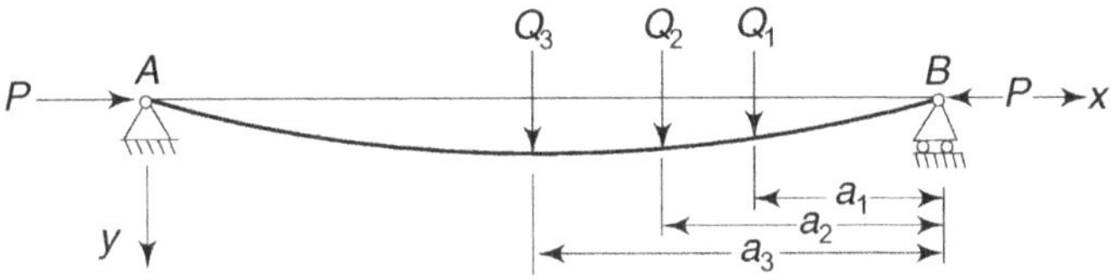

Fig. 10.8 *Beam-column with several lateral loads*

At some section left of Q_3, let y_1 be the deflection due to Q_1 alone with P, y_2 the deflection at the same point due to Q_2 alone with P, and y_3 the deflection due to Q_3 alone with P. From Eq. (10.5) with each Q and P [(similar to Eq. (10.8)], we get the following:

$$EI\frac{d^2y_1}{dx^2} = -\frac{Q_1a_1}{L}x - Py_1$$

$$EI\frac{d^2y_2}{dx^2} = -\frac{Q_2a_2}{L}x - Py_2$$

$$EI\frac{d^2y_3}{dx^2} = -\frac{Q_3a_3}{L}x - Py_3$$

By adding these equations

$$EI\frac{d^2(y_1+y_2+y_3)}{dx^2} = -\frac{Q_1a_1}{L}x - \frac{Q_2a_2}{L}x - \frac{Q_3a_3}{L} - P(y_1+y_2+y_3) \tag{10.15}$$

If Q_1, Q_2 and Q_3 are acting together with P, then the bending moment at section x is

$$M = \frac{Q_1a_1}{L}x + \frac{Q_2a_2}{L}x + \frac{Q_3a_3}{L}x + P(y_1+y_2+y_3)$$

From Eq. (10.5), therefore, we get

$$EI\frac{d^2(y_1+y_2+y_3)}{dx^2} = -\frac{Q_1a_1}{L}x - \frac{Q_2a_2}{L}x - \frac{Q_3a_3}{L}x - P(y_1+y_2+y_3)$$

This is identical to Eq. (10.15). Therefore, when there are several loads acting on a bar with an axial force P, the resultant deflection can be obtained by the superposition of the deflection produced by each lateral load acting in combination with axial force P.

Let a beam be acted upon by n lateral loads $Q_1, Q_2, \ldots, Q_m, Q_{m+1}, \ldots, Q_n$ at distances $a_1, a_2, \ldots, a_m, a_{m+1}, \ldots, a_n$ respectively from the right hand support. At some point x, which lies between Q_m, and Q_{m+1}, the total deflection is obtained from Eq. (10.10) as

$$y = \frac{\sin kx}{Pk\sin kL}\sum_{i=1}^{m} Q_i \sin ka_i - \frac{x}{PL}\sum_{i=1}^{m} Q_i a_i$$

$$+\frac{\sin k(L-x)}{Pk\sin kL}\sum_{i=m+1}^{m} Q_i \sin k(L-a_i) - \frac{L-x}{PL}\sum_{i=m+1}^{m} Q_i(L-a_i) \qquad (10.16)$$

In the above equation, we have made use of Eq. (10.10a) for loads $Q_1, Q_2, \ldots, Q_m$ lying to the right of x and Eq. (10.10b) for loads $Q_{m+1}, Q_{m+2}, \ldots, Q_n$ lying to the left of x.

10.6 CONTINUOUS LATERAL LOAD

The result obtained for a single load and the method of superposition can be used to solve the problem of a beam subjected to a continuously distributed load and an axial force, shown in Fig. 10.9. Let q be the intensity of loading. At distance a from the right hand support, the load on an elementary length Δa is $q\,\Delta a$.

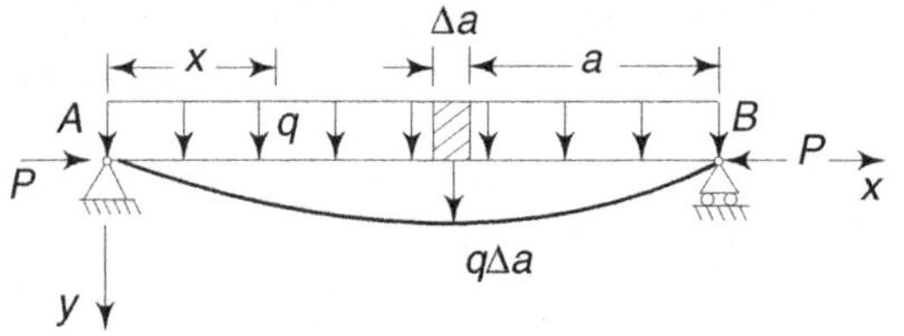

Fig. 10.9 *Beam-column with continuous lateral load*

At distance x from A, the deflection due to the load $q\,\Delta a$ from Eq. (10.10a) is

$$\Delta y = \frac{q\,\Delta a \sin ka}{Pk\sin kL}\sin kx - \frac{q\,a\,\Delta a}{PL}x$$

Assuming a to vary from 0 to $(L - x)$ from the right hand support, the deflection due to this part of the load, from Eq. (10.10a) and the principle of superposition, is

$$y_1 = \frac{\sin kx}{Pk\sin kL}\int_0^{L-x} q\sin ka\,da - \frac{x}{PL}\int_0^{L-x} qa\,da$$

Similarly, using Eq. (10.10b) and the principle of superposition, the deflection at x due to the loading to the left of x, i.e. for a varying from $(L - x)$ to L, is

$$y_2 = \frac{\sin k(L-x)}{Pk\sin kL}\int_{L-x}^{L} q\sin k(L-a)\,da - \frac{L-x}{PL}\int_{L-x}^{L} q(L-a)\,da$$

The total deflection y due to complete loading is $y = y_1 + y_2$. Summing the above two quantities and integrating with q as constant, we obtain the result, with $u = \frac{1}{2} kL$ as

$$y = \frac{qL^4}{16EIu^4}\left[\frac{\cos(u - 2ux/L)}{\cos u} - 1\right] - \frac{qL^2}{8EI\,u^2}(L - x)x \tag{10.17}$$

Example 10.2 *Determine the deflection y, using the general differential equation for a beam-column given by Eq. (10.7), for a beam uniformly loaded laterally and subjected to an axial force P.*

Solution The general differential Eq. (10.7) is

$$EI\frac{d^4y}{dx^4} + P\frac{d^2y}{dx^2} = q$$

where q is a constant. The general solution of this equation is

$$y = A\sin kx + B\cos kx + Cx + D + \frac{qx^2}{2P}$$

where A, B, C and D are constants. The boundary conditions are that the deflection and bending moment are zero at $x = 0$ and $x = L$, i.e.

$$y = 0 \quad \text{and} \quad \frac{d^2y}{dx^2} = 0 \quad \text{at } x = 0 \text{ and at } x = L$$

These give

$$B = -D = \frac{q}{k^2P}; \qquad A = \frac{q}{k^2P}\,\frac{1-\cos kL}{\sin kL}; \qquad C = -\frac{qL}{2P}$$

Therefore,

$$y = \frac{q}{k^2P}\,\frac{1-\cos kL}{\sin kL}\sin kx + \frac{q}{k^2P}(\cos kx - 1) - \frac{qL}{2P}x + \frac{qx^2}{2P}$$

Putting

$$u = \frac{kL}{2} \quad \text{and} \quad P = \frac{4u^2EI}{L^2}$$

the above equation can be written as

$$y = \frac{qL^4}{16EIu^4}\left[\frac{1-\cos 2u}{\sin 2u}\sin\frac{2ux}{L} + \cos\frac{2ux}{L} - 1\right] - \frac{qL^2}{8EIu^2}(L-x)x$$

$$= \frac{qL^4}{16EIu^4}\left[\frac{\sin f - \cos 2u\sin f + \sin 2u\cos f}{\sin 2u} - 1\right] - \frac{qL^2}{8EIu^2}(L-x)x$$

where, we have put $f = 2ux/L$. Simplifying, we obtain

$$y = \frac{qL^4}{16EIu^4}\left[\frac{\cos(u - f)}{\cos u} - 1\right] - \frac{qL^2}{8EIu^2}(L-x)x$$

as in Eq. (10.17).

10.7 BEAM-COLUMN WITH END COUPLE

Consider the beam shown in Fig. 10.10, where a moment M_b is applied at support B.

Fig. 10.10 *Beam-column with one end couple*

The solution to this can be obtained from the equation for the deflection curve due to a single concentrated load. For this purpose, we assume that the distance a where the load Q is applied is made to approach zero, however, keeping the product $Qa = M_b$ constant. In this manner, we obtain moment M_b acting at support B. From Eq. (10.10a)

$$y = \frac{Q \sin ka}{Pk \sin kL} \sin kx - \frac{Qa}{PL} x$$

Now, the limit of $Q \sin ka$ as $a \to 0$ and $Q \to \infty$, so that $Qa = M_b$ remains constant is

$$Lt\, Q\left(ka - \frac{k^3 a^3}{3!} + \ldots\right) = k(Qa) = M_b k$$

Hence, $$y = \frac{M_b k}{Pk \sin kL} \sin kx - \frac{M_b}{PL} x$$

or $$y = \frac{M_b}{P}\left(\frac{\sin kx}{\sin kL} - \frac{x}{L}\right) \tag{10.18}$$

If two couples M_a and M_b are applied at the ends A and B of the bar, as shown in Fig. 10.11, the equation for the deflection curve can be obtained by applying the modified principle of superposition.

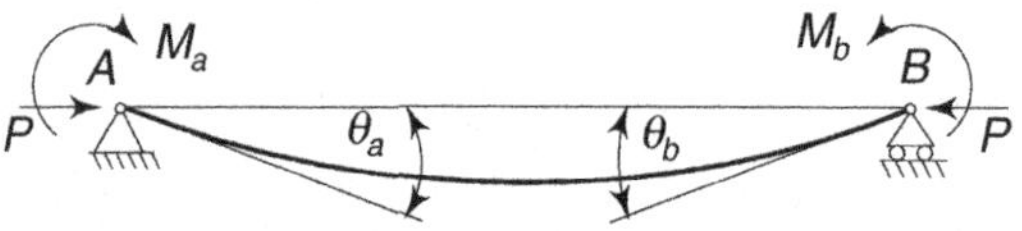

Fig. 10.11 *Beam-column with two end couples*

Equation (10.18) gives the deflection produced by M_b. In this equation, if we substitute M_a for M_b and $(L - x)$ for x, we obtain the deflection produced by M_a. Adding these results, we get the deflection curve for M_a and M_b acting together. Thus,

$$y = \frac{M_b}{P}\left(\frac{\sin kx}{\sin kL} - \frac{x}{L}\right) + \frac{M_a}{P}\left[\frac{\sin k(L - x)}{\sin kL} - \frac{L - x}{L}\right] \tag{10.19}$$

The slopes θ_a and θ_b at A and B can be obtained by differentiating the above expression and putting $x = 0$ and $x = L$, i.e.

$$\theta_a = \left(\frac{dy}{dx}\right) at\ x = 0, \quad \text{and} \quad \theta_b = -\left(\frac{dy}{dx}\right) at\ x = L$$

The negative sign in θ_b expression is because of the sign convention adopted [Fig. (10.11)]. The slopes are

$$\theta_a = \frac{M_a L}{3EI}\psi(u) + \frac{M_b L}{6EI}\phi(u) \tag{10.20a}$$

$$\theta_b = \frac{M_b L}{3EI}\psi(u) + \frac{M_a L}{6EI}\phi(u) \tag{10.20b}$$

where $$\phi(u) = \frac{3}{u}\left(\frac{1}{\sin 2u} - \frac{1}{2u}\right)$$

$$\psi(u) = \frac{3}{2u}\left(\frac{1}{2u} - \frac{1}{\tan 2u}\right)$$

and $$u = \frac{1}{2}kL = \frac{1}{2}L\left(\frac{P}{EI}\right)^{1/2}$$

Example 10.3 *A beam-column carries a triangular load, as shown in Fig. 10.12. Find the slopes at the ends of the column.*

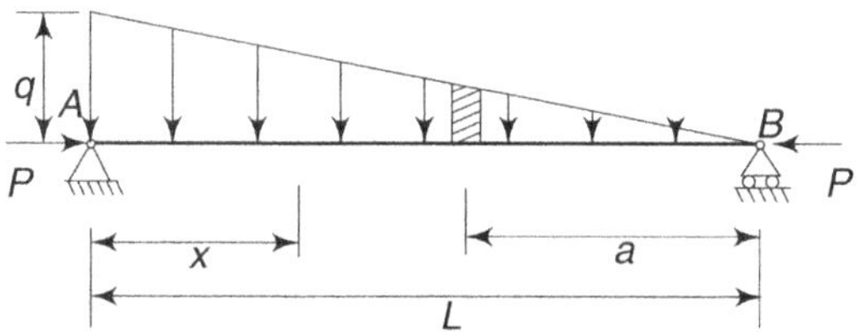

Fig. 10.12 *Example 10.3*

Solution We make use of Eq. (10.16). The loading Q_i will now be equal to $\frac{q}{L}$ ada acting at distance a from B. Replacing the summation by integration, the equation becomes

$$y = \frac{\sin kx}{Pk \sin kL}\int_0^{L-x}\frac{qa}{L}\sin ka\,da - \frac{x}{PL}\int_0^{L-x}\frac{qa^2}{L}\,da$$

$$+ \frac{\sin k(L-x)}{Pk \sin kL}\int_{L-x}^{L}\frac{qa}{L}\sin k(L-a)\,da - \frac{L-x}{PL}\int_{L-x}^{L}\frac{qa}{L}(L-a)\,da$$

Now, we make use of the formula

$$\frac{d}{d\alpha}\int_{U(\alpha)}^{V(\alpha)} F(\alpha, x)\,dx = \int_{U(\alpha)}^{V(\alpha)}\frac{dF(\alpha, x)}{d\alpha}\,dx + F(\alpha, V)\frac{dV}{d\alpha} - F(\alpha, U)\frac{dU}{d\alpha}$$

Then,

$$\frac{dy}{dx} = \frac{k\cos kx}{Pk\sin kL}\int_0^{L-x}\frac{qa}{L}\sin ka\,da - \frac{\sin kx}{Pk\sin kL}\,\frac{q(L-x)}{L}\sin k(L-x)$$

$$-\frac{1}{PL}\int_0^{L-x}\frac{qa^2}{L}\,da + \frac{x}{PL}\,\frac{q(L-x)^2}{L} - \frac{k\cos k(L-x)}{Pk\sin kL}\times$$

$$\int_{L-x}^{L} \frac{qa}{L} \sin k(L-a)da + \frac{\sin k(L-x)}{Pk \sin kL} \cdot \frac{q(L-x)}{L} \sin kx$$

$$+ \frac{1}{PL} \int_{L-x}^{L} \frac{qa}{L}(L-a)\, da - \frac{L-x}{PL} \frac{q(L-x)}{L} x$$

$$\therefore \qquad \left\{\frac{dy}{dx}\right\}_{x=0} = \frac{q}{PL \sin kL} \int_0^L a \sin ka\, da - \frac{q}{PL^2} \int_0^L a^2\, da$$

$$= \frac{q}{3Pk^2 L \tan kL}(3 \tan kL - 3kL - k^2 L^2 \tan kL)$$

Similarly,

$$\left\{\frac{dy}{dx}\right\}_{x=L} = \frac{q}{6Pk^2 L \sin kL}(6 \sin kL - 6kL - k^2 L^2 \sin kL)$$

II GENERAL TREATMENT OF COLUMN STABILITY PROBLEMS (As an Eigenvalue Problem)

10.8 GENERAL DIFFERENTIAL EQUATION AND SPECIFIC EXAMPLES

The general differential equation derived for a beam column, given by Eq. (10.7), can be used as a general equation to determine the critical loads of buckled bars. If the column is not subjected to lateral loads, then the general equation becomes, with $q = 0$,

$$EI \frac{d^4 y}{dx^4} + P \frac{d^2 y}{dx^2} = 0 \tag{10.21}$$

If EI varies along x, then the general equation can be derived on the same lines as in Sec. 10.3, giving

$$\frac{d^2}{dx^2}\left(EI \frac{d^2 y}{dx^2}\right) + P \frac{d^2 y}{dx^2} = 0 \tag{10.22}$$

Equations (10.21) and (10.22) are the equilibrium equations of a slightly buckled beam subjected to axial load only. Hence, the axial load will represent the critical load. The boundary conditions, i.e. the end conditions, can be quite general. Hence, these equations represent the general differential equations for a column.

For the present, we shall assume that EI is constant along x and use Eq. (10.21). On using the notation $k^2 = \dfrac{P}{EI}$, Eq. (10.21) becomes

$$\frac{d^4 y}{dx^4} + k^2 \frac{d^2 y}{dx^2} = 0 \tag{10.23}$$

The general solution of this equation is

$$y = A \sin kx + B \cos kx + Cx + D \tag{10.24}$$

The constants are determined from the end conditions of the bar. We can consider the following particular cases:

Column with Hinged Ends (Fundamental Case) In the case of a bar with hinged ends (Fig. 10.11), the deflection and bending moments are zero at the two ends, i.e.

$$y = 0, \quad \text{and} \quad \frac{d^2y}{dx^2} = 0 \quad \text{at } x = 0 \quad \text{and } x = L$$

These conditions give

$$B = 0, \qquad C = 0, \qquad D = 0, \qquad \text{and } \sin kL = 0$$

Hence, $kL = n\pi$

The deflection curve is then obtained as

$$y = A \sin kx = A \sin \frac{n\pi x}{L} \tag{10.25}$$

where A is undetermined, i.e. in determining the load that keeps the column in a slightly buckled form, the amplitude of the deflection remains undetermined. For $n = 1$, $n = 2$ and $n = 3$, the shapes of the buckled bar are as shown in Fig. 10.13.

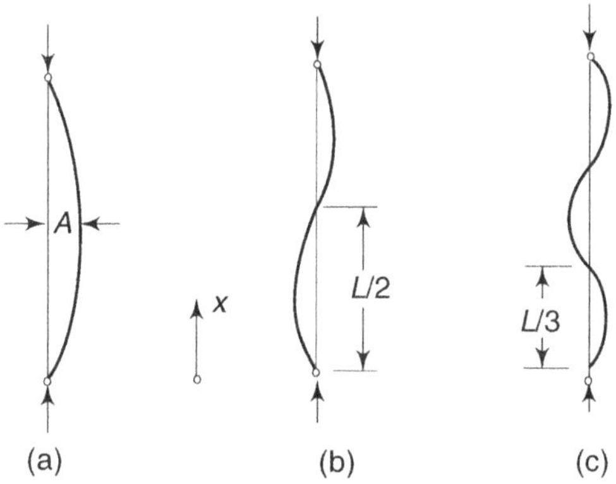

Fig. 10.13 *Various modes of buckling of a column with two ends hinged*

The corresponding loads are obtained from the equation

$$k = \frac{n\pi}{L} = \left(\frac{P}{EI}\right)^{1/2} \quad \text{or} \quad P_{cr} = \frac{n^2\pi^2 EI}{L^2}$$

Column with One End Fixed and the Other End Free The end conditions are:

at fixed end (i.e. at $x = 0$), $y = 0$ and $dy/dx = 0$

at free end (at $x = L$), moment and shear force are zero, i.e. $d^2y/dx^2 = 0$ and from Eq. (10.6)

$$EI \frac{d^3y}{dx^3} + P\frac{dy}{dx} = 0$$

From these, the constants are determined as

$$B = -D, \qquad C = -Ak,$$

$$A \sin kL + B \cos kL = 0, \qquad C = 0$$

Hence, $\quad A = C = 0 \quad \text{and} \quad \cos kL = 0$

or $$kL = (2n-1)\frac{\pi}{2}$$

The deflection curve is therefore

$$y = B(1 - \cos kx)$$

or $$y = B\left[1 - \cos(2n-1)\frac{\pi x}{2L}\right] \tag{10.26}$$

With $n = 1$, we obtain

$$kL = \frac{\pi}{2} \quad \text{or,} \quad P_{cr} = \frac{\pi^2 EI}{4L^2}$$

This is the smallest load that can keep the column in a slightly buckled shape. When $n = 2, 3$, etc., we get the other critical loads as

$$P_{cr} = \frac{9\pi^2 EI}{4L^2}, \qquad P_{cr} = \frac{25\pi^2 EI}{4L^2}, \ldots, \text{etc.}$$

The corresponding deflection curves are shown in Fig. 10.14.

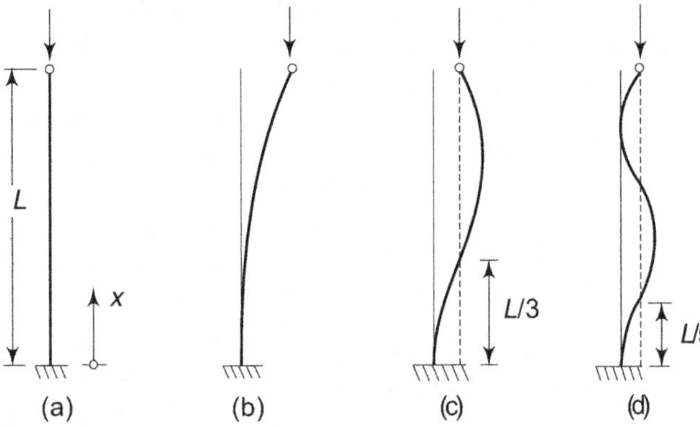

Fig. 10.14 *Various modes of buckling of a column with one end fixed and the other end free*

Column with One End Fixed and Other End Pinned This case is shown in Fig. 10.15 and was discussed in Sec. 10.1. Since the top end is pinned, a lateral force R is necessary to keep the column in that position.

The end conditions are

$$y = 0, \quad \frac{dy}{dx} = 0 \quad \text{at } x = 0$$

$$y = 0, \quad \frac{d^2y}{dx^2} = 0 \quad \text{at } x = L$$

With these, the general solution yields the following equations:

$$B + D = 0$$
$$Ak + C = 0$$

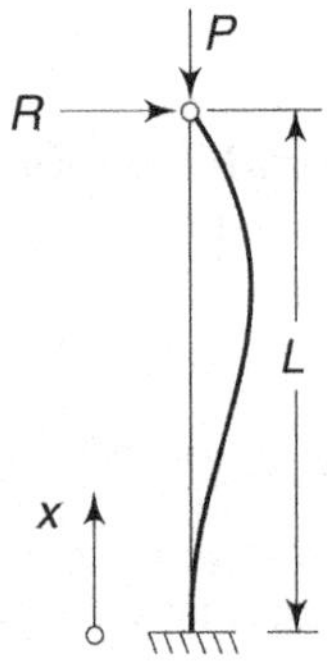

Fig. 10.15 *Column with one end fixed and other end pinned*

$$CL + D = 0$$
$$A \sin kL + B \cos kL = 0$$

A trivial solution for the above set of equations is $A = B = C = D = 0$, which means that the deflection curve is a straight line (i.e. $y = 0$). For the existence of a non-trivial solution, the determinant of the coefficients should be equal to zero. The determinant is

$$\begin{vmatrix} 0 & 1 & 0 & 1 \\ k & 0 & 1 & 0 \\ 0 & 0 & L & 1 \\ \sin kL & \cos kL & 0 & 0 \end{vmatrix} = -\sin kL + kL \cos kL$$

For the existence of a non-trivial solution, the above quantity should be equal to zero, i.e.

$$-\sin kL + kL \cos kL = 0$$

or
$$\tan kL = kL \tag{10.27}$$

The load which keeps the column in a slightly buckled form should, therefore, satisfy the above transcendental equation. The smallest root of this equation is $kL = 4.493$ and the corresponding critical load is

$$P_{cr} = \frac{20.19\, EI}{L^2} = \frac{\pi^2 EI}{(0.699L)^2}$$

Column with Ends Fixed For a column with both ends fixed, the boundary conditions are

$$y = 0, \quad \frac{dy}{dx} = 0 \quad \text{for } x = 0$$

$$y = 0, \quad \frac{dy}{dx} = 0 \quad \text{for } x = L$$

Substituting these in Eq. (10.24), we get

$$B + D = 0$$
$$Ak + C = 0$$
$$A \sin kL + B \cos kL + CL + D = 0$$
$$Ak \cos kL - Bk \sin kL + C = 0$$

For the existence of a non-trivial solution, the following determinant should be equal to zero

$$\begin{vmatrix} 0 & 1 & 0 & 1 \\ k & 0 & 1 & 0 \\ \sin kL & \cos kL & L & 1 \\ k \cos kL & -k \sin kL & 1 & 0 \end{vmatrix} = 0$$

i.e.
$$2(\cos kL - 1) + kL \sin kL = 0$$

or
$$\sin\frac{kL}{2}\left(\frac{kL}{2}\cos\frac{kL}{2}-\sin\frac{kL}{2}\right)=0 \tag{10.28}$$

One solution is

$$\sin\frac{kL}{2}=0$$

i.e.
$$kL = 2n\pi \text{ and hence } P_{cr} = \frac{4n^2\pi^2 EI}{L^2}$$

Noting that $\sin kL = 0$ and $\cos kL = 1$, whenever $\sin kL/2 = 0$, the constants are found as

$$A = 0, \qquad C = 0, \qquad B = -D$$

and the deflection curve is therefore

$$y = B\left(\cos\frac{2n\pi x}{L}-1\right) \tag{10.29}$$

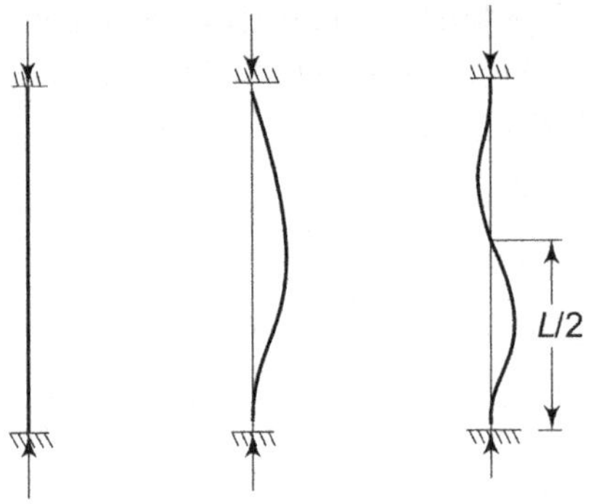

Fig. 10.16 *Modes of buckling of a column with both ends fixed*

A second solution to Eq. (10.28) is

$$\frac{kL}{2}\cos\frac{kL}{2}-\sin\frac{kL}{2}=0$$

or
$$\tan\frac{kL}{2}=\frac{kL}{2}$$

The lowest root of this transcendental equation is $kL/2 = 4.493$ and hence

$$P_{cr} = \frac{8.18\pi^2 EI}{L^2}$$

The deflection curves corresponding to these two critical loads are shown in Fig. 10.16.

Column with Load Passing Through a Fixed Point Consider a column with one end fixed and the other end loaded in such a manner that the load passes through a fixed point (Fig. 10.17). The load may be applied through the tension of a cable passing through the fixed point O.

During buckling, because the force P becomes inclined, a shearing force is developed at the top end. This shearing force is equal to the horizontal component of the inclined force P. Since the deflection is assumed to be very small, the vertical component will be almost equal to P and the horizontal component is

$$V = -P\frac{\delta}{c}$$

From Eq. (10.6)

$$EI\frac{d^3y}{dx^3}+P\frac{dy}{dx}=P\frac{\delta}{c}$$

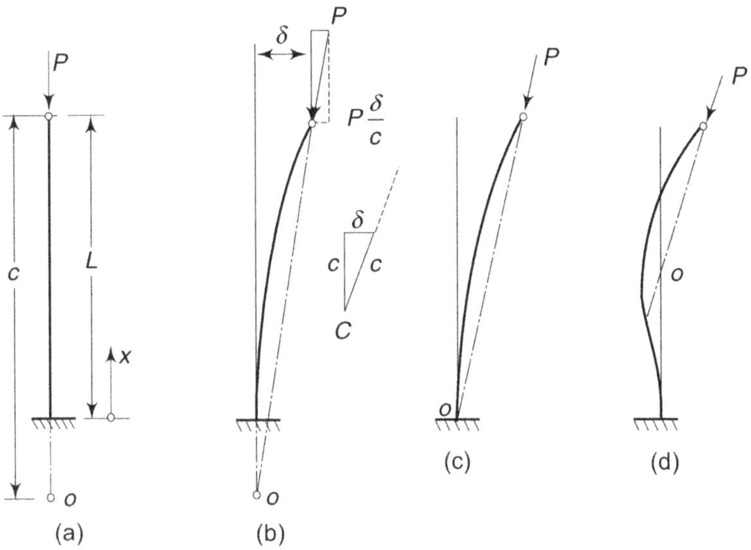

Fig. 10.17 *Column with load passing through a fixed point*

or $$\frac{d^3y}{dx^3} + k^2\frac{dy}{dx} = \frac{k^2\delta}{c}$$

This is one of the boundary conditions. The other conditions are

$$y = 0 \quad \text{and} \quad \frac{dy}{dx} = 0 \quad \text{at } x = 0$$

$$\frac{d^2y}{dx^2} = 0 \quad \text{at } x = L$$

Substituting these in the general solution given by Eq. (10.24)

$$B + D = 0$$
$$Ak + C = 0$$
$$C = \frac{\delta}{c}$$
$$A \sin kL + B \cos kL = 0$$

Solving these equations, the constants are obtained as

$$A = -\frac{\delta}{kc}, \qquad B = \frac{\delta}{kc}\tan kL = -D, \qquad C = \frac{\delta}{c}$$

Substituting these, the deflection curve is obtained as

$$y = \frac{\delta}{kc}\left[\tan kL\,(\cos kL - 1) + kL - \sin kL\right]$$

or $$\tan kL = kL\left(1 - \frac{c}{L}\right) \tag{10.30}$$

The above equation gives the value of the critical load for any given ratio of c/L. For three specific values of c/L, the values of kL and $\frac{L^2 P_{cr}}{\pi^2 EI}$ are as follows:

$\frac{c}{L} = 0$	1	∞
$kL = 4.493$	π	$\frac{\pi}{2}$
$\frac{L^2 P_{cr}}{\pi^2 E I} = 2.05$	1	0.25

When $c = 0$, we get the case of a column pinned at the top and fixed at the bottom, which is the case discussed in (iii). When $c = L$, the critical load is obtained as

$$kL = \pi \quad \text{and} \quad P_{cr} = \frac{\pi^2 EI}{L^2}$$

which is the same as the one obtained for case (i), i.e. the fundamental case. This can be explained by the fact that when the line of action of P passes through the base point, the moment at the base is zero and the end behaves like a hinged end. The moment at the top end is also zero and consequently, the column acts as a hinged-end column. When c approaches infinity, the column behaves like it did in case (ii), where the load is aways vertical.

10.9 BUCKLING PROBLEM AS A CHARACTERISTIC VALUE (EIGENVALUE) PROBLEM

In Sec. 10.8, the buckling problem was discussed, starting with the general differential equation of equilibrium of a slightly buckled column subjected to axial load only. The specific examples analysed, bring out some general characteristic features of the differential equation, which will be discussed now. These features give us a better insight into the problem and provide a basis for the application of energy methods to buckling problems.

The general differential equation of equilibrium of a slightly buckled column with general boundary conditions was given by Eq. (10.22) as

$$\frac{d^2}{dx^2}\left(EI_x \frac{d^2 y}{dx^2}\right) + P\frac{d^2 y}{dx^2} = 0 \tag{10.31a}$$

It is assumed that the moment of inertia I_x can vary along the axis of the column. We can write

$$I_x = Ip(x)$$

where I is a constant moment of inertia and $p(x)$ is a dimensionless function of x. The differential equation then becomes

$$\frac{d^2}{dx^2}\left[EIp(x)\frac{d^2 y}{dx^2}\right] + P\frac{d^2 y}{dx^2} = 0 \tag{10.31b}$$

Dividing by EI and using the notation $k^2 = \frac{P}{EI}$, the above equation can be written as

$$\frac{d^2}{dy^2}\left[p(x)\frac{d^2y}{dx^2}\right] + k^2\frac{d^2y}{dx^2} = 0 \tag{10.32}$$

This is a homogeneous differential equation with the following general solution:

$$y = C_1\phi_1(k, x) + C_2\phi_2(k, x) + C_3\frac{x}{L} + C_4 \tag{10.33}$$

where C_1, C_2, C_3 and C_4 are constants and $\phi_1(k, x)$ and $\phi_2(k, x)$ are transcendental functions of x and k. When EI_x is constant, the general solution has the form given by Eq. (10.24). The constants have the dimensions of length and are determined from the boundary conditions. The most frequently encountered boundary conditions are

For freely supported end $\quad y = 0$ and $\frac{d^2y}{dx^2} = 0$

For a fixed end $\quad y = 0$ and $\frac{dy}{dx} = 0 \quad (10.34)$

For a free end $\quad \frac{d^2y}{dx^2} = 0$ and shear $= 0$

The zero shear force condition is represented by an equation similar to Eq. (10.6) as

$$\frac{d}{dx}\left[p(x)\frac{d^2y}{dx^2}\right] + k^2\frac{dy}{dx} = 0$$

These boundary conditions are linear and homogeneous equations and will, therefore, be referred to as homogeneous boundary conditions. Substituting these homogeneous boundary conditions for a specific case in Eq. (10.33), we get a set of four linear homogeneous equations with the following general form

$$\begin{aligned}
\alpha_{11}C_1 + \alpha_{21}C_2 + \alpha_{31}C_3 + \alpha_{41}C_4 &= 0\\
\alpha_{12}C_1 + \alpha_{22}C_2 + \alpha_{32}C_3 + \alpha_{42}C_4 &= 0\\
\alpha_{13}C_1 + \alpha_{23}C_2 + \alpha_{33}C_3 + \alpha_{43}C_4 &= 0\\
\alpha_{14}C_1 + \alpha_{24}C_2 + \alpha_{34}C_3 + \alpha_{44}C_4 &= 0
\end{aligned} \tag{10.35}$$

Some of these coefficients αs are transcendental functions of the parameter k while others are constants. We have come across these kinds of equations in Sec. 10.8. A trivial solution of Eq. (10.35) is that $C_1 = C_2 = C_3 = C_4 = 0$, representing a straight undeflected column ($y = 0$). However, a non-trivial solution, in which we are interested, exists if the determinant of the coefficients is zero, i.e.

$$\Delta = \begin{vmatrix} \alpha_{11} & \alpha_{21} & \alpha_{31} & \alpha_{41} \\ \alpha_{12} & \alpha_{22} & \alpha_{32} & \alpha_{42} \\ \alpha_{13} & \alpha_{23} & \alpha_{33} & \alpha_{43} \\ \alpha_{14} & \alpha_{24} & \alpha_{34} & \alpha_{44} \end{vmatrix} = 0 \tag{10.36}$$

The expansion of the determinant is Δ and equating it to zero yields an equation for the parameter k, which is the only unknown in this equation. In general, Eq. (10.36) is a transcendental equation providing an infinite number of roots k_i ($i = 1, 2,...,$). These are called the characteristic roots or values of the parameter k and for each k_i, there is a corresponding critical load P_i given by

$$k_i^2 = \frac{P_i}{EI}$$

Introducing one of the characteristic values k_i into the system of Eq. (10.35), we get four equations to determine the four constants C_{1i}, C_{2i}, C_{3i} and C_{4i}. However, as these equations are homogeneous, only the ratios of these constants can be determined. Let

$$\frac{C_{2i}}{C_{1i}} = \bar{C}_{2i}, \qquad \frac{C_{3i}}{C_{1i}} = \bar{C}_{3i}, \qquad \frac{C_{4i}}{C_{1i}} = \bar{C}_{4i}$$

Substituting these in Eq. (10.33), we get the deflection curve corresponding to the load P_i as

$$y_i = C_{1i}\left[\phi_1(k_i, x) + \bar{C}_{2i}\,\phi_2(k_i, x) + \bar{C}_{3i}\,\phi_3(k_i, x) + \bar{C}_{4i}\,\phi_4(k_i, x)\right] \tag{10.37}$$

The constant C_{1i} remains undetermined. y_i is called characteristic function of the homogeneous differential Eq. (10.32) associated with the set of particular boundary conditions of the case under consideration.

The above analysis of the homogeneous differential Eq. (10.32) shows that there exists a set of values of the parameter k for which a deflected configuration of the column is possible. For each value of the parameter k a corresponding critical load to keep the column in that buckled shope is obtained from Eq. (10.37). The amplitude (i.e. the magnitude) of deflection however remains indeterminate. There is a close similarity between the analysis of a buckling problem and the analysis of a vibration problem connected with small oscillations. The relationship between the two groups of problems is as follows:

Problem	**Equation $\Delta = 0$**	**Characteristic values**	**Characteristic functions**
Buckling	Stability criterion	Buckling loads	Buckling modes
Vibrations	Frequency equation	Frequencies	Principal modes of vibration

10.10 THE ORTHOGONALITY RELATIONS

The characteristic functions y_i satisfying the homogeneous differential equation have an important property which play an important role in the energy methods. Consider the homogeneous differential equation

$$\frac{d^2}{dy^2}\left[p(x)\frac{d^2y}{dx^2}\right] + k^2\frac{d^2y}{dx^2} = 0$$

This is satisfied by any characteristic function y_i and the associated characteristic value of the parameter k_i, i.e.

$$\frac{d^2}{dy^2}\left[p(x)\frac{d^2y_i}{dx^2}\right]+k_i^2\frac{d^2y_i}{dx^2}=0$$

Multiply this equation by any other characteristic functions y_k and integrate over the length L of the column, obtaining

$$\int_0^L \frac{d^2}{dy^2}\left[p(x)\frac{d^2y_i}{dx^2}\right]y_k\,dx + k_i^2\int_0^L \frac{d^2y_i}{dx^2}y_k\,dx = 0 \tag{10.38}$$

Integrating the first term by parts, we obtain

$$\int_0^L \frac{d^2}{dy^2}\left[p(x)\frac{d^2y_i}{dx^2}\right]y_k\,dx$$

$$=\frac{d}{dy}\left[p(x)\frac{d^2y_i}{dx^2}\right]y_k\Bigg|_0^L-\int_0^L\frac{d}{dy}\left[p(x)\frac{d^2y_i}{dx^2}\right]\frac{dy_k}{dx}\,dx$$

Integrating once again by parts

$$=\frac{d}{dy}\left[p(x)\frac{d^2y_i}{dx^2}\right]y_k\Bigg|_0^L-\left[p(x)\frac{d^2y_i}{dx^2}\right]\frac{dy_k}{dx}\Bigg|_0^L+\int_0^L\left[p(x)\frac{d^2y_i}{dx^2}\right]\frac{d^2y_k}{dx^2}\,dx$$

Similarly, the second term in Eq. (10.38) yields

$$k_i^2\int_0^L\frac{d^2y_i}{dx^2}y_k\,dx = k_i^2\frac{dy_i}{dx}y_k\Bigg|_0^L-k_i^2\int\frac{dy_i}{dx}\frac{dy_k}{dx}\,dx$$

Substituting these in Eq. (10.38), we obtain

$$\left[\left\{\frac{d}{dy}p(x)\frac{d^2y_i}{dx^2}+k_i^2\frac{dy_i}{dx}\right\}y_k\right]\Bigg|_0^L-\left[p(x)\frac{d^2y_i}{dx^2}\frac{dy_k}{dx}\right]\Bigg|_0^L$$

$$+\int_0^L\left[p(x)\frac{d^2y_i}{dx^2}\frac{d^2y_k}{dx^2}\right]dx-k_i^2\int_0^L\frac{dy_i}{dx}\frac{dy_k}{dx}\,dx=0$$

The first term within the brackets vanish for any combination of the homogeneous boundary conditions given by Eq. (10.34). Consequently,

$$\int_0^L p(x)\frac{d^2y_i}{dx^2}\frac{d^2y_k}{dx^2}\,dx-k_i^2\int_0^L\frac{dy_i}{dx}\frac{dy_k}{dx}\,dx=0 \tag{10.39}$$

Since the above equation is valid for each combination of two characteristic functions, we can interchange y_i and y_k and obtain

$$\int_0^L p(x)\frac{d^2y_k}{dx^2}\frac{d^2y_i}{dx^2}\,dx-k_k^2\int_0^L\frac{dy_k}{dx}\frac{dy_i}{dx}\,dx=0$$

Subtracting one from the other

$$\left(k_i^2 - k_k^2\right)\int_0^L \frac{dy_i}{dx}\frac{dy_k}{dx}\,dx = 0 \tag{10.40}$$

If i is different from k then in general $k_i^2 - k_k^2$ will be different from zero, and consequently,

$$\int_0^L \frac{dy_i}{dx}\frac{dy_k}{dx}\,dx = 0 \tag{10.41}$$

Using Eq. (10.41) in Eq. (10.39), we observe that

$$\int_0^L p(x)\frac{d^2y_i}{dx^2}\frac{d^2y_k}{dx^2}\,dx = 0 \tag{10.42}$$

If $i = k$, the integral

$$\int_0^L \left(\frac{dy_i}{dx}\right)^2 dx$$

cannot be equal to zero since the integrand is always a positive quantity and in Eq. (10.39), as k_i^2 is also different from zero, we observe that

$$\int_0^L p(x)\left(\frac{d^2y_i}{dx^2}\right)^2 dx \neq 0 \quad \text{and} \quad \int_0^L \left(\frac{dy_i}{dx}\right)^2 dx \neq 0 \tag{10.43}$$

Equations (10.41) and (10.42) express the fundamental properties of the characteristic solutions of the differential Eq. (10.32) and these are known as the orthogonality relations of the characteristic functions y_i. A family of functions consisting of all of Eq. (10.32) with prescribed boundary conditions is said to constitute a complete system of orthogonal functions.

If I_x is independent of x, then in Eq. (10.32) $p(x) = 1$ and consequently, Eqs (10.41) and (10.42) reduce to

$$\int_0^L \frac{dy_i}{dx}\frac{dy_k}{dx}\,dx = \int_0^L \frac{d^2y_i}{dx^2}\frac{d^2y_k}{dx^2}\,dx = 0 \tag{10.44}$$

For example, the sequence of functions

$$y_i = \sin i\frac{\pi x}{L} \quad (i = 1, 2, \ldots, \infty)$$

which are the terms of a Fourier expansion, form a complete system of orthogonal functions satisfying conditions given by Eq. (10.44). We may recall that for a hinged column, $\left(\text{i.e. } y = \frac{d^2y}{dx^2} = 0 \text{ for } x = 0 \text{ and } x = L\right)$, the functions

$$y_n = \sin\frac{n\pi x}{L}$$

are the characteristic solutions which satisfy the orthogonality conditions. The advantage of representing a deflection curve by a series like

$$y = a_1 \sin\frac{\pi x}{L} + a_2 \sin\frac{2\pi x}{L} + a_3 \sin\frac{3\pi x}{L} + \ldots$$

will be demonstrated in Sec. 10.19.

III ENERGY METHODS FOR BUCKLING PROBLEMS

10.11 THEOREM OF STATIONARY POTENTIAL ENERGY

The energy method of analysing the problems of elastic stability is based on an extremum principle of mechanics. Consider an elastic body subjected to external surface and body forces. Let the body be in equilibrium. During the application of these forces, the body deforms and consequently, these forces do a certain amount of work W. The internal forces which are set up inside the elastic body also do work during the deformation process and this is stored as elastic strain energy. When external forces are applied gradually and no dissipation of energy takes place due to friction etc. the work done by the external forces should be equal to the internal elastic energy U, i.e.

$$W = U \tag{10.45}$$

Let portions of the body be given small virtual displacements. These are small displacements that are consistent with the constraints imposed on the body. For example, if a point of the body is fixed, then the virtual displacement there is zero. If a point of the body is constrained to lie on the surface of another body, then the virtual displacement there should be tangential to the surface of the contacting body. These virtual displacements being very small, the changes necessary in the external forces to bring about these virtual displacements will also be very small and will vanish in the limit. The work done by external surface and body forces P_i during these virtual displacements is

$$\delta W = \Sigma P_i \delta\, \Delta_i + \text{higher order terms} \tag{10.46}$$

where $\delta\Delta_i$ are the work absorbing components of the virtual displacements. It is convenient to define a potential V of the external forces in such a manner that the work done during virtual displacements is equal to $-\delta V$, i.e. a decrease in potential energy in the form of an equation

$$-\delta V = \Sigma P_i \delta\Delta_i = \delta W \tag{10.47}$$

In the above equation, we have neglected the higher order terms of Eq. (10.46). If a part of the body is subjected to distributed external forces, then over that part, the summation must be replaced by a surface integral.

From Eq. (10.47)

$$-\delta V - \delta W = 0$$

Using Eq. (10.45), the above equation can be written as

$$\delta (U + V) = 0 \tag{10.48}$$

The expression $U + V$ is known as the total potential of the system. Consequently, Eq. (10.48) can be stated as follows:

The first-order change in the total potential energy must vanish for every virtual displacement when an elastic body is in equilibrium.

First-order change means a change in which only those terms that contain first-order terms are considered. Terms of higher order, as in Eq. (10.46), are ignored. Equation (10.48) is also stated as that in which the quantity $U + V$ assumes a stationary value, i.e.

$$U + V = \text{stationary} \tag{10.49}$$

It is shown in books on elasticity that for stable equilibrium, any virtual displacement will cause a positive change in the total potential energy of the system, which means that for a system in stable equilibrium, the total potential energy is minimum.

Example 10.4 *Figure 10.18 shows a three-bar truss, the point D of which is subjected to P units of force. Applying the principle of minimum potential energy, determine the vertical and horizontal displacements of D due to the load. The members have equal cross-sectional areas.*

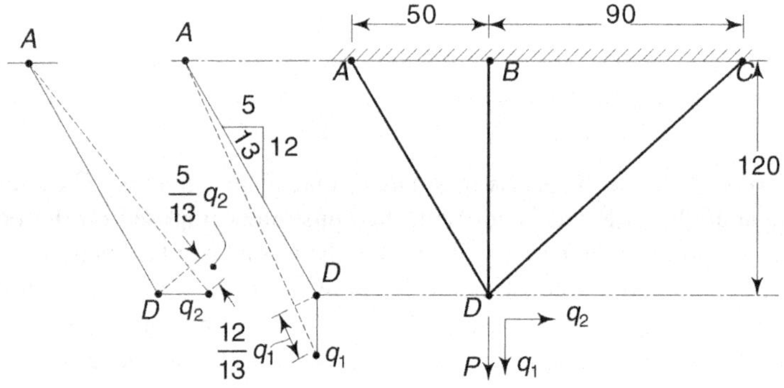

Fig. 10.18 *Example 10.4*

Solution Let the point D have a vertical displacement q_1 and a horizontal displacement q_2. The elongation caused in each member due to these displacements can be calculated geometrically. This is shown in the figure for member AD. The total extension of each member is

$$\text{member } AD \quad \frac{12}{13}q_1 + \frac{5}{13}q_2$$

$$\text{member } BD \quad q_1$$

$$\text{member } CD \quad \frac{4}{5}q_1 - \frac{3}{5}q_2$$

If A is the cross-sectional area and E is the Young's modulus, then the elastic strain energy stored in a member of length L is

$$U = \frac{EA}{2L}\delta^2$$

where δ is the elongation. Hence, for the three members, the strain energies are

$$\text{for } AD \qquad U_1 = \frac{EA}{2(130)}\left(\frac{12}{13}q_1 + \frac{5}{13}q_2\right)^2$$

$$\text{for } BD \qquad U_2 = \frac{EA}{2(120)}q_1^2$$

$$\text{for } CD \qquad U_3 = \frac{EA}{2(150)}\left(\frac{4}{5}q_1 - \frac{3}{5}q_2\right)^2$$

The total elastic strain energy is the sum of the above three quantities. Hence,

$$U = EA\left(958q_1^2 - 47q_1q_2 + 177q_2^2\right) \times 10^{-5}$$

Taking the undeformed position as the datum, the potential energy in the deformed configuration is

$$V = -Pq_1$$

Hence, the total potential energy is

$$U + V = EA\left(958q_1^2 - 47q_1q_2 + 177q_2^2\right) \times 10^{-5} - pq_1$$

For equilibrium position, the first-order variation of the above quantity should be equal to zero, i.e.

$$\delta(U+V) = EA\left[\,958(2q_1\,\delta q_1) - 47(q_1\delta q_2 + q_2\delta q_1) + 177(2q_2\delta q_2)\,\right] \times 10^{-5} - P\delta q_1 = 0$$

or $$\left(1916q_1 - 47q_2 - \frac{P}{EA} \times 10^5\right)\delta q_1 + \left(-47q_1 + 354q_2\right)\delta q_2 = 0$$

Since δq_1 and δq_2 are arbitrary virtual displacements, the quantities inside the parentheses should vanish individually. Thus,

$$1916q_1 - 47q_2 = \frac{P}{EA} \times 10^5$$

$$-47q_1 + 354q_2 = 0$$

Solving these two equations, we obtain

$$q_1 = 52.36\frac{P}{EA} \quad \text{and} \quad q_2 = 6.95\frac{P}{EA}$$

It should be observed that we have not made use of any equation of statics in solving the problem.

10.12 COMPARISON WITH THE PRINCIPLE OF CONSERVATION OF ENERGY

It is important to realise that the principle of the minimum total potential is different from the law of conservation of energy. The latter principle states that in an equilibrium condition, the work done by all external forces during the loading process is equal to the internal elastic strain energy stored, i.e. Eq. (10.49) becomes

$$U - W = 0$$

If the loading is done gradually, the work done is equal to

$$W = \Sigma \frac{1}{2} P_i y_i$$

where y_i is the work absorbing component of the deflection at P_i. On the other hand, the virtual work done is

$$\delta W = \Sigma P_i \,\Delta y_i$$

There is no 1/2 factor in this case since the forces $p_i s$ are acting with full magnitude during the virtual displacements $\Delta y_i\, s$.

10.13 ENERGY AND STABILITY CONSIDERATIONS

In Example 10.4, we have demonstrated the use of the theorem of stationary potential energy in solving a statically indeterminate problem. Now we shall show, with reference to a specific problem, how energy considerations can be used to analyse stability problems. Consider a vertical bar hinged at one end and supported at the other end by a spring, as shown in Fig. 10.19. It is assumed that the bar is infinitely rigid. It carries a centrally applied load P.

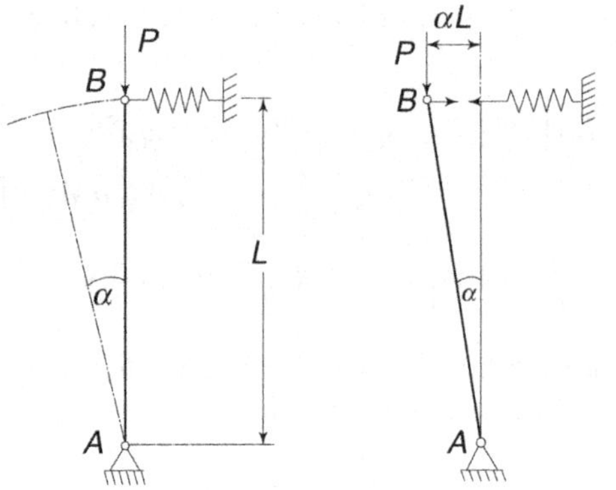

Fig. 10.19 *Vertical bar hinged at one end and supported by spring at the other end*

Let the bar be displaced through a small angle α. Because of this displacement, the load P is lowered by the amount

$$L - L\cos\alpha = L(1-\cos\alpha) \approx L\frac{\alpha^2}{2}$$

The decrease in potential energy is equal to the work done by P, i.e. $\frac{1}{2}PL\,\alpha^2$. At the same time, the spring elongates by an amount αL and the energy stored due to this is $\frac{1}{2}S(\alpha L)^2$ where S is the spring constant. If the decrease in the potential energy is greater than the energy stored in the spring, i.e. if

$$\frac{1}{2}PL\alpha^2 > \frac{1}{2}S\alpha^2L^2$$

then the system is unstable. On the other hand, if

$$\frac{1}{2}PL\alpha^2 < \frac{1}{2}S\alpha^2L^2$$

then the system is stable. If

$$\frac{1}{2}PL\alpha^2 = \frac{1}{2}S\alpha^2L^2$$

i.e. $$P_{cr} = SL$$

then we get the value of the critical load which keeps the column in equilibrium in a slightly displaced configuration.

The same conclusion can be obtained by applying the principles of statics. For equilibrium in the displaced position, the moment about A should be zero. The end B of the column is subjected to a vertical load P and a horizontal force $S\alpha L$. For moment equilibrium about A,

$$P\alpha L = S\alpha L^2 \quad \text{or} \quad P_{cr} = SL$$

An analysis of stability problems in column buckling, using the above concept, will be taken up again in Sec. 10.17.

10.14 APPLICATION TO BUCKLING PROBLEMS

We shall now discuss the application of the minimum total energy principle to column buckling problems.

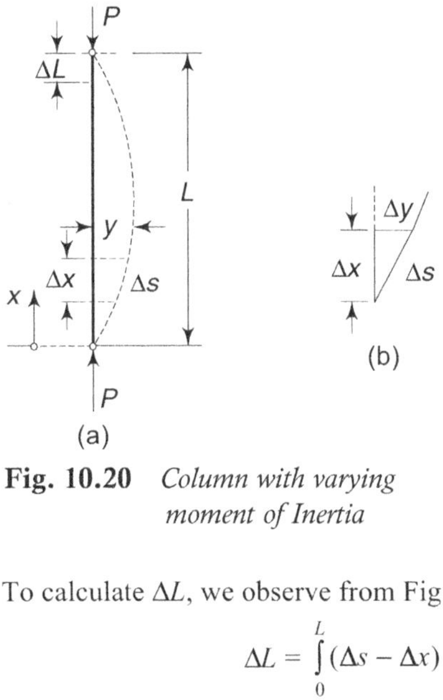

Fig. 10.20 *Column with varying moment of Inertia*

Consider the column shown in Fig. 10.20, carrying an axial load P. Let the moment of inertia I_x be variable. In calculating the strain energy, we shall consider only the bending energy. From the straight equilibrium configuration, let the column be moved to a neighbouring bent configuration.

Let the buckled form be expressed by $y = f(x)$. The elastic strain energy is

$$U = \frac{1}{2} E \int_0^L I_x \left(\frac{d^2 y}{dx^2}\right)^2 dx \tag{10.50}$$

Taking the undeflected position as datum, the potential energy in the buckled form is

$$V = -P\Delta L$$

To calculate ΔL, we observe from Fig. (10.20(b))

$$\Delta L = \int_0^L (\Delta s - \Delta x)$$

But

$$\Delta S = (\Delta x^2 + \Delta y^2)^{1/2} \approx \Delta x + \frac{1}{2}\left(\frac{\Delta y}{\Delta x}\right)^2 \Delta x$$

Hence,

$$P\Delta L = \frac{1}{2} P \int_0^L \left(\frac{dy}{dx}\right)^2 dx \tag{10.51}$$

The total potential energy is, therefore, given by

$$U + V = \frac{1}{2} E \int_0^L I_x \left(\frac{d^2 y}{dx^2}\right)^2 dx - \frac{1}{2} P \left(\frac{dy}{dx}\right)^2 dx \tag{10.52}$$

For equilibrium, the variation of the above quantity should vanish, i.e.

$$\delta(U + V) = \delta\left[\frac{1}{2} E \int_0^L I_x \left(\frac{d^2 y}{dx^2}\right)^2 dx - \frac{1}{2} P \int_0^L \left(\frac{dy}{dx}\right)^2 dx\right] = 0 \tag{10.53}$$

The above equation permits the determination of the function $y = f(x)$ by applying the technique of the calculus of variations. While the mathematical procedure will not be discussed here, it may be mentioned that the final result agrees with the differential equation given by Eq. (10.31a). While this differential equation can be derived from static equilibrium considerations as was done in the derivation of Eq. (10.31a), the merit of the energy criterion for the solution of stability problems becomes evident in the Rayleigh–Ritz method discussed in the next section.

10.15 THE RAYLEIGH–RITZ METHOD

A direct solution to the extremum problem stated by Eq. (10.49) is obtained by the Rayleigh–Ritz method, revealing the importance of the energy criterion. We shall demonstrate the method with respect to the buckling problem discussed in the previous section. The deflection of the buckled column is expressed in the form of a finite series:

$$y = a_1\phi_1 + a_2\phi_2 + \dots + a_n\phi_n \tag{10.54}$$

The ϕ terms are a set of arbitrarily chosen functions of x, such that each term satisfies the prescribed boundary conditions of the column. These are called co-ordinate functions. The coefficients a correspond to a set of parameters, as yet undetermined. With the value of y as given by Eq. (10.54), the elastic strain energy and the potential energy can be calculated, using Eqs (10.50) and (10.51). These lead to an expression involving the n parameters a, and having the form

$$U + V = F_1(a_1, a_2, \dots, a_n) - PF_2(a_1, a_2, \dots, a_n) \tag{10.55}$$

in which F_1 and F_2 are quadratic forms of the parameters a. If y, as given by Eq. (10.54), is to be a solution of the problem, then the parameters a must be chosen so as to make the total potential energy an extremum [Eq. (10.55)]. The problem has, therefore, been reduced to the familiar maximum–minimum problem involving the parameters $a_1, a_2, \dots, a_n$. Hence, the conditions become

$$\frac{\partial(U+V)}{\partial a_i} = 0, \quad (i = 1, 2, \dots, n) \tag{10.56}$$

i.e.

$$\frac{\partial F_1}{\partial a_1} - P\frac{\partial F_2}{\partial a_1} = 0$$

$$\frac{\partial F_1}{\partial a_2} - P\frac{\partial F_2}{\partial a_2} = 0 \tag{10.57}$$

$$\frac{\partial F_1}{\partial a_n} - P\frac{\partial F_2}{\partial a_n} = 0$$

The above set of equations involve only linear functions since these are derivatives of the quadratic expressions involved in $U + V$. Since Eq. (10.57) is a set of homogeneous equations, for the existence of a non-trivial solution, the determinant of the coefficients should be equal to zero, i.e.

$$\Delta = 0 \tag{10.58}$$

This is an equation of degree n in the unknown P and is the stability condition from which P can be determined. The smallest of the roots gives the critical load P_{cr}.

Introducing $P = P_{cr}$ in Eq. (10.57), a set of n linear homogeneous equations is obtained from which the ratios of the parameters a can be determined. Calling

$$\frac{a_2}{a_1} = \alpha_2, \qquad \frac{a_3}{a_1} = \alpha_3, \ldots, \qquad \frac{a_n}{a_1} = \alpha_n$$

the buckling mode is obtained from Eq. (10.54) as

$$y = a_1(\phi_1\alpha_1 + \phi_2\alpha_2 + \ldots + \phi_n\alpha_n) \tag{10.59}$$

The importance of the Rayleigh–Ritz method lies in the fact that it offers a method of obtaining an approximate solution to the buckling problem. The method, in many cases, involves less labour than is involved in solving the differential equation and the associated eigenvalue (i.e. characteristic value problem) . In the majority of cases, a few terms of the series in Eq.(10.54) give a sufficiently accurate result. Success or failure in applying the Rayleigh–Ritz method to any problem depends largely on the proper choice of the coordinate functions. In the majority of cases, satisfactory results can be obtained only when the coordinate functions chosen form a system of orthogonal functions discussed in Sec. 10.10. This is the reason why Fourier Series play such an important role in the applications of the Rayleigh–Ritz method.

Example 10.5 *Consider a pin-ended column subjected to an axial compressive load P, as shown in Fig. 10.20. Assume that the buckled shape is given by*

$$y = a \sin \frac{\pi x}{L}$$

where a is an unknown parameter. The coordinate function chosen satisfies the boundary conditions which are

$$y = 0 \quad \text{at } x = 0 \quad \text{and} \quad \text{at } x = L$$

$$\frac{d^2y}{dx^2} = 0 \quad \text{at } x = 0 \quad \text{and} \quad \text{at } x = L$$

Solution From Eq. (10.50), the strain energy is obtained as

$$U = \frac{1}{2} EI \int_0^L \left(\frac{d^2y}{dx^2}\right)^2 dx$$

$$= \frac{1}{2} EI \int_0^L a^2 \left(\frac{\pi}{L}\right)^4 \sin^2 \frac{\pi x}{L} dx$$

$$= \frac{1}{4} \pi^4 a^2 \left(\frac{EI}{L^3}\right)$$

From Eq. (10.51), the potential energy is obtained as

$$V = -\frac{1}{2} P \int_0^L \left(\frac{dy}{dx}\right)^2 dx$$

$$= -\frac{1}{2} P \int_0^L a^2 \left(\frac{\pi}{L}\right)^2 \cos^2 \frac{\pi x}{L} dx$$

$$= -\frac{1}{4} P\pi^2 \left(\frac{a^2}{L}\right)$$

Thus, the total potential energy is

$$U + V = \frac{1}{4} \pi^4 a^2 \frac{EI}{L^3} - \frac{1}{4} P\pi^2 \frac{a^2}{L}$$

For this to be an extremum, we must have

$$\frac{1}{2} \pi^4 a \frac{EI}{L^3} - \frac{1}{2} P\pi^2 \frac{a}{L} = 0$$

or
$$\frac{1}{2} \pi^2 \frac{a}{L} \left(\pi^2 \frac{EI}{L^2} - P\right) = 0$$

The non-trivial solution is obtained when

$$P = P_{cr} = \frac{\pi^2 EI}{L^2}$$

We have been able to obtain an exact solution since the coordinate function we used happens to give the exact deflected shape for the column.

Example 10.6 *Consider a column fixed at one end and free at the other end (Fig. 10.21). It is subjected to a compressive load P at the free end. Determine the approximate critical load assuming the deflection curve as*

$$y = a_1 \left(\frac{x}{L}\right)^2 + a_2 \left(\frac{x}{L}\right)^3$$

Fig. 10.21 *Example 10.6*

Solution The boundary conditions are

$$y = 0 \quad \text{at } x = 0, \qquad \frac{dy}{dx} = 0 \quad \text{at } x = 0$$

and
$$\frac{d^2 y}{dx^2} = 0 \quad \text{at } x = L$$

(i) Let us ignore the last condition for the time being. The first two conditions are satisfied by the coordinate function. The strain energy is equal to

$$U = \frac{1}{2} EI \int_0^L \left(\frac{d^2 y}{dx^2}\right)^2 dx$$

$$= \frac{1}{2} EI \int_0^L \left(\frac{2a_1}{L^2} + \frac{6a_2 x}{L^3}\right)^2 dx$$

$$= \frac{2EI}{L^3}\left(a_1^2 + 3a_1 a_2 + 3a_2^2\right)$$

The potential energy is to equal to

$$V = -\frac{1}{2} P \int_0^L \left(\frac{dy}{dx}\right)^2 dx$$

$$= -\frac{1}{2} P \int_0^L \left(\frac{2a_1 x}{L^2} + \frac{3a_2 x^2}{L^3}\right)^2 dx$$

$$= -\frac{P}{30L}\left(20a_1^2 + 45a_1 a_2 + 27a_2^2\right)$$

Hence, the total potential energy is

$$U + V = a_1^2\left(\frac{2EI}{L^3} - \frac{2P}{3L}\right) + a_1 a_2\left(\frac{6EI}{L^3} - \frac{3P}{2L}\right) + a_2^2\left(\frac{6EI}{L^3} - \frac{9P}{10L}\right)$$

For an extremum we should have

$$\frac{\partial(U+V)}{\partial a_1} = \left(\frac{4EI}{L^3} - \frac{4P}{3L}\right) a_1 + \left(\frac{6EI}{L^3} - \frac{3P}{2L}\right) a_2 = 0$$

$$\frac{\partial(U+V)}{\partial a_2} = \left(\frac{6EI}{L^3} - \frac{3P}{2L}\right) a_1 + \left(\frac{12EI}{L^3} - \frac{9P}{5L}\right) a_2 = 0$$

For the existence of a non-trivial solution, the determinant D of the coefficients should be equal to zero. Hence,

$$\Delta = \left(\frac{4EI}{L^3} - \frac{4P}{3L}\right)\left(\frac{12EI}{L^3} - \frac{9P}{5L}\right) - \left(\frac{6EI}{L^3} - \frac{3P}{2L}\right)^2 = 0$$

or

$$3P^2L^4 - 104PL^2EI + 240E^2I^2 = 0$$

Therefore,

$$P = 2.49\frac{EI}{L^2} \quad \text{or} \quad 32.18\frac{EI}{L^2}$$

The smaller value is

$$P_{cr} = 2.49\frac{EI}{L^2}$$

Compared to the exact value $\frac{\pi^2 EI}{4L^2}$, the error is only +0.92%.

(ii) In the above analysis, we have ignored the third boundary condition, i.e. at $x = L, \dfrac{d^2y}{dx^2} = 0$. If we use this condition

$$\frac{d^2y}{dx^2}(\text{at } x = L) = \frac{2a_1}{L^2} + \frac{6a_2L}{L^3} = 0$$

or

$$a_1 = -3a_2$$

Using this

$$y = a_1\left(\frac{x}{L}\right)^2 - \frac{1}{3}a_1\left(\frac{x}{L}\right)^3$$

Substituting $a_1 = -3a_2$ in the expressions for U and V

$$U = \frac{2EI}{L^2}\left(a_1^2 - a_1^2 + \frac{a_1^2}{3}\right) = \frac{2}{3}\frac{EI}{L^3}a_1^2$$

and

$$V = -\frac{P}{L}\left(\frac{2}{3}a_1^2 - \frac{1}{2}a_1^2 + \frac{1}{10}a_1^2\right) = -\frac{4}{15}\frac{P}{L}a_1^2$$

Therefore,

$$U + V = \left(\frac{2}{3}\frac{EI}{L^3} - \frac{4}{15}\frac{P}{L}\right)a_1^2$$

For an extremum, the quantity inside the parentheses should be equal to zero, i.e.

$$\frac{4}{15}\frac{P}{L} = \frac{3}{4}\frac{EP}{L^3}$$

or

$$P_{cr} = 2.5\frac{EI}{L^2}$$

which is almost identical with the previous solution but the solution has been obtained with comparative ease.

10.16 TIMOSHENKO'S CONCEPT OF SOLVING BUCKLING PROBLEMS

Consider a straight column subjected to an axial load P. If P is less than the critical load, then the column is in stable equilibrium, which means that if the column is slightly displaced from its straight equilibrium position by any transverse disturbing force, it will return to its vertical position as soon as the disturbing force is removed. In terms of energy, this means that when P is less than P_{cr}, in the slightly bent configuration, the elastic strain energy stored in the bent column is greater than the work done by the axial load in moving through a distance ΔL, i.e.

$$U - W > 0 \tag{10.60}$$

where U is the strain energy and $W = P\,\Delta L$. On the other hand, when P exceeds P_{cr}, if the column is slightly displaced, the work done by the external load P will exceed the strain energy in bending and the equilibrium becomes unstable. Consequently, the condition

$$U - W = 0 \tag{10.61}$$

characterises the state when the equilibrium configuration changes from stable to unstable.

Following the same procedure as in the Rayleigh–Ritz method, we can assume that the buckled column curve can be expressed by Eq. (10.54) as

$$y = a_1\phi_1 + a_2\phi_2 + \ldots + a_n\phi_n$$

The ϕ terms are functions of x so that each term satisfies the boundary conditions of the column. The constants a_1, a_2,..., define the amplitudes of the terms. The strain energy is given by

$$U = \frac{1}{2} EI \int_0^L \left(\frac{d^2y}{dx^2}\right)^2 dx = F_1\,(a_1, a_2, \ldots, a_n)$$

The work done by the external force during deformation is from Eq. (10.51)

$$W = \frac{1}{2} P \int_0^L \left(\frac{dy}{dx}\right)^2 dx = PF_2\,(a_1, a_2, \ldots, a_n)$$

Using Eq. (10.61)

$$P = EI \frac{\int_0^L (d^2y/dx^2)^2\,dx}{\int_0^L (dy/dx)^2\,dx} = \frac{F_1\,(a_1, a_2, \ldots, a_n)}{F_2\,(a_1, a_2, \ldots, a_n)} \tag{10.62}$$

Observing that for a pin-ended column or a column with one end free

$$M = -EI\frac{d^2y}{dx^2}$$

and that $\qquad M = Py$

and $\qquad EI\int\left(\frac{d^2y}{dx^2}\right)^2 dx = \frac{P^2}{EI}\int y^2\,dx$

Eq. (10.62) can also be written as

$$P = \frac{EI\int_0^L (dy/dx)^2\,dx}{\int_0^L y^2\,dx} \tag{10.63}$$

Since we need the minimum value for the load P, the critical load is obtained when the expression in Eq. (10.62) or Eq. (10.63) is made a minimum. This requires that the derivatives of Eq. (10.62) or Eq. (10.63) with respect to each coefficient a_i must vanish. This yields

$$\frac{\partial F_1}{\partial a_i} - P\frac{\partial F_2}{\partial a_i} = 0, \quad (i = 1, 2, \ldots) \tag{10.64}$$

These are identical to Eq. (10.57). Since there are n homogeneous equations, a non-trivial solution exists when the determinant of the coefficients is equal to zero. This discussion is identical to that given in Sec. 10.15.

There is a fundamental difference between Eqs (10.62) and (10.63) though they appear to be equivalent. The elastic strain energy is obtained from the expression

$$U = \frac{1}{2EI}\int_0^L M^2\, dx$$

If we take the deflection curve as $y = y\,(x)$, then M could be expressed in two ways

$$M = -EI\frac{d^2 y}{dx^2} \tag{10.65}$$

or

$$M = Py$$

where P is the axial force acting on a pin-ended column or a column with one end free end the other end fixed. If we use the first expression in Eq. (10.65), we get the strain energy for any assumed form of the column. If we use the second expression, we take the external force also into account and consequently, the final result obtained for P_{cr} using Eq. (10.63) gives a slightly more accurate result. Equation (10.62) is generally referred to as the Rayleigh–Ritz formula and Eq. (10.63) as the Timoshenko formula.

10.17 COLUMNS WITH VARIABLE CROSS-SECTIONS

So far, in the examples considered, we have treated the moment of inertia I as independent of x. We shall consider a few problems where I varies with x. The energy method will be found to be very suitable to obtain fairly good solutions.

Example 10.7 *Consider a column with the moment of inertia of the cross-sectional area varying according to the equation*

$$I = I_0\left(1 + \sin\frac{\pi x}{L}\right)$$

Solution The column is hinged at both ends. Assume that the deflection curve can be represented by the series

$$y = \Sigma a_n \sin\frac{n\pi x}{L}$$

Since the deflection curve must be symmetrical with respect to the middle point of the column (because the moment of inertia is symmetrical about the middle point), the even parameters in the above series vanish. The deflection equation then becomes

$$y = a_1 \sin\frac{\pi x}{L} + a_3 \sin\frac{3\pi x}{L} + a_5 \sin\frac{5\pi x}{L} + \ldots$$

We shall consider only two terms of the series. Thus

$$y = a_1 \sin\frac{\pi x}{L} + a_3 \sin\frac{3\pi x}{L}$$

$$U = \frac{1}{2}\int_0^L EI\left(\frac{d^2y}{dx^2}\right)^2 dx$$

$$= \frac{1}{2}\int_0^L EI_0\left(1+\sin\frac{\pi x}{L}\right)\left[-a_1\left(\frac{\pi}{L}\right)^2 \sin\frac{\pi x}{L} - a_3\left(\frac{3\pi}{L}\right)^2 \sin\frac{3\pi x}{L}\right]^2 dx$$

$$= \frac{1}{2} EI_0\left(\frac{\pi}{L}\right)^3\left[\left(\frac{4}{3}+\frac{\pi}{2}\right)a_1^2 - \frac{24}{5}a_1a_3 + \left(\frac{2916}{35}+\frac{81\pi}{2}\right)a_3^2\right]$$

and

$$\Delta L = \frac{1}{2}\int_0^L\left(\frac{dy}{dx}\right)^2 dx = \frac{1}{2}\left(\frac{\pi}{L}\right)^2\left(\frac{L}{2}a_1^2 + \frac{L}{2}\times 9a_3^2\right)$$

Substituting in Eq. (10.62)

$$P = EI_0\,\frac{\pi}{L^2}\,\frac{(8/3+\pi)a_1^2 - (48/5)\,a_1a_3 + (5832/35) + 81\,\pi)a_3^2}{a_1^2 + 9a_3^2} = \frac{F_1}{F_2}$$

For minimum P, we should have from Eq. (10.64)

$$\frac{\partial}{\partial a_1}(F_1 - PF_2) = 0 \quad \text{and} \quad \frac{\partial}{\partial a_2}(F_1 - PF_2) = 0$$

Thus,

$$2\left(\frac{8}{3}+\pi-P^*\right)a_1 - \frac{48}{5}a_3 = 0$$

and

$$-\frac{48}{5}a_1 + 2\left(\frac{5832}{35}+81\pi-9P^*\right)a_3 = 0$$

where $P^* = \dfrac{PL^2}{\pi EI_0}$

For the existence of a non-trivial solution, the determinant of the coefficients should be equal to zero. This gives

$$\Delta = 9P^{*2} - \left(\frac{5832}{35}+81\pi+24+9\pi\right)P^*$$

$$+\left(\frac{8}{3}+\pi\right)\left(\frac{5832}{35}+81\pi\right) - \left(\frac{24}{5}\right)^2 = 0$$

Solving,

$$P^* = 5.746 \quad \text{or} \quad P = 18.05\,\frac{EI_0}{L^2}$$

10.18 USE OF TRIGONOMETRIC SERIES

In many instances, it will be useful to represent the deflection curve in the form of a trigonometric series. We have discussed in Sec. 10.11 that the functions satisfying Eqs (10.31a) and (10.31b) also satisfy orthogonality conditions. The trigonometric series which we shall consider now is made up of such functions. Let the deflection curve be represented by

$$y = a_1 \sin\frac{\pi x}{L} + a_2 \sin\frac{2\pi x}{L} + \ldots + a_n \sin\frac{n\pi x}{L} + \ldots \tag{10.66}$$

By properly determining the coefficients $a_1, a_2, \ldots$, the above series can be made to represent any deflection curve. These coefficients may be calculated by a consideration of the strain energy of the beam or the column. The strain energy is given by

$$U = \frac{1}{2} EI \int_0^L \left(\frac{d^2 y}{dx^2}\right)^2 dx$$

Now, $$\frac{d^2 y}{dx^2} = -a_1 \frac{\pi^2}{L^2}\sin\frac{\pi x}{L} - a_2 \frac{2^2\pi^2}{L^2}\sin\frac{2\pi x}{L} - a_3\frac{3^2\pi^2}{L^2}\sin\frac{3\pi x}{L} - \ldots$$

Hence, the square of the above expression will involve terms of two kinds

$$a_n^2 \frac{n^4\pi^4}{L^4}\sin^2\frac{n\pi x}{L}$$

and $$2a_m a_n \frac{n^2 m^2 \pi^4}{L^4}\sin\frac{n\pi x}{L}\sin\frac{m\pi x}{L}$$

By direct integration it can be seen that

$$\int_0^L \sin^2\frac{n\pi x}{L}\,dx = \frac{L}{2}, \quad \text{and} \quad \int_0^L \sin\frac{n\pi x}{L}\sin\frac{m\pi x}{L}\,dx = 0, \qquad \text{for } n \neq m$$

These are the orthogonality relations expressed by Eqs (10.42) and (10.43). Hence, in the expression for strain energy, terms containing products like $a_m\, a_n$ vanish and only terms like a_n^2 remain. Then

$$U = \frac{1}{2} EI\left[a_1^2 \frac{\pi^4}{L^4}\frac{L}{2} + a_2^2\frac{2^4\pi^4}{L^4}\frac{L}{2} + a_3^2\frac{3^4\pi^4}{L^4}\frac{L}{2} + \ldots\right]$$

$$= \frac{EI\pi^4}{4L^3}\left(1\,a_1^2 + 2^4 a_2^2 + 3^4 a_3^2 + \ldots\right)$$

$$= \frac{EI\pi^4}{4L^3}\sum_{n=1}^{\infty} n^4 a_n^2 \tag{10.67}$$

Similarly, if we consider the expression

$$\Delta L = \frac{1}{2}\int_0^L \left(\frac{dy}{dx}\right)^2 dx$$

we find the integrand to consist of two kinds of terms

$$a_n^2 \frac{n^2\pi^2}{L^2} \cos^2 \frac{n\pi x}{L}$$

and
$$2a_m a_n \frac{mn\pi^2}{L^2} \cos \frac{m\pi x}{L} \cos \frac{n\pi x}{L}$$

By direct integration it can be shown that

$$\int_0^L \cos^2 \frac{n\pi x}{L} = \frac{L}{2}$$

$$\text{and} \int_0^L \cos^2 \frac{n\pi x}{L} \cos \frac{m\pi x}{L} dx = 0 \quad \text{for } n \neq m$$

Consequently,

$$\Delta L = \frac{\pi^2}{4L} \sum_{n=1}^{\infty} n^2 a_n^2 \tag{10.68}$$

With these expressions for U and ΔL, we can consider the following example.

Example 10.8 *A beam column is subjected to an axial force P and a lateral force Q at x = c (Fig. 10.22). Determine the deflection curve using the energy method.*

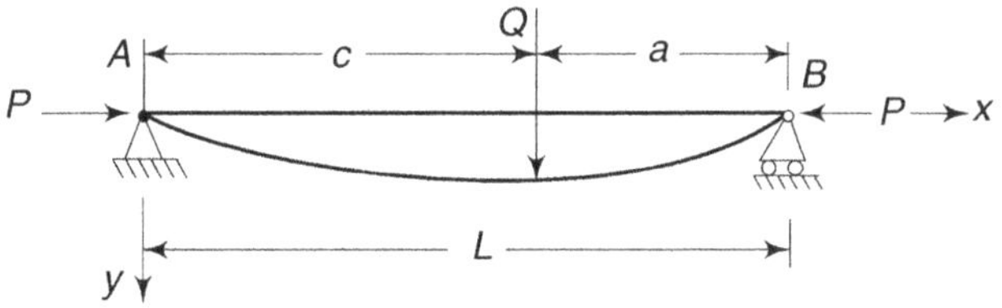

Fig. 10.22 *Example 10.8*

Solution Let the deflection curve be

$$y = a_1 \sin \frac{\pi x}{L} + a_2 \sin \frac{2\pi x}{L} + \ldots$$

Let a virtual displacement δy_n be given. This virtual displacement is obtained by changing one of the terms a_n sin $(n\pi x/L)$ to $(a_n + \delta a_n)$ sin $(n\pi x/L)$. In other words, the deflection curve δa_n sin $(n\pi x/L)$ is superimposed on the original deflection curve. The work done by the external forces Q and P is

$$\delta W = Q\, \delta a_n \sin \frac{n\pi c}{L} + P\delta\,(\Delta L)$$

Using Eq. (10.68)

$$\delta W = Q\, \delta a_n \sin \frac{n\pi c}{L} + P \frac{\pi^2}{4L} 2n^2 a_n \delta_n$$

The increase in strain energy is

$$\delta U = \frac{\partial U}{\partial a_n} \delta a_n$$

From Eq. (10.67)

$$\delta U = \frac{EI\pi^4}{2L^3} n^4 a_n \, \delta a_n$$

Since the increase in strain energy should be equal to the work done, we have

$$Q \sin \frac{n\pi c}{L} \delta a_n + P \frac{\pi^2}{2L} n^2 a_n \, \delta a_n = \frac{EI\pi^4}{2L^3} n^4 a_n \, \delta a_n$$

from which,

$$a_n = \frac{2QL^3}{EI\pi^4} \frac{1}{\left(n^2 - \dfrac{PL^2}{EI\pi^2}\right)} \sin \frac{n\pi c}{L}$$

If we use the notation

$$\beta = \frac{PL^2}{EI\pi^2}$$

then, $$a_n = \frac{2QL^3}{EI\pi^4} \frac{1}{n^2 (n^2 - \beta)} \sin \frac{n\pi c}{L}$$

The deflection curve is, therefore, given by

$$y = \frac{2QL^3}{EI\pi^4} \sum_{n=1}^{\infty} \frac{1}{n^2 (n^2 - \beta)} \sin \frac{n\pi c}{L} \sin \frac{n\pi x}{L}$$

Example 10.9 *Using an infinite series, determine the deflection curve for the beam column shown in Fig. 10.23.*

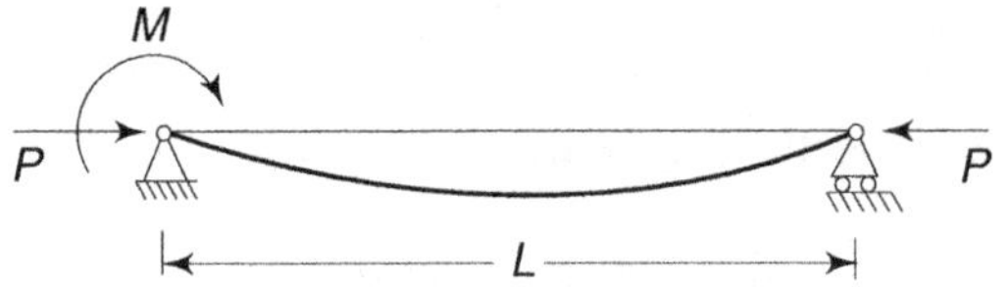

Fig. 10.23 *Example 10.9*

Solution In the solution of the previous example consider c to be very small.

Then $$\sin \frac{n\pi c}{L} \approx \frac{n\pi c}{L}$$

and $$y = \frac{2L^3}{EI\pi^4} \frac{\pi}{L} Qc \sum_{n=1}^{\infty} \frac{n}{n^2 (n^2 - \beta)} \sin \frac{n\pi x}{L}$$

Let $c \to 0$ and $Q \to \infty$, such that $Qc = M =$ constant.

Then, $$y = \frac{2ML^2}{EI\pi^3} \sum_{n=1}^{\infty} \frac{1}{n(n^2 - \beta)} \sin \frac{n\pi x}{L}$$

Problems

10.1 A uniformly loaded beam is built-in at one end and simply supported at the other end. It is subjected to an axial force P. Determine the moment M_b at the built-in end (Fig. 10.24).

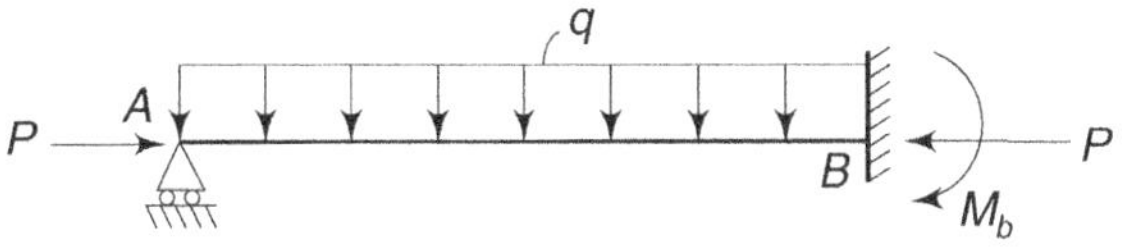

Fig. 10.24 *Problem 10.1*

$$\left[\begin{aligned} &\textit{Ans.} \quad M_b = +\frac{qL^2}{8}\frac{\beta(u)}{\phi(u)} \\ &\text{where, } \beta(u) = \frac{3}{u^3}(\tan u - u) \\ &\phi(u) = \frac{3}{2u}\left(\frac{1}{2u} - \frac{1}{\tan 2u}\right) \end{aligned}\right]$$

10.2 A beam of uniform cross-section has an initial curvature given by the equation

$$y_0 = \delta \sin \frac{\pi x}{L}$$

It is subjected to end couples M_a and M_b and to an axial force P (Fig. 10.25). Determine the deflection curve.

Fig. 10.25 *Problem 10.2*

$$\left[\textit{Ans.}\ y = -\frac{M_a}{P}\left[\frac{L-x}{L} - \frac{\sin k(L-x)}{\sin kL}\right] + \frac{M_b}{P}\left[\frac{\sin kx}{\sin kL} - \frac{x}{L}\right] + \frac{\delta\pi^2}{\pi^2 - k^2L^2}\sin\frac{\pi x}{L}\right]$$

10.3 The initial shape of a bar can be approximated by the series

$$y = \delta_1 \sin \frac{\pi x}{L} + \delta_2 \sin \frac{2\pi x}{L} + \ldots$$

If the bar is simply supported and subjected to axial force P only, show that the deflection curve due to P is given by

$$y_1 = \alpha\left(\frac{\delta_1}{1-\alpha}\sin\frac{\pi x}{L} + \frac{\delta_2}{2^2-\alpha}\sin\frac{2\pi x}{L} + \ldots\right), \text{ where } \alpha = \frac{k^2L^2}{\pi^2}$$

10.4 For a column with one end built-in and the other end free and carrying an axial load P, it is assumed that the deflection curve has the form

$$y = \frac{\delta x^2}{L^2}$$

where L is the length of the column and x is measured from the fixed end. Using the energy method, determine the critical load.

$$\left[Ans.\ P_{cr} = 2.5 \frac{EI}{L^2} \right]$$

10.5 The deflection curve for a pin-ended column is represented by a polynomial as

$$y = ax^4 + bx^3 + cx^2 + dx + e$$

Determine the critical load by the energy method.

$$\left[Ans.\ P_{cr} = 9.88 \frac{EI}{L^2} \right]$$

10.6 A prismatic bar with hinged ends (Fig. 10.26) is subjected to the action of a uniformly distributed axial load of intensity q and an axial compressive force P. Find the critical value of P by assuming, for the deflection curve, the equation

$$y = \delta \sin \frac{\pi x}{L}$$

$$\left[Ans.\ P_{cr} = \frac{\pi^2 EI}{L^2} - \frac{qL}{2} \right]$$

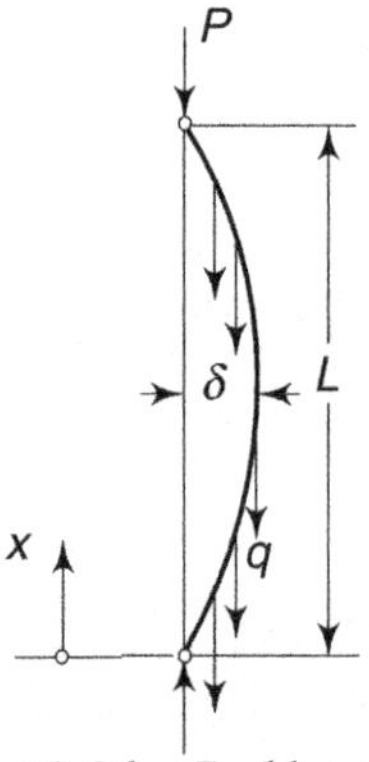

Fig. 10.26 *Problem 10.6*

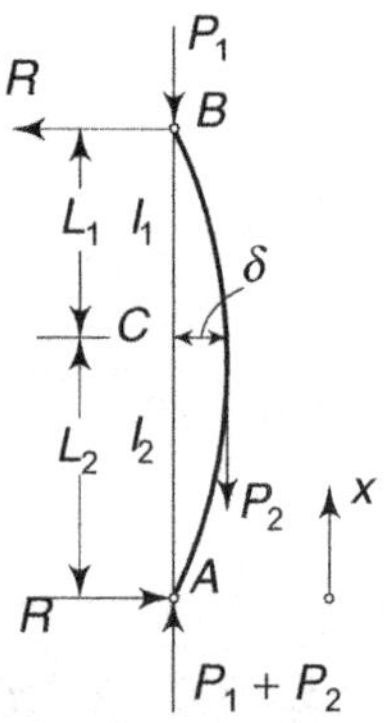

Fig. 10.27 *Problem 10.7*

10.7 Determine the critical load $(P_1 + P_2)$ by the energy method for the case shown in Fig. 10.27. The column has a moment of inertia I_1 for half the length and moment of inertia I_2 for the other half.

Assume the deflection curve in the form

$$y = \delta \sin \frac{\pi x}{L}$$

[*Ans.* $(P_1 + P_2)_{cr} =$

$$\frac{(\pi^2 EI_2 / L^2)(m+1)}{m + \frac{m}{6}\left(\frac{m-1}{m}\right)^2 - \frac{8}{\pi^2}(m-1) + n\left[\frac{1}{m} + \frac{m}{6}\left(\frac{m-1}{m}\right)^2 + \frac{8}{\pi^2}\frac{m-1}{m}\right]}$$

where $m = \frac{P_1 + P_2}{P_1}$ and $n = \frac{I_2}{I_1}$]

CHAPTER

11 Introduction to Composite Materials

11.1 INTRODUCTION

Till now, we have been considering materials that are homogeneous and isotropic. These qualifications are with respect to their elastic properties. If the properties are the same at every point in the body, then it is said to be homogeneous. Isotropy implies that the properties are independent of directions. Materials like steel, copper, aluminium, etc. are both homogeneous and isotropic. However, wood is homogeneous but is not isotropic since its strength along the fibres is greater than its strength in a direction transverse to the fibres. Materials that are not isotropic are called anisotropic materials. At any point of such a body, the elastic properties are different for different directions. Directions for which the elastic properties are the same are said to be elastically equivalent. Generally speaking, one would like to use materials that are suitable to specific applications. For example, a cable wire or rope that is used for hauling purposes needs to have the required tensile strength in the direction of the cable. A structure built up of bricks is good to carry compressive loads. A reinforced concrete beam with steel reinforcement at the bottom is good to carry bending loads which will induce compressive stresses in concrete and tensile stresses in the steel reinforcement. This is an example of a composite material that is designed for a specific operation. On the other hand, consider a bundle of glass fibres or carbon filaments, which is useless when used as an engineering structure. It has no shape and no defined hard surface for machining purposes. The bundle can resist tensile forces, but it is useless for compressive, bending and torsional forces. But, when the same bundle of filaments or fibres is dipped into a bath of resin, drained and allowed to harden, it behaves as a new material possessing properties that are comparable to those of steel or other metals, and can resist forces in tension, compression and bending. It has a definite shape, a durable surface and it can be machined. Such a material is called a composite material.

Generally speaking, composites are produced when two or more materials are joined to give a combination of properties that cannot be attained in the original materials. Composites can be placed into three categories—particulate, fibre and laminar—based on the shapes of the constituent materials. Concrete, a mixture of cement and gravel, is a particulate composite; fiberglass, containing glass fibres embedded in a polymer, is a fibre-reinforced composite; and plywood, having alternating layers of wood veneer, is a laminar composite. In this chapter, we shall focus our attention mainly on fibre-reinforced composites and laminates.

11.2 STRESS–STRAIN RELATIONS

In Chapter 3, it was stated that for a linearly elastic body, the stresses are linearly related to the strains and are given by

$$\begin{aligned}
\sigma_x &= a_{11}\,\varepsilon_x + a_{12}\,\varepsilon_y + a_{13}\,\varepsilon_z + a_{14}\,\gamma_{xy} + a_{15}\,\gamma_{yz} + a_{16}\,\gamma_{zx} \\
\sigma_y &= a_{21}\,\varepsilon_x + a_{22}\,\varepsilon_y + a_{23}\,\varepsilon_z + a_{24}\,\gamma_{xy} + a_{25}\,\gamma_{yz} + a_{26}\,\gamma_{zx} \\
\sigma_z &= a_{31}\,\varepsilon_x + a_{32}\,\varepsilon_y + a_{33}\,\varepsilon_z + a_{34}\,\gamma_{xy} + a_{35}\,\gamma_{yz} + a_{36}\,\gamma_{zx} \\
\tau_{xy} &= a_{41}\,\varepsilon_x + a_{42}\,\varepsilon_y + a_{43}\,\varepsilon_z + a_{44}\,\gamma_{xy} + a_{45}\,\gamma_{yz} + a_{46}\,\gamma_{zx} \\
\tau_{yz} &= a_{51}\,\varepsilon_x + a_{52}\,\varepsilon_y + a_{53}\,\varepsilon_z + a_{54}\,\gamma_{xy} + a_{55}\,\gamma_{yz} + a_{56}\,\gamma_{zx} \\
\tau_{zx} &= a_{61}\,\varepsilon_x + a_{62}\,\varepsilon_y + a_{63}\,\varepsilon_z + a_{64}\,\gamma_{xy} + a_{65}\,\gamma_{yz} + a_{66}\,\gamma_{zx}
\end{aligned} \tag{11.1}$$

Assuming that the sixth-order determinant of the coefficients a_{ij}s in Eq. (11.1) is not zero, one can solve for $\varepsilon_x, \varepsilon_y, \ldots, \gamma_{zx}$ in terms of $\sigma_x, \sigma_y, \ldots, \tau_{zx}$. The expressions for the strain components will then be

$$\begin{aligned}
\varepsilon_x &= b_{11}\,\sigma_x + b_{12}\,\sigma_y + b_{13}\,\sigma_z + b_{14}\,\tau_{xy} + b_{15}\,\tau_{yz} + b_{16}\,\tau_{zx} \\
\varepsilon_y &= b_{21}\,\sigma_x + b_{22}\,\sigma_y + b_{23}\,\sigma_z + b_{24}\,\tau_{xy} + b_{25}\,\tau_{yz} + b_{26}\,\tau_{zx} \\
\varepsilon_z &= b_{31}\,\sigma_x + b_{32}\,\sigma_y + b_{33}\,\sigma_z + b_{34}\,\tau_{xy} + b_{35}\,\tau_{yz} + b_{36}\,\tau_{zx} \\
\gamma_{xy} &= b_{41}\,\sigma_x + b_{42}\,\sigma_y + b_{43}\,\sigma_z + b_{44}\,\tau_{xy} + b_{45}\,\tau_{yz} + b_{46}\,\tau_{zx} \\
\gamma_{yz} &= b_{51}\,\sigma_x + b_{52}\,\sigma_y + b_{53}\,\sigma_z + b_{54}\,\tau_{xy} + b_{55}\,\tau_{yz} + b_{56}\,\tau_{zx} \\
\gamma_{zx} &= b_{61}\,\sigma_x + b_{62}\,\sigma_y + b_{63}\,\sigma_z + b_{64}\,\tau_{xy} + b_{65}\,\tau_{yz} + b_{66}\,\tau_{zx}
\end{aligned} \tag{11.2}$$

where the coefficients b_{ij}s are related to a_{ij}s. The stress–strain relations given by Eq. (11.1) contain altogether 36 elastic constants. However, this number can be reduced based on the material properties. Let us assume that there exists an elastic potential V such that

$$\sigma_x = \frac{\partial V}{\partial \varepsilon_x}, \qquad \sigma_y = \frac{\partial V}{\partial \varepsilon_y}, \qquad \ldots, \qquad \tau_{zx} = \frac{\partial V}{\partial \gamma_{zx}} \tag{11.3}$$

The physical meaning of the potential V will become clear soon. Assuming the existence of such a potential, from Eq. (11.3), one obtains

$$\frac{\partial \sigma_x}{\partial \varepsilon_y} = \frac{\partial^2 V}{\partial \varepsilon_x \partial \varepsilon_y} = \frac{\partial \sigma_y}{\partial \varepsilon_x}, \qquad \frac{\partial \sigma_x}{\partial \gamma_{xy}} = \frac{\partial^2 V}{\partial \varepsilon_x\, \partial \gamma_{xy}} = \frac{\partial \tau_{xy}}{\partial \varepsilon_x}, \qquad \text{etc.} \tag{11.4}$$

From Eqs. (11.4) and (11.1), one immediately gets

$$a_{12} = a_{21};\; a_{31} = a_{13}, \ldots, a_{65} = a_{56}$$

And, in general, based on the existence of such a potential,

$$a_{ij} = a_{ji} \quad (i, j = 1, 2, 3, \ldots, 6) \tag{11.5a}$$

Consequently, in Eq. (11.2)

$$b_{ij} = b_{ji} \quad (i, j = 1, 2, 3, \ldots, 6) \tag{11.5b}$$

As a result of this, i.e. $a_{ij} = a_{ji}$ and $b_{ij} = b_{ji}$, the 36 elastic constants a_{ij}s in Eq. (11.1) or the b_{ij}s in Eq. (11.2) get reduced to 21. In other words, since $a_{12} = a_{21}$, $a_{13} = a_{31}, \ldots, a_{16} = a_{61}$ and similarly, $b_{12} = b_{21}$, $b_{13} = b_{31}, \ldots, b_{16} = b_{61}$, the number of independent elastic constants in Eqs (11.1) and (11.2) gets reduced by15, resulting in 21 constants. So, for a general case of anisotropy, the number of independent elastic constants is 21. However, because of the symmetry of the material properties, it is claimed by material scientists that the number of independent elastic constants does not exceed 18.

Equation (11.6) gives the expression for V which can be verified by differentiating with respect to ε_x, ε_y, ε_z, etc., and comparing with Eq. (11.1). Thus,

$$\begin{aligned} V = {} & \frac{1}{2} a_{11} \varepsilon_x^2 + a_{12} \varepsilon_x \varepsilon_y + a_{13} \varepsilon_x \varepsilon_z + a_{14} \varepsilon_x \gamma_{xy} + a_{15} \varepsilon_x \gamma_{yz} + a_{16} \varepsilon_x \gamma_{zx} \\ & + \frac{1}{2} a_{22} \varepsilon_y^2 + a_{23} \varepsilon_y \varepsilon_z + a_{24} \varepsilon_y \gamma_{xy} + a_{25} \varepsilon_y \gamma_{yz} + a_{26} \varepsilon_y \gamma_{zx} \\ & + \frac{1}{2} a_{33} \varepsilon_z^2 + a_{34} \varepsilon_z \gamma_{xy} + a_{35} \varepsilon_z \gamma_{yz} + a_{36} \varepsilon_z \gamma_{zx} \\ & + \frac{1}{2} a_{44} \gamma_{xy}^2 + a_{45} \gamma_{xy} \gamma_{yz} + a_{46} \gamma_{xy} \gamma_{zx} \\ & + \frac{1}{2} a_{55} \gamma_{yz}^2 + a_{56} \gamma_{yz} \gamma_{zx} \\ & + \frac{1}{2} a_{66} \gamma^2 zx \end{aligned} \tag{11.6}$$

By differentiating Eq. (11.6) with respect to ε_x, ε_y, ε_z, etc., one gets expressions for σ_x, σ_y, σ_z, etc., thus verifying Eq. (11.3). The terms of Eq. (11.6) can be grouped to give for V an equivalent expression as

$$\begin{aligned} V = {} & \frac{1}{2} (a_{11} \varepsilon_x + a_{12} \varepsilon_y + a_{13} \varepsilon_z + a_{14} \gamma_{xy} + a_{15} \gamma_{yz} + a_{16} \gamma_{zx}) \, \varepsilon_x \\ & + \frac{1}{2} (a_{12} \varepsilon_x + a_{22} \varepsilon_y + a_{23} \varepsilon_z + \ldots + a_{16} \gamma_{zx}) \, \varepsilon_y \\ & + \frac{1}{2} (a_{13} \varepsilon_x + a_{23} \varepsilon_y + a_{33} \varepsilon_z + \ldots + a_{36} \gamma_{zx}) \, \varepsilon_z \\ & + \frac{1}{2} (a_{14} \varepsilon_x + a_{24} \varepsilon_y + a_{34} \varepsilon_z + a_{44} \gamma_{xy} + \ldots) \, \gamma_{xy} + \ldots \end{aligned} \tag{11.7}$$

Using Eq. (11.1), the elastic potential assumes the form

$$V = \frac{1}{2} (\sigma_x \varepsilon_x + \sigma_y \varepsilon_y + \sigma_z \varepsilon_z + \tau_{xy} \gamma_{xy} + \tau_{yz} \gamma_{yz} + \tau_{xz} \gamma_{zx}) \tag{11.8}$$

This is nothing but the elastic energy per unit volume at any point of the body, when forces are applied uniformly, i.e without dynamic effects.

Consider a body for which Eqs (11.1) and (11.2) are valid and let a uniform stress $\sigma_x = p$ be applied along the x-axis. The remaining stress components are zero. Then, the strain components as given by Eq. (11.2) are constant for every point, i.e

$$\begin{aligned} \varepsilon_x &= b_{11}p \quad & \varepsilon_y &= b_{21}p \quad & \varepsilon_z &= b_{31}p \\ \gamma_{xy} &= b_{41}p \quad & \gamma_{yz} &= b_{51}p \quad & \gamma_{zx} &= b_{61}\,p \end{aligned} \tag{11.9}$$

It becomes obvious that small segments passing through different points and parallel to x-axis get extended by the same amount. In general, all segments parallel to a given direction n and drawn through different points, undergo equal elongations. A homogeneous anisotropic body exhibiting this property is said to be rectilinearly anisotropic. For such a body, all parallel directions are elastically equivalent. Equations (11.9) show that elements of the same size in the form of rectangular parallelepipeds with respective parallel faces deform identically, no matter where they are located. However, it should be noted that, in general, rectangular parallelepipeds deform into oblique parallelepipeds having no right angles between the faces. The constants a_{ij}s appearing in Eq. (11.1) are sometimes called material constants or components of modulus. The constants b_{ij}s appearing in Eq. (11.2) are called components of compliance. In other words, material constants or components of modulus are used to determine stresses from strains, and components of compliance are used to get strains from stresses.

11.3 BASIC CASES OF ELASTIC SYMMETRY

The number of elastic constants involved either in Eq. (11.1) or Eq. (11.2) is 36, and these get reduced to 21 independent elastic constants under the assumption that an elastic potential V (equivalent to strain energy density function) exists. It was also stated that even in the most general case, the number of independent elastic constants (according to material scientists) does not exceed 18. For an isotropic body, it has been shown earlier that the number of independent elastic constants is only 2, these being the Lame's constants λ and μ, Eq. (3.4); or the engineering constants, E the Young's modulus and μ the Poisson's ratio. For an anisotropic body, the number of independent elastic constants get reduced depending on the type of symmetry that exists.

When a surface of revolution is rotated through any angle about the axis of revolution, the position of every point on the surface, but not on the axis, gets changed, but the position of the figure as a whole is not changed. In other words, the surface can be made to coincide with itself after an operation which changes the positions of some of its points. Any geometrical figure which can be brought into coincidence with itself after an operation which changes the position of any of its points is said to possess symmetry. A body which can be brought into coincidence with itself by a rotation about an axis, is said to possess an axis of symmetry. A body which after rotation can be brought into coincidence with itself by reflection in a plane, is said to possess a *plane of symmetry.*

Transversely Isotropic Consider a fibre-reinforced body in which the filaments are fairly long and are all oriented in the same direction, Fig. 11.1. Let the z-axis be parallel to the fibre elements and let the x and y axes lie in a plane perpendicular to the element orientation. If the fibres are uniformly distributed in the matrix, then it is obvious that the elastic properties at any point in the x–y plane, which is the

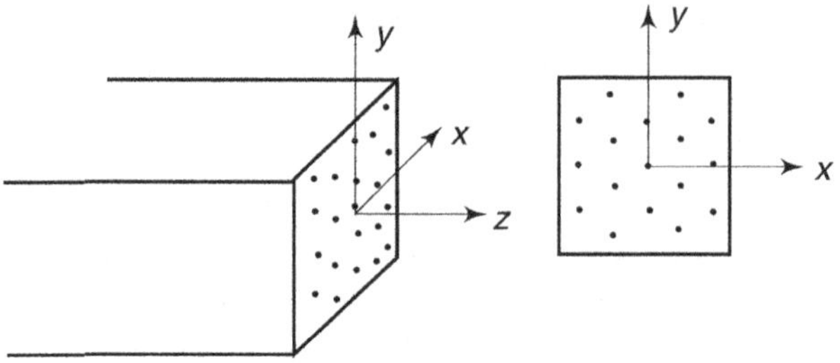

Fig. 11.1 *Transversely isotropic composite*

plane of symmetry, are independent of the directions in that plane. A body of this nature is said to be transversely isotropic. The axis normal to the plane of elastic symmetry is some times called the principal direction. It is assumed that no debonding between the fibres and matrix occurs when the body is stressed.

If a uniform force p is applied in the z direction, such that $\sigma_z = p$ and all other stress components are zero, then any rectangular parallelepiped with faces parallel to x, y and z planes will deform into a rectangular parallelepiped with equal lateral contraction (or extension) in x and y directions. Equations (11.2) can therefore be written as (using double subscripts for ε_x, ε_y and ε_z),

$$\varepsilon_{zz} = b_{33}\sigma_z, \qquad \varepsilon_{xx} = b_{13}\sigma_z, \qquad \varepsilon_{yy} = b_{23}\sigma_z = b_{13}\sigma_z, \qquad \gamma_{xy} = \gamma_{yz} = \gamma_{zx} = 0$$

Because of transverse isotropy, $b_{13} = b_{23}$ (i.e. transverse strains are equal). For a uniform stress σ_x, with the remaining stress components being zero, the strain components will be

$$\varepsilon_{xx} = b_{11}\sigma_x, \qquad \varepsilon_{yy} = b_{21}\sigma_x, \qquad \varepsilon_{zz} = b_{31}\sigma_x, \qquad \gamma_{xy} = \gamma_{yz} = \gamma_{zx} = 0$$

One should observe that the transverse strains ε_{yy} and ε_{zz} will not be equal.

A similar set of equations can be written with stress σ_y, and other stress components being zero. Now consider the shearing stresses. For a shearing stress τ_{xy} in the xy plane, the deformation of a rectangular parallelepiped will only be in the xy plane. Thus, for τ_{xy} alone (with other stress components being zero),

$$\varepsilon_{xx} = \varepsilon_{yy} = \varepsilon_{zz} = 0, \qquad \gamma_{xy} = b_{44}\tau_{xy}, \qquad \gamma_{yz} = \gamma_{zx} = 0$$

Now consider with τ_{yz} alone. The strain components will be

$$\varepsilon_{xx} = \varepsilon_{yy} = \varepsilon_{zz} = 0, \qquad \gamma_{yz} = b_{55}\tau_{yz}, \qquad \gamma_{zx} = \gamma_{xy} = 0$$

Similarly, with τ_{zx} alone, all strain components other than $\gamma_{zx} = b_{66}\,\tau_{zx}$ will be zero. In general, with all stress components σ_{ij} and τ_{ij} acting, the strain components ε_{ij} can be obtained by superposition of all the above expressions. Thus,

$$\begin{aligned}
\varepsilon_{xx} &= b_{11}\sigma_x + b_{12}\sigma_y + b_{13}\sigma_z \\
\varepsilon_{yy} &= b_{12}\sigma_x + b_{11}\sigma_y + b_{13}\sigma_z \\
\varepsilon_{zz} &= b_{31}\sigma_x + b_{31}\sigma_y + b_{33}\sigma_z \\
\gamma_{xy} &= b_{44}\tau_{xy} \\
\gamma_{yz} &= b_{55}\tau_{yz} \\
\gamma_{zx} &= b_{66}\tau_{zx}
\end{aligned} \tag{11.10}$$

In the above set of equations, there are eight constants. However, because of the reciprocal relations, $b_{13} = b_{31}$, and as a result, the number gets reduced to seven elastic constants. Also, as a result of the plane of symmetry, $b_{55} = b_{66}$. Further, the elastic constants in the plane of isotropy, i.e. b_{11}, b_{12} and b_{44} are also related. One can see this if Eq. (11.10) are written using the familiar engineering constants E,v and G, as follows.

$$\begin{aligned}
\varepsilon_{xx} &= \frac{1}{E_{xx}}\sigma_x - \frac{v_{xy}}{E_{yy}}\sigma_y - \frac{v_{xz}}{E_{zz}}\sigma_{zz} \\
\varepsilon_{yy} &= -\frac{v_{yx}}{E_{xx}} + \frac{1}{E_{yy}}\sigma_y - \frac{v_{yz}}{E_{zz}}\sigma_{zz} \\
\varepsilon_{zz} &= \frac{v_{zx}}{E_{xx}}\sigma_x - \frac{v_{zy}}{E_{yy}}\sigma_y + \frac{1}{E_{zz}}\sigma_{zz} \\
\gamma_{xy} &= \frac{1}{G_{xy}}\tau_{xy} \\
\gamma_{yz} &= \frac{1}{G_{yz}}\tau_{yz} \\
\gamma_{zx} &= \frac{1}{G_{yz}}\tau_{zx}
\end{aligned} \tag{11.11}$$

In these equations, v_{xy} is the Poisson's ratio in x direction due to a stress in y direction, i.e the ratio of lateral contraction in x direction to axial extension in y direction. This contraction is indicated by a negative sign. Slight variations exist in the subscript representation of Poisson's ratio from book to book. Also, the strain–stress equations are written in different ways. What is important is the reciprocal identity, i.e $b_{ij} = b_{ji}$.

Since xy plane is a plane of isotropy, $E_{xx} = E_{yy}$; $v_{yx} = v_{xy}$; $v_{zx} = v_{zy}$; $G_{yz} = G_{zx}$. Also, because of reciprocal identity,

$$\frac{v_{yz}}{E_{zz}} = \frac{v_{zy}}{E_{yy}} \quad \text{and} \quad \frac{v_{xz}}{E_{zz}} = \frac{v_{zx}}{E_{xx}} \tag{11.12a}$$

Further, in the plane of isotropy, one has from Eq. (3.14)

$$G = \frac{E}{2\,(1+v)}$$

$$\text{i.e.} \qquad G_{xy} = \frac{E_{xx}}{2(1+v_{xy})} = \frac{1}{2\left(\dfrac{1}{E_{xx}} + \dfrac{v_{xy}}{E_{xx}}\right)} \tag{11.12b}$$

In Eq. (11.11)

$$E_{xx} = \frac{1}{b_{11}}, \qquad -\frac{\nu_{xy}}{E_{xx}} = b_{12}, \qquad G_{xy} = \frac{1}{b_{44}}$$

Substituting these in Eq. (11.12b), one gets

$$\frac{1}{b_{44}} = \frac{1}{2} \cdot \frac{1}{(b_{11} - b_{12})}$$

i.e.

$$b_{44} = 2(b_{11} - b_{12}) \tag{11.13}$$

As a result of this, the number of independent elastic constants in Eq. (11.11) are only five; these being E_{xx}, E_{zz}, ν_{xy}, ν_{xz}, G_{yz}.

Orthotropic Body Let the fibres in a composite be aligned along the x and y axes and let these be uniformly distributed. z-axis is taken normal to this plane, Fig. 11.2. The planes normal to x, y and z axes are planes of symmetry (by reflection) and the body is said to be orthogonally anisotropic or orthotropic. The axes x, y and z are the principal directions.

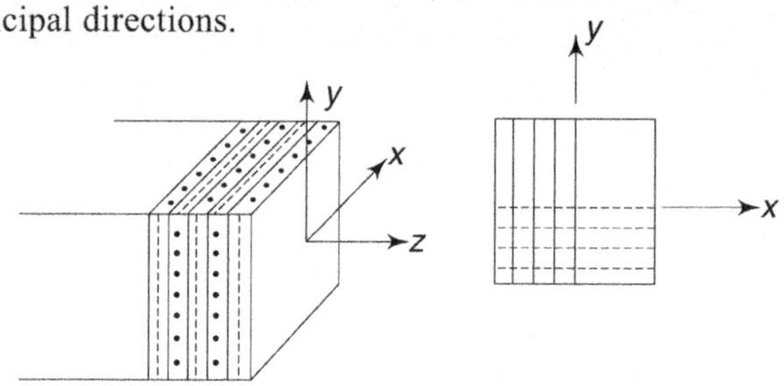

Fig. 11.2 *Orthotropic composite*

To be quite general, the fibres parallel to x-axis may be different from the fibres parallel to y-axis.

Consequently, for an axial force in z direction, the lateral contractions (or extensions) in x and y directions will be different. Similarly, the elasticity moduli corresponding to these two directions will also be different. But, when the thickness in z direction is large, the properties in x and y directions tend to become equal. However, retaining the difference, the strain–stress relations will be (recalling that $b_{ij} = b_{ji}$):

$$\begin{aligned}
\varepsilon_{xx} &= b_{11}\sigma_x + b_{12}\sigma_y + b_{13}\sigma_z \\
\varepsilon_{yy} &= b_{12}\sigma_x + b_{22}\sigma_y + b_{23}\sigma_z \\
\varepsilon_{zz} &= b_{13}\sigma_x + b_{23}\sigma_y + b_{33}\sigma_z \\
\gamma_{xy} &= b_{44}\tau_{xy} \\
\gamma_{yz} &= b_{55}\tau_{yz} \\
\gamma_{zx} &= b_{66}\tau_{zx}
\end{aligned} \tag{11.14}$$

Observe that because of orthogonal symmetries, the shear stresses cause deformations only in their respective planes. There are thus altogether nine independent elastic constants. One can rewrite Eq. (11.14) using the familiar engineering constants (ν_{ij} is the Poisson's effect in i direction due to a force in j direction):

$$\varepsilon_{xx} = \frac{1}{E_{xx}}\sigma_x - \frac{\nu_{xy}}{E_{yy}}\sigma_y - \frac{\nu_{xz}}{E_{zz}}\sigma_z$$

$$\varepsilon_{yy} = -\frac{\nu_{yx}}{E_{xx}}\sigma_x + \frac{1}{E_{yy}}\sigma_y - \frac{\nu_{yz}}{E_{zz}}\sigma_z$$

$$\varepsilon_{zz} = -\frac{\nu_{zx}}{E_{xx}}\sigma_x - \frac{\nu_{zy}}{E_{yy}}\sigma_y + \frac{1}{E_{zz}}\sigma_z$$

$$\gamma_{xy} = \frac{1}{G_{xy}}\tau_{xy} \tag{11.15}$$

$$\gamma_{yz} = \frac{1}{G_{yz}}\tau_{yz}$$

$$\gamma_{zx} = \frac{1}{G_{zx}}\tau_{zx}$$

In the above equations, the following conditions hold good because of the reciprocal identity:

$$\frac{\nu_{xy}}{E_{yy}} = \frac{\nu_{yx}}{E_{xx}}$$

i.e $$E_{xx}\,\nu_{xy} = E_{yy}\,\nu_{yx} \tag{11.16}$$

and, $$E_{xx}\,\nu_{xz} = E_{zz}\,\nu_{zx}, \qquad E_{yy}\,\nu_{yz} = E_{zz}\,\nu_{zy}$$

It should be observed that $\nu_{yx} \neq \nu_{xy}$, $\nu_{zx} \neq \nu_{zx}$ and $\nu_{zy} \neq \nu_{yz}$, unlike in the case of an isotropic body. If the fibres in the x and y directions are identical in their elastic properties and assuming that they are uniformly interwoven, the following additional relations hold good among the elastic constants.

$$b_{11} = b_{22}, \qquad b_{13} = b_{23}, \qquad b_{55} = b_{66} \tag{11.17}$$

Consequently, the number of elastic constants gets reduced to six.

11.4 LAMINATES

So far, our discussion has been quite general, in the sense that the composite element that was being considered was an element in a three-dimensional or a bulk material having special properties. However, composites are manufactured keeping in view certain specific applications. For example, a plywood with veneers oriented in different directions is essentially a laminate designed to meet specific requirements. Laminates, which are essentially thick sheets, are produced not only to be used as such or in a moulded form as corrugated sheets but also to produce bulk materials (by cementing one sheet on top of another, or wrapping one sheet after another about a mandrel). So, an analysis of composite laminates becomes important. Let xy plane represent the midplane of a composite laminate, with z-axis normal to the plane.

Unidirectional Laminates Let the composite consist of fibres all aligned parallel to x-axis. Such a composite will obviously be stronger in the x direction than in the y direction. We assume that the laminate will be subjected to a plane state of stress. At any point, the rectangular stress components will be σ_x, σ_y and τ_{xy}.

The strain components will be ε_x, ε_y, ε_z and γ_{xy}, their values being (remembering that $b_{12} = b_{21}$):

$$\begin{aligned}
\varepsilon_{xx} &= b_{11}\sigma_{xx} + b_{12}\sigma_{yy} \\
\varepsilon_{yy} &= b_{12}\sigma_{xx} + b_{22}\sigma_{yy} \\
\varepsilon_{zz} &= b_{31}\sigma_{xx} + b_{32}\sigma_{yy} \\
\gamma_{xy} &= b_{44}\tau_{xy}
\end{aligned} \tag{11.18}$$

We shall be using double subscripts for stresses and strains in order to be consistent with a_{ij}s, b_{ij}s and τ_{ij}s.

There are six independent elastic constants called compliance coefficients. Using engineering constants E and ν,

$$\begin{aligned}
\varepsilon_{xx} &= \frac{1}{E_{xx}}\sigma_{xx} - \frac{\nu_{xy}}{E_{yy}}\sigma_{yy} \\
\varepsilon_{yy} &= -\frac{\nu_{yx}}{E_{xx}}\sigma_{xx} + \frac{1}{E_{yy}}\sigma_{yy} \\
\varepsilon_{zz} &= -\frac{\nu_{zx}}{E_{xx}}\sigma_{xx} - \frac{\nu_{zy}}{E_{yy}}\sigma_{yy} \\
\gamma_{xy} &= \frac{1}{G_{xy}}\tau_{xy}
\end{aligned} \tag{11.19}$$

ν_{xy} is the Poisson's effect (ratio) in x direction due to a stress in y direction, ν_{zx} is the Poisson's ratio in z direction due to a stress in x direction, etc. A negative sign is used to indicate lateral contraction for an extensional stress at right angles. As in Eq. (11.16), because of reciprocal identity,

$$\frac{\nu_{xy}}{E_{yy}} = \frac{\nu_{yx}}{E_{xx}} \tag{11.20}$$

Table 11.1 gives typical values of Young's moduli in x and y directions, shear modulus, volume fraction of fibre V_f, Poisson's ratio, and specific gravity for selected unidirectional (along x-axis) composites.

Table 11.1 *Fibres Along xx-axis*

Material	E_{xx} (GPa)	E_{yy} (GPa)	G_{xy} (GPa)	ν_{yx}	V_f	Sp. gravity
Graphite + Epoxy	181	10.3	7.17	0.28	0.70	1.6
Boron + Epoxy	204	18.5	5.59	0.23	0.5	2.0
Graphite + Epoxy	138	8.96	7.1	0.3	0.66	1.6
Glass + Epoxy	38.6	8.27	4.14	0.26	0.45	1.8
Kevlar + Epoxy	76	5.5	2.3	0.34	0.60	1.46

Similar to Eq. (11.18), one can express the stress components in terms of strain components. Thus, for a plane state of stress with $\sigma_{zz} = 0$,

$$\sigma_{xx} = a_{11}\varepsilon_{xx} + a_{12}\varepsilon_{yy}$$
$$\sigma_{yy} = a_{12}\varepsilon_{xx} + a_{22}\varepsilon_{yy} \tag{11.21}$$
$$\tau_{xy} = a_{44}\gamma_{xy}$$

Or, one can solve Eq. (11.19) to obtain expressions for σ_{xx}, σ_{yy} and τ_{xy} in terms of the engineering constants and the strain components. These come out as

$$\sigma_{xx} = \frac{E_{xx}}{1-\nu_{xy}\,\nu_{yx}}(\varepsilon_{xx} + \nu_{xy}\varepsilon_{yy})$$
$$\sigma_{yy} = \frac{E_{yy}}{1-\nu_{xy}\nu_{yx}}(\nu_{yx}\,\varepsilon_{xx} + \varepsilon_{yy}) \tag{11.22}$$
$$\tau_{xy} = G_{xy}\gamma_{xy}$$

The reciprocal identity given by Eq. (11.20) holds good

Off-axis Loading in writing the strain–stress equations, the axes x and y were chosen along the principal directions, i.e. along and perpendicular to the fibre directions. If the laminate is stressed such that the rectangular stress components for these axes can easily be determined, then one can directly use Eq. (11.19), and in practice, through experiments or otherwise, the elastic constants (like E_{xx}, E_{yy} ν_{yx}, etc.) along the principal directions can be determined. However, if the loading is in an arbitrary direction, say x' and y' directions that are oriented at an angle θ to x and y axes, then, it is desirable to get $\varepsilon_{x'x'}$, $\varepsilon_{y'y'}$, etc. using the elastic constants transformed to these new axes. Consider Fig.11.3, where the principal directions are x and y and the arbitrary loading directions are x' and y'. The axes x' and y' are rotated through an angle θ counter-clockwise.

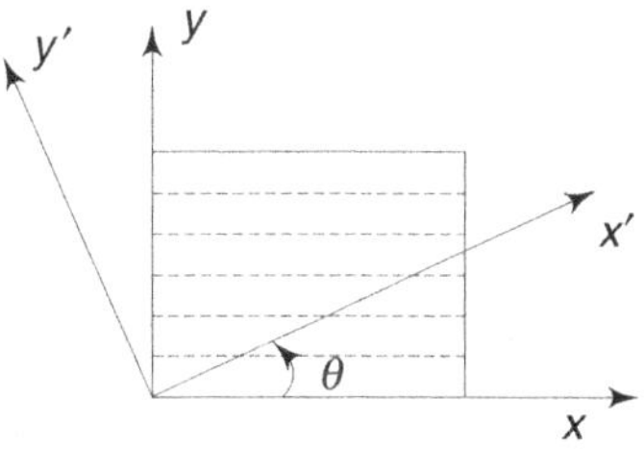

Fig. 11.3 *Off-axis loading*

The positive x'-axis makes an angle θ with fibre direction (i.e. x-axis) and the angle is positive when it is measured counter-clockwise.

Let the stresses applied be $\sigma_{x'x'}$, $\sigma_{y'y'}$ and $\tau_{x'y'}$ and let the corresponding strain components be, $\varepsilon_{x'x'}$, $\varepsilon_{y'y'}$, and $\gamma_{x'y'}$. The procedure to get $\varepsilon_{i'j'}$, in terms of $\sigma_{i'j'}$ is as follows:

$$\sigma_{ij} = f_1(\sigma_{i'j'}, \theta) \quad \text{according to Eqs (1.59) and (1.60)}$$
$$\varepsilon_{ij} = f_2(\sigma_{ij}, b_{ij}) \quad \text{according to Eq. (11.18)}$$
$$\varepsilon_{i'j'} = f_3(\varepsilon_{ij}, \theta) \quad \text{according to Eqs (2.20) and (2.36a)}$$

$$\therefore \quad \varepsilon_{i'j'} = f_3\left[f_2(\sigma_{ij}, b_{ij}), \theta\right]$$
$$= f_3\left\{f_2\left[f_1(\sigma_{i'j'}, \theta), b_{ij}\ \theta\right]\right\}$$

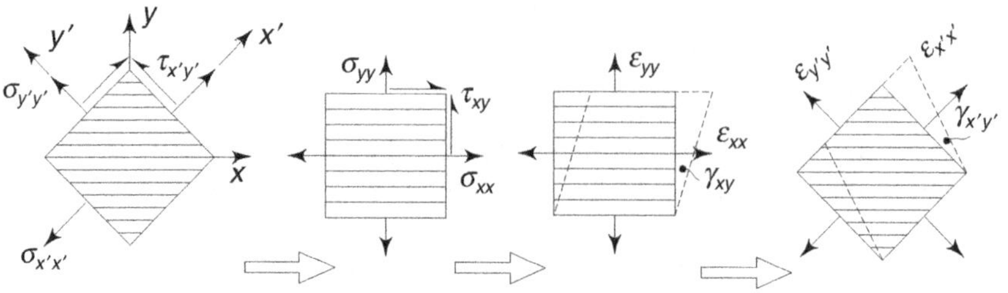

Fig.11.4 *Off-axis stresses and strains*

The above transformations are illustrated in Fig. 11.4.

To transform $\sigma_{i'j'}$ into σ_{ij}, we make use of Eqs (1.59) and (1.60). σ_{xx} is the normal stress on the x plane, and the normal to this plane which is the x-axis, makes angles $-\theta$ and $(90 + \theta)$ with x' and y' axes. Hence, $n_x' = \cos\theta$ and $n_y' = -\sin\theta$. Similarly, for the y-axis, $n_x' = \cos(90 - \theta) = \sin\theta$ and $n_y' = \cos\theta$.

$$\begin{aligned}\sigma_{xx} &= \sigma_{x'x'}\cos^2\theta + \sigma_{y'y'}\sin^2\theta - 2\tau_{x'y'}\sin\theta\cos\theta \\ \sigma_{yy} &= \sigma_{x'x'}\sin^2\theta + \sigma_{y'y'}\cos^2\theta + 2\tau_{x'y'}\sin\theta\cos\theta \\ \tau_{xy} &= +\frac{1}{2}(\sigma_{x'x'} - \sigma_{y'y'})\sin 2\theta + \tau_{x'y'}\cos 2\theta\end{aligned} \tag{11.23}$$

From Eq. (11.18)

$$\begin{aligned}\varepsilon_{xx} &= b_{11}(\sigma_{x'x'}\cos^2\theta + \sigma_{y'y'}\sin^2\theta - \tau_{x'y'}\sin 2\theta) \\ &\quad + b_{12}(\sigma_{x'x'}\sin^2\theta + \sigma_{y'y'}\cos^2\theta + \tau_{x'y'}\sin 2\theta) \\ \varepsilon_{yy} &= b_{12}(\sigma_{x'x'}\cos^2\theta + \sigma_{y'y'}\sin^2\theta - \tau_{x'y'}\sin 2\theta) \\ &\quad + b_{22}(\sigma_{x'x'}\sin^2\theta + \sigma_{y'y'}\cos^2\theta + \tau_{x'y'}\sin 2\theta) \\ \varepsilon_{zz} &= b_{31}(\sigma_{x'x'}\cos^2\theta + \sigma_{y'y'}\sin^2\theta - \tau_{x'y'}\sin 2\theta) \\ &\quad + b_{32}(\sigma_{x'x'}\sin^2\theta + \sigma_{y'y'}\cos^2\theta + \tau_{x'y'}\sin 2\theta) \\ \gamma_{xy} &= +\frac{1}{2}b_{44}\left[(\sigma_{x'x'} - \sigma_{y'y'})\sin 2\theta + 2\tau_{x'y'}\cos 2\theta\right]\end{aligned} \tag{11.24}$$

To obtain $\varepsilon_{i'j'}$ in terms of ε_{ij} we make use of Eqs (2.20) and (2.36). In using Eq. (2.36a) for the shear strain, one ignores in the denominator, quantities of higher order compared to unity. For x'-axis, $n_x = \cos\theta$, $n_y = \sin\theta$, and for y'-axis, $n_x = \cos(90 + \theta) = -\sin\theta$, $n_y = \cos\theta$.

$$\begin{aligned}\varepsilon_{x'x'} &= \varepsilon_{xx}\cos^2\theta + \varepsilon_{yy}\sin^2\theta + \gamma_{xy}\sin\theta\cos\theta \\ \varepsilon_{y'y'} &= \varepsilon_{xx}\sin^2\theta + \varepsilon_{yy}\cos^2\theta + \gamma_{xy}\sin\theta\cos\theta \\ \gamma_{x'y'} &= -2\varepsilon_{xx}\cos\theta\sin\theta + 2\varepsilon_{yy}\sin\theta\cos\theta + \gamma_{xy}(\cos^2\theta - \sin^2\theta)\end{aligned} \tag{11.25}$$

Substituting for ε_{xx}, ε_{yy} and γ_{xy} from Eq. (11.24),

$$\begin{aligned}\varepsilon_{x'x'} &= \cos^2\theta\Big[b_{11}(\sigma_{x'x'}\cos^2\theta + \sigma_{y'y'}\sin^2\theta - \tau_{x'y'}\sin 2\theta) \\ &\quad + b_{12}(\sigma_{x'x'}\sin^2\theta + \sigma_{y'y'}\cos^2\theta + \tau_{x'y'}\sin 2\theta)\Big] \\ &\quad + \sin^2\theta\Big[b_{12}(\sigma_{x'x'}\cos^2\theta + \sigma_{y'y'}\sin^2\theta - \tau_{x'y'}\sin 2\theta)\end{aligned}$$

$$+ b_{22}(\sigma_{x'x'} \sin^2\theta + \sigma_{y'y'} \cos^2\theta + \tau_{x'y'} \sin 2\theta)\Big]$$

$$+\frac{1}{4}\sin 2\theta\Big[b_{44}(\sigma_{x'x'} - \sigma_{y'y'})\sin 2\theta + 2b_{44}\,\tau_{x'y'}\cos 2\theta\Big]$$

Grouping the coefficients of $\sigma_{x'x'}$, $\sigma_{y'y'}$ and $\tau_{x'y'}$, we get

$$\varepsilon_{x'x'} = \sigma_{x'x'}\left(b_{11}\cos^4\theta + 2b_{12}\cos^2\theta\sin^2\theta + b_{22}\sin^4\theta + \frac{1}{4}b_{44}\sin^2 2\theta\right)$$

$$+ \sigma_{y'y'}\Big(b_{11}\cos^2\theta\sin^2\theta + b_{12}\cos^4\theta + b_{12}\sin^4\theta$$

$$+ b_{22}\cos^2\theta\sin^2\theta - \frac{1}{4}b_{44}\sin^2 2\theta\Big)$$

$$+ \tau_{x'y'}\Big(-b_{11}\cos^2\theta\sin 2\theta + b_{12}\cos^2\theta\sin 2\theta - b_{12}\sin^2\theta\sin 2\theta$$

$$+ b_{22}\sin 2\theta\sin^2\theta + \frac{1}{2}b_{44}\sin 2\theta\cos 2\theta\Big)$$

Using the notation $n = \cos\theta$ and $m = \sin\theta$, one can rewrite the expression for $\varepsilon_{x'x'}$ as

$$\varepsilon_{x'x'} = \left(b_{11}n^4 + 2b_{12}n^2m^2 + b_{22}m^4 + b_{44}n^2m^2\right)\sigma_{x'x'}$$

$$+\left(b_{11}n^2m^2 + b_{12}n^4 + b_{12}m^4 + b_{22}n^2m^2 - b_{44}n^2m^2\right)\sigma_{y'y'}$$

$$+\left[-2b_{11}n^3m + 2b_{12}n^3m - 2b_{12}nm^3 + 2b_{22}nm^3 + b_{44}nm\,(n^2 - m^2)\right]\tau_{x'y'}$$

Substituting for b_{11}, b_{22}, b_{12} and b_{44} from equations (11.18) and (11.19),

$$\varepsilon_{x'x'} = \left[\frac{1}{E_{xx}}n^4 + \frac{1}{E_{yy}}m^4 - \left(\frac{2\nu_{xy}}{E_{yy}} - \frac{1}{G_{xy}}\right)n^2m^2\right]\sigma_{x'x'}$$

$$+\left[\left(\frac{1}{E_{xx}} + \frac{1}{E_{yy}} - \frac{1}{G_{xy}}\right)n^2m^2 - \frac{\nu_{xy}}{E_{yy}}(n^4 + m^4)\right]\sigma_{y'y'}$$

$$+\left[-2\left(\frac{1}{E_{xx}}n^2 - \frac{1}{E_{yy}}m^2\right)nm - (n^2 - m^2)\,nm\left(\frac{2\nu_{xy}}{E_{yy}} - \frac{1}{G_{xy}}\right)\right]\tau_{x'y'}$$

Observing that

$$n^4 + m^4 = n^4 + m^4 + 2n^2m^2 - 2n^2m^2$$

$$= (n^2 + m^2)^2 - 2n^2m^2 = 1 - 2n^2m^2,$$

the expression for $\varepsilon_{x'x'}$ can be simplified as

$$\varepsilon_{x'x'} = \left[\frac{n^4}{E_{xx}} + \frac{m^4}{E_{yy}} + \left(\frac{1}{G_{xy}} - \frac{2\nu_{xy}}{E_{yy}}\right)n^2m^2\right]\sigma_{x'x'}$$

$$+\left[\left(\frac{1}{E_{xx}} + \frac{1}{E_{yy}} - \frac{1}{G_{xy}} + \frac{2\nu_{xy}}{E_{yy}}\right)n^2m^2 - \frac{\nu_{xy}}{E_{yy}}\right]\sigma_{y'y'} \qquad (11.26)$$

$$+\left[2\left(\frac{m^2}{E_{yy}}-\frac{n^2}{E_{xx}}\right)+\left(\frac{1}{G_{xy}}-\frac{2\nu_{xy}}{E_{yy}}\right)(n^2-m^2)\right]nm\tau_{x'y'}$$

Similarly, expressions for $\varepsilon_{y'y'}$ and $\varepsilon_{x'y'}$, can be obtained.

For an anisotropic body, it is preferable to use compliance coefficients while expressing strains in terms of stresses. Thus, one can write

$$\begin{aligned}
\varepsilon_{x'x'} &= b'_{11}\sigma_{x'x'} + b'_{12}\sigma_{y'y'} + b'_{14}\tau_{x'y'}\\
\varepsilon_{y'y'} &= b'_{12}\sigma_{x'x'} + b'_{22}\sigma_{y'y'} + b'_{24}\tau_{x'y'}\\
\varepsilon_{z'z'} &= b'_{31}\sigma_{x'x'} + b'_{32}\sigma_{y'y'} + b'_{34}\tau_{x'y'}\\
\gamma_{x'y'} &= b'_{14}\sigma_{x'x'} + b'_{24}\sigma_{y'y'} + b'_{44}\tau_{x'y'}
\end{aligned} \tag{11.27}$$

From Eq. (11.26),

$$\begin{aligned}
b'_{11} &= \frac{n^4}{E_{xx}} + \frac{m^4}{E_{yy}} + \left(\frac{1}{G_{xy}} - \frac{2\nu_{xy}}{E_{yy}}\right)n^2m^2\\
b'_{12} &= \left(\frac{1}{E_{xx}} + \frac{1}{E_{yy}} - \frac{1}{G_{xy}} + \frac{2\nu_{xy}}{E_{yy}}\right)n^2m^2 - \frac{\nu_{xy}}{E_{yy}}\\
b'_{14} &= \left[2\left(\frac{m^2}{E_{yy}} - \frac{n^2}{E_{xx}}\right) + \left(\frac{1}{G_{xy}} - \frac{2\nu_{xy}}{E_{yy}}\right)(n^2-m^2)\right]nm
\end{aligned} \tag{11.28}$$

Proceeding on the lines for getting $\varepsilon_{x'x'}$, one can get expressions for $\varepsilon_{y'y'}$ and $\gamma_{x'y'}$. The compliance coefficients for these will come out as

$$\begin{aligned}
b'_{22} &= \frac{m^4}{E_{xx}} + \frac{n^4}{E_{yy}} + \left(\frac{1}{G_{xy}} - \frac{2\nu_{xy}}{E_{yy}}\right)n^2m^2\\
b'_{24} &= \left[2\left(\frac{n^4}{E_{yy}} - \frac{m^2}{E_{xx}}\right) - \left(\frac{1}{G_{xy}} - \frac{2\nu_{xy}}{E_{yy}}\right)(n^2-m^2)\right]nm\\
b'_{44} &= 4\left(\frac{1}{E_{xx}} + \frac{1}{E_{yy}} + \frac{2\nu_{xy}}{E_{yy}} - \frac{1}{G_{xy}}\right)n^2m^2 + \frac{1}{G_{xy}}
\end{aligned} \tag{11.29}$$

Since changes in thickness of the laminate are not of much concern, the compliance coefficients for $\varepsilon_{z'z'}$ have not been written. Hence, for a unidirectionally reinforced laminate, if the material constants (i.e., Es and νs) along the principal directions are known, one can get from Eqs (11.28) and (11.29), the compliance coefficients for any arbitrarily oriented axes x' and y'. One should remember that positive x'-axis makes an angle $+\theta$ counter-clockwise with the fibre axis, i.e. $+x$ axis. Equations (11.28) and (11.29) can be written in a compact from using the notations of Eqs (11.18) and (11.19). Thus,

$$\begin{aligned}
b'_{11} &= b_{11}n^4 + b_{22}m^4 + 2b_{12}n^2m^2 + b_{44}n^2m^2\\
b'_{12} &= b_{11}n^2m^2 + b_{22}n^2m^2 + b_{12}(n^4 + m^4) - b_{44}n^2m^2
\end{aligned}$$

$$
\begin{aligned}
b'_{14} &= -2b_{11}n^3m + 2b_{22}nm^3 + 2b_{12}(n^2 - m^2)\,nm + b_{44}(n^2 - m^2)\,nm \\
b'_{22} &= b_{11}m^4 + b_{22}n^4 + 2b_{12}n^2m^2 + b_{44}n^2m^2 \\
b'_{24} &= -2b_{11}nm^3 + 2b_{22}n^3m - 2b_{12}(n^2 - m^2)\,nm - b_{44}(n^2 - m^2)\,nm \\
b'_{44} &= 4b_{11}n^2m^2 + 4b_{22}n^2m^2 - 8b_{12}n^2m^2 + b_{44}(n^2 - m^2)^2
\end{aligned} \tag{11.30a}
$$

where

$$
b_{11} = \frac{1}{E_{xx}}, \quad b_{22} = \frac{1}{E_{yy}}, \quad b_{12} = -\frac{\nu_{xy}}{E_{yy}} = -\frac{\nu_{yx}}{E_{xx}}, \quad b_{44} = \frac{1}{G_{xy}} \tag{11.30b}
$$

Equations (11.30a) can be further modified for ease of applications when we take up for analysis multidirectional composites. Towards this, consider the following trigonometric identities:

$$
n^4 = \frac{1}{8}(3 + 4\cos 2\theta + \cos 4\theta), \qquad n^3m = \frac{1}{8}(2\sin 2\theta + \sin 4\theta) \tag{11.31a}
$$

$$
n^2m^2 = \frac{1}{8}(1 - \cos 4\theta), \qquad nm^3 = \frac{1}{8}(2\sin 2\theta - \sin 4\theta) \tag{11.31b}
$$

$$
m^4 = \frac{1}{8}(3 - 4\cos 2\theta + \cos 4\theta) \tag{11.31c}
$$

Substituting these for b'_{11} in Eq. (11.30a)

$$
\begin{aligned}
b'_{11} &= \frac{1}{8}(3 + 4\cos 2\theta + \cos 4\theta)\,b_{11} + \frac{1}{8}(3 - 4\cos 2\theta + \cos 4\theta)\,b_{22} \\
&\quad + \frac{1}{8}(1 - \cos 4\theta)(2b_{12} + b_{44}) \\
&= \frac{1}{8}(3b_{11} + 3b_{22} + 2b_{12} + b_{44}) + \frac{1}{2}(b_{11} - b_{22})\cos 2\theta \\
&\quad + \frac{1}{8}(b_{11} + b_{22} - 2b_{12} - b_{44})\cos 4\theta \\
&= P_1 + P_2\cos 2\theta + P_3\cos 4\theta
\end{aligned} \tag{11.32a}
$$

Similarly, substituting for other components in Eq. (11.30a),

$$
\begin{aligned}
b'_{12} &= \frac{1}{8}(b_{11} + b_{22} + 6b_{12} - b_{44}) - \frac{1}{8}(b_{11} + b_{22} - 2b_{12} - b_{44})\cos 4\theta \\
&= P_4 - P_3\cos 4\theta \\
b'_{14} &= -P_2\sin 2\theta - 2P_3\sin 4\theta \\
b'_{22} &= P_1 - P_2\cos 2\theta + P_3\cos 4\theta \\
b'_{24} &= -P_2\sin 2\theta + 2P_3\sin 4\theta \\
b'_{44} &= \frac{1}{2}(b_{11} + b_{22} - 2b_{12} + b_{44}) - \frac{1}{2}(b_{11} + b_{22} - 2b_{12} + b_{44})\cos 4\theta \\
&= P_5 - 4P_3\cos 4\theta
\end{aligned} \tag{11.32b}
$$

In the above expressions for b'_{ij}

$$P_1 = \frac{1}{8}(3b_{11} + 3b_{22} + 2b_{12} + b_{44})$$

$$P_2 = \frac{1}{2}(b_{11} - b_{22})$$

$$P_3 = \frac{1}{8}(b_{11} + b_{22} - 2b_{12} - b_{44}) \tag{11.33}$$

$$P_4 = \frac{1}{8}(b_{11} + b_{22} + 6b_{12} - b_{44})$$

$$P_5 = \frac{1}{2}(b_{11} + b_{22} - 2b_{12} - b_{44})$$

These equations are useful in two ways. Firstly, the quantities P_is are material properties of the composite.

Once these quantities are determined, they can be used for any off-axis loading direction. Secondly, as mentioned before, when we take up multi-direction composites, these equations become useful.

Example 11.1 *Consider a graphite-epoxy laminate whose elastic constants along and perpendicular to the fibres are as follows.*

$E_{xx} = 181$ GPa, $E_{yy} = 10.3$ GPa, $G_{xy} = 7.17$ GPa, $\nu_{yx} = 0.28$,

$\nu_{xy} = 0.01594$

Obtain the compliance coefficients appropriate to $x'y'$ axes which are at (a) +30° (counter-clockwise) to xy axes and (b) +90° to xy axes

Solution (a) $\theta = 30°$, $n = \cos 30° = 0.866$, $m = \sin 30° = 0.5$

From Eqs. (11.28) and (11.29),

$$b'_{11} = [3.107 + 6.068 + (139.5 - 3.094) \times 0.1875] \times 10^{-12}$$
$$= (34.75) \times 10^{-3}\,(\text{GPa})^{-1}$$

$$b'_{12} = [(5.525 + 97.09 - 139.47 + 3.094) \times 0.1875 - 1.547] \times 10^{-12}$$
$$= -(7.88 \times 10^{-3})\,(\text{GPa})^{-1}$$

$$b'_{14} = [2(24.3 - 4.143) + (139.47 - 3.094)(0.75 - 0.25)] \times 0.433 \times 10^{-12}$$
$$= (40.3 + 68.188) \times 0.433 \times 10^{-12} = 46.98 \times 10^{-3}\,(\text{GPa})^{-1}$$

$$b'_{22} = [0.345 + 54.61 + (139.47 - 3.094) \times 0.1875] \times 10^{-12}$$
$$= 80.53 \times 10^{-3}\,(\text{GPa})^{-1}$$

$$b'_{24} = [2(72.81 - 1.381) - (139.47 - 3.094)(0.75 - 0.25)] \times 0.433 \times 10^{-12}$$
$$= (142.86 - 68.188) \times 0.433 \times 10^{-12} = 32.33 \times 10^{-3}\,(\text{GPa})^{-1}$$

$$b'_{44} = [4(5.525 + 97.09 + 3.094 - 139.47) \times 0.1875 + 139.47] \times 10^{-12}$$
$$= 114.15 \times 10^{-3}\,(\text{GPa})^{-1}$$

(b) $\theta = 90°$, $n = 0$, $m = 1$,

From Eqs (11.27) and (11.28),

$$b'_{11} = \frac{1}{E_{yy}} = 97.09 \times 10^{-3}\ (\text{GPa})^{-1}$$

$$b'_{12} = -\frac{\nu_{xy}}{E_{xx}} = -1.55 \times 10^{-3}\ (\text{GPa})^{-1}$$

$$b'_{14} = 0$$

$$b'_{22} = \frac{1}{E_{xx}} = 5.525 \times 10^{-3}\ (\text{GPa})^{-1}$$

$$b'_{24} = 0$$

$$b'_{44} = \frac{1}{G_{xy}} = 139.5 \times 10^{-3}\ (\text{GPa})^{-1}$$

It should be observed that $x'y'$ frame is obtained through rotation of the xy frame by 90° counter-clockwise. Consequently, the values of the elastic constants get switched since the x'-axis will be along the y-axis, and the y'-axis will be along the x-axis (but in the opposite direction). Thus, $E_{x'x'} = E_{yy}$, $= E_{y'y'} = E_{xx}$ and $G_{x'y'} = G_{xy}$ as the results show.

Example 11.2 *At a point in a laminate the following stress state exists:*

$$\sigma_{x'x'} = 100\ MPa, \qquad \sigma_{y'y'} = 30\ MPa,$$

$$\tau_{x'y'} = 30\ MPa$$

The laminate is unidirectionally reinforced and the fibre orientation is 30° to x'-axis, as shown in Fig. 11.5. The elastic constants along the principal directions of the laminate are

$$E_{xx} = 100\ GPa, \qquad E_{yy} = 10\ GPa,$$

$$G_{xy} = 5\ GPa, \qquad \nu_{yx} = 5.25$$

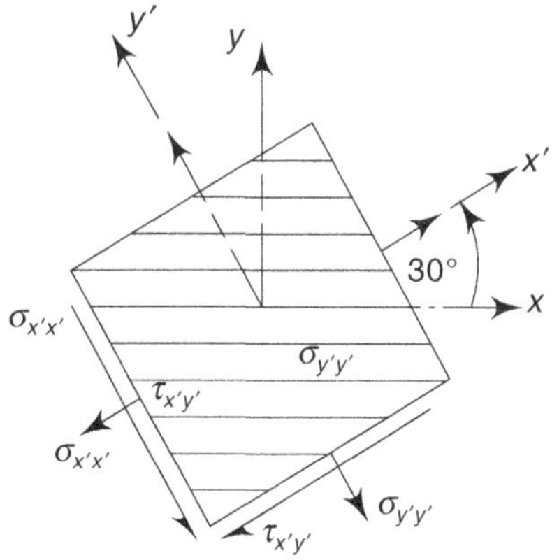

Fig. 11.5 *Example 11.2*

Determine the principal stresses, principal strains and their orientations in the plane of the laminate.

Solution From Eq. (1.61), the principal stresses are

$$\sigma_{1,2} = \frac{1}{2}(\sigma_{x'x'} + \sigma_{y'y'}) \pm \sqrt{\left[\left(\frac{\sigma_{x'x'} - \sigma_{y'y'}}{2}\right)^2 + \tau_{x'y'}^2\right]}$$

$$= \frac{1}{2}(130) \pm \sqrt{\left[(35)^2 + 30^2\right]} = 111 \quad \text{or} \quad 19$$

$$\therefore \qquad \sigma_1 = 111\ \text{MPa} \quad \text{and}\ \sigma_2 = 19\ \text{MPa}\ (\text{check:}\ \ \sigma_{x'x'} + \sigma_{y'y'} = \sigma_1 + \sigma_2)$$

From Eq. (1.62),

$$\tan 2\phi' = \frac{2\tau_{x'y'}}{\sigma_{x'x'} - \sigma_{y'y'}}$$

$$= \frac{60}{70} = 0.8571$$

$$\therefore \qquad \phi_1' = 20.3^\circ \quad \text{and} \quad \phi_2' = 110.3^\circ$$

From Strength of Materials, the algebraically maximum principal stress, which in our present case is $\sigma_1 = 111$ MPa, lies within the principal 45° angle. Thus, $\sigma_1 = 111$ MPa makes an angle of 20.3° with x'-axis, and $\sigma_2 = 19$ MPa makes an angle of 110.3° with x'-axis (counter clockwise).

To determine the principal strains, the required rectangular components can be obtained either with respect to $x'y'$ or with respect to xy axes. To determine the components with respect to $x'y'$ axes, we need the corresponding compliance coefficients. To obtain the strain components with respect to xy axes, we need to transform the given stress components to these axes, and then use Eq. (11.19). Let us transform the given stress components to xy axes. From Eq. (11.23),

$$\sigma_{xx} = (100 \times 0.75) + (30 \times 0.25) - (2 \times 30 \times 0.433) = 56.52 \text{ MPa}$$

$$\sigma_{yy} = (100.0 \times 0.25) + (30 \times 0.75) + (2 \times 30 \times 0.433) = 73.48 \text{ MPa}$$

$$(\text{check } \sigma_{xx} + \sigma_{yy} = \sigma_{x'x'} + \sigma_{y'y'})$$

$$\tau_{xy} = \frac{1}{2}(70) \times 0.866 + (30 \times 0.5) = 45.31 \text{ MPa}$$

From Eq. (11.19) and using the reciprocal identity, Eq. (11.20),

$$\varepsilon_{xx} = \left(\frac{56.52}{100} - \frac{0.25}{100} \times 73.48\right) \times 10^{-3} = 0.3815 \times 10^{-3}$$

$$\varepsilon_{yy} = \left(-\frac{0.25}{100} \times 56.52 + \frac{73.48}{10}\right) \times 10^{-3} = 7.207 \times 10^{-3}$$

$$\gamma_{xy} = \frac{45.31}{5} \times 10^{-3} = 9.062 \times 10^{-3}$$

The principal strains corresponding to these are, from Eq. (2.50),

$$\varepsilon_{1,2} = \frac{\varepsilon_{xx} + \varepsilon_{yy}}{2} \pm \sqrt{\left[\left(\frac{\varepsilon_{xx} - \varepsilon_{yy}}{2}\right)^2 + \left(\frac{\gamma_{xy}}{2}\right)^2\right]}$$

$$= \frac{1}{2}(0.3815 + 7.207) \times 10^{-3} \pm \sqrt{\left[(3.413)^2 + (4.531)^2\right]} \times 10^{-3}$$

$$= (3.7942 \times 10^{-3}) \pm (5.6726)$$

$$= 9.4668 \times 10^{-3} \quad \text{or} \quad -1.8784 \times 10^{-3}$$

$$(\text{check: } \varepsilon_{xx} + \varepsilon_{yy} = \varepsilon_1 + \varepsilon_2)$$

From Eq. (2.51), the directions of these principal strains are

$$\tan 2\phi^* = \frac{\gamma_{xy}}{\varepsilon_{xx} - \varepsilon_{yy}} = \frac{9.062}{0.3815 - 7.207} = -1.3277$$

$$\therefore \qquad \phi_1^* = -26.5^\circ \qquad \text{and} \qquad \phi_2^* = +63.5^\circ$$

These angles are with respect to the x-axis. By subtracting 30°, we get the orientations of the principal strain axes with respect to x'-axis. Thus,

$$\phi_1^* = -56.5^\circ \qquad \text{and} \qquad \phi_2^* = 33.5^\circ.$$

Unlike an isotropic case, in general, the principal stress axes do not coincide with the principal strain axes in an anisotropic body. In this example, the principal stress axes are at 40.6° and 130.6°, while the principal strain axes are at 33.5° and 123.5° (i.e. – 56.5°) to x'-axis.

Example 11.3 *In Example 11.2, the directions of the principal strains were obtained by transforming the applied stresses to the principal direction axes. Show that the same results can be obtained by using the loading or the stress axes as reference and obtaining the corresponding compliance coefficients.*

Solution $\theta = 30^\circ$, $\quad n \cos 30^\circ = 0.866$, $\quad m = \sin 30^\circ = 0.5$

From Equations (11.28) and (11.29)

$$b'_{11} = [5.6243 + 6.25 + (200 - 5) \times 0.1875] \times 10^{-12} = 48.4368 \times 10^{-12}$$

$$b'_{12} = [(10 + 100 - 200 + 5) \times 0.1875 - 2.5] \times 10^{-12} = -18.4375 \times 10^{-12}$$

$$b'_{14} = [2(25 - 7.4996) + (200 - 5) \times 0.5] \times 0.433 \times 10^{-12} = 57.3728 \times 10^{-12}$$

$$b'_{22} = [0.625 + 56.24 + (200 - 5) \times 0.1875] \times 10^{-12} = 93.4275 \times 10^{-12}$$

$$b'_{24} = [2(74.996 - 2.5) - (200 - 5) \times 0.5] \times 0.433 \times 10^{-12} = 20.564 \times 10^{-12}$$

$$b'_{44} = [4(10 + 100 + 5 - 200) \times 0.1875 + 200] \times 10^{-12} = 136.25 \times 10^{-12}$$

From Eq. (11.25),

$$\varepsilon_{x'x'} = [(48.4368 \times 100) - (18.4375 \times 30) + (57.3728 \times 30)] \times 10^{-6}$$
$$= 0.006012$$

$$\varepsilon_{y'y'} = [-(18.4375 \times 100) + (93.4275 \times 30) + (20.564 \times 30)] \times 10^{-6}$$
$$= 0.001576$$

$$(\text{check: } \varepsilon_{x'x'} + \varepsilon_{y'y'} = \varepsilon_{xx} + \varepsilon_{yy})$$

$$\gamma_{x'y'} = [(57.3728 \times 100) + (20.564 \times 30) + (136.25 \times 30)] \times 10^{-6}$$
$$= 0.010434$$

The principal strains are

$$\varepsilon_{1,2} = \frac{\varepsilon_{x'x'} + \varepsilon_{y'y'}}{2} \pm \sqrt{\left[\left(\frac{\varepsilon_{x'x'} - \varepsilon_{y'y'}}{2}\right)^2 + \left(\frac{\gamma_{x'y'}}{2}\right)^2\right]}$$

$$= \frac{1}{2}(6.012 + 1.576) \times 10^{-3} \pm \sqrt{\left[\left(\frac{6.012 - 1.576}{2}\right)^2 + (5.217)^2\right]} \times 10^{-3}$$

$$= (3.794 \pm 5.669) \times 10^{-3} = 0.00946; - 0.00187$$

Off-axis Components of Modulus In the previous discussions we obtained the off-axis components of compliances b'_{ij}s from Eqs (11.28) and (11.29). The motivation for considering this first is that in practice, composites are used to comply with situations where stresses or loads are prescribed which usually are not along principal directions. To estimate the deformations using Eq. (11.27), one needs compliance coefficients from known elastic constants along principal directions. On the other hand, when we need to analyse multidirectional fibre composites we need to know the stress values for given off-axis strain values. To get this, one can follow a similar procedure as was adopted earlier, i.e obtain $\sigma_{i'j'}$ in terms of $\varepsilon_{i'j'}$s.

$$\varepsilon_{ij} = f_1(\varepsilon_{i'j'}, \theta) \quad \text{from Eqs (2.20) and (2.36a)}$$

$$\sigma_{ij} = f_2(\varepsilon_{ij}, a_{ij}) \quad \text{from Eq. (11.21)}$$

$$\sigma_{i'j'} = f_3(\sigma_{ij}, \theta) \quad \text{from Eqs (1.59) and (1.60)}$$

$$= f_3\left[f_2(\varepsilon_{ij}, a_{ij}), \theta\right]$$

$$= f_3\left\{f_2[f_1(\varepsilon_{i'j'}, \theta), a_{ij}], \theta\right\}$$

Alternatively, one can solve for $\sigma_{x'x'}$, $\sigma_{y'y'}$ and $\tau_{x'y'}$, from Eq. (11.27). For this, we need expressions for $\varepsilon_{x'x'}$, $\varepsilon_{y'y'}$ and $\gamma_{x'y'}$.

$$\varepsilon_{x'x'} = b'_{11}\sigma_{x'x'} + b'_{12}\sigma_{y'y'} + b'_{14}\tau_{x'y'}$$

$$\varepsilon_{y'y'} = b'_{12}\,\sigma_{x'x'} + b'_{22}\,\sigma_{y'y'} + b'_{24}\,\tau_{x'y'}$$

$$\gamma_{x'y'} = b'_{41}\,\sigma_{x'x'} + b'_{42}\,\sigma_{y'y'} + b'_{44}\,\tau_{x'y'}$$

In Eq. (a), $b'_{41} = b'_{14}$ and $b'_{42} = b'_{24}$. The determinant of the coefficients in Eq. (a) is

$$\Delta = b'_{11}(b'_{22}\,b'_{44} - b'^{2}_{24}) - b'_{12}(b'_{12}\,b'_{44} - b'_{24}\,b'_{14}) + b'_{14}(b'_{12}\,b'_{24} - b'_{22}\,b'_{14})$$

$$= b'_{11}\,b'_{22}\,b'_{44} + 2b'_{12}\,b'_{24}\,b'_{14} - b'_{22}\,b'^{2}_{14} - b'_{11}\,b'^{2}_{24} - b'_{44}\,b'^{2}_{12}$$

Hence, the solutions for $\sigma_{x'x'}$, $\sigma_{y'y'}$ and $\tau_{x'y'}$, from Eq. (a), are

$$\sigma_{x'x'} = \frac{1}{\Delta}[(b'_{22}\,b'_{44} - b'^{2}_{24})\,\varepsilon_{x'x'} - (b'_{12}\,b'_{44} - b'_{14}\,b'_{24})\,\varepsilon_{y'y'}$$
$$+ (b'_{12}\,b'_{24} - b'_{14}\,b'_{22})\,\gamma_{x'y'}]$$

$$\sigma_{y'y'} = \frac{1}{\Delta}[(b'_{12}\,b'_{44} - b'_{24}\,b'_{14})\,\varepsilon_{x'x'} - (b'_{11}\,b'_{44} - b'^{2}_{14})\,\varepsilon_{y'y'}$$
$$+ (b'_{11}\,b'_{24} - b'_{14}\,b'_{12})\,\gamma_{x'y'}]$$

$$\tau_{x'y'} = \frac{1}{\Delta}[(b'_{12}b'_{42} - b'_{22}b'_{14})\,\varepsilon_{x'x'} - (b'_{11}b'_{24} - b'_{12}b'_{14})\,\varepsilon_{y'y'}$$

$$+ (b'_{11}b'_{22} - b'^2_{12})\, \gamma_{x'y'}] \tag{11.34}$$

From the general stress–strain equations, for a laminate under plane state of stress with an off-axis coordinate system, one has, similar to Eq. (11.27),

$$\begin{aligned} \sigma_{x'x'} &= a'_{11}\, \varepsilon_{x'x'} + a'_{12}\, \varepsilon'_{y'y'} + a'_{14}\, \gamma_{x'y'} \\ \sigma_{y'y'} &= a'_{12}\, \varepsilon_{x'x'} + a'_{22}\, \varepsilon'_{y'y'} + a'_{24}\, \gamma_{x'y'} \\ \tau_{x'y'} &= a'_{14}\, \varepsilon_{x'x'} + a'_{24}\, \varepsilon'_{y'y'} + a'_{44}\, \gamma_{x'y'} \end{aligned} \tag{11.35}$$

Comparing the coefficients $\varepsilon_{i'j'}$ in Eqs (11.34) and (11.35), one gets

$$\begin{aligned} a'_{11} &= \frac{1}{\Delta}(b'_{22}b'_{44} - b'^2_{24}) \\ a'_{12} &= -\frac{1}{\Delta}(b'_{12}b'_{44} - b'_{14}b'_{24}) \\ a'_{14} &= \frac{1}{\Delta}(b'_{12}b'_{24} - b'_{14}b'_{22}) \\ a'_{22} &= -\frac{1}{\Delta}(b'_{11}b'_{44} - b'^2_{14}) \\ a'_{24} &= \frac{1}{\Delta}(b'_{11}b'_{24} - b'_{14}b'_{12}) \\ a'_{44} &= \frac{1}{\Delta}(b'_{11}b'_{22} - b'^2_{12}) \end{aligned} \tag{11.36}$$

Application of Eq. (11.36) to get a'_{ij} s involves calculations of b'_{ij}s. Instead, one can follow the procedure adopted earlier. This is shown schematically in Fig. 11.6. The results are the following:

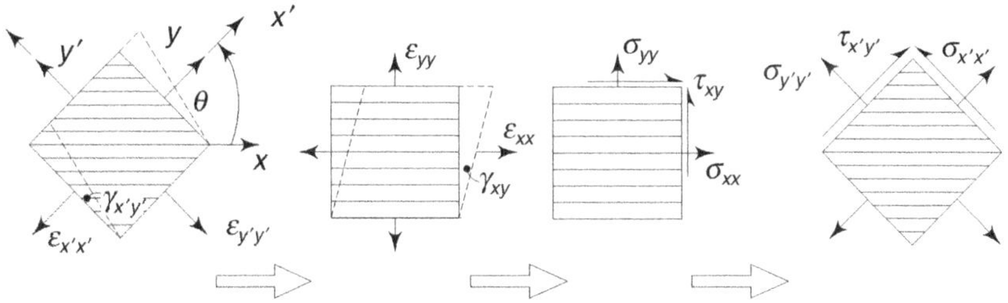

Fig. 11.6 *Off-axis components of stresses and strains*

$$\begin{aligned} a'_{11} &= a_{11}n^4 + a_{22}m^4 + 2a_{12}n^2m^2 + 4a_{44}n^2m^2 \\ a'_{12} &= a_{11}n^2m^2 + a_{22}\, n^2m^2 + a_{12}\,(n^4 + m^4) - 4a_{44}\, n^2m^2 \\ a'_{14} &= -a_{11}n^3m + a_{22}nm^3 + a_{12}\,(n^2 - m^2)nm + 2a_{44}\,(n^2 - m^2)nm \\ a'_{24} &= -a_{11}nm^3 + a_{22}n^3m - a_{12}\,(n^2 - m^2)nm - 2a_{44}\,(n^2 - m^2)nm \\ a'_{22} &= a_{11}m^4 + a_{22}n^4 + 2a_{12}n^2m^2 + 4a_{44}n^2m^2 \end{aligned} \tag{11.37a}$$

$$a'_{44} = a_{11}n^2m^2 + a_{22}n^2m^2 - 2a_{12}n^2m^2 + a_{44}(n^2 - m^2)^2$$

where, from Eqs (11.21) and (11.22),

$$a_{11} = \frac{E_{xx}}{1 - \nu_{yx}\,\nu_{xy}}, \quad a_{22} = \frac{E_{yy}}{1 - \nu_{yx}\,\nu_{xy}}, \quad a_{12} = + \frac{\nu_{yx}\,E_{xx}}{1 - \nu_{yx}\,\nu_{xy}}, \quad a_{44} = G_{xy} \tag{11.37b}$$

Equations (11.37a) can be recast as was done in the case of compliance coefficients, i.e. Eqs (11.32) and (11.33). Using the trigonometric identities given by Eq. (11.31),

$$\begin{aligned} a'_{11} &= \frac{1}{8}(3 + 4\cos 2\theta + \cos 4\theta)\, a_{11} + \frac{1}{8}(3 - 4\cos 2\theta + \cos 4\theta)\, a_{22} \\ &\quad + \frac{1}{8}(1 - \cos 4\theta)(2a_{12} + 4a_{44}) \\ &= \frac{1}{8}(3a_{11} + 3a_{22} + 2a_{12} + 4a_{44}) + \frac{1}{2}(a_{11} - a_{22})\cos 2\theta \\ &\quad + \frac{1}{8}(a_{11} + a_{22} - 2a_{12} - 4a_{44})\cos 4\theta \\ &= Q_1 + Q_2\cos 2\theta + Q_3\cos 4\theta \end{aligned} \tag{11.38a}$$

Similarly, for other components we get,

$$\begin{aligned} a'_{22} &= Q_1 - Q_2\cos 2\theta + Q_3\cos 4\theta \\ a'_{12} &= Q_4 - Q_3\cos 4\theta \\ a'_{14} &= -\frac{1}{2}Q_2\sin 2\theta - Q_3\sin 4\theta \\ a'_{24} &= -\frac{1}{2}Q_2\sin 2\theta + Q_3\sin 4\theta \\ a'_{44} &= Q_5 - Q_3\cos 4\theta \end{aligned} \tag{11.38b}$$

where,

$$\begin{aligned} Q_1 &= \frac{1}{8}(3a_{11} + 3a_{22} + 2a_{12} + 4a_{44}) \\ Q_2 &= \frac{1}{2}(a_{11} - a_{22}) \\ Q_3 &= \frac{1}{8}(a_{11} + a_{22} - 2a_{12} - 4a_{44}) \\ Q_4 &= \frac{1}{8}(a_{11} + a_{22} + 6a_{12} - 4a_{44}) \\ Q_5 &= \frac{1}{8}(a_{11} + a_{22} - 2a_{12} + 4a_{44}) \end{aligned} \tag{11.39}$$

The Q_is involve only material properties and once they are determined, the compliance coefficients for any off-axis direction can be determined using Eq. (11.38b).

Example 11.4 *A unidirectionally reinforced composite of 'Toray' filament and 'Nameo' resin has the following moduli and Poisson's ratio.*

$E_{xx} = 181\ GPa, \quad E_{yy} = 10.3\ GPa, \quad \nu_{xy} = 0.0159, \qquad G_{xy} = 7.17\ GPa,$

$(1 - \nu_{xy}\nu_{yx})^{-1} = 1.0045$

Estimate the components of moduli for an off-axis orientation of
(a) $\theta = +30°$ and (b) $\theta = +45°$.

(a) for $\theta = +30°$, $n = \cos\theta = 0.866$ and $m = \sin\theta = 0.5$.

Soluion From Eq. (11.37b),

$$a_{11} = \frac{E_{xx}}{1 - \nu_{yx}\nu_{xy}} = 181 \times 1.0045 = 181.8 \text{ GPa}$$

$$a_{22} = \frac{E_{yy}}{1 - \nu_{yx}\nu_{xy}} = 10.3 \times 1.0045 = 10.34 \text{ GPa}$$

$$a_{12} = \frac{\nu_{yx} E_{xx}}{1 - \nu_{yx}\nu_{xy}} = 0.0159 \times 181 \times 1.0045 = 2.891 \text{ GPa}$$

$$a_{44} = G_{xy} = 7.17 \text{ GPa}$$

Further, $n = 0.866, \qquad m = 0.5, \qquad nm = 0.433$

$$n^2 = 0.750, \quad n^4 = 0.562, \quad m^2 = 0.25, \quad m^4 = 0.0625$$

$$n^3 m = 0.3248, \quad nm^3 = 0.1083, \quad n^2 m^2 = 0.1875$$

Substituting in Eq. (11.37a),

$$a'_{11} = (181.8 \times 0.562) + (10.34 \times 0.625) + (2 \times 2.891 \times 0.1875) + (4 \times 7.17 \times 0.1875) = 109.2 \text{ GPa}$$

$$a'_{12} = (181.8 \times 0.1875) + (10.34 \times 0.1875) + (2.891)(0.6245) - (4 \times 7.17 \times 0.1875) = 32.45 \text{GPa}$$

$$a'_{14} = -(181.8 \times 0.3248) + (10.34 \times 0.1083) + (2.891 \times 0.5 \times 0.433) + (2 \times 7.17 \times 0.5 \times 0.433) = -54.19 \text{ Gpa}$$

$$a'_{22} = 23.64 \text{ GPa}, \quad a'_{24} = -20.05 \text{ GPa}, \quad a'_{44} = 36.78 \text{ GPa};$$

Similarly, for (b) with $\theta = +45°$

$a'_{11} = 56.6$ GPa, $a'_{12} = 42.32$ GPa, $a'_{14} = -42.87$ GPa, $a'_{22} = 46.59$ GPa;
$a'_{24} = -42.87$ GPa; $a'_{44} = 46.59$ GPa

Multi-directional Laminates Multi-directional laminates can be formed by cementing plies with different fibre orientations. The effective in-plane modulus of laminate plies is found to be simply the arithmetic mean of the moduli of the constituent plies. Laminates with midplane symmetry will behave like homogeneous anisotropic plates. A multi-directional composite laminate is defined by a code which describes the stacking sequence of the ply groups. For example, the code

$$[0_2/90_2/45/-45_3]_S \tag{11.40}$$

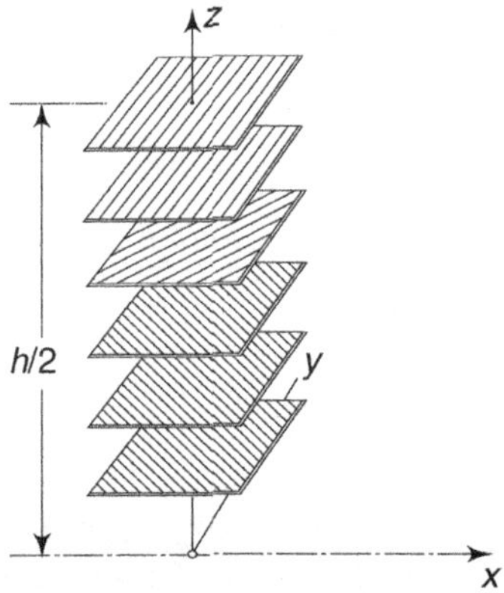

Fig. 11.7 *Multidirectional laminate—Schematic representation*

means the following:

The thickness of the laminate is h. Starting from the bottom of the laminate, at $z = -\frac{h}{2}$, the first ply group has two plies of 0° orientation, followed by the next group with two 90° plies, followed by one 45° ply, and finally the last group with three −45° plies. For a symmetric laminate, the ascending order from the bottom face is identical to the descending order from the top face, i.e $z = +\frac{h}{2}$. The subscript S denotes that the laminate is symmetric with respect to the midplane, i.e $z = 0$. The upper half of the laminate is the same as the lower half except that the stacking sequence is reversed to maintain midplane symmetry.

A subscript T is used to describe the total laminate without resorting to describe symmetry or otherwise. For example, the laminate described by the code given in Eq. (11.40) can be written with the following code also:

$$\left[0_2/90_2/45/-45_3/-45_3/45/90_2/0_2\right]_T \tag{11.41a}$$

or

$$\left[0_2/90_2/45/-45_6/45/90_2/0_2\right]_T \tag{11.41b}$$

where the middle six ply groups with the same orientations have been grouped together. Figure 11.7 shows the laminate schematically.

Inplane Stress–Strain Relations In deriving the stress–strain relations for a multi-directional laminate, the following assumptions are made:

(i) The laminate is symmetric, i.e.

$$\theta(z) = \theta(-z) \tag{11.42a}$$

and

$$a_{ij}(z) = a_{ij}(-z) \tag{11.42b}$$

Hence, both the ply orientation and the ply material modulus are symmetric with respect to the midplane of the laminate.

(ii) The strain is uniformly the same across the thickness of the laminate, i.e.

$$\begin{aligned} \varepsilon_{xx}(z) &= \overset{*}{\varepsilon}_{xx} \\ \varepsilon_{yy}(z) &= \overset{*}{\varepsilon}_{yy} \\ \gamma_{xy}(z) &= \overset{*}{\gamma}_{xy} \end{aligned} \tag{11.43}$$

The above assumption is fairly reasonable when the total laminate thickness is small and bonding between plies is good. x and y axes are arbitrary axes with reference to which the strains are prescribed. These axes may not in general coincide with any fibre axes.

Because of different orientations of the plies, the components of moduli for any given direction are not the same for each ply. Hence, for a given uniform strain, the stresses vary from ply to ply, and it is useful to discuss in terms of average stresses across the thickness of the laminate. Thus,

$$\bar{\sigma}_{yy} = \frac{1}{h}\int_{-h/2}^{h/2} \sigma_{yy}\, dz \tag{11.44}$$

$$\bar{\tau}_{xy} = \frac{1}{h}\int_{-h/2}^{h/2} \tau_{xy}\, dz$$

Now, from Eq. (11.35) for any ply, remembering that x and y are arbitrary axes,

$$\sigma_{xx} = a_{11}\,\varepsilon_{xx} + a_{12}\,\varepsilon_{yy} + a_{14}\,\gamma_{xy}$$

$$\sigma_{yy} = a_{12}\,\varepsilon_{xx} + a_{22}\,\varepsilon_{yy} + a_{24}\,\gamma_{xy}$$

$$\tau_{xy} = a_{14}\,\varepsilon_{xx} + a_{24}\,\varepsilon_{yy} + a_{44}\,\gamma_{xy}$$

Since the strains are the same in the plies,

$$\bar{\sigma}_{xx} = \frac{1}{h}\int_{-h/2}^{h/2}\left[a_{11}\,\varepsilon^*_{xx} + a_{12}\varepsilon^*_{yy} + a_{14}\gamma^*_{xy}\right]dz$$

$$= \frac{1}{h}\left[\varepsilon^*_{xx}\int a_{11}\,dz + \varepsilon^*_{yy}\int a_{12}\,dz + \gamma^*_{xy}\int a_{14}\,dz\right]$$

$$= \frac{1}{h}\left[A_{11}\,\varepsilon^*_{xx} + A_{12}\varepsilon^*_{yy} + A_{14}\gamma^*_{xy}\right] \tag{11.45a}$$

where,

$$A_{11} = \int_{-h/2}^{h/2} a_{11}\,dz; \qquad A_{12} = \int_{-h/2}^{h/2} a_{12}\,dz; \qquad A_{14} = \int_{-h/2}^{h/2} a_{14}\,dz \tag{11.45b}$$

Similarly,

$$\bar{\sigma}_{yy} = \frac{1}{h}\left[A_{12}\,\varepsilon^*_{xx} + A_{22}\varepsilon^*_{yy} + A_{24}\,\gamma^*_{xy}\right] \tag{11.45c}$$

$$\bar{\tau}_{xy} = \frac{1}{h}\left[A_{14}\,\varepsilon^*_{xx} + A_{24}\varepsilon^*_{yy} + A_{44}\,\gamma^*_{xy}\right] \tag{11.45d}$$

where,

$$A_{22} = \int_{-h/2}^{h/2} a_{22}\,dz; \qquad A_{24} = \int_{-h/2}^{h/2} a_{24}\,dz; \qquad A_{44} = \int_{-h/2}^{h/2} a_{44}\,dz \tag{11.45e}$$

$\bar{\sigma}_{xx}$, $\bar{\sigma}_{yy}$ and $\bar{\tau}_{xy}$, are the average stresses across the thickness of the laminate, i.e. stresses per unit thickness. Hence, for a laminate of thickness h, the stress resultants are

$$N_{xx} = h\bar{\sigma}_{xx}, \qquad N_{yy} = h\bar{\sigma}_{yy}, \qquad N_{xy} = h\bar{\tau}_{xy}$$

Substituting for $\bar{\sigma}_{xx}, \bar{\sigma}_{yy}$ and $\bar{\tau}_{xy}$

$$N_{xx} = A_{11}\,\varepsilon^*_{xx} + A_{12}\,\varepsilon^*_{xy} + A_{14}\,\gamma^*_{xy}$$

$$N_{yy} = A_{12}\,\varepsilon^*_{xx} + A_{22}\,\varepsilon^*_{yy} + A_{24}\,\gamma^*_{xy} \tag{11.46}$$

$$N_{xy} = A_{14}\,\varepsilon^*_{xx} + A_{24}\,\varepsilon^*_{yy} + A_{44}\,\gamma^*_{xy}$$

The quantities A_{11}, A_{22}, etc. are the integrated values (across the thickness) of the off-axis components of moduli of the laminate.

Evaluation of In-plane Moduli The resultant values of the moduli components are obtained by integrating the moduli component values across the thickness. In practice, when different plies of finite thicknesses are bonded to get the laminate, the values of a_{ij}s change in discrete steps from ply to ply and they are not continuous functions of z as indicated by Eqs (11.45b and e). However, continuing the integration sign as used earlier,

$$A_{11} = \int_{-h/2}^{h/2} a_{11}\, dz$$

$$= \int (Q_1 + Q_2 \cos 2\theta + Q_3 \cos 4\theta)\, dz$$

$$= Q_1 \int dz + Q_2 \int \cos 2\theta\, dz + Q_3 \int \cos 4\theta\, dz$$

The Qs for a laminate are constant because of our assumption that the laminate consists of plies of the same kind having identical material constants, but in the bonding process, the plies are put with their fibre axes oriented differently, i.e. θ changes from ply to ply, but the Qs are the same for each ply. Thus,

$$A_{11} = Q_1 h + Q_2 V_1 + Q_3 V_2 \qquad (11.47a)$$

where

$$V_1 = \int_{-h/2}^{h/2} \cos 2\theta\, dz, \quad \text{and} \quad V_2 = \int_{-h/2}^{h/2} \cos 4\theta\, dz \qquad (11.47b)$$

Similarly, from Eqs (11.45 b and e),

$$A_{22} = Q_1 h - Q_2 V_1 + Q_3 V_2$$

$$A_{12} = Q_4 h - Q_3 V_2$$

$$A_{14} = -\frac{1}{2} Q_2 V_3 - Q_3 V_4 \qquad (11.47c)$$

$$A_{24} = -\frac{1}{2} Q_2 V_3 - Q_3 V_4$$

$$A_{44} = Q_5 h - Q_3 V_2$$

where $$V_3 = \int_{-h/2}^{h/2} \sin 2\theta\, dz, \quad \text{and} \quad V_4 = \int_{-h/2}^{h/2} \sin 4\theta\, dz \qquad (11.47d)$$

It was assumed that the laminate consists of even number of plies and are symmetrically positioned, i.e. the positioning sequence from the bottom of the laminate, i.e. from $z = -\frac{h}{2}$ to $z = 0$ is the same as from $z = +\frac{h}{2}$ to $z = 0$, the mid-plane. Hence the limits for the integrals can be changed to $z = 0$ and $z = +\frac{h}{2}$, and the quantities multiplied by 2. Also, it was explicitly stated that the orientations θ change in finite steps, from ply to ply. So, the integration sign can be changed to summation sign, i.e.

$$V_1 = 2\sum_{i=1}^{n} \cos 2\theta_i h_i \tag{11.48}$$

where θ_i is the fibre orientation of the ith ply whose thickness is h_i, and the total number of plies in the laminate is $2n$, so that the number of plies from $z = 0$ to $z = \frac{h}{2}$ is n. The summation in Eq. (11.48) is over all the plies from $z = 0$ to $z = \frac{h}{2}$, i.e. n. Let the laminate be composed of $2k_1$ number of plies with fibre orientation θ_1, $2k_2$ number of plies with fibre orientation θ_2, and $2k_i$ number with θ_i orientation. Then,

$$V_1 = 2k_1h_1 \cos 2\theta_1 + 2k_2h_2 \cos 2\theta_2 + \ldots + 2k_ih_i \cos 2\theta_i + \ldots \tag{11.49}$$

Also, $\quad 2k_1 + 2k_2 + \ldots + 2k_i + \ldots = 2n$

and $\quad 2k_1h_1 + 2k_2h_2 + \ldots + 2k_ih_i + \ldots = h$

or $\quad 2k_1 \frac{h_1}{h} + 2k_2 \frac{h_2}{h} + \ldots + 2k_i \frac{h_i}{h} + \ldots = 1$

It is easily seen that $(2k_ih_i/h)$ is the volume fraction of plies with fibre orientation i in the laminate. If the volume fractions are indicated by v_is, then Eq. (11.49) can be written as

$$V_1^* = \frac{V_1}{h} = v_1 \cos 2\theta_1 + v_2 \cos 2\theta_2 + \ldots + v_i \cos 2\theta_i + \ldots \tag{11.50}$$

where, $\quad v_1 + v_2 + \ldots + v_i + \ldots = 1$

Equation (11.50) is simply the rule mixtures equation which will be discussed later. Thus, Eqs (11.47b and d) can be rewritten as

$$\begin{aligned} V_1^* &= \frac{V_1}{h} = v_1 \cos 2\theta_1 + v_2 \cos 2\theta_2 + \ldots \\ V_2^* &= \frac{V_2}{h} = v_1 \cos 4\theta_1 + v_2 \cos 4\theta_2 + \ldots \\ V_3^* &= \frac{V_3}{h} = v_1 \sin 2\theta_1 + v_2 \sin 2\theta_2 + \ldots \\ V_4^* &= \frac{V_4}{h} = v_1 \sin 4\theta_1 + v_2 \sin 4\theta_2 + \ldots \end{aligned} \tag{11.51}$$

If the thickness of each ply is the same, say t, then on the basis of Eq. (11.49), one can write

$$\begin{aligned} V_1 &= t(2k_1 \cos 2\theta_1 + 2k_2 \cos 2\theta_2 + \ldots) \\ V_2 &= t(2k_1 \cos 4\theta_1 + 2k_2 \cos 4\theta_2 + \ldots) \\ V_3 &= t(2k_1 \sin 2\theta_1 + 2k_2 \sin 2\theta_2 + \ldots) \\ V_4 &= t(2k_1 \sin 4\theta_1 + 2k_2 \sin 4\theta_2 + \ldots) \end{aligned} \tag{11.52}$$

where, as mentioned earlier, $2k_1$ is the number of plies in the laminate with θ_1 orientation of fibres, $2k_2$ is the number of plies in the laminate with θ_2 orientations, etc.

Using either Eq. (11.51) or Eq. (11.52), one can easily compute the in-plane moduli of multi-directional laminates with any ply orientation. The information needed is orientation and volume fraction (or the number of plies) of each ply group. Using Eqs (11.51) or (11.52), the values of V_i^*s (or V_is) can be determined. Since the plies are identical, the values of Q_is are the same for each ply and these can be evaluated from Eq. (11.39). The values of a_{ij}s needed in Eq. (11.39) are obtained from Eq. (11.37b). Finally, Eq. (11.47) gives the values of A_{ij}s. The units of A_{ij}s are Pa m or Nm^{-1}.

As an illustration of the steps involved consider the following case:

Cross-ply composites are commonly used in practice when uniform strength is required in both x and y directions. The laminate consists of plies with fibres oriented at $\theta = 0°$ and $\theta = 90°$ The laminate is symmetric. Let v_0 be the volume fraction of plies with $\theta = 0°$ orientation, and v_{90} be the volume fraction of plies with fibres at $\theta = 90°$ orientation. In general, v_0 and v_{90} are not equal.

$$\theta_1 = 0°, \quad \cos 2\theta = \cos 4\theta = 1, \quad \sin 2\theta = \sin 4\theta = 0$$
$$\theta_2 = 90°, \quad \cos 2\theta = -1, \quad \cos 4\theta = +1, \quad \sin 2\theta = \sin 4\theta = 0$$

From Eq. (11.51),

$$V_1 = v_0 h - v_{90}\, h = (v_0 - v_{90})h$$
$$V_2 = (v_0 + v_{90})h = h$$
$$V_3 = V_4 \equiv 0$$

From Eqs (11.47a and c),

$$A_{11} = \left[Q_1 + Q_2(v_0 - v_{90}) + Q_3\right]h$$
$$A_{22} = \left[Q_1 - Q_2(v_0 - v_{90}) + Q_3\right]h$$
$$A_{44} = [Q_5 - Q_3]h \tag{11.53}$$
$$A_{12} = [Q_4 - Q_3]h$$
$$A_{14} = A_{24} \equiv 0$$

A laminate of this type is called an orthotropic laminate, Fig. 11.8. A three-dimensional body formed by cross-ply laminates was said to be orthogonally anisotropic or orthotropic.

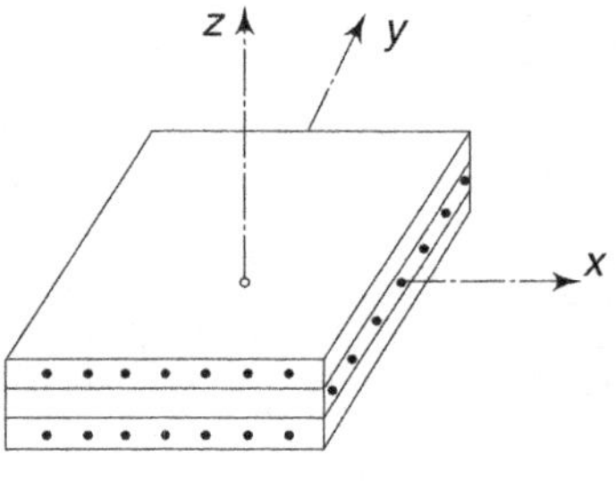

Fig. 11.8 *Cross-ply laminate*

It can easily be seen from the figure that because of the difference in fibre densities in x and y directions, E_{xx} and E_{yy} can be different. Further, a shear stress τ_{xy} produces only shear strain in the xy plane, and does not cause any linear strain either in x direction or in y direction. These are being reflected in A_{14} and A_{24}, both being zero. Further, the moduli components A_{11} and A_{22} vary linearly with $(v_0 - v_{90})$. When v_0 or v_{90}, is zero, we get unidirectional composite laminates. When $v_0 = v_{90}$, the volume fractions are equal and $A_{11} = A_{22}$.

Example 11.5 *Estimate the in-plane moduli and compliances for a cross-ply laminate formed by using unidirectional composites with 'Toray' filament and 'Namco' resin. The modulus data for this ply was given in Example 11.4 and is repeated here:*

$$E_{xx} = 181\,GPa, \quad E_{yy} = 10.3\,GPa, \quad \nu_{xy} = 0.0159, \quad G_{xy} = 7.17\,GPa,$$
$$(1 - \nu_{yx}\nu_{xy})^{-1} = 1.0045$$

Solution From Example 11.4, the values of a_{ij}s for any ply are

$$a_{11} = 181.8 \text{ GPa}, \quad a_{22} = 10.34 \text{ GPa},$$
$$a_{12} = 2.891 \text{ GPa}, \quad a_{44} = 7.17 \text{ GPa}$$

From Eq. (11.39),

$$Q_1 = \frac{1}{8}(3a_{11} + 3a_{22} + 2a_{12} + 4a_{44})$$
$$= \frac{1}{8}[(3 \times 181.8) + (3 \times 10.34) + (2 \times 2.891) + (4 \times 7.17)]$$
$$= 76.36 \text{ GPa}$$
$$Q_2 = \frac{1}{2}(a_{11} - a_{22}) = \frac{1}{2}(181.8 - 10.34) = 85.73 \text{ GPa}$$
$$Q_3 = \frac{1}{8}(a_{11} + a_{22} - 2a_{12} - 4a_{44}) \tag{b}$$
$$= \frac{1}{8}[181.8 + 10.34 - (2 \times 2.891) - (4 \times 7.17)] = 19.71 \text{ GPa}$$
$$Q_4 = \frac{1}{8}(a_{11} + a_{22} + 6a_{12} - 4a_{44})$$
$$= \frac{1}{8}[181.8 + 10.34 + (6 \times 2.891) - (4 \times 7.17)] = 22.6 \text{ GPa}$$
$$Q_5 = \frac{1}{8}(a_{11} + a_{22} - 2a_{12} + 4a_{44}) = 26.88 \text{ GPa}$$

Substituting these in the expressions for A_{ij}s from Eq. (11.53),

$$\frac{1}{h}A_{11} = 76.36 + (\nu_0 - \nu_{90})\,85.73 + 19.71$$
$$\frac{1}{h}A_{22} = 76.36 - (\nu_0 - \nu_{90})\,85.73 + 19.71 \tag{c}$$
$$\frac{1}{h}A_{44} = 26.88 - 19.71 = 7.17$$
$$\frac{1}{h}A_{12} = 22.60 - 19.71 = 2.89$$
$$A_{14} = A_{24} = 0$$

The average values of the compliance coefficients are obtained by the inversion of Eq. (11.46). If Δ is the determinant of the A_{ij}s in Eq. (11.46), then

$$\Delta = A_{11}(A_{22}A_{44} - A^2_{24}) - A_{12}(A_{12}A_{44} - A_{24}A_{14}) + A_{14}(A_{12}A_{24} - A_{22}A_{14})$$

Corresponding to Eq. (11.46), one can write for ε^*_{ij}

$$\begin{aligned}\varepsilon_{xx}^{*} &= \bar{b}_{11}N_{xx} + \bar{b}_{12}N_{yy} + \bar{b}_{14}N_{xy} \\ \varepsilon_{yy}^{*} &= \bar{b}_{12}N_{xx} + \bar{b}_{22}N_{yy} + \bar{b}_{24}N_{xy} \\ \gamma_{xy}^{*} &= \bar{b}_{41}N_{xx} + \bar{b}_{42}N_{yy} + \bar{b}_{44}N_{xy}\end{aligned} \tag{11.54}$$

Solving Eq. (11.46) for ε_{ij}^{*} s and comparing with the coefficients in Eqs (11.54), one gets

$$\begin{aligned}\bar{b}_{11} &= \frac{1}{\Delta}(A_{22}A_{44} - A_{24}^2) \\ \bar{b}_{12} &= -\frac{1}{\Delta}(A_{12}A_{44} - A_{14}A_{24}) \\ \bar{b}_{14} &= \frac{1}{\Delta}(A_{12}A_{24} - A_{14}A_{22}) \\ \bar{b}_{22} &= \frac{1}{\Delta}(A_{11}A_{44} - A_{14}^2) \\ \bar{b}_{24} &= \frac{1}{\Delta}(A_{11}A_{24} - A_{14}A_{12}) \\ \bar{b}_{44} &= \frac{1}{\Delta}(A_{11}A_{22} - A_{12}^2)\end{aligned} \tag{11.55}$$

To find the values of A_{ij}s we need the values of ν_0 and ν_{90}. Assuming $\nu_0 = \nu_{90} = 0.5$,

$$A_{11} = 96.07h, \quad A_{22} = 96.07h, \quad A_{44} = 7.17h, \quad A_{12} = 2.89h,$$

$$A_{14} = A_{24} = 0$$

Substituting these,

$$\Delta = 96.07\,(96.07 \times 7.17) - 2.89\,(2.89 \times 7.17) = 66.12 \times 10^{30} h^3 \,(Pa)^3$$

$$\frac{1}{\Delta} = 0.0151 \times 10^{-30} h^{-3}$$

$$\begin{aligned}\therefore \quad \bar{b}_{11}h &= 0.0151 \times 10^{-30}\,(96.07 \times 7.17) \times 10^{18} = 10.40 \times 10^{-12}\,(Pa)^{-1} \\ \bar{b}_{12}h &= -0.0151 \times 10^{-30}\,(2.89 \times 7.17) \times 10^{18} \\ &= -0.313 \times 10^{-12}\,(Pa)^{-1} \\ \bar{b}_{22}\,h &= 0.0151 \times 10^{-30}\,(96.07 \times 7.17) \times 10^{18} = 10.40 \times 10^{-12}\,(Pa)^{-1} \\ \bar{b}_{44}h &= 0.0151 \times 10^{-30}\,(96.07 \times 96.07 - 2.89^2) \times 10^{18} \\ &= 138.3 \times 10^{-12}\,(Pa)^{-1}\end{aligned} \tag{d}$$

Angle-ply Laminates Another class of laminates that are commonly used in practice are the angle-ply laminates. In these laminates, there are only two ply orientations which have the same magnitudes but are of opposite signs. The laminate is said to be balanced when there are equal number of plies with positive and negative orientations. Hence, for a balanced angle-ply laminate assuming complete symmetry, one has, from Eqs (11.42a and b), the following:

$$\theta_1 = +\beta, \quad \theta_2 = -\beta, \quad v_1 = v_2 = \frac{1}{2} \tag{11.56}$$

Substituting these in Eq. (11.51),

$$\begin{aligned} V_1^* &= \frac{1}{2}(\cos 2\beta + \cos 2\beta) = \cos 2\beta \\ V_2^* &= \cos 4\beta \\ V_3^* &= V_4^* = 0 \end{aligned} \tag{11.57}$$

From Eqs (11.47a and c),

$$\begin{aligned} A_{11} &= \left[Q_1 + Q_2 \cos 2\beta + Q_3 \cos 4\beta\right] h \\ A_{22} &= \left[Q_1 - Q_2 \cos 2\beta + Q_3 \cos 4\beta\right] h \\ A_{12} &= \left[Q_4 - Q_3 \cos 4\beta\right] h \\ A_{44} &= \left[Q_5 - Q_3 \cos 4\beta\right] h \\ A_{14} &= A_{24} = 0 \end{aligned} \tag{11.58}$$

As an example, consider a balanced symmetric angle-ply laminate with $\beta = 45°$. For such a laminate,

$$A_{11} = A_{22} = (Q_1 - Q_3)\, h, \quad A_{12} = (Q_4 + Q_3)\, h, \quad A_{44} = (Q_5 + Q_3)\, h$$

$$A_{14} = A_{24} = 0$$

For the composite considered in Example 11.5, the values of Q_is are,

$$Q_1 = 76.36 \text{ GPa}, \quad Q_2 = 85.73 \text{ GPa}, \quad Q_3 = 19.71 \text{ GPa},$$

$$Q_4 = 22.6 \text{ GPa}, \quad Q_5 = 26.88 \text{ GPa}$$

For a laminate formed from these composites,

$$\begin{aligned} A_{11} &= A_{22} = (76.36 - 19.71)h = 56.65\, h \text{ GPa} \\ A_{12} &= (22.6 + 19.71)h = 42.31\, h \text{ GPa} \\ A_{44} &= (26.88 + 19.71)\, h = 46.59\, h \text{ GPa} \\ A_{14} &= A_{24} = 0 \end{aligned}$$

The components of compliance are obtained from Eq. (11.55)

$$\begin{aligned} \bar{b}_{11}h &= \bar{b}_{22}h = 39.91 \times 10^{-12}\,(Pa)^{-1} \\ \bar{b}_{12}h &= -29.81 \times 10^{-12}\,(Pa)^{-1} \\ \bar{b}_{44}h &= 21.46 \times 10^{-12}\,(Pa)^{-1} \\ \bar{b}_{14} &= \bar{b}_{24} = 0 \end{aligned}$$

The corresponding Engineering constants are

$$\bar{E}_{xx} = \frac{1}{\bar{b}_{11}h} = 25.05 \text{ GPa}$$

$$\bar{E}_{yy} = \frac{1}{\bar{b}_{22}h} = 25.05 \text{ GPa}$$

$$\bar{G}_{xy} = \frac{1}{\bar{b}_{44}h} = 46.59 \text{ GPa}$$

$$\bar{\nu}_{xy} = -\bar{b}_{12}\bar{E}_{xx}\,h = 0.746$$

The reason for working out the values of the Engineering constants is to show that in the case of a composite, the value of Poisson's ratio can be greater than 0.5. In this particular case, the Poisson's ratio in the x direction is 0.746.

11.5 PLY STRESS AND PLY STRAIN

The stress analysis of symmetrical laminates discussed in the previous section was based on the fundamental assumption that all the plies in the laminate experienced uniform strains. As the fibre orientations in the plies are different, for a given strain, the stresses induced in individual plies will be different. Also, under the same assumption of uniform strain, a given load or stress gets distributed according to the stiffness of each ply. As an example, consider a laminate having the code $[0_4/90_4]$s. This is a symmetric cross-ply laminate having a total of 16 plies. Let the plies be of the same composite material that we have been discussing so far, i.e. the values of $h\bar{b}_{ij}$ s are as given in Eq. (d). Let the thickness of each ply be 130×10^{-6} m, and let the laminate be subjected to a uniaxial stress resultant $N_{xx} = 1$ MN/m.

Thickness of laminate = $h = 16 \times 130 \times 10^{-6} = 2.08 \times 10^{-3}$ m

The compliance coefficients for the laminate from Eq. (d) are

$$\bar{b}_{11} = \frac{1}{h} \times 10.40 \times 10^{-12} = \frac{1}{2.08} \times 10.40 \times 10^{-9} = 5 \times 10^{-9}\ (\text{N/m})^{-1}$$

$$\bar{b}_{12} = -\frac{1}{h} \times 0.313 \times 10^{-12} = -\frac{1}{2.08} \times 0.313 \times 10^{-9} = -0.15 \times 10^{-9}\ (\text{N/m})^{-1}$$

$$\bar{b}_{14} = 0$$

From Eq. (11.54), the strains are

$$\varepsilon^*_{xx} = \bar{b}_{11} N_{xx} = 5 \times 10^{-9} \times 10^6 = 5 \times 10^{-3}$$

$$\varepsilon^*_{yy} = \bar{b}_{12} N_{xx} = -0.15 \times 10^{-9} \times 10^6 = -0.15 \times 10^{-3}$$

These are the strains experienced by each ply. For the ply, the stress–strain equations are

$$\sigma_{xx} = a_{11}\varepsilon_{xx} + a_{12}\varepsilon_{yy}$$

$$\sigma_{yy} = a_{21}\varepsilon_{xx} + a_{22}\varepsilon_{yy}$$

From Example 11.4, for 0° fibre orientation,

$$a_{11} = 181.8\ \text{GPa}, \quad a_{22} = 10.34\ \text{GPa}, \quad a_{21} = 2.891\ \text{GPa}$$

$$\therefore \quad \sigma_{xx} = (181.8 \times 5 \times 10^6) - (2.891 \times 0.15 \times 10^6) = 908.6\ \text{MPa.}$$

$$\sigma_{yy} = (2.891 \times 5 \times 10^6) - (10.34 \times 0.15 \times 10^6) = 12.9\ \text{MPa}$$

$$\tau_{xy} = 0$$

For the 90° fibre orientation,

$$a_{11} = 10.34\ \text{GPa}, \quad a_{22} = 181.8\ \text{GPa}, \quad a_{21} = 2.891\ \text{GPa}$$

$$\therefore \quad \sigma_{xx} = (10.34 \times 5 \times 10^6) - (2.891 \times 0.15 \times 10^6) = 51.3\ \text{MPa}$$

$$\sigma_{yy} = (2.891 \times 5 \times 10^6) - (181.8 \times 0.15 \times 10^6) = -12.9\ \text{MPa}$$

$$\tau_{xy} = 0$$

It should be observed that for 0° fibre orientation ply group, σ_{xx} and σ_{yy} are the stresses along and perpendicular to the fibres, whereas for 90° ply group, σ_{xx} and σ_{yy} are stresses perpendicular and parallel to the fibres.

From these results, the average stresses in x and y directions are

$$\bar{\sigma}_{xx} = \frac{1}{2}(908.6 + 51.3) = 479.95 \text{ MPa}$$

$$\bar{\sigma}_{y} = 0, \qquad \bar{\tau}_{xy} = 0$$

The resultant stress in x direction is

$$\bar{\sigma}_{xx}h = (479.95 \times 10^6 \text{ Pa}) \times (2.08 \times 10^{-3} \text{ m}) = 998 \times 10^3 \text{ N/m} \simeq 1 \text{ MN/m}$$

This checks with the resultant applied stress N_{xx}.

One of the reasons in estimating the stresses and strains in individual plies is to check whether they meet the failure criteria. Failure criteria will be discussed in the next section. For example, in the present case if the maximum strain criterion is applied with the limit that

$$\varepsilon_{\max} \text{ along fibre} \leq 10 \times 10^{-3}$$

$$\varepsilon_{\max} \text{ perpendicular to fibre } \leq 4.5 \times 10^{-3}$$

then, for the 90° ply group, $\overset{*}{\varepsilon}_{xx} = 5 \times 10^{-3}$ is the strain perpendicular to the fibres and this is greater than 4.5×10^{-3}, which is the limit. Hence, based on the maximum strain criterion, failure would have occurred in the 90° ply group, when the resultant applied stress reached a value

$$N_{xx(\max)} = \frac{4.5}{5} = 0.9 \text{ MN/m}$$

11.6 FAILURE CRITERIA OF COMPOSITE MATERIALS

It is obvious from the discussions so far that the failure theories for composite materials would be quite different and more complex compared to theories of failure for an isotropic solid. In this section, we shall briefly consider some of the failure criteria found suitable for composites. We shall restrict our discussion to orthotropic materials in general and to laminates in particular. As in the case of isotropic materials, the maximum stress theory and the maximum strain theory are the basic theories that are considered first.

Maximum Stress Theory The maximum stress theory assumes that failure occurs when any of the stresses in the principal material axes reach a critical value. There are three possible modes of failure, and the conditions for these are

$$\begin{aligned} \sigma_{11} &= \overset{*}{\sigma}_{11} \\ \sigma_{22} &= \overset{*}{\sigma}_{22} \\ \tau_{12} &= \overset{*}{\tau}_{12} \end{aligned} \tag{11.59}$$

The stresses σ_{ij} are referred to the principal directions 1 and 2. $\overset{*}{\sigma}_{11}$ is the ultimate tensile or compressive stress in direction 1, $\overset{*}{\sigma}_{22}$ is the ultimate tensile or compressive stress in direction 2 and $\overset{*}{\tau}_{12}$ is the ultimate shear stress acting on plane 1 (i.e. plane with normal in direction 1) in direction 2. The values of $\overset{*}{\sigma}_{11}$, $\overset{*}{\sigma}_{22}$ and $\overset{*}{\tau}_{12}$ are obtained experimentally for a given composite.

If the load were to be applied at an angle θ to the fibre axis direction 1, Fig. 11.5, and if x', y' are the corresponding frame of reference for the applied stresses, then, from the transformation equations

$$\begin{aligned} \sigma_{11} &= \sigma_{x'x'} \cos^2 \theta \\ \sigma_{22} &= \sigma_{x'x'} \sin^2 \theta \\ \tau_{12} &= +\sigma_{x'x'} \sin \theta \cos \theta \end{aligned} \tag{11.60}$$

Combining Eqs (11.59) and (11.60), according to the maximum stress theory, failure occurs when $\sigma_{x'x'}$ assumes the smallest of the following three values

$$\begin{aligned} \sigma_{x'x'} &= \frac{\overset{*}{\sigma}_{11}}{\cos^2 \theta} \\ \sigma_{x'x'} &= \frac{\overset{*}{\sigma}_{22}}{\sin^2 \theta} \\ \sigma_{x'x'} &= \frac{\overset{*}{\tau}_{12}}{\sin \theta \cos \theta} \end{aligned} \tag{11.61}$$

Instead of $\sigma_{x'x'}$ alone, if the stresses acting are $\sigma_{x'x'}$, $\sigma_{y'y'}$ and $\tau_{x'y'}$, then from Eq. (11.23), the stresses in the principal directions 1 and 2 are

$$\begin{aligned} \sigma_{11} &= \sigma_{x'x'} \cos^2 \theta + \sigma_{y'y'} \sin^2 \theta - \tau_{x'y'} \sin 2\theta \\ \sigma_{22} &= \sigma_{x'x'} \sin^2 \theta + \sigma_{y'y'} \cos^2 \theta + \tau_{x'y'} \sin 2\theta \\ \tau_{12} &= \frac{1}{2}(\sigma_{x'x'} - \sigma_{y'y'}) \sin 2\theta + \tau_{x'y'} \cos 2\theta \end{aligned} \tag{11.62}$$

If the applied stresses $\sigma_{x'x'}$, $\sigma_{y'y'}$ and $\tau_{x'y'}$ either individually or in combination cause σ_{11} or σ_{22} or σ_{12} to exceed their maximum allowable values, failure occurs.

Maximum Strain Theory According to this theory, failure occurs when the strain along any principal direction assumes a critical value, i.e when

$$\begin{aligned} \varepsilon_{11} &= \overset{*}{\varepsilon}_{11} \\ \varepsilon_{22} &= \overset{*}{\varepsilon}_{22} \\ \gamma_{12} &= \overset{*}{\gamma}_{12} \end{aligned} \tag{11.63}$$

where $\overset{*}{\varepsilon}_{11}$ is the maximum tensile or compressive strain in direction 1, $\overset{*}{\varepsilon}_{22}$ is the maximum tensile or compressive strain in direction 2, and $\overset{*}{\gamma}_{12}$ is the maximum shear strain in plane 1-2. If E_{11}, E_{22} and G_{12} are the material constants, then, according to the maximum strain theory, failure occurs when any of the following conditions hold:

$$\varepsilon_{11} = \frac{\sigma_{11}}{E_{11}} - \nu_{12}\frac{\sigma_{22}}{E_{22}} \geq \varepsilon_{11}^{*}$$

$$\varepsilon_{22} = -\nu_{12}\frac{\sigma_{11}}{E_{11}} + \frac{\sigma_{22}}{E_{22}} \geq \varepsilon_{22}^{*} \tag{11.64}$$

$$\gamma_{12} = \frac{\tau_{11}}{G_{12}} \geq \gamma_{12}^{*}$$

If $\sigma_{x'x'}$, $\sigma_{y'y'}$ and $\tau_{x'y'}$ are the stresses applied, then the values of σ_{11}, σ_{22} and, τ_{12} are obtained from Eq. (11.23), which can then be substituted into Eq. (11.62).

Distortion Energy Theory While the maximum stress and maximum strain theories are easy to apply, they have limitations since experiments do not completely support them. Another theory which is commonly used in design processes is the energy of distortion theory, which sometimes is called the Tsai–Hill theory. This theory is similar to the distortion energy or the deviatoric stress theory applied to isotropic solids. For an isotropic solid, Eq. (4.12) gives the distortion energy as

$$U^* = \frac{1}{12G}\left[(\sigma_1 - \sigma_2)^2 + (\sigma_2 - \sigma_3)^2 + (\sigma_3 - \sigma_1)^2\right]$$

where σ_1, σ_2 and σ_3 are the principal stresses and G is the shear modulus. For an orthotropic solid, this expression is generalised and written as

$$2f(\sigma_{ij}) = F(\sigma_{11} - \sigma_{22})^2 + G(\sigma_{22} - \sigma_{33})^2 + H(\sigma_{33} - \sigma_{11})^2$$

$$+ 2L\,\tau_{12}^2 + 2M\,\tau_{23}^2 + 2N\,\tau_{31}^2 = 1 \tag{11.65}$$

where 1, 2 and 3 are the principal directions of symmetry and F, G, H, L, M and N are parameters characterising the anisotropy of the material. In the stress-space, Eq. (11.65) represents a six-dimensional surface. The critical values of σ_{ij}s and τ_{ij}s will give the limits to this yield surface. If the applied stresses lie within the surface, then no failure occurs. The values of the parameters are obtained from tests conducted on a sample of the composite. Let $\sigma_{11}^*, \sigma_{22}^*, \sigma_{33}^*$ be the normal or yield strengths in the directions of anisotropic symmetry. Then, with σ_{11}^* alone, Eq. (11.65) gives

$$F\sigma_{11}^{*2} + H\sigma_{11}^{*2} = 1$$

$$F + H = \frac{1}{\sigma_{11}^{*2}} \tag{11.66a}$$

Similarly, with σ_2^* and σ_3^* individually applied, one gets

$$F + G = \frac{1}{\sigma_{22}^{*2}} \tag{11.66b}$$

$$G + H = \frac{1}{\sigma_{33}^{*2}} \tag{11.66c}$$

On the same lines, for τ_{12}, τ_{23} and τ_{31}, one gets

$$2L = \frac{1}{\overset{*2}{\tau_{12}}}, \quad 2M = \frac{1}{\overset{*2}{\tau_{23}}}, \quad 2N = \frac{1}{\overset{*2}{\tau_{31}}} \tag{11.66d}$$

where $\overset{*}{\tau}_{ij}$s are the yield strengths in shear.

From Eqs (11.66a–c), one can solve for F, G and H. The solutions are

$$F = \frac{1}{2}\left(\frac{1}{\overset{*2}{\sigma_{11}}} + \frac{1}{\overset{*2}{\sigma_{22}}} - \frac{1}{\overset{*2}{\sigma_{33}}}\right)$$

$$G = \frac{1}{2}\left(-\frac{1}{\overset{*2}{\sigma_{11}}} + \frac{1}{\overset{*2}{\sigma_{22}}} + \frac{1}{\overset{*2}{\sigma_{33}}}\right) \tag{11.67}$$

$$H = \frac{1}{2}\left(\frac{1}{\overset{*2}{\sigma_{11}}} - \frac{1}{\overset{*2}{\sigma_{22}}} + \frac{1}{\overset{*2}{\sigma_{33}}}\right)$$

Equation (11.65) with values for F, G, H, L, M and N substituted from Eqs (11.66d) and (11.67) describes a failure surface in a six-dimensional space. So long as the point $(\sigma_{11}, \sigma_{22}, \sigma_{23}, \tau_{12}, \tau_{23}, \tau_{31})$ lies within this surface, no failure occurs. If the point happens to fall either on the surface or outside the surface, failure occurs.

Consider now a unidirectionally reinforced composite as shown in Fig. 11.1. Let x-axis be along the fibre direction instead of z as shown in that figure. Then, the plane yz will be a transverse plane of isotropy, and for this plane, the transverse yield strengths $\overset{*}{\sigma}_{yy}$ and $\overset{*}{\sigma}_{zz}$ will be equal to each other. In our notation, this means that $\overset{*}{\sigma}_{22}$ and $\overset{*}{\sigma}_{33}$ are equal. Also, for this body, $\overset{*}{\tau}_{12}$ and $\overset{*}{\tau}_{13}$ are equal, i.e. $L = N$. Hence, for an orthotropic body, substituting the present values for F, G, H, etc., Eq. (11.65) becomes

$$\frac{1}{2\overset{*2}{\sigma_{11}}}\left[(\sigma_{11} - \sigma_{22})^2 - (\sigma_{22} - \sigma_{33})^2 + (\sigma_{33} - \sigma_{11})^2\right] + \frac{1}{\overset{*2}{\sigma_{22}}}(\sigma_{22} - \sigma_{33})^2$$

$$+ \frac{1}{\overset{*2}{\tau_{12}}}\left(\tau_{12}^2 + \tau_{31}^2\right) + \frac{1}{\overset{*2}{\tau_{23}}}(\tau_{23})^2 = 1 \tag{11.68}$$

In the case of a laminate with unidirectional reinforcements, if the state of stress is a plane state, then one has $\sigma_{33} = \tau_{31} = \tau_{23} \equiv 0$. Equation (11.68) then reduces to

$$\left(\frac{\sigma_1}{\overset{*}{\sigma}_{11}}\right)^2 - \left(\frac{\sigma_{11}}{\overset{*}{\sigma}_{11}}\right)\left(\frac{\sigma_{22}}{\overset{*}{\sigma}_{11}}\right) + \left(\frac{\sigma_{22}}{\overset{*}{\sigma}_{22}}\right)^2 + \left(\frac{\tau_{12}}{\overset{*}{\tau}_{12}}\right)^2 = 1 \tag{11.69}$$

Equation (11.69) describes a failure envelope and so long as the point $(\sigma_{11}, \sigma_{22}, \tau_{12})$ lies within the surface no failure occurs. If the unidirectional laminate is subjected to a stress $\sigma_{x'x'}$ at an angle θ to x-axis, then from Eqs (11.60) and (11.69) failure occurs when

$$\sigma_{x'x'} = \left[\frac{\cos^4 \theta}{\sigma_{11}^{*2}} + \left(\frac{1}{\tau_{12}^{*2}} - \frac{1}{\sigma_{11}^{*2}} \right) \sin^2 \theta \cos^2 \theta + \frac{\sin^4 \theta}{\sigma_{22}^{*2}} \right]^{-\frac{1}{2}} \tag{11.70}$$

Example 11.6 *For a class of E-glass-epoxy composite with unidirectional reinforcement, the following data apply:*

$$E_{11} = 53.8\ GPa, \qquad E_{22} = 17.9\ GPa$$

$$\nu_{12} = 0.25 \qquad G_{12} = 8.6\ GPa$$

$$\sigma_{11}^{*}\ (tens) = 1304\ MPa \qquad \sigma_{11}^{*}\ (comp) = 1034\ MPa$$

$$\sigma_{22}^{*}\ (tens) = 27.64\ MPa \qquad \sigma_{22}^{*}\ (comp) = 138\ MPa$$

$$\tau_{12}^{*} = 55.2\ MPa$$

Determine the minimum value of $\sigma_{x'x'}$ applied at an angle of 30° to the fibre axis to cause failure according to (a) maximum stress theory (tension and compression), (b) maximum strain theory (tension) and (c) distortion energy theory (tension).

Solution (a) Maximum Stress Theory

(i) Tension: From Eq. (11.61),

$$\sigma_{x'x'} = \frac{\sigma_{11}^{*}}{\cos^2 \theta} = \frac{1304}{0.74} = 1378.7 \text{ MPa}$$

$$\sigma_{x'x'} = \frac{\sigma_{22}^{*}}{\sin^2 \theta} = \frac{27.6}{0.25} = 110.4 \text{ MPa}$$

$$\sigma_{x'x'} = \frac{\tau_{12}^{*}}{\sin \theta \cos \theta} = \frac{55.2}{0.433} = 127.5 \text{ MPa}$$

Failure occurs when $\sigma_{x'x'} \geq 110.4$ MPa (tension).

(ii) Compression: From Eq. (11.61)

$$\sigma_{x'x'} = \frac{\sigma_{11}^{*}}{\cos^2 \theta} = \frac{1034}{0.75} = 1378.7 \text{ MPa}$$

$$\sigma_{x'x'} = \frac{\sigma_{22}^{*}}{\sin^2 \theta} = \frac{138}{0.25} = 552 \text{ MPa}$$

$$\sigma_{x'x'} = \frac{\tau_{12}^{*}}{\sin \theta \cos \theta} = \frac{55.2}{0.433} = 127.5 \text{ MPa}$$

Failure occurs when $\sigma_{x'x'} \geq 127.5$ MPa (Compression)

(b) Maximum Strain Theory: From Eq. (11.60)

$$\sigma_{11} = \sigma_{x'x'} \cos^2 \theta, \qquad \sigma_{22} = \sigma_{x'x'} \sin^2 \theta, \tag{e}$$

$$\tau_{12} = \frac{1}{2}\sigma_{x'x'} \sin 2\theta$$

Further, from Eq. (11.64), the maximum tensile strain in direction 1 when yield (or failure) stress σ_{11}^* is applied is

$$\varepsilon_{11}^* = \frac{\sigma_{11}^*}{E_{11}} = \frac{1.034}{53.8} = 0.01922$$

Similarly, the maximum tensile strain in direction 2 when yielding (or failure) occurs is

$$\varepsilon_{22}^* = \frac{\sigma_{22}^*}{E_{22}} = \frac{0.0276}{17.9} = 0.001542$$

Further, the shear strain at the time of yielding is

$$\gamma_{12}^* = \frac{\tau_{12}^*}{G_{12}} = \frac{0.0552}{8.6} = 0.00642$$

From Eqs (11.64) and (e)

$$\varepsilon_{11} = \frac{\sigma_{x'x'}}{E_{11}} \cos^2\theta - \nu_{12}\frac{\sigma_{x'x'}}{E_{22}} \sin^2\theta$$

$$= \sigma_{x'x'}\left(\frac{\cos^2\theta}{E_{11}} - \nu_{12}\frac{\sin^2\theta}{E_{22}}\right)$$

$$\varepsilon_{22} = \sigma_{x'x'}\left(-\nu_{21}\frac{\cos^2\theta}{E_{11}} + \frac{\sin^2\theta}{E_{22}}\right)$$

$$\gamma_{12} = \sigma_{x'x'}\left(\frac{1}{2}\frac{\sin 2\theta}{G_{12}}\right)$$

Substituting the critical values for the strains and solving for $\sigma_{x'x'}$ $\left(\text{with the reciprocal identity } \frac{\nu_{12}}{E_{22}} = \frac{\nu_{21}}{E_{11}}\right)$

$$\sigma_{x'x'} = \varepsilon_{11}^*\left[\frac{\cos^2\theta}{E_{11}} - \nu_{12}\frac{\sin^2\theta}{E_{22}}\right]$$

$$= 0.01922\,(0.01394 - 0.003492)^{-1}$$

$$= 1.8395 \text{ GPa} = 1839 \text{ MPa}$$

$$\sigma_{x'x'} = \varepsilon_{22}^*\left[-\nu_{12}\frac{\cos^2\theta}{E_{22}} + \frac{\sin^2\theta}{E_{22}}\right]^{-1}$$

$$= 0.001542\,(-0.01047 + 0.01397)^{-1}$$

$$= 0.4406 \text{ GPa} = 441 \text{ MPa}$$

$$\sigma_{x'x'} = \gamma_{12}^{*}\left(\frac{1}{2}\frac{\sin 2\theta}{G_{12}}\right)^{-1}$$

$$= 0.00642\,(0.04511)^{-1}$$

$$= 0.1275 \text{ GPa} = 142 \text{ MPa}$$

Based on the minimum of the three values, the critical stress value is 142 MPa.

(c) Distortion Energy Theory: From Eq. (11.70)

$$\sigma_{x'x'} = 10^8 \times \left[0.0052612 + (3.2819 - 0.00935) \times 0.1875 + 0.82047\right]^{-1/2}$$
$$= 83 \text{ MPa}$$

11.7 MICROMECHANICS OF COMPOSITES

In this section, the micromechanical aspects of fibrous composites based on the rule of mixtures, where the sharing of the loads by the matrix and the fibre is dependent on the volume-weighted averages of the component properties, will be examined.

Consider a composite of mass m_c and volume v_c. The total mass of the composites is the sum of the matrix mass m_m and the reinforcing fibre mass m_f, i.e.

$$m_c = m_m + m_f \tag{11.71}$$

The subscripts c, m and f refer to composite, matrix and fibre, respectively. The volume v_c of the composite is given by

$$v_c = v_m + v_f + v_v \tag{11.72}$$

where v_v is the volume of voids that the composite element may contain. Dividing Eq. (11.71) by m_c and Eq. (11.72) by v_c, one gets

$$M_m + M_f = 1 \tag{11.73}$$

and

$$V_m + V_f + V_v = 1 \tag{11.74}$$

where the Ms and Vs stand for mass and volume fractions.

Consider a rectangular, unidirectional composite rod, Fig. 11.9, subjected to a force P_c in the direction of the fibres. Assume that the rod extends uniformly with no delaminations between the matrix and the fibres.

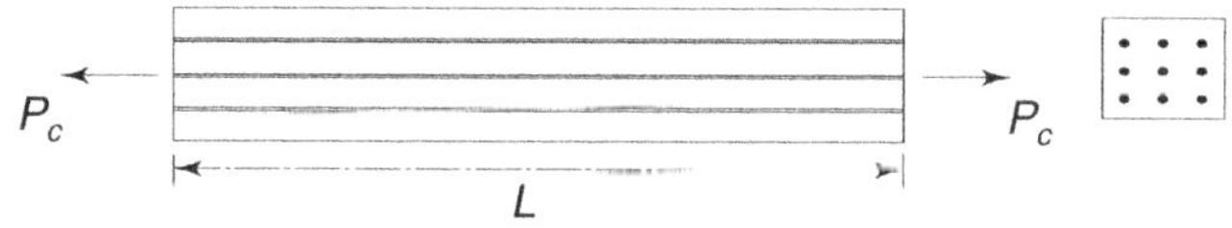

Fig. 11.9 *Unidirectional composite rod*

Assume that transverse sections that were plane before loading remain plane after loading. This means that the strain in the matrix and the strains in the fibres are the same. Thus,

$$\varepsilon_{cl} = \varepsilon_m = \varepsilon_f = \frac{\Delta L}{L} \tag{11.75}$$

where ε_{cl} indicates the strain in the composite in the longitudinal direction.

The situation depicted by Eq. (11.75) is known as the isostrain situation. It is further assumed that the Poisson's ratios of the matrix and the fibres are equal. If E_m and E_f are the Young's moduli for the matrix and the fibre, then the stresses are

$$\sigma_m = E_m\, \varepsilon_{cl}, \quad \sigma_f = E_f\, \varepsilon_{cl}$$

If A_c is the total cross-sectional area of the composite, then

$$P_c = P_m + P_f$$

i.e.

$$\sigma_c A_c = \sigma_m A_m + \sigma_f A_f \tag{11.76a}$$

$$= (A_m E_m + A_f E_f)\, \varepsilon_{cl} \tag{11.76b}$$

or

$$\frac{\sigma_c}{\varepsilon_{cl}} = E_{cl} = E_m \frac{A_m}{A_c} + E_f \frac{A_f}{A_c} \tag{11.77}$$

Since the lengths of the composite, the matrix and the fibres are all equal,

$$v_m = A_m L, \qquad v_f = A_f L, \qquad v_c = A_c L$$

and

$$\frac{A_m}{A_c} = \frac{v_m}{v_c} = V_m, \qquad \frac{A_f}{A_c} = \frac{v_f}{v_c} = V_f \tag{11.78}$$

Eq. (11.77) becomes

$$E_{cl} = E_m\, V_m + E_f V_f = E_{11} \tag{11.79}$$

E_{11} is the Young's modulus for the composite in the fibre direction. This is called the rule of mixtures for the Young's modulus in the fibre direction. From Eqs (11.76a) and (11.78), one can obtain an expression for the composite strength in the fibre direction as

$$\sigma_{cl} = \sigma_m\, V_m + \sigma_f\, V_f \tag{11.80}$$

If the composite is loaded in the transverse direction, and if it is assumed once again that there is no separation between the fibres and the matrix, then one can group the fibres together as one phase material that is continuous, and the matrix as one group, Fig. (11.10).

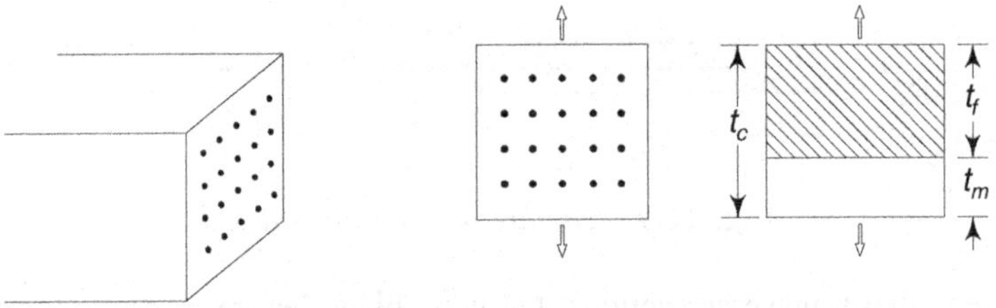

Fig. 11.10 *Two phases of unidirectional composite rod*

If the applied load is uniformly distributed across the transverse faces, then the transverse stresses in the two phases are equal, i.e.

$$\sigma_{ct} = \sigma_m = \sigma_f \tag{11.81}$$

The total transverse displacement is

$$\Delta t_c = \Delta t_m + \Delta t_f \tag{11.82}$$

t_m and t_f are the equivalent gauge lengths of the matrix and the fibre respectively, when each is considered as one phase material. If L is the length of the member as shown in Fig. 11.9, then

$$Lt_m = v_m, \qquad Lt_f = v_f, \qquad Lt_c = v_c \tag{11.83}$$

Dividing Eq. (11.82) by t_c

$$\varepsilon_{ct} = \frac{\Delta t_c}{t_c} = \frac{\Delta t_m}{t_c} + \frac{\Delta t_f}{t_c}$$

$$= \frac{\Delta t_m}{t_m} \cdot \frac{t_m}{t_c} + \frac{\Delta t_f}{t_f} \cdot \frac{t_f}{t_c}$$

$$= \varepsilon_m \frac{v_m}{v_c} + \varepsilon_f \frac{v_f}{v_c}$$

i.e. $$\varepsilon_{ct} = \varepsilon_m V_m + \varepsilon_f V_f \tag{11.84}$$

Now, $$\varepsilon_{ct} = \frac{\sigma_{ct}}{E_{ct}}, \qquad \varepsilon_m = \frac{\sigma_m}{E_m}, \qquad \varepsilon_f = \frac{\sigma_f}{E_f}$$

Using Eq. (11.81), Eq. (11.84) becomes

$$\frac{\sigma_{ct}}{E_{ct}} = \frac{\sigma_{ct}}{E_m} V_m + \frac{\sigma_{ct}}{E_f} V_f$$

or, $$\frac{1}{E_{ct}} = \frac{V_m}{E_m} + \frac{V_f}{E_f} \tag{11.85}$$

Equation (11.85) gives the Young's modulus for the composite in a direction transverse to the fibre direction according to the rules of mixtures. It should be observed that equations (11.79) and (11.85) for the values of the Young's moduli in the axial direction (i.e. in the direction of the fibres) and the transverse direction are obtained under the assumption that the Poisson's ratios for the matrix and the fibres are equal. If the Poisson's ratios are different, then the analysis becomes complicated. Some aspects of this will be discussed subsequently.

When a composite cylindrical rod of uniform cross-section is subjected to a force P_c, assuming that cross-sections remain plane, the stresses in the fibre and matrix, and the linear strain in the rod are given by

$$\varepsilon_{cl} = \frac{\sigma_f}{E_f} = \frac{\sigma_m}{E_m}$$

$$= \frac{P_c}{A_c E_{cl}}$$

$$= \frac{P_c}{A_c}\left[\frac{1}{E_m V_m + E_f V_f}\right], \quad \text{from Eq. (11.79)} \tag{11.86}$$

$$\sigma_f = E_f \, \varepsilon_{cl}, \qquad \sigma_m = E_m \, \varepsilon_{cl} \tag{11.87}$$

The determination of the shear modulus G_c for the composite in terms of shear moduli for the fibre and matrix is not simple. However, under some simple assumptions, an expression can be obtained as indicated next.

Assume, as shown in Fig. 11.11, that the composite can be considered to be a combination of two continuous phase materials, one that of fibre and the other that of matrix.

As shown in Fig. 11.11(b), the shear stresses on the complementary faces are equal. Consequently, if the shear moduli for the matrix and the composite are not equal, there will be some discontinuity in the shear strains as shown in Fig. 11.11(c). Ignoring this discontinuity,

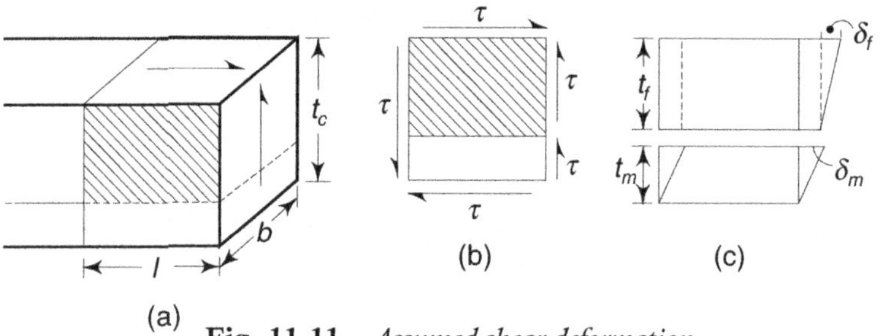

Fig. 11.11 *Assumed shear deformation*

the shear strain in the fibre is $\left(\dfrac{\delta_f}{t_f}\right)$

the shear strain in the matrix is $\left(\dfrac{\delta_m}{t_m}\right)$

and the total shear strain in the composite is $\left(\dfrac{\delta_f + \delta_m}{t_c}\right)$

The shear modulus for the composite is

$$G_c = \frac{\tau}{(\delta_f + \delta_m)/t_c} = \frac{\tau t_c}{\delta_f + \delta_m} \tag{11.88}$$

If G_f and G_m are the shear moduli for the fibre and matrix, then

$$G_f = \frac{\tau}{(\delta_f / t_f)} \quad \text{and} \quad G_m = \frac{\tau}{(\delta_m / t_m)}$$

i.e.

$$d_f = \frac{\tau t_f}{G_f} \quad \text{and} \quad \delta_m = \frac{\tau t_m}{G_m}$$

Substituting these in Eq. (11.88),

$$G_c = \frac{\tau t_c}{(\tau t_f / G_f) + (\tau t_m / G_m)}$$

$$= \frac{t_c}{(t_f / G_f) + (t_m / G_m)}$$

or

$$G_c = \frac{G_f \, G_m}{V_f G_m + V_m G_f} \tag{11.89}$$

Equation (11.89) gives the composite shear modulus in terms of the constituent shear moduli. In obtaining an expression for the composite elastic modulus E_{11} in the fibre direction, it was assumed that the Poisson's ratios for the fibre and matrix were equal. If the ratios happen to be different, one can get the composite Poisson's ratio in terms of the matrix and fibre ratios under some simple assumptions. For this, consider Fig. 11.9 and Eq. (11.79). The longitudinal strain for the composite is

$$\varepsilon_{cl} = \frac{\sigma_{cl}}{E_{cl}} = \frac{\sigma_{cl}}{E_m V_m + E_f V_f}$$

If ν_f and ν_m are the Poisson's ratios for the constituents, then the change in the transverse dimension t_c is

$$\delta t_c = t_m \nu_m \varepsilon_{cl} + t_f \nu_f \varepsilon_{cl}$$

This is under the assumption that there are no transverse stresses when the bar is subjected to uniaxial tension. The transverse strain is therefore

$$\varepsilon_{ct} = \frac{\delta t_c}{t_c} = \frac{\varepsilon_{cl}}{t_c} (t_m \nu_m + t_f \nu_f)$$

$$= \varepsilon_{cl} (V_m \nu_m + V_f \nu_f) \tag{11.90a}$$

using Eq. (11.83). Hence, the Poisson's ratio for the composite is

$$\nu_c = \frac{\varepsilon_{ct}}{\varepsilon_{cl}} = V_m \nu_m + V_f \nu_f \tag{11.90b}$$

It should be observed that the transverse strain ε_{cl} as given by Eq. (11.90a) is negative when ε_{cl}, the longitudinal strain, is positive.

Among the several important properties of composites, the specific strength and specific modulus are the special characteristics. These are defined as follows:

$$\text{Specific strength} = \frac{\sigma}{\rho} \tag{11.91}$$

$$\text{Specific modulus} = \frac{E}{\rho}$$

where σ is the yield or tensile strength, ρ is the density and E is the modulus of clasticity. Pıoperties of some typical fibres are given in Table 11.2. The highest specific modulus is usually found in materials having a low atomic number and covalent bonding, such as carbon and boron. One should be careful about the units involved in Eq. (11.91). In the metric system the yield strength σ will be expressed in kgf/cm^2 and the density in kgf/cm^3. Thus, the specific strength will be expressed in

$$\frac{\text{kgf}}{\text{cm}^2}\cdot\frac{\text{cm}^3}{\text{kg}}: \quad \frac{\text{kg/cm}}{\text{s}^2\text{cm}^2}\cdot\frac{\text{cm}^3}{\text{kg}}: \quad \frac{\text{cm}^2}{\text{s}^2}$$

In SI units also, the specific strength or the specific modulus will be expressed in m^2/s^2.

Table 11.2

Material	Density (Mg/m^3)	Tensile strength (MPa)	Elasticity modulus (GPa)
Polymers			
Kevlar	1.44	4480	124
Polyethelene	1.14	3300	172
Metals			
Be	1.83	1275	303
Boron	2.36	3450	379
Glass			
E-glass	2.55	3450	72.4
S-glass	2.50	4480	86.9
R-glass	2.76	4137	85
Carbon			
High strength	1.75	5650	276
High modulus	1.90	1860	531

Example 11.7 *One of the important light weight composites used for high temperature applications is borasic-reinforced aluminium containing 40% by volume of fibres. Estimate the density, modulus of elasticity and tensile strength parallel to the fibre axis. Also estimate the modulus of elasticity perpendicular to the fibres. The following data is given:*

Material	*Density (kg/m^3)*	*E (GPa)*	*Tensile strength (MPa)*
Fibres	2.36×10^3	380	2760
Aluminium	2.70×10^3	70	35

Solution A cubic metre of composite consists of 0.4 m^3 of fibres and 0.6 m^3 of aluminium. Hence, from Eq. (11.71), the density ρ_c (kg/m^3) of the composite is

$$\rho_c = 0.6(2.70\times10^3) + 0.4(2.36\times10^3)$$

$$= 2.56\times10^3\ \text{kg/m}^3$$

From Eq. (11.79),

$$E_{cl} = 70(0.6) + 380(0.4) = 194\ \text{GPa}$$

From Eq. (11.80),

$$\sigma_{cl} = 35(0.6) + 2760\,(0.4) = 1125\ \text{MPa}$$

In a direction perpendicular to the fibre axis, from Eq. (11.85),

$$\frac{1}{E_{ct}} = \frac{0.6}{70} + \frac{0.4}{380} = 9.624 \times 10^{-3}$$

$$\therefore \quad E_{ct} = 103.9 \text{ GPa}$$

Example 11.8 *A glass fibre reinforced nylon composite contains E-glass ibres 30% by volume. Calculate the percentage of load carried by the fibres when the composite is loaded. The moduli of elasticity of the constituents are E (glass) = 72 GPa, E (nylon) = 2.8 GPa.*

Solution Assuming isostrain condition,

$$\varepsilon_{cl} = \varepsilon_m = \varepsilon_f$$

But,

$$\varepsilon_m = \frac{\sigma_m}{E_m} \quad \text{and} \quad \varepsilon_f = \frac{\sigma_f}{E_f}$$

$$\therefore \quad \frac{\sigma_m}{E_m} = \frac{\sigma_f}{E_f}$$

i.e

$$\frac{\sigma_f}{\sigma_m} = \frac{E_f}{E_m} = \frac{72}{2.8} = 25.71$$

The load carried by the composite is

$$F_c = F_m + F_f$$

Hence, the fraction of the load carried by the fibre is

$$\frac{F_f}{F_m + F_f} = \frac{\sigma_f A_f}{\sigma_m A_m + \sigma_f A_f}$$

$$= \frac{\sigma_f v_f}{\sigma_m v_m + \sigma_f v_f}, \quad \text{using Eq. (11.78)}$$

$$= \frac{\sigma_f (0.3)}{\sigma_m (0.7) + \sigma_f (0.3)}$$

$$= \frac{0.3}{0.7(\sigma_m / \sigma_f) + 0.3}$$

$$= \frac{0.3}{0.7(1/25.71) + 0.3} = 0.92$$

Hence, the fibres carry 92% of the applied load.

Example 11.9 *An important part of a structure which currently is being made of an aluminium alloy having a modulus of elasticity of 60 GPa is to be replaced by a composite material consisting of E-glass fibres in nylon matrix. It is desired that while weight reduction is important, the specific modulus of the composite should not be lower than that of the current material. The direction of loading in the composite will be in the fibre direction. The density of aluminium alloy used is* 2.8×10^3 *kgf* / m^3.

Solution The specific modulus of the aluminium alloy is

$$\frac{60 \text{ GPa}}{2.8 \times 10^3 \text{ kg m}^{-3}} = \frac{60 \times 10^9 \text{ Nm}^3}{2.8 \times 10^3 \text{ kg m}^2}$$

$$= 20.69 \times 10^6 \text{ m}^2 \text{ sec}^{-2}$$

From Table 11.2, the density of E-glass is 2.55×10^3 kg/m^3, and its modulus is 72 GPa. For nylon, the corresponding values are 1.14×10^3 kg/m^3 and 2.8 GPa. If we use 60% by volume of glass fibres in the composite, then the density and modulus of the composite will be

$$\rho_c = (0.6) \times 2.55 \times 10^3 + (0.4) \times 1.14 \times 10^3 = 1.986 \times 10^3 \text{ kg/m}^3.$$

$$E_c = (0.6) \times 72 + (0.4) \times 2.8$$

$$= 44.32 \text{ GPa}$$

∴ specific modulus of the composite is

$$\frac{44.32 \times 10^9 \text{ Nm}^{-2}}{1.986 \times 10^3 \text{ kg m}^{-3}} = 22.9 \times 10^6 \text{ Nkg}^{-1}\text{m} = 22.9 \times 10^6 \text{ m}^2 \text{ sec}^{-2}$$

While the specific modulus is marginally increased by 10%, the density is reduced by 29% of the original values.

Example 11.10 *A microlaminate is produced using five sheets of 0.4 mm thick aluminium and four sheets of 0.2 mm epoxy which is reinforced with unidirectionally oriented Kevlar fibres. The volume fraction of Kevlar fibres in these intermediate epoxy sheets is 55%. Calculate the modulus of elasticity of the microlaminate parallel and perpendicular to the fibre alignment.*

Solution In each epoxy sheet of 0.2 mm thickness, the fibre content is 55%. Thus, in a 1 mm × 1 mm sheet size, the fibre content is

$$(0.2 \times 0.55) = 0.11 \text{ mm}^3$$

and that of pure epoxy content is

$$(0.2 \times 0.45) = 0.09 \text{ mm}^3$$

Since there are four such fibre reinforced epoxy sheets, the total fibre content is 0.44 mm^3, and that of pure epoxy is 0.36 mm^3.

A microlaminate of size 1 mm × 1 mm has a total volume equal to

$$V_c = (5 \times 0.4) + (4 \times 0.2) = 2.88 \text{ mm}^3$$

Out of this, the aluminium content is 2 mm^3, the pure epoxy content is 0.36 mm^3, and that of fibres is 0.44 mm^3. Hence, the modulus along the fibre according to the rule of mixtures is

$$E_{cl} = \frac{1}{2.8}[(2 \times 70) + (0.36 \times 3) + (0.44 \times 124)]$$

$$= 69.87 \text{ GPa}$$

To evaluate the modulus perpendicular to the fibre orientation, we have to do it in two steps. The aluminium sheets being isotropic, its modulus will be direction independent. However, for the reinforced epoxy, we have to use Eq. (11.85). For each of the reinforced epoxy sheets, if E'_{cl} is the modulus in a direction transverse to fibre orientation, then

$$\frac{1}{E'_{cl}} = \frac{V_m}{E_m} + \frac{V_f}{E_f}$$

$$= \frac{1}{0.2}\left(\frac{0.09}{3} + \frac{0.11}{124}\right) = 0.1544\,(\text{GPa})^{-1}$$

$\therefore$ $$E_{ct} = 6.477 \text{ GPa}$$

Now, the laminate consists of aluminium (volume content = 2 mm^3, E_{al} = 70 GPa), and fibre reinforced epoxy (volume content = 0.8 mm^3, E'_{ct} = 6.477 GPa). Hence, for the laminate, the modulus will be

$$\frac{1}{E_{ct}} = \frac{1}{2.8}\left[\frac{2}{70} + \frac{0.8}{6.477}\right] = 0.0543$$

$\therefore$ $$E_{ct} = 18.4 \text{ GPa}$$

In getting the above answer, we have used Eq. (11.85) assuming isostress conditions, and this gives a low modulus value. However, for the reinforced epoxy, a modulus in the transverse direction has already been determined as E'_{ct}. So, if the bonding is good between the aluminium sheets and the reinforced epoxy sheets, one can use the isostrain condition and obtain a modulus value as

$$E_{ct} = \frac{1}{2.8}[(2 \times 70) + (0.8 \times 6.477)] = 51.8 \text{ GPa}$$

The actual value will however be in between these two values.

Example 11.11 *It is desired to design a tensile member made of a uni-directional composite material. The structure is to carry a load of 2.2 kN and is to be 3 m long having a circular cross-section. The matrix is to be epoxy with a yield strength of 80 MPa. The yield strength of the composite should not exceed the yield strength of the epoxy. This is to make sure that if the fibres break, the epoxy will be able to carry the load without any catastrophic failure. Assume a modulus of 3.5 GPa for the epoxy. It is also required that the composite member should not stretch more than 2.5 mm.*

Solution If the member is made entirely of epoxy without any fibres, then

$$\varepsilon_{max} = \frac{25 \text{ mm}}{3000 \text{ mm}} = 0.88 \times 10^{\;3}$$

$$\sigma_{max} = \varepsilon_{max} \times E$$

$$= 0.83 \times 10^{-3} \times 3.5 \times 10^{9}$$

$$= 2.92 \times 10^{6} \text{ Nm}^{-2}$$

$$\text{Area of section} = \frac{2.2 \times 10^3}{2.92 \times 10^6 \text{ Nm}^{-2}}$$

$$= 0.753 \times 10^{-3} \text{ m}^2$$

$\therefore$ Diameter of the member $= d = 31$ mm

Assuming a specific weight of 1.25×10^3 kgf m^{-3}, the weight of the tensile member will be

$$W(\text{epoxy}) = (1.25 \times 10^3)\,(0.753 \times 10^{-3})\,(3) = 2.83 \text{ kgf}$$

For the composite, the maximum strain permitted is still 0.83×10^{-3}. The maximum yield strength for the composite is 80 MPa. Hence, the minimum modulus for the composite will be

$$E_c\,(\text{minm}) = \frac{\sigma}{\varepsilon} = \frac{80 \times 10^6 \text{ Nm}^{-2}}{0.83 \times 10^{-3}} = 96.4 \times 10^9 \text{ Nm}^{-2}$$

From Table 11.2, the moduli of glass fibres are less than the minimum required. So one has to look for a fibre having a higher modulus. High-modulus carbon having a modulus of 531 GPa, and a density of 1.90 Mgm^{-3} meets our requirement. If V_f is the volume fraction of the carbon fibre in the composite, the modulus of the composite will be

$$E_c = V_f\,(531) + (1 - V_f) \times 3.5 = 96.4$$

This should be equal to or greater than 96.4. Thus,

$$V_f\,(531) + (1 - V_f) \times 3.5 = 96.4$$

or

$$V_f = 0.176$$

The volume fraction of the carbon is 0.176 and that of epoxy is 0.824. A composite of this nature will have a modulus not less than 96.4 GPa.

If the structure is made of such a composite, and if the fibres break when a load of 2.2 kN is applied, then the epoxy alone should be able to carry the load. If A_c is the total area of section, then 0.824 A_c is the area of epoxy and the stress on this should not exceed 80 MPa. Thus,

$$0.824\,A_c \times 80 \times 10^6 \text{ Nm}^{-2} = 2.2 \times 10^3 \text{ N}$$

or

$$A_c = \frac{2.2 \times 10^3 \text{ N}}{0.824 \times 80 \times 10^6 \text{ Nm}^{-2}}$$

$$= 0.0333 \times 10^{-3} \text{ m}^2 = 33.4 \text{ mm}^2$$

The diameter of the composite is 6.5 mm.

$$\text{Volume} = 33.4 \times 10^{-6} \text{ m}^2 \times 3 \text{ m} = 10 \times 10^{-5} \text{ m}^3$$

$$\text{Weight} = [(1.9 \times 0.176) + (1.25 \times 0.824)] \times 10 \times 10^{-5} = 0.137 \text{ kgf}$$

Therefore, the carbon fibre reinforced structure is less than one-quarter the diameter of pure epoxy structure, and one-twentieth the weight of pure epoxy.

11.8 PRESSURE VESSELS

Let the thickness of the vessel be small compared to the radius of the vessel, so that the curvature effects on the fibres can be neglected. The problem concerned is with the orientation of the fibres for optimum strength. In the netting theory, it is assumed that only the fibres take the load and that too in the directions of the fibres only. The strength of the fibre in its transverse direction is taken as zero. The contribution of the matrix to the strength is ignored.

Consider a cylindrical pressure vessel with closed ends, as shown in Fig. 11.12(a), subjected to an internal pressure p. The longitudinal and hoop stresses are

$$\sigma_z = \frac{pa}{2h}, \qquad \sigma_\theta = \frac{pa}{h} \tag{11.92}$$

where a is the radius of the vessel and h is the thickness of the vessel. Assume a helical winding as shown in Fig. 11.12, and let us consider the stresses along the fibre orientation. If σ is the stress along the fibre orientation, then the stress in the z direction is $\sigma \cos^2 \phi$ and that in the hoop direction is $\sigma \sin^2 \phi$. For equilibrium,

$$\sigma \cos^2 \phi = \frac{pa}{2h} \quad \text{and} \quad \sigma \sin^2 \phi = \frac{pa}{h} \tag{11.93}$$

From these two,

$$\tan^2 \phi = 2 \quad \text{or} \quad \phi \cong 55° \tag{11.94}$$

Hence, the optimum orientation of the fibre does not coincide with the principal stress direction. The shear stress shown in Fig. 11.12(c) is balanced by the shear stress caused by the fibre in the $-\phi$ direction. In practice, the fibres are not made to run in the optimum directions as given by Eq. (11.94), because such a pattern cannot be used to form the end domes. Generally, a small winding angle ϕ is used to form both the cylindrical portion and the end domes, and then an overlay of fibres in the circumferential direction is put to resist the hoop stresses. Thus, in practice, the fibres run approximately in the principal stress directions.

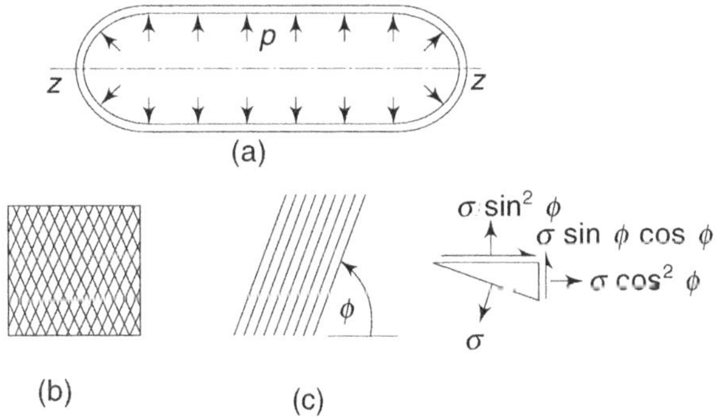

(b) (c)

Fig. 11.12 *Composite pressure vessel*

11.9 TRANSVERSE STRESSES

In the previous sections, it was assumed that the Poisson's ratios of the matrix material and of the fibres were equal. When the ratios are different, one can expect forces between the surfaces of contact because of different contractile tendencies. To see this, consider a cylindrical composite member having a single fibre as a core, Fig. 11.13. Let a be the radius of the fibre and b the outer radius of the matrix. Let the composite cylinder be subjected to a uniaxial load in the z direction.

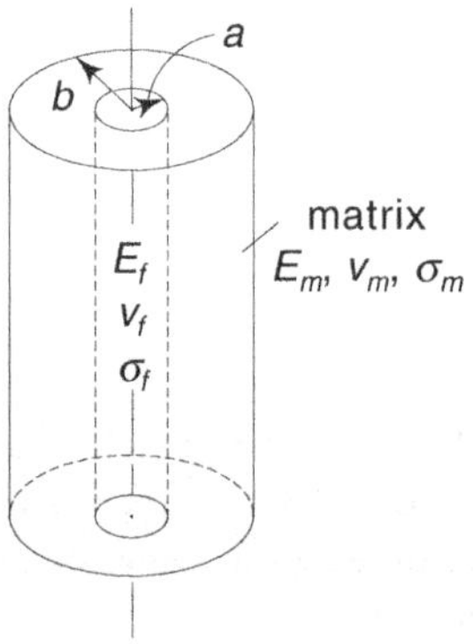

Fig. 11.13 *Cylinder with a single fibre*

Let ε_r, ε_θ, ε_z be the strains and σ_r, σ_θ, σ_z the stress components in the polar coordinate system. Then, the general Hooke's law with Young's modulus in the longitudinal direction as E and Poisson's ratio as ν, is

$$\begin{vmatrix} \varepsilon_r & 0 & 0 \\ 0 & \varepsilon_\theta & 0 \\ 0 & 0 & \varepsilon_z \end{vmatrix} = \frac{1+\nu}{E} \begin{vmatrix} \sigma_r & 0 & 0 \\ 0 & \sigma_\theta & 0 \\ 0 & 0 & \sigma_z \end{vmatrix} -\frac{\nu}{E}(\sigma_r + \sigma_\theta + \sigma_z) \begin{vmatrix} 1 & 0 & 0 \\ 0 & 1 & 0 \\ 0 & 0 & 1 \end{vmatrix} \tag{11.95}$$

The equation of equilibrium is

$$\frac{d\sigma_r}{dr} + \frac{\sigma_r - \sigma_\theta}{r} = 0$$

This is the only equilibrium equation for the case under consideration. Let the strain in the z direction be constant. The strain-displacement equations are

$$\varepsilon_r = \frac{du_r}{dr}, \qquad \varepsilon_\theta = \frac{u_r}{r}, \qquad \varepsilon_z = \text{constant} \tag{11.96}$$

where u_r is the radial displacement. Equation (11.95) can be solved for σ_r and σ_θ in terms of ε_r, ε_θ and ε_z. The results are

$$\sigma_r = K[(1-\nu)\,\varepsilon_r + \nu(\varepsilon_\theta + \varepsilon_z)]$$

$$\sigma_\theta = K[(1-\nu)\,\varepsilon_\theta + \nu(\varepsilon_r + \varepsilon_z)] \tag{11.97a}$$

where
$$K = \frac{E}{(1+\nu)(1-2\nu)} \tag{11.97b}$$

Substituting for ε_r and ε_θ from Eq. (11.96), one gets

$$\sigma_r = K\left[(1-\nu)\frac{du_r}{dr} + \nu\frac{u_r}{r} + \nu\varepsilon_z\right] \tag{11.98}$$

$$\sigma_\theta = K\left[\nu\frac{du_r}{dr} + (1-\nu)\frac{u_r}{r} + \nu\varepsilon_z\right]$$

Substituting into the equilibrium equation, the result appears as

$$\frac{d^2u_r}{dr^2} + \frac{1}{r}\frac{du_r}{dr} - \frac{u_r}{r^2} = 0 \tag{11.99}$$

The solution of the above differential equation is

$$u_r = C_1 r + \frac{C_2}{r} \tag{11.100}$$

where C_1 and C_2 are constants to be determined from the boundary conditions. Equation (11.100) is valid for both matrix and fibre. Representing the fibre equation by subscript f, and the matrix equation by m, Eq. (11.100) becomes

$$u_{rf} = C_{1f}\, r + \frac{C_{2f}}{r} \tag{11.101}$$

$$u_{rm} = C_{1m}\, r + \frac{C_{2m}}{r} \tag{11.102}$$

The boundary conditions to determine the constants are

(i) At the free surface $r = b$, $s_{rm} = 0$

(ii) At the interface $r = a$, because of continuity, $u_{rf} = u_{rm}$ and $\sigma_{rf} = \sigma_{rm}$

(iii) At $r = 0$, $u_{rf} = 0$ This gives $C_{2f} = 0$

Applying the above boundary conditions, the following equations are obtained:

(i) $$(1-\nu_m)\left(C_{1m} - \frac{C_{2m}}{b^2}\right) + \nu_m\left(C_{1m} - \frac{C_{2m}}{b^2}\right) + \nu_m\varepsilon_z = 0$$

or, $$C_{1m} + \frac{1+2\nu_m}{b^2}C_{2m} = -\nu_m\,\varepsilon_z \tag{f}$$

(ii) $$(1-\nu_m)\left(C_{1m} - \frac{C_{2m}}{a^2}\right) + \nu_m\left(C_{1m} + \frac{C_{2m}}{a^2}\right) + \nu_m\,\varepsilon_z$$

$$= (1-\nu_f)\,C_{1f} + \nu_f\, C_{1f} + \nu_f\,\varepsilon_z$$

or, $$C_{1m} + \frac{1+2\nu_m}{a^2}C_{2m} + \nu_m\,\varepsilon_z = C_{1f} + \nu_f\,\varepsilon_z$$

i.e. $$C_{1m} + \frac{1+2\nu_m}{a^2}C_{2m} - C_{1f} = (\nu_f - \nu_m)\,\varepsilon_z \tag{g}$$

Also, $$C_{1m}a + \frac{C_{2m}}{a^2} - C_{if}\; a = 0 \tag{h}$$

Equations (f) – (h) can be solved for the constants. The stress ($-p$), at the interface is obtained as

$$p = \frac{2\varepsilon_z(\nu_m - \nu_f)V_m}{(V_f/2K_m) + (V_m/2K_f) + (1/G_m)}$$

where K is given by Eq. (11.97b).
The stress components in the fibre are

$$\sigma_{rf} = \sigma_{\theta f} = -p$$

$$\sigma_{zf} = E_f\varepsilon_z - 2\nu_f p$$

The stress components in the matrix are

$$\sigma_{rm} = p\left(\frac{a^2}{b^2 - a^2}\right)\left(1 - \frac{b^2}{r^2}\right)$$

$$\sigma_{\theta m} = p\left(\frac{a^2}{b^2 - a^2}\right)\left(1 + \frac{b^2}{r^2}\right) \qquad (11.103)$$

$$\sigma_{zm} = E_m\varepsilon_z + 2\nu_m\, p\left(\frac{a^2}{b^2 - a^2}\right)$$

Problems

11.1 A particular laminate has the following elastic constants along the principal axes x-y:

$$E_{xx} = 200\text{ GPa}, \quad E_{yy} = 20\text{ GPa}, \quad G_{xy} = 10\text{ GPa}, \quad \nu_{yx} = 0.25$$

At a point in the laminate, the following state of stress exists:

$$\sigma_{x'x'} = 200\text{ MPa}, \quad \sigma_{y'y'} = 20\text{ MPa}, \quad \tau_{x'y'} = 20\text{ MPa}$$

The x'-axis makes an angle of 30° with the fibre axis, counter-clockwise. Calculate the principal stresses, the principal strains and their orientations.

[*Ans.* $\sigma_{1,2} = 202.2$ MPa; 7.8 MPa
$\phi' = 6.25°$ and $96.25°$
$\varepsilon_{1,2} = 7.207 \times 10^{-3}; -2.255 \times 10^{-3}$
$\phi^* = -33.6°$ and $56.4°$]

11.2 For a graphite-epoxy laminate having uniaxial reinforcements (parallel to x-axis), the following elastic constants apply:

$$E_{xx} = 181\text{ GPa}, \quad E_{yy} = 10.3\text{ GPa}, \quad G_{xy} = 7.17\text{ GPa}, \quad \nu_{yx} = 0.28$$

Obtain the off-axis compliance coefficients when the axes are rotated by (a) +45° and (b) +60°. Express the results in $(10^{12}\text{ Pa})^{-1}$ units.

$$\left[\begin{array}{ll} Ans. & (a)\ b'_{11} = 59.75;\ b'_{22} = 59.75;\ b'_{12} = -9.99 \\ & b'_{44} = 105.7;\ b'_{14} = -45.78;\ b'_{24} = -45.78 \\ & (b)\ b'_{11} = 80.53;\ b'_{22} = 34.75;\ b'_{12} = -7.88 \\ & b'_{44} = 114.2\ b'_{14} = -32.34;\ b'_{24} = -46.96 \end{array}\right]$$

11.3 Estimate the components of moduli and compliances for a cross-ply laminate formed from composites consisting of Toray filament and Namco resin. The modulus data are

$$E_{xx} = 181\,\text{GPa}, \quad E_{yy} = 10.3\,\text{GPa}, \quad \nu_{yx} = 0.159, \quad G_{xy} = 7.17\,\text{GPa},$$

$$(1 - \nu_{yx}\,\nu_{xy})^{-1} = 1.0045$$

The laminate code is (a) $[0_2/90]$s (b) $[0_4/90]$. Assume that the composites are of uniform thicknesses. Express

$$\left[\begin{array}{ll} Ans. & (a)\ A_{11} = 124.65h;\ \ A_{22} = 67.49h; \\ & A_{44} = 7.17h;\ \ A_{12} = 2.89h; \\ & b_{11}h = 8.03;\ \ b_{22}h = 14.82; \\ & b_{44}h = 139.47;\ \ b_{12}h = -0.344; \\ & (b)\ A_{11} = 147.51h;\ \ A_{22} = 44.63h; \\ & A_{44} = 7.17h;\ \ A_{12} = 2.89h \\ & b_{11}h = 6.78;\ \ b_{22}h = 22.43; \\ & b_{44}h = 139.47;\ \ b_{12}h = -0.440 \end{array}\right]$$

11.4 A laminate is formed from angle-ply composite plies having elastic constants given in Example 11.5. Estimate the components of moduli and compliances for the laminate described by the following codes: (a) $\phi = \pm 30°$ and (b) $\phi = \pm 60°$.

$$\left[\begin{array}{ll} Ans. & (a)\ A_{11} = 109.3h;\ \ A_{22} = 23.6h; \\ & A_{12} = 32.46\ h \\ & A_{44} = 36.73h;\ \ A_{14} = A_{24} = 0 \\ & h\bar{b}_{11} = 15.42;\ \ h\bar{b}_{22} = 71.36; \\ & h\bar{b}_{12} = -21.8; \\ & h\bar{b}_{44} = 27.22;\ \ h\bar{b}_{14} = \bar{b}_{24} = 0 \end{array}\right]$$

$$\left[\begin{array}{l}\text{(b) } A_{11} = 23.6h;\quad A_{22} = 109.3h;\\ A_{12} = 32.46\,h\\ A_{44} = 36.73h;\quad A_{14} = A_{24} = 0\\ h\bar{b}_{11} = 71.36;\quad h\bar{b}_{22} = 15.42;\\ h\bar{b}_{12} = -21.18;\quad h\bar{b}_{44} = 27.22;\\ \bar{b}_{14} = \bar{b}_{24} = 0\end{array}\right]$$

11.5 For the laminates of Problem 11.4, estimate the average values of the engineering constants (Es and νs) corresponding to x and y axes.

$$\left[\begin{array}{ll}\textit{Ans.}\ \text{(a)} & \bar{E}_{xx} = 64.9 \text{ GPa}; \bar{E}_{yy} = 14 \text{ GPa};\\ & \bar{G}_{xy} = 36.7 \text{ GPa}; \bar{\nu}_{xy} = 1.376;\\ \text{(b)} & \bar{E}_{xx} = 14 \text{ GPa}; \bar{E}_{yy} = 64.9 \text{ GPa};\\ & \bar{G}_{xy} = 36.7 \text{ GPa}; \bar{\nu}_{xy} = 0.297\end{array}\right]$$

11.6 For the laminate described in Example 11.6, determine the minimum failure stresses $\sigma_{x'x'}$ applied at $\theta = 45°$ and $\theta = 60°$ to the fibre axis according to (a) maxmium stress theory in tension and compression; (b) maximum strain theory in tension only; (c) distortion energy theory in tension and compression. Use the data given in Example 11.6.

$$\left[\begin{array}{lll}\textit{Ans.}\ \text{(a)} & \text{Tension:} & \theta = 45°, \sigma_{x'x'} = 55.2 \text{ MPa}\\ & & \theta = 60°, \sigma_{x'x'} = 36.8 \text{ MPa}\\ & \text{Compression:} & \theta = 45°, \sigma_{x'x'} = 110.4 \text{ MPa}\\ & & \theta = 60°, \sigma_{x'x'} = 127.5 \text{ MPa}\\ \text{(b)} & \text{Tension:} & \theta = 45°, \sigma_{x'x'} = 73.6 \text{ MPa}\\ & & \theta = 60°, \sigma_{x'x'} = 40 \text{ MPa}\\ \text{(c)} & \text{Tension:} & \theta = 45°, \sigma_{x'x'} = 49.4 \text{ MPa}\\ & & \theta = 60°, \sigma_{x'x'} = 35.3 \text{ MPa}\\ & \text{Compression:} & \theta = 45°, \sigma_{x'x'} = 102 \text{ MPa}\\ & & \theta = 60°, \sigma_{x'x'} = 105 \text{ MPa}\end{array}\right]$$

11.7 A cemented carbide cutting tool used for machining contains 75% by weight tungsten carbide (WC), 15% by weight titanium carbide (TiC), 5% by weight TaC, and 5% by weight cobalt (Co). Estimate the density of the composite, given the following densities for the constituents:

$$\rho_{wc} = 15.77 \text{ Mgm}^{-3}, \qquad \rho_{Tic} = 4.94 \text{ Mgm}^{-3},$$
$$\rho_{Tac} = 14.5 \text{ Mgm}^{-3}, \qquad \rho_{co} = 8.90 \text{ Mgm}^{-3},$$

[*Ans.* 11.50 Mgm^{-3}]

11.8 An electrical contact material is produced by infiltrating copper into a porous tungsten-carbide (WC) compact. The density of WC is 15.77 Mgm^{-3} and that of the final composite is 12.3 Mgm^{-3}. Assuming that all of the pores are filled with copper, and given ρ_c = 8.94 Mgm^{-3}, calculate

(i) the volume fraction of copper in the composite,

(ii) the volume fraction of pores in WC compact before infiltration and

(iii) the original density of WC compact.

[*Ans.* (a) 0.507; (b) 0.507; (c) 7.775]

11.9 An epoxy matrix is reinforced with 40% by volume E-glass fibres to produce a 20 mm diametre composite to carry a load of 25 kN. Calculate the stress acting on the fibre elements. The modulus of epoxy is 3 Gpa and that of glass fibre is 72.4 Gpa.

[*Ans.* 187.3 MPa]

11.10 In the design problem of Example 11.11, if one uses high strength carbon instead of the high modulus carbon, what will be the changes as compared to the pure epoxy member?

[*Ans.* Diameter = 7.3 mm; Weight = 0.179 kgf]

11.11 If Kevlar fibres are used instead of carbon fibres in Example 11.11, show that the volume fraction of fibre needed would be 0.8, and the diameter of the member would be 13.1 mm, and the weight 0.57 kgf.

CHAPTER

12 Introduction to Stress Concentration and Fracture Mechanics

I STRESS CONCENTRATION

12.1 INTRODUCTION

While analysing the stresses induced in members subjected to tension, compression, torsion, and bending, it is generally assumed that members do not have abrupt changes in their cross-sections. In the case of a tapered member under tension or compression, the cross-section changes uniformly. But, abrupt changes in the cross-sections of load-bearing members cannot be avoided. Shafts subjected to torsion will have shoulders to take up thrusts, and key-ways for pulleys and gears. Oil grooves, holes, notches, etc., are common. In such cases, the analysis of stresses and strains become complicated. Elementary equations derived under the assumption of no abrupt changes in the geometry of the section are no longer valid. Sectional discontinuities are called *stress raisers*, and the distribution of stresses in the neighbourshood of such regions are higher than in other regions. They are called regions of *stress concentration*. Generally, stress concentration is a highly localized effect. Figures 12.1(a) and (b) show members with stepped cross-sections under tension and torsion respectively. Let the members be circular in their cross-sections. In the case of the member under tension, let A_1, A_2, and A_3 be respectively the cross-sectional areas of the parts A, B, and C. If P is the axial tensile force, the stresses in the parts according to elementary analysis are $\frac{P}{A_1}$, $\frac{P}{A_2}$, and $\frac{P}{A_3}$. However, these values are valid in regions for removed from sectional discontinuities including the region where the load P is applied. The corners where the discontinuities occur are regions of stress concentration. These are shown by dots. Similarly, in the case of the torsion member, the shear stresses by elementary analysis are $\frac{Tr}{I_a}$ and $\frac{Tr}{I_b}$, where I_a and I_b are the polar moments of inertia of the parts A and B. As before, these average stress values are valid in regions far removed from geometrical discontinuities. At points of discontinuities and nearabout, the stress values are high.

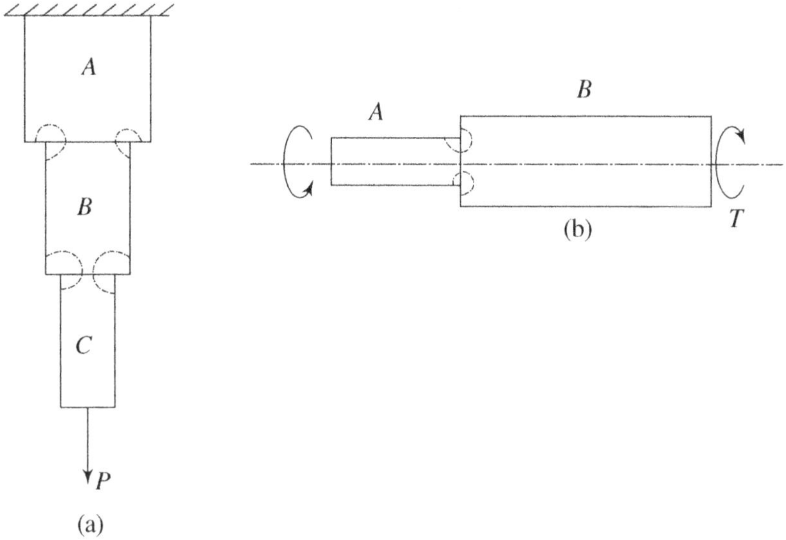

Fig. 12.1 *Stepped cross-sections*

12.2 MEMBERS UNDER TENSION

Figure 12.2 shows a two-dimensional member having two semi-circular grooves and subjected to tensile loading.

The distribution of normal stresses across the section *mn* is shown qualitatively in the figure. At points *m* and *n*, the stress magnitudes are high and they fall rapidly to a uniform value as shown. Ignoring stress concentration, the average or the nominal stress across the section *mn* is

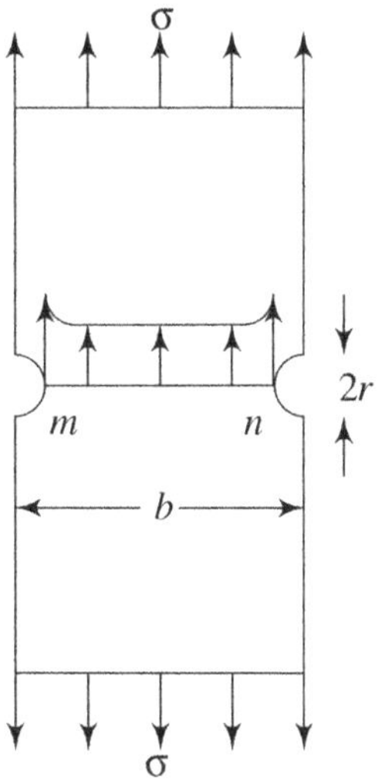

Fig. 12.2 *Plate with semicircular grooves*

$$\sigma_0 = \frac{\sigma bt}{(b-2r)t} = \frac{\sigma b}{(b-2r)}$$

where b is the width and t the thickness of the plate. At points m and n, the stresses are maximum, and let their values be σ_{max}. The ratio of σ_{max} to the nominal or average stress σ_0 is called the *stress-concentration factor* K_t; i.e.,

$$K_t = \frac{\sigma_{max}}{\sigma_0} = \frac{\sigma_{max}(b-2r)}{\sigma b}.$$

The subscript t in K_t represents that this stress concentration factor is obtained theoretically or experimentally and does not depend on the mechanical properties (within the elastic limit) of the plate material. Sometimes, instead of using the area across *mn*, the area away from discontinuity is used to calculate the nominal stress. In the present case, this will be

$$\sigma_0' = \frac{\sigma bt}{bt} = \sigma$$

and $$K_t' = \frac{\sigma_{max}}{\sigma}.$$

So, while referring to design tables, one should be careful about the meaning of the stress concentration factor.

With reference to Figures 12.1(a) and (b), it was said that the regions where the cross-sections abruptly change are zones of high stress concentration. To reduce stresses, these regions are smoothened by fillets as shown in Fig. 12.3.

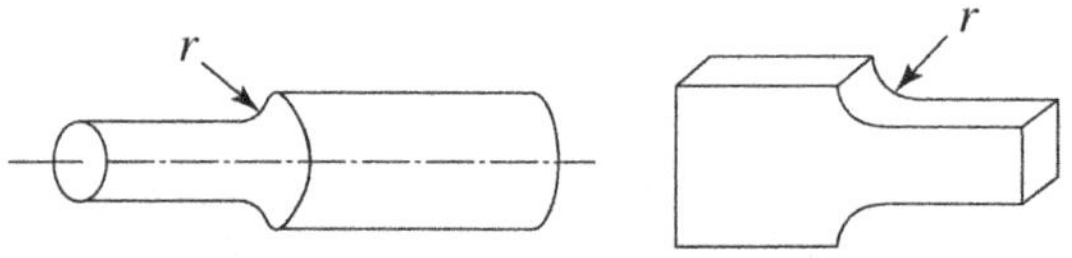

Fig. 12.3 *Members with fillets*

Figure 12.4 dispalys qualitatively how the stress concentration factor in plates varies with the ratio $\frac{r}{d}$, where r is the radius of the groove or the fillet and d is the width of the plate near the groove or the fillet. Determination of stress concentration factors purely from theoretical analysis for sectional discontinuities of several shapes is difficult and complicated. The majority of data for design purposes are obtained experimentally.

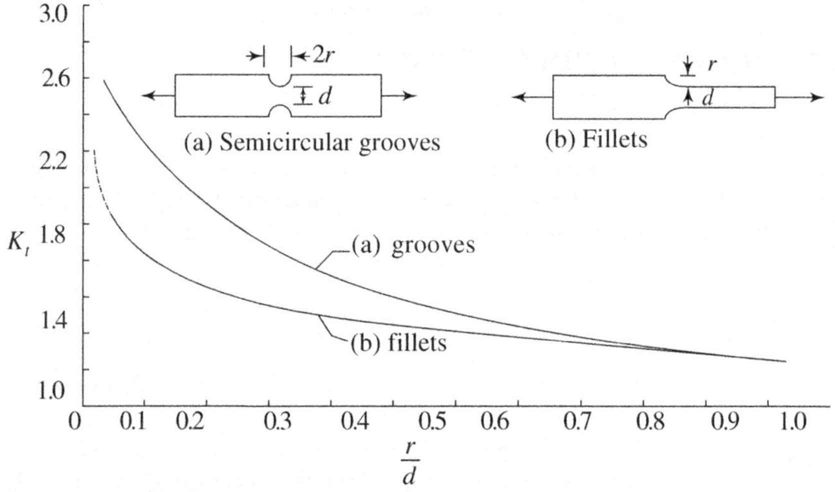

Fig. 12.4 *Stress concentration factor for grooves and fillets*

The case of a very wide plate with hyperbolic grooves has been solved theoretically and the solution shows that the stress concentration factor near the roots of the grooves can be represented approximately by the formula

$$K_t = \sqrt{0.8\frac{d}{2r} + 1.2} - 0.1 \tag{a}$$

where d is the width of the plate at the grooves, and r is the radius of curvature at the bottom of the groove. Poisson's ratio is taken as 0.3 in the foregoing equation.

In the case of a circular member of large diameter with hyperbolic grooves and subjected to tension, the maximum stress occurs again at the bottom of the grooves. The stress concentration factor is given by

$$K_t = \sqrt{0.5\frac{d}{2r} + 0.85} + 0.08 \qquad \text{(b)}$$

Comparing Eq. (a) with Eq. (b), it is seen that the stress concentration factor in the case of a cylinder under tension is smaller than the stress concentration factor for a plate under tension. For example, with $\frac{d}{2r} = 10$ in both cases, $K_t = 2.93$ in the case of the plate, and $K_t = 2.5$ in the case of the cylinder.

(a) Plate with a Circular Hole Figure 12.5 shows a plate of width w and thickness t with a small circular hole of radius c. The plate is subjected to a tensile stress σ at a distance far removed from the hole. The width w is assumed to be large compared to c, the radius of the hole. This problem has an exact solution given by the theory of elasticity. The detailed solution, which is fairly simple, is given in the Appendix at the end of this chapter. An approximate solution can also be obtained using the energy method and curved beam theory discussed in chapters 5 and 6. For this, consider a *large circle* drawn concentric with the hole and having a radius b. Since this circle is far removed from the hole, it can be assumed that the stress condition around the circumference of the circle is not affected by the presence of the hole.

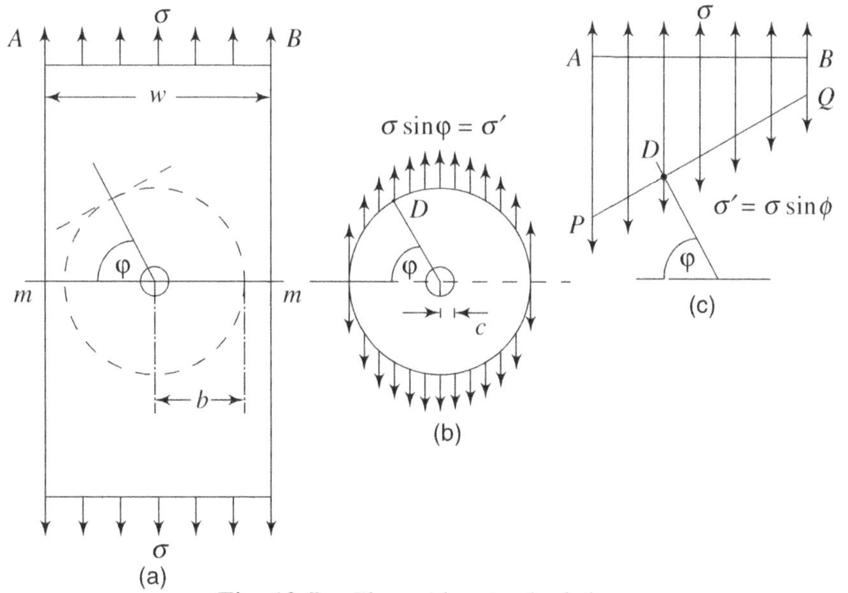

Fig. 12.5 *Plate with a circular hole*

To determine the stress distribution around the circumference, consider a tangential plane PQ at point D, Fig. 12.5 (c). The radius vector makes an angle φ with mm. The area of the section across PQ is $\frac{wt}{\sin\varphi}$. Hence, the stress across PQ is

$$\sigma' = \frac{\sigma wt}{(wt/\sin\varphi)} = \sigma \sin\varphi \cdot$$

This stress distribution, which is a function of φ, is shown in Fig. 12.5 (b).

The problem is now reduced to a thick circular ring of thickness t with inner radius c, outer radius b, and subjected to loading $\sigma \sin \varphi$ around the periphery as shown in Figures 12.5 (b) and 12.6 .

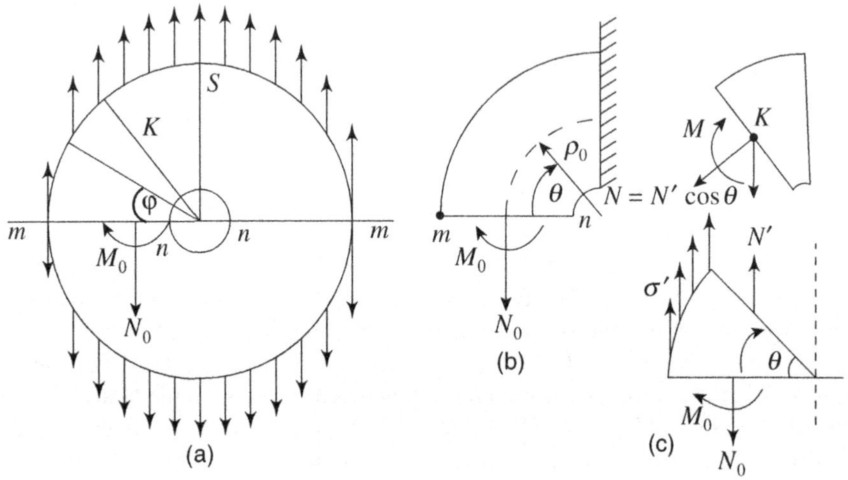

Fig. 12.6 *Thick ring subjected to periferal loading*

Consider a qudrant *mn* of the ring across the section *mm*, Fig. 12.6(b). The reactive forces across *mn* consist of a longitudinal force N_0 and a moment M_0 which maintain the slope there as zero. The value of N_0 is obtained by integrating $\sigma \sin \varphi$ from 0 to $\pi/2$; i.e.,

$$N_0 = \int_0^{\pi/2} \sigma \sin \varphi \, bt \, d\varphi = \sigma bt$$

The strain energy method (similar to the method in Example 6.8) is used to obtain M_0. Cosider a section of the quadrant at angle θ, Fig. 12.6 (c). The face of this section is subjected to the following moments:

moment due to $M_0 = M_0$

moment due to $N_0 = -N_0\left(\frac{b+c}{2} - \rho_0 \cos\theta\right) = -N_0 \frac{b+c}{2}(1-\cos\theta)$

moment due to distributed forces $= \int_0^{\theta} \sigma t \sin \phi \, bd\, \phi\left(bco\phi - \frac{b+c}{2}\cos\theta\right)$

$$= \sigma bt \int_0^{\theta} \sin \phi (b\cos\phi - \frac{b+c}{2}\cos\theta) d\phi$$

$$= \sigma bt \left. \frac{1}{2} b \sin^2 \phi + \frac{b+c}{2}\cos\theta\cos\phi \right|_0^{\theta}$$

$$= \sigma bt \left[\frac{1}{2} b \sin^2\theta + \frac{b+c}{2}\cos\theta(\cos\theta - 1)\right]$$

The total moment

$$M = M_0 - \sigma bt\frac{b+c}{2}(1-\cos\theta)+\frac{1}{2}\sigma b^2 t\sin^2\theta+\sigma bt\frac{b+c}{2}\cos\theta(\cos\theta-1)$$

$$= M_0 - \sigma bt\frac{b+c}{2}(1-\cos\theta)(1+\cos\theta)+\frac{1}{2}\sigma b^2 t\sin^2\theta$$

i.e.,
$$M = M_0 - \frac{1}{2}\sigma btc\sin^2\theta.$$

The vertical force N' on the face at θ is obtained from statics; i.e.,

$$N' + \int_0^\theta \sigma\sin\phi t \; bd\phi = N_0$$

or
$$N' = -\sigma bt(\cos\theta - 1) + \sigma bt = \sigma bt\cos\theta$$

The face at section θ is subjected to moment M, normal force $N = N'\cos\theta$ $= \sigma bt\cos^2\theta$, and shear force $V = N'\sin\theta = \sigma\, bt\cos\theta\sin\theta$. Observing that the direction of M is opposite to the one shown in Fig. 6.30, the total strain energy V for the quadrant from Eq. (6.49) is,

$$U = \int_0^{\pi/2}\left(\frac{\alpha V^2}{2AG} + \frac{N^2}{2AE} + \frac{M^2}{2AeE\rho_0} - \frac{MN}{AE\rho_0}\right)\rho_0 d\theta$$

Here, ρ_0 is the radius of the centre line, A is the cross-sectional area, e is the distance of the neutral axis of the curved member from the centroid of the section, and E is young's modulus. Since there is no change of slope across mn,

$$\frac{\partial U}{\partial M_0} = 0 = \int_0^{\pi/2}\left(\frac{M}{AEe} - \frac{N}{AE}\right)\frac{\partial M}{\partial M_0}d\theta$$

Substituting for M and N, and observing that $\dfrac{\partial M}{\partial M_0} = 0,$

$$\int_0^{\pi/2}\left[\frac{1}{e}\left(M_0 - \frac{1}{2}\sigma btc\sin^2\theta\right) - \sigma bt\cos^2\theta\right]d\theta = 0$$

i.e.,
$$\left.\left\{\frac{1}{e}\left[M_0\theta - \frac{1}{2}\sigma btc\left(\frac{\theta}{2} - \frac{1}{4}\sin 2\theta\right)\right] - \sigma bt\left(\frac{\theta}{2} + \frac{1}{4}\sin 2\theta\right)\right\}\right|_0^{\pi/2} = 0$$

i.e.,
$$\frac{1}{e}\left(M_0\frac{\pi}{2} - \frac{1}{2}\sigma btc\frac{\pi}{4}\right) - \sigma bt\frac{\pi}{4} = 0$$

∴
$$M_0 = \frac{2}{\pi}\left(\frac{1}{2}\sigma btc\frac{\pi}{4} + e\sigma bt\frac{\pi}{4}\right)$$

$$= \frac{\sigma bt}{2}\left(\frac{c}{2} + e\right)$$

The normal stress at the point n of the section mn is σ_1 due to the moment M_0, plus σ_2 due to the longitudinal force N_0. From Eq. (6.35), and since M_0 is opposite to one in Fig. 6.20,

$$\sigma_1 = \frac{M_0}{Ae}\frac{y}{(r_0 - y)}$$

In this equation; making reference to Fig. 6.20,

$$M_0 = \frac{\sigma bt}{2}\left(\frac{c}{2}+e\right)$$

$$A = (b-c)t$$

$$y = \frac{b-c}{2} - e$$

$$r_0 = \rho_0 - e = \frac{b-c}{2} + c - e = \frac{b+c}{2} - e$$

Substituting these, simplifying, and expressing in the form of ratios, one gets

$$\sigma_1 = \sigma\left[\frac{b}{4c}\left(\frac{c}{2e}+1\right)\left(1-\frac{2e/c}{b/c-1}\right)\right]$$

Similarly,

$$\sigma_2 = \left[\frac{b/c}{b/c-1}\right]$$

$$e = \rho_0 - \frac{(b-c)}{\log(b/c)} = \frac{b+c}{2} - \frac{(b-c)}{\log(b/c)}$$

or

$$\frac{e}{c} = \frac{b/c+1}{2} - \frac{b/c-1}{\log(b/c)}.$$

Let $b/c = 5$ as an example. Then,

$$\frac{e}{c} = \frac{5+1}{2} - \frac{5-1}{\log 5} = 0.5147$$

$$\sigma_1 = \sigma\left[\frac{5}{4}\left(\frac{1}{1.0294}+1\right)\left(1-\frac{1.0294}{4}\right)\right] = 1.83\sigma$$

$$\sigma_2 = \sigma\left[\frac{5}{5-1}\right] = 1.25\sigma.$$

$$\therefore \quad \sigma_{max} = \sigma_1 + \sigma_2 = (1.83+1.25)\sigma = 3.08\sigma.$$

Table 12.1 gives the values of σ_1, σ_2, and σ_{max} for several values of *b/c*.

Table 12.1

b/c=	*3*	*4*	*5*	*6*	*7*	*8*
σ_1=	1.50	1.33	1.25	1.20	1.14	1.11
σ_2=	2.33	1.93	1.83	1.83	1.95	2.19
σ_{max} =	3.83	3.26	3.08	3.03	3.09	3.30

Comparing the values in the table with the exact solution $\sigma_{max} = 3\sigma$ for a very small hole, it can be seen that for *b/c* between 5 and 8, the results of the approximate calculation agree closely with the exact solution. When $b/c < 5$, the hole cannot be considered as small. Consequently, the distribution of stress on the outer perifery of the bigger circle is no longer what was assumed. It is also seen from the table, when $b/c > 8$, the approximate value deviates substantially from the exact value, though the hole is small. The reason for this is that the stress calculated for the curved beam according to the elementary theory is not accurate enough.

From the exact theory, the stress σ_r at a distance r from the centre across the section *mm* is given by

$$\sigma_r = \frac{1}{2}\sigma\left(2 + \frac{c^2}{r^2} + \frac{3c^4}{r^4}\right)$$

where σ is the uniform tensile stress across the ends of the plate, Fig. 12.5 (a). When $r = c$, i.e., at the point n of the hole, the tensile stress $\sigma_r = 3\sigma$ as stated earlier. When r increases, the stress falls down rapidly as shown in Fig. 12.7(a). At point $r = 2c$, the stress is

$$\sigma_{r=2c} = \frac{1}{2}\sigma\left(2 + \frac{1}{4} + \frac{3}{16}\right) = 1.22\sigma$$

The exact theory also tells that at the point s i.e., when φ in Fig. 12.7 is equal to $\frac{\pi}{2}$, the stress is compressive and is equal to σ. This means that when the plate is subjected to uniform tensile stress σ, at the boundary, the point n at the hole experiences a tensile stress of magnitude 3σ, and the point s at the hole experiences a compressive stress of magnitude σ.

Instead of the stress σ at the boundary being tensile, if it is compressive as shown in Fig. 12.7 (b), the sign of the stresses around the hole become reversed; i.e., the point n will experience a stress of magnitude -3σ, and the point s will experience a tensile stress of magnitude σ. This is important if the material is brittle like glass. Brittle materials are strong in compression and weak in tension. Hence, as shown in Fig. 12.7 (b) , when a glass plate is subjected to compressive stress σ at the boundary, due to tensile stress, cracks develop at points s.

Figure 12.8 (a) shows a plate subjected to a biaxial state of stress σ_x and σ_y, where both stresses are tensile. Due to σ_x, the stresses at points *n-n* are $-\sigma_x$ each, and those to σ_y are $3\sigma_y$ each. The combined stresses at points *n-n* are each $(-\sigma_x+3\sigma_y)$. Similarly, at the points *s-s*, the combined stresses due to σ_x and σ_y are each $(3\sigma_x - \sigma_y)$. A thin tube with a hole and subjected to torsion is shown in Fig. 12.8(b). If the hole is small

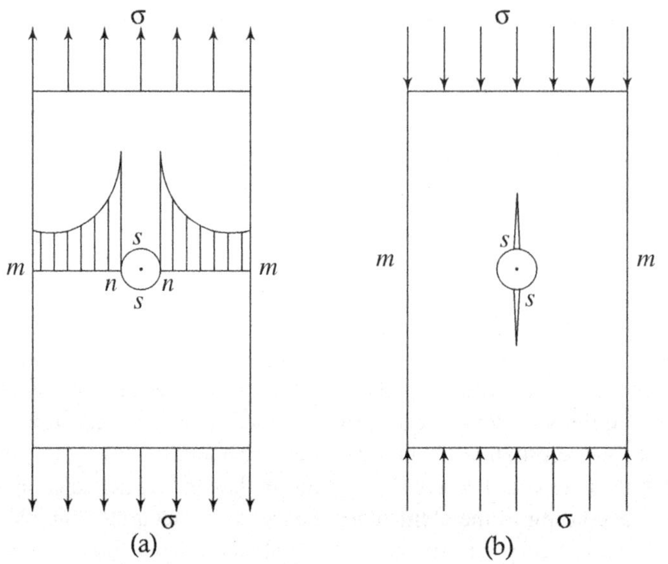

Fig. 12.7 *(a) Plate under tensile stress; (b) Plate under compressive stress*

compared to the radius of the tube and is far removed from the ends, the area around the hole can be considered to be subjected to a biaxial state of stress with $+\sigma_x$ and $-\sigma_y$. These are equal in magnitude. Due to σ_x, the stresses at n and s are respectively $-\sigma_x$ and $+3\sigma_x$. Due to σ_y, the stresses at n and s are respectively $-3\sigma_y$ and $+\sigma_y$. The net stresses are therefore:

$$\text{at } n: -\sigma_x - 3\sigma_y = -4\sigma, \text{ since } |\sigma_x| = |\sigma_y| = \sigma.$$

$$\text{at } s: 3\sigma_x + \sigma_y = +4\sigma.$$

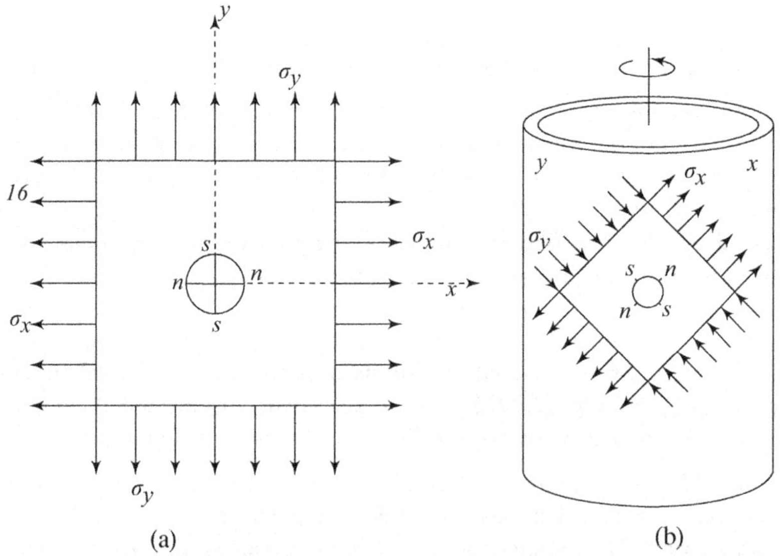

Fig. 12.8 *(a) Sheet subjected to biaxial stress state; (b) Thin tube subjected to pure torsion*

Hence, when a thin tube with a hole, is subjected to pure torsion in the direction shown in Fig. 12.8(b), at points such as *s*, there will be tensile stresses which are four times the shear stress in the tube.

In the previous discussions, it was assumed that the hole was small compared to the width of the plate and was far from the loaded ends. The problem of a hole in a plate of finite width has also been solved theoretically. Referring to Fig. 12.9, if the radius *c* of the hole is equal to $\frac{b}{2}$, where *b* is half-width of the plate; i.e., distance of the straight edge from the centre of the hole, and the plate is subjected to a uniform tensile stress σ at the ends, then σ_θ at points *n* and *m* are

$$\sigma_\theta \text{ at } n = 4.3\sigma$$
$$\sigma_\theta \text{ at } m = 0.75\sigma.$$

Hence, for a finite plate with a hole, the stress σ_θ at *n* is more than that for a large plate (theoretically, the width $2b \to \infty$) with a hole.

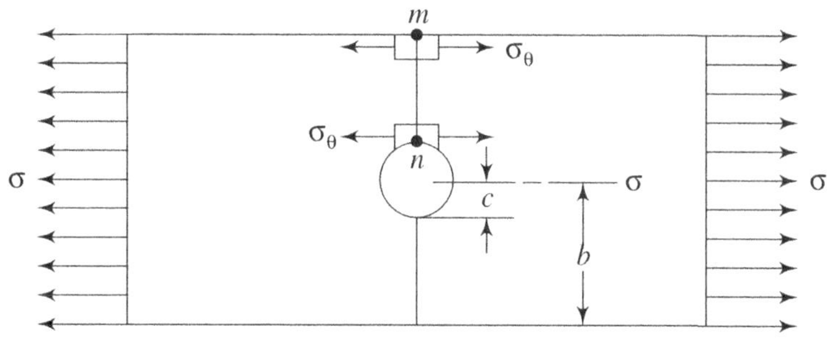

Fig. 12.9 *Finite plate with a hole*

(b) Plate with an Elliptical Hole Figure 12.10 shows a plate of large width (theoretically infinite) with a hole which is elliptical in shape, and the plate is subjected to uniform tension σ at the ends in the direction of the minor axis of the ellipse.

The exact analysis of the problem gives the magnitude of the stress at point *n* of the major axis of the ellipse as

$$\sigma^* = \sigma\left(1 + 2\frac{a}{b}\right)$$

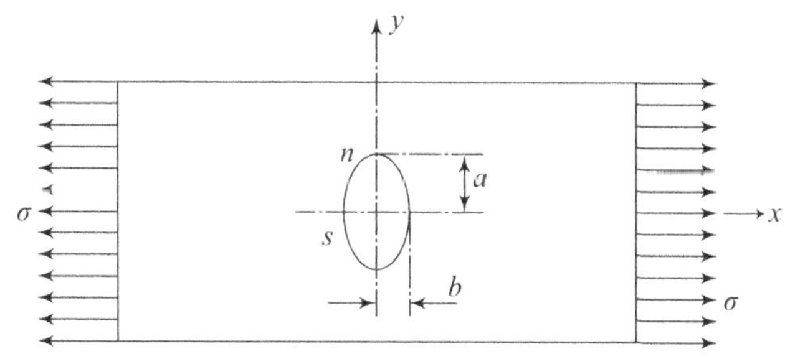

Fig. 12.10 *Plate with an elliptical hole under tension*

where a is the semi-major axis and b is the semi-minor axis of the ellipse. As the equation shows, the stress σ_θ at the ends of the major axis keeps increasing as the ellipse becomes more and more slender. In the limit, when b tends to zero, tends to infinity. When $a = b$, the ellipse degenerates into a circle, and $\sigma_\theta = 3$. This agrees with the previous discussion of a hole in a wide plate. In the case of the elliptical hole, the least value of stress occurs at the ends of the minor axis, point s and its value is $-\sigma$.

When the uniaxial tension σ is along the major axis of the ellipse, the maximum value of the stress σ_θ occurs at the tips s of the minor axis, and its value is

$$\sigma^* = \sigma\left(1 + 2\frac{b}{a}\right)$$

In this case, when the ellipse becomes very narrow, i.e., $b \longrightarrow 0$, the value of σ_θ tends to σ, and the narrow slit is along the direction of the external loading.

When the plate with an elliptical hole is subjected to pure shear τ parallel to the x and y axes, it is eqvivalent to subjecting the plate to a tensile stress $\sigma = \tau$ at $\pi/4$ and a compressive stress $-\sigma$ at $3\pi/4$ to the x-axis, Fig. 12.11.

The solution from the theory of elasticity shows that the stresses at the tips of both major and minor axes, i.e., points n and s respectively are both zero. The value of the maximum stress is

$$\sigma^* = \sigma_2 \frac{(a+b)^2}{ab}$$

and the minimum stress is

$$\sigma^* = -\sigma\frac{(a+b)^2}{ab}.$$

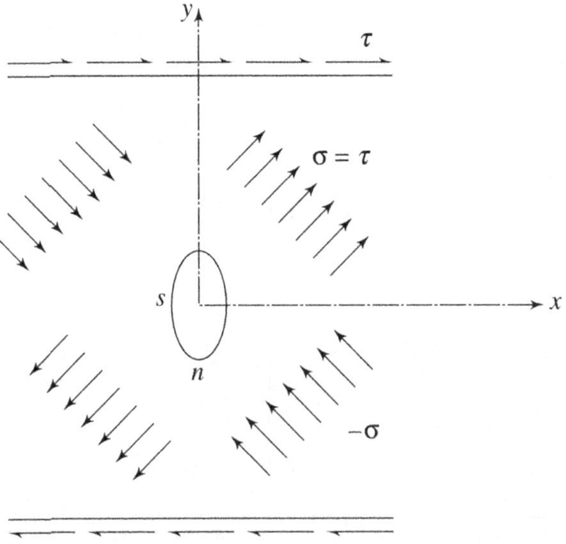

Fig. 12.11 *Plate with an elliptical hole under shear*

These occur at points whose location depends on the ratio $\frac{b}{a}$. When the ellipse becomes very narrow, the value of σ^* becomes very high, and the points where they occur are close to the tips of the major axis.

It becomes clear why cracks perpendicular to the direction of tensile loading tend to spread. Since the maximum stress in the case of a circular hole is finite (stress concentration factor being 3), to prevent spreading of cracks, small holes are drilled at the ends of a crack. Plates with semicircular grooves subjected to tension as shown in Fig. 12.2, also experience stress concentration as stated earlier. Experiments reveal that the stresses at points m and n, are nearly three times the stress at the ends of the plate as the radius r of the groove is very small in comparison with the width d of the minimum section. This is seen in Fig. 12.4, where the curve tends to 3 as $\frac{r}{d}$ tends to zero.

All of the foregoing conclusions regarding stress distribution assume that the maximum stresses are within the elastic limits of the materials under test. Beyond the elastic limit, the distribution of stresses depend on the ductility of the material. A ductile material can be stretched considerably beyond the elastic limit without a great increase in stress, since the stresses tend to get distributed more and more uniformly as the member gets stretched. This is the reason why in the case of ductile materials, holes, notches and grooves do not affect the ultimate strength of the material.

In the case of brittle materials however, the stress concentration caused by grooves and fillets remain up to the point of breaking. There are no redistribution of stresses. This is the reason why brittle members with grooves or fillets show a lower ultimate strength compared to members with no geometrical changes. But, in the case of glass, which is a brittle material, fine surface scratches do not produce any noticeable weakening effect, though at the bottom of fine scratches, the stress magnitudes should be quite high. As an explanation to this, it is stated that common glass, in its natural state has many microscopic cracks and defects and that a few additional ones deliberately caused do not substantially affect the strength.

12.3 MEMBERS UNDER TORSION

Similar to members in tension or compression, geometrical discontinuities or irregularities in members under torsion act as stress raisers. In discussing torsion problems, the hydrodynamical analogy is useful. This analogy compares the torsional stresses in a bar of uniform cross-section with that of the motion of a frictionless fluid circulating in a shell having the same cross-section as that of the torsion member. Figure 12.12 shows the cross-section of a shell in which an *ideal fluid* is circulating. An ideal fluid is characterized by two qualities; (a) *incompressibility*, and (b) *frictionlessness*. At point A in Fig. 12.12, let V_x and V_y be the components in the x and y directions respectively of the velocity of the circulating fluid.

In the case of deformable solids, the volumetric strain, i.e., change in volume per unit volume is given by Eq. 2.34, i.e.,

$$\Delta = \frac{\Delta V}{V} = \frac{\partial u_x}{\partial x} + \frac{\partial u_y}{\partial y} + \frac{\partial u_z}{\partial z} \tag{12.1}$$

where u_x, u_y, and u_z, are the displacements at a point in the x, y, and z directions. For a two-dimensional body, this becomes

$$\Delta = \frac{\partial u_x}{\partial x} + \frac{\partial u_y}{\partial y} \tag{12.2}$$

If the body under consideration is incompressible, then

$$\Delta = \frac{\partial u_x}{\partial x} + \frac{\partial u_y}{\partial y} = 0 \tag{12.3}$$

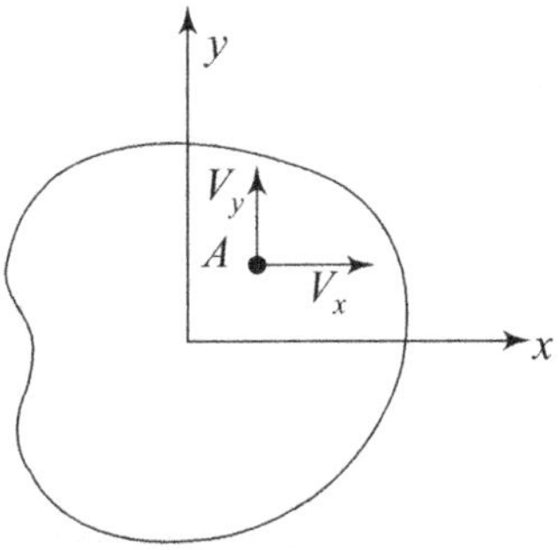

Fig. 12.12 *Circulating ideal fluid in a shell*

In the case of a fluid in motion, the *continuity equation* is the mathematical expression of the *conservation of mass*. If ρ is the density, for a two-dimensional flow field as in Fig. 12.12, the conservation of mass gives

$$\frac{d\rho}{dt} + \rho\left(\frac{\partial v_x}{\partial x} + \frac{\partial v_y}{\partial y}\right) = 0 \tag{12.4}$$

where ρ is the density of the fluid. The terms inside the brackets represent the volumetric strain. If the flow is *steady*, the density ρ is independent of time, and the conservation of mass equation becomes

$$\frac{\partial v_x}{\partial x} + \frac{\partial v_y}{\partial y} = 0 \cdot \tag{12.5}$$

Then the fluid is said to be incompressible.

In Chapter 2, dealing with the analysis of strain, Eq. 2.25 gave $\omega_{yx} = \omega_z$ as *rigid body rotation* about the z-axis without strain or deformation. If the rigid body rotation is uniform every where,

then, $$\omega_{yx} = \frac{1}{2}\left(\frac{\partial u_y}{\partial x} - \frac{\partial u_x}{\partial y}\right) = \omega_z = \text{constant}$$

i.e., $$\frac{\partial u_y}{\partial x} - \frac{\partial u_x}{\partial y} = \text{constant}. \tag{12.6}$$

Similarly, in the case of a fluid, the *vorticity* or *rotation* is given by the expression

$$\omega_{yx} = \frac{\partial v_y}{\partial x} - \frac{\partial v_x}{\partial y} \tag{12.7}$$

The condition of uniform vorticity is therefore,

$$\frac{\partial v_y}{\partial x} - \frac{\partial v_x}{\partial y} = \text{constant}. \tag{12.8}$$

Hence, an incompressible fluid circulating with uniform vorticity in a shell is expressed by

$$\frac{\partial v_x}{\partial x} + \frac{\partial v_y}{\partial y} = 0 \tag{12.9a}$$

and
$$\frac{\partial v_y}{\partial x} - \frac{\partial v_x}{\partial y} = \text{constant} \tag{12.9b}$$

Now define a *stream function* ϕ such that

$$v_x = \frac{\partial \phi}{\partial y}, \text{ and } v_y = -\frac{\partial \phi}{\partial x} \tag{12.10}$$

Such a function satisfies Eq. 12.9 (a). In order to satisfy Eq. 12.9 (b), we should have

$$\frac{\partial^2 \phi}{\partial x^2} + \frac{\partial^2 \phi}{\partial y^2} = \text{constant.} \tag{12.11}$$

This stream function coincides with Eq. 7.21 for the forsion stress function, or Prandtl's torsion stress function. From torsion stress function and Eq. 7.19,

$$\tau_{zx} = \frac{\partial \phi}{\partial y}, \text{ and } \tau_{zy} = -\frac{\partial \phi}{\partial x} \tag{12.12}$$

From the stream function and Eq. 12.10,

$$v_x = \frac{\partial \phi}{\partial y}, \text{ and } v_y = -\frac{\partial \phi}{\partial x} \tag{12.13}$$

This means that the velocity components v_x and v_y correspond to shear stress components τ_{zx} and τ_{zy} respectively.

Consider Fig. 12.13 which shows a shaft with a small eccentric hole. Let the shaft be subjected to torsion.

The effect of this hole on the stress distribution is similar to the velocity distribution of a circulating fluid in a shell with a solid cylinder of the same diameter as the hole. Such a cylinder obviously alters the velocity distribution in the neighbourhood of the obstruction. According to hydrodynamic analysis, the velocities of the circulating fluid in the front and rear points of the solid cylinder are zero, while at points m and n, the velocities are doubled. Analogously therefore, when the shaft with a small circular hole is subjected to torsion, the shear stresses in the immediate neighbourhood of the hole will be twice of what it would be in the absence of the hole.

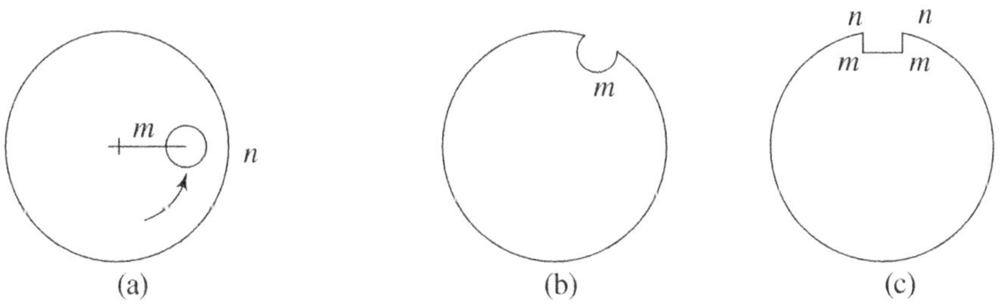

Fig. 12.13 *(a) Shaft with a circular hole; (b) Shaft with a semicircular groove; (c) Shaft with a key way*

Figure 12.13 (b) shows a shaft with a *semicircular groove* at the periphery. Based on the hydrodynamic analogy, the shear stress at the bottom of the groove, point *m*, is about twice the shearing stress at the surface of the shaft far away from the groove. In the case of a *key way* with sharp corners, Fig. 12.13 (c), the hydrodynamic analogy indicates a zero velocity of the circulating fluid at the corners protruding or projecting outwards, points *n-n*. Hence, the shearing stresses at these corresponding points in the torsion problem are zero. The corners *m-m* are called reentrant corners. At these points, the velocities of the circulating fluid are theoretically infinite. In the corresponding torsion problem, the shearing stresses at these points are also very high. This means that even a small torque will induce permanent set at these points. The stress concentration can however be reduced by rounding the corners *n-n*. Generally speaking, reentrant corners are points of high stress-concentration, and protruding corners experience zero stresses. Figure 12.14 illustrates protruding corners or projecting corners, and vertices of reentrant corners. Some of these are sharp and some are rounded corners.

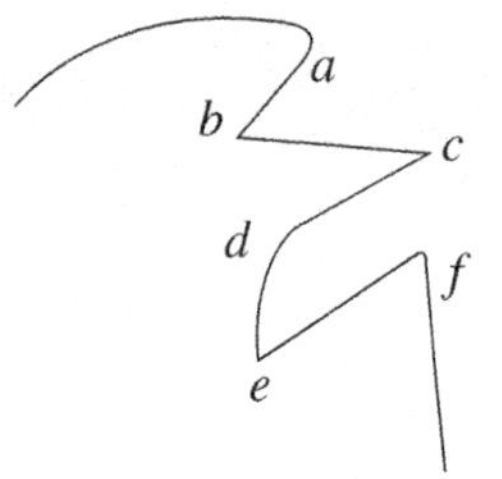

Fig. 12.14 *Reentrant (b, e) and protruding (a, c, f) corners*

The hydrodynamic analogy explains the effects of a small *hole of elliptical corss-section* or of a *groove with a semi-elliptic cross-section* in a shaft under torsion. Let the principal axes be *a* and *b*. If the principal axis *a* is along the radial direction of the shaft, then the shearing stresses at the ends of the major axis *a* are increased in the proportion $\left[1+\left(\frac{a}{b}\right)\right]:1$. Thus, the maximum stress induced depends on the ratio $\frac{a}{b}$. When *a* and *b* become equal, the ellipse tends to become a circle; i.e., a hole in the shaft, and the discussion can be applied. When *b* becomes very small, the ellipse resembles a crack in the radial direction, and the shearing stresses at the tips of this crack become very high. This explains why shafts with radial cracks are weak in torsion. Figures 12.15 (a) and (b) illustrate these.

Circular shafts with abrupt changes in diameters are subjected to high stress concentrations under torsion. If the diameter changes gradually, then one may use the elementary analysis to get the values of the stresses. To reduce the

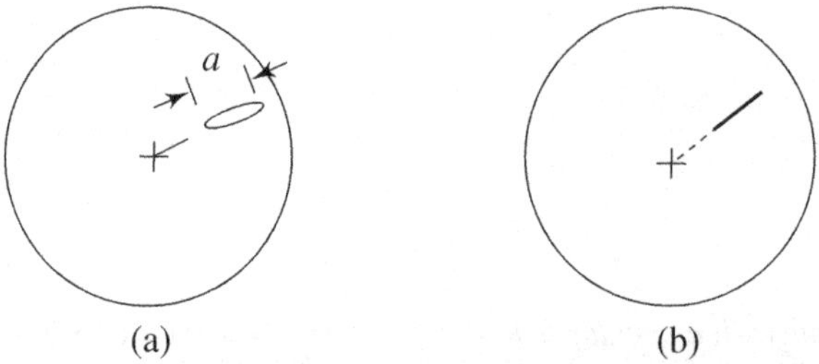

Fig. 12.15 *(a) Shaft with an elliptical hole; (b) shaft with a radial crack*

occurrence of high stresses, fillets or shoulders are provided in stepped shafts, Fig. 12.16. The magnitude of the maximum stress depends on the ratios ρ/d and D/d, where ρ is the radius of the fillet, and d and D are the two diameters of the circular shaft.

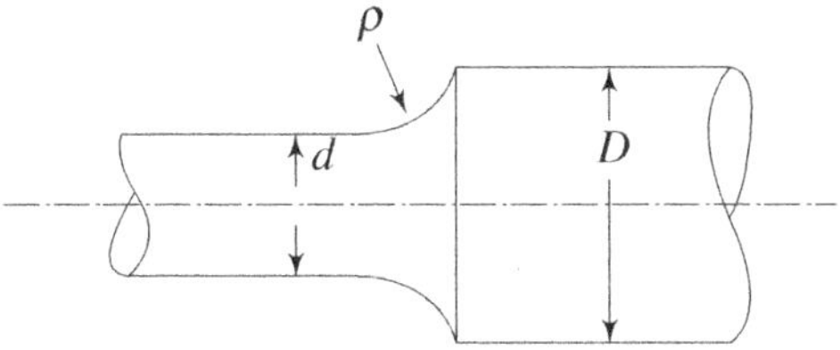

Fig. 12.16 *Shaft with variable diameter*

Figure 12.17 illustrates the stress concentration factors K_t as a function of ρ/d for two values of D/d.

The stress concentration factor K_t is equal to the ratio of the maximum shear stress τ_{max} occurring at the fillet to the stress τ_0 occurring in the shaft with the smaller diameter, i.e., d. The value of τ_0 is given by

$$\tau_0 = \frac{Td}{2J} = \frac{16T}{\pi d^3}$$

where T is the torque applied and J is the polar moment of inertia of the smaller shaft. Thus,

$$K_t = \frac{\tau_{max}}{\tau_0} = \tau_{max}\frac{\pi d^3}{16T}.$$

These localized high stresses may not be dangerous for ductile materials subjected to static loading. However, when these structural members or machine components are subjected to fluctuating loads, as in the case of turbine rotors and crankshafts, these stress concentrations will have pronounced effects.

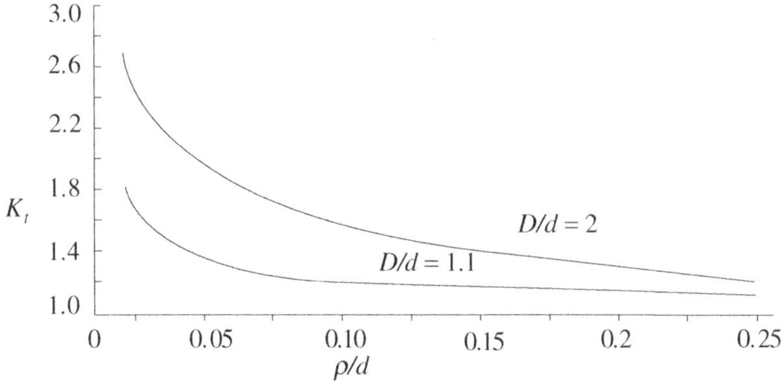

Fig. 12.17 *Variation of stress concentration factor*

12.4 MEMBERS UNDER BENDING

Equations obtained for normal and shearing stresses in the case of prismatic beams are very often applied to cases of beams of variable cross-section. If the changes in the sections of the beam are not abrupt and are gradual, the solutions obtained by the application of elementary analysis are fairly satisfactory. If the changes are abrupt, then, as in the previous cases of tension and torsion, the maximum stress values will be greater than those obtained from elementary formulas. The maximum stress can be expressed as

$$\sigma_{\max} = K_t \sigma$$

in which σ is the stress at the point under consideration as obtained from the prismatic beam formula, and K_t is the stress concentration factor. Only in limited number of cases, the values of K_t have been obtained using the equations of the theory of elasticity. For example, for a circular shaft with a hyperbolic groove, Fig. 12.18(a), the stress concentration factor in the case of pure bending is obtained as

$$K_t = \frac{3}{4N}\left[1+\sqrt{\frac{d}{2r}+1}\right]\left[\frac{3d}{2r}+4+v-(1-2v)\sqrt{\frac{d}{2r}+1}\right] \tag{12.14a}$$

where

$$N = 3\left(\frac{d}{2r}+1\right)+(1+4v)\sqrt{\frac{d}{2r}+1}+\frac{1+v}{1+\sqrt{\frac{d}{2r}+1}} \tag{12.14b}$$

where d is the diameter of the minimum cross-section and r is the smallest radius of curvature at the bottom of the groove. v is the Poisson's ratio for the material.

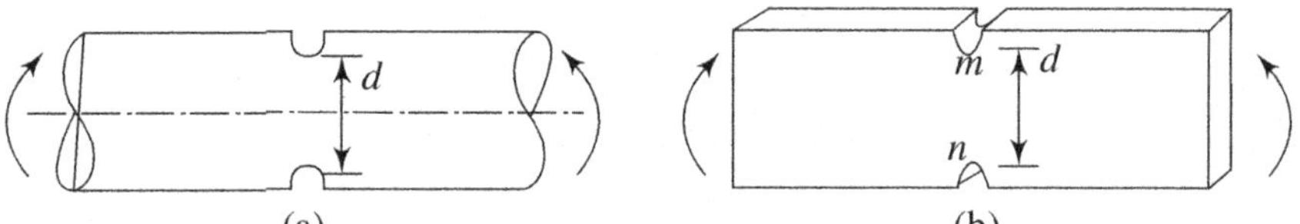

Fig. 12.18 *Shaft and plate with hyperbolic grooves under bending*

When $\frac{d}{2r}$ is fairly large, Eq. (12.14a) can be replaced with sufficient accuracy by the following approximate equation

$$K_t = \frac{3}{4}\sqrt{\frac{d}{2r}} \tag{12.14c}$$

similar to the circular shaft, a large plate with hyperbolic notches subjected to pure bending has also been rigorously analysed for stress distribution near the notches. The stress concentration factor near the roots fo the notch, m and n, Fig.12.18(b), can approximately be represented by

$$K_t = 0.08+\sqrt{0.355\frac{d}{r}+0.85} \tag{12.14d}$$

where d is the minimum width of the plate and r, the radius of curvature at the bottom of the groove.

As in the case of tension, Sec. 12.2, a circular shaft with hyperbolic notches subjected to bending has a smaller stress concentration factor at the roots than a wide plate with hyperbolic notches under bending.

12.5 NOTCH SENSITIVITY

It was stated earlier in this chapter that when the sectional geometry of a member under stress has geometrical discontinuities like grooves, fillets, holes, keyways, etc., at these zones, stresses higher than the nominal stress values are induced. The value σ_{max} of stress at these highly stressed zones was obtained by multiplying the nominal stress value σ_0 by a factor K_t called the stress concentration factor; i.e.,

$$\sigma_{max} = K_t \sigma_0 \tag{a}$$

However, there are some materials that are not very sensitive to notches, grooves, etc. For such materials, a lower stress concentration factor can be used for design purpose. In line with Eq.(a), for these materials, the maximum stress value is

$$\sigma_{max} = K_f \sigma_0 \tag{b}$$

where K_f is a reduced value of K_t and σ_0 is the nominal stress value. *Notch sensitivity* q is defined by the equation

$$q = \frac{K_f - 1}{K_t - 1} \tag{12.15}$$

where q is usually between zero and unity. Equation (12.15) shows that if $q = 0$, then $K_f = 1$, and the material under consideration has no sensitivity to notches at all. On the other hand, if $q = 1$, then $K_f = K_t$ and the material has full notch sensitivity. For design purposes, the factor K_t is obtained first for a given geometry either from theoretical considerations or experimental results. This factor K_t is independent of the material. Next, for the material under consideration, find q from design charts. With these, the value of K_f is obtained from the equation

$$K_f = 1 + q\,(K_t - 1) \tag{12.16}$$

Figure 12.19 shows how the notch sensitivity factor varies with the notch radius for two materials, aluminium alloy and steel whose σ_{ult} = 0.7 GPa. The notch

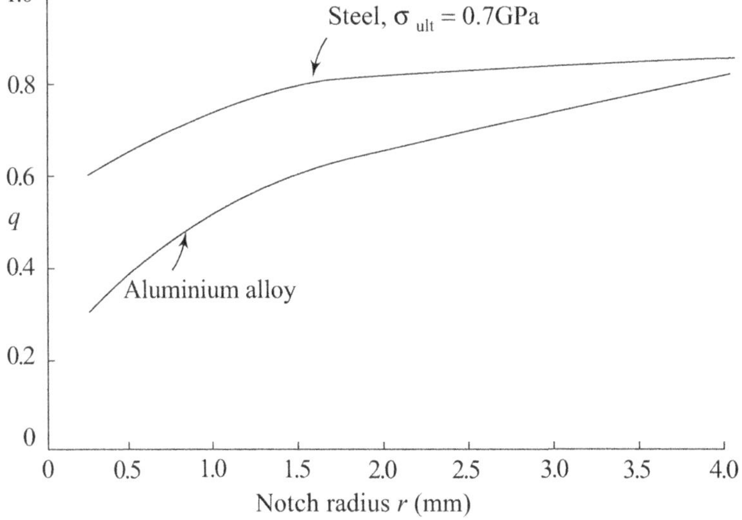

Fig. 12.19 *Variation of q with notch radius*

sensitivity factor curves involve considerable scatter and because of this many design calculations involve only the stress concentration factor K_t.

12.6 CONTACT STRESSES

Stresses developed during the pressing actions of two bodies need careful attention since the occurrence of such cases are very frequent. Gears, ball-and-roller bearings, wheel on rails, etc., are familiar examples. When bodies with curved surfaces come into contact without any pressure or forces between them, the geometry of contact is in general either a point or a line. When pressure is applied between the contacting bodies, the point or line contact become area contacts. Since the areas of contact are small, the stresses developed will be high. Typical failures due to these high contact stresses are seen as cracks, pits or flaking in the surface material.

The general analysis of contact stresses involve bodies having double radius of curvature; that is, when the radius in the plane of rolling is different from the radius in a perpendicular plane. Figure 12.20 illstrates a body having a double radius of curvature.

In the present discussion, only two special cases will be considered, i.e., contacting spheres and contacting cylinders, because of their importance.The stresses developed are generally referred to as *Hertzian stresses*, named after the scientist who developed the theory.

(a) Two Spheres in Contact Consider two spheres of diameters d_1 and d_2 brought into contact. Initially, when no pressure is applied, the spheres experience point contacts. When a force F is applied, a circular area of contact is developed due to axial symmetry. Let this contact area have a radius a, and let E_1, ν_1 and E_2, ν_2 be the respective elastic constants of the two spheres. According to Hertzian analysis, the radius a of the contact surface is given by

$$a = \sqrt[3]{\frac{3F}{8}\left[\frac{\left(1-\nu_1^2\right)/E_1 + \left(1-\nu_2^2\right)/E_2}{1/d_1 + 1/d_2}\right]} \tag{12.17}$$

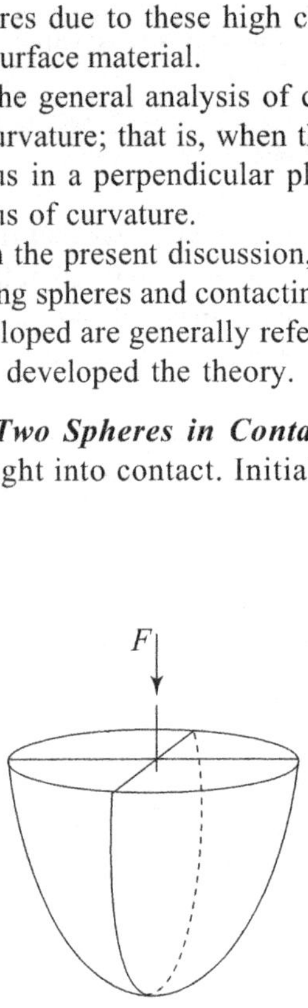

Fig. 12.20 *Body having a double radius of curvature*

If the spheres are extremely rigid with $E_1 \to \infty$ and $E_2 \to \infty$, the area of contact will be a point as the expression reveals. The stresses at all points within the area of contact in the two spheres are *not uniform*. They have a semi-elliptical distribution. Figure 12.21(a) shows two spheres in contact and the frame of reference xyz. The axis of z is downword and the force F acts along the z-axis. Figure 12.21(b) shows the stress distribution in the spheres and in the area of contact; and this is shown in Fig. 12.21(c), separately.

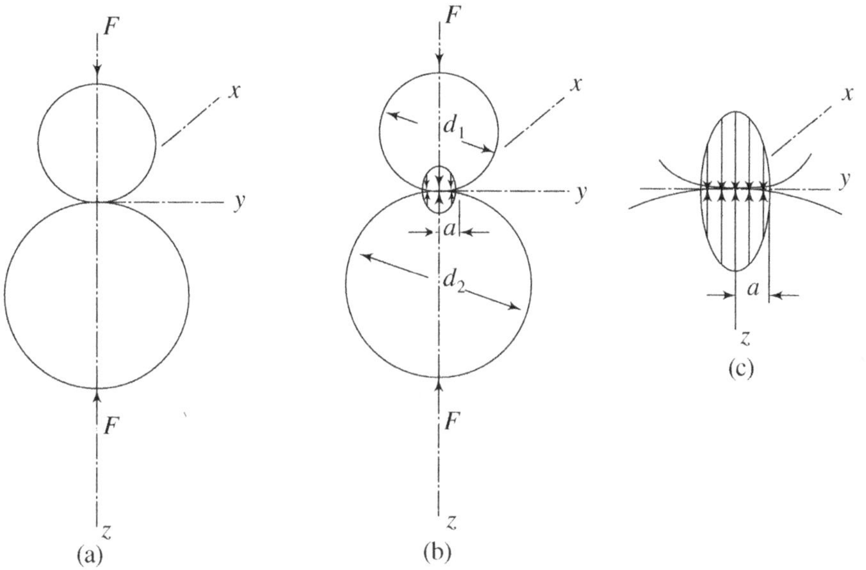

Fig. 12.21 *(a) Geometry of spheres; (b) Stress distribution within the area of contact; (c) Enlarged sketch*

The maximum pressure p_{max} occurs at the centre of the contact area, and its magnitude is given as

$$p_{max} = \frac{3F}{2\pi a^2} \tag{12.18}$$

Equations (12.17) and (12.18) are general expressions in the sense that they are valid for a sphere in contact with a plane surface, or a sphere inside another spherical surface. For a sphere of diameter d_1 in contacte with a plane, $d_2 = \infty$. For a sphere d_1 in contact within another internal spherical surface, d_2 is negative. These cases are shown in Fig. 12.22.

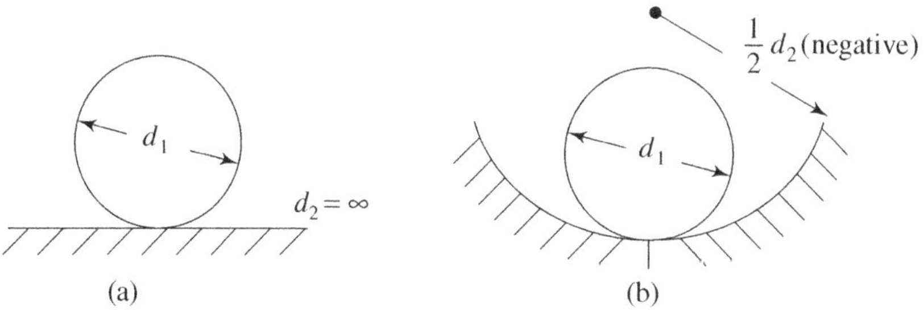

Fig. 12.22 *(a) Sphere in contact with a plane; (b) Sphere inside another spherical surface*

All points within the spheres experience stresses, but the stresses along the z-axis are maximum; i.e., in any diametrical plane at z the stresses at point (0, 0, z)

are maximum. Their values are

$$\sigma_x = \sigma_y = -p_{max}\left\{\left[1 - \frac{z}{a}\tan^{-1}\left(\frac{1}{z/a}\right)\right](1+v) - \frac{1}{2\left(1+z^2/a^2\right)}\right\} \tag{12.19}$$

$$\sigma_z = -\,p_{max}\left[\frac{1}{1+z^2/a^2}\right] \tag{12.20}$$

where p_{max} is the numerical value as given by Eq. (12.18). These are the principal stresses at any point z along the z-axis.

The average stress at the area of contact is $F/\left(\pi a^2\right)$. Hence, the maximum pressure p_{max} as given by Eq. (12.18) which occurs at the centre of the contact area, is $1\frac{1}{2}$ times the average stress. Assuming both the spheres have the same elastic properties, and taking $v = 0.3$, the maximum pressure which is compressive, is

$$p_{max} = \frac{3}{2}\frac{F}{\pi a^2} = 0.388\left[\sqrt[3]{FE^2\frac{(r_1+r_2)^2}{r_1^2 r_2^2}}\right] \tag{12.21}$$

where r_1 and r_2 are the radii of the two spheres. Let the sphere with radius r_1 be pressed on to a plane surface which has the same elastic properties as that of the sphere. Putting $r_2 = \infty$, the radius of the contact area, and the maximum pressure are

$$a = 1.109\left[\sqrt[3]{\frac{Fr_1}{E}}\right], \quad p_{max} = 0.388\left[\sqrt[3]{\frac{FE^2}{r_1^2}}\right] \tag{12.22}$$

Equations (12.17) to (12.22) are valid for both spheres; but appropriate value for Poisson's ratio corresponding to the sphere considered, need to be used.

The Mohr's circles for the state of stress described by equations (12.19) and (12.20) consist of a point and circle.

Further as $\sigma_x = \sigma_y, \tau_{xy} = 0$; and

$$\tau_{xz} = \tau_{yz} = \frac{\sigma_x - \sigma_z}{2} = \frac{\sigma_y - \sigma_z}{2} \tag{12.23}$$

One can plot the values σ_x and σ_z along z to display their variations as a functions of the distance from z = 0. In measuring the distances along the z-axis, the radius a of the surface of contact is taken as the unit. For the stresses, p_{max} is taken as the unit. Figure 12.23 shows graphically the plots of σ_z and $\sigma_x = \sigma_y$. The plot of $\tau_{xz} = \tau_{yz}$, Eq. (12.23), is also shown. All the normal stresses are compressive in nature, and v is taken as equal to 0.3.

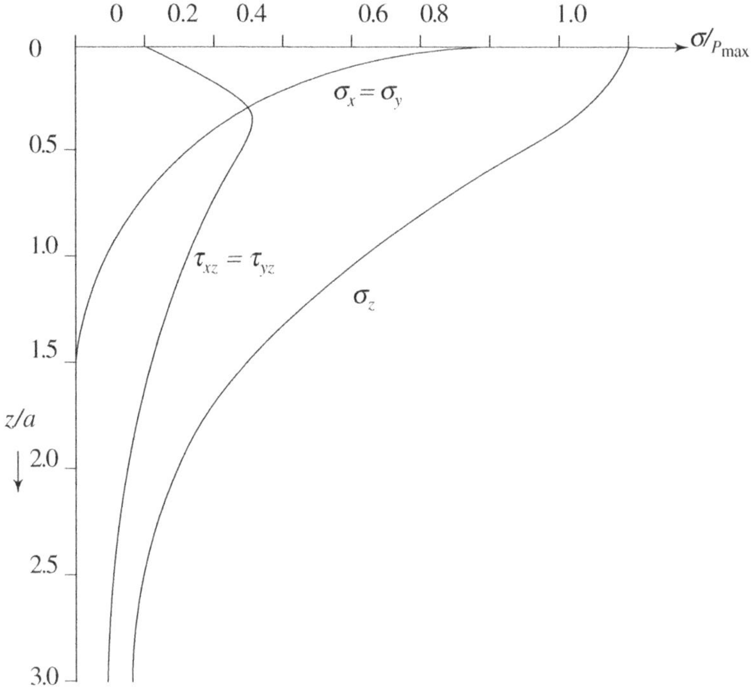

Fig. 12.23 *Plot of* $\sigma_{max}, \sigma_x, \sigma_y, \tau_{xz}$ *and* τ_{yz}

At the central point of contact, i.e., at $x = y = z = 0$, the values of $\sigma_x = \sigma_y$ are from Eq. (12.19),

$$\sigma_x = \sigma_y = -p_{max}\left[(1+\nu) - \frac{1}{2}\right] = -\frac{1+2\nu}{2}\,p_{max} \tag{12.24a}$$

From Eq. (12.20),

$$\sigma_z = -p_{max}$$

The maximum shear stress at the point from Eq. (12.24) is

$$\tau_{xz} = \tau_{yz} = \frac{1}{2}(\sigma_x - \sigma_z) = \frac{1}{2}\left(-\frac{1+2\nu}{2} + 1\right)p_{max} = \frac{1-2\nu}{4}\,p_{max} \tag{12.24b}$$

With $\quad \nu = 0.3, \quad \tau_{xz} = \tau_{yz} = 0.1p_{max}$ (12.24c)

This being too small, it dose not cause any yielding of materials such as steel, which depend on shear stresses for yielding. In fact, the maximum shear stress occurs *inside the sphere* at approximately half the distance of the radius of the contact area. This point must be considered as the weakest point in such materials as steel. The maximum shearing stress at this point, for $\nu = 0.3$ is about $0.31p_{max}$. It is suggested that cracks originate at this point below the surface and progresses to the surface. The lubricant, which is under pressure, enters the fine crack and wedges the chip loose.

(b) Two Cylinders in Contact Now consider two cylinders, Fig. 12.24, pressing against each other. Before the application of force F, there will be a line of contract l. After the application of the pressing force, the bodies deform and the line of contact becomes a narrow rectangle of width $2b$ and length l. The pressure distribution within the area of contact is once again semi-elliptical as in the case of the two spheres.

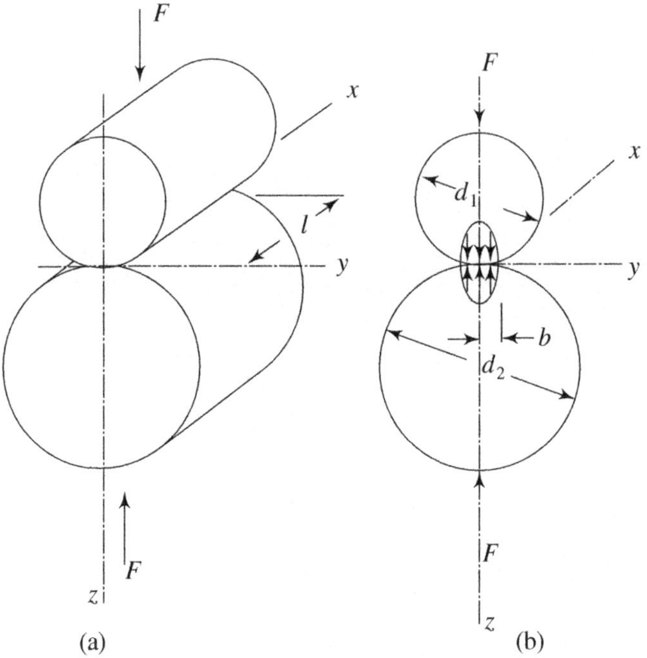

Fig. 12.24 *(a) Two cylinders in contact; (b) Pressure distribution in the contact area*

If d_1 and d_2 are the diameters of the cylinders, and if E_1, ν_1, and E_2, ν_2 are the respective elastic constants, then the half-width b of the rectangle area of contact is given by

$$b = \sqrt{\left[\frac{2F}{\pi l}\frac{\left(1-\nu_1^2\right)/E_1+\left(1-\nu_2^2\right)/E_2}{\dfrac{1}{d_1}+\dfrac{1}{d_2}}\right]} \tag{12.25}$$

The maximum pressure; i.e., compressive stress σ_{max} which occurs along the middle line of the contact area is given by

$$p_{max} = \frac{2F}{\pi b l} \tag{12.26}$$

Equations (12.25) and (12.26) are general and are applicable to both cylinders. If a cylinder of diameter d_1 presses on a plate, then d_2 becomes infinite in Eq. (12.25). If the cylinder is in contact with a hollow cylinder of diameter d_2, then d_2 is negative. A wheel pressing on a rail is a case where $d_2 = \infty$.

The state of stress along the z-axis for two cylinders is given by

$$\sigma_x = -2\nu p_{\max}\left[\sqrt{\left(1+\frac{z^2}{b^2}\right)} - \frac{z}{b}\right] \tag{12.27}$$

$$\sigma_y = -p_{\max}\left[\left(2-\frac{1}{1+z^2/b^2}\right)\sqrt{1+z^2/b^2} - 2\frac{z}{b}\right] \tag{12.28}$$

$$\sigma_z = -p_{\max}\frac{1}{\sqrt{1+z^2/b^2}} \tag{12.29}$$

If the elastic properties of the two cylinders are identical, then equations (12.25) and (12.26) reduce to the following.

$$b = 1.128\left[\sqrt{\frac{F}{El}\frac{(1-\nu)^2 d_1 d_2}{(d_1+d_2)}}\right];\ p_{\max} = 0.564\left[\sqrt{\frac{EF}{l}\frac{d_1+d_2}{(1-\nu^2)d_1 d_2}}\right] \tag{12.30}$$

Figure 12.25 is a plot of σ_x, σ_y, and σ_z as a function of depth form the centre of the contact area. The unit of depth is b, the half-width of the contract area, and the unit of stress is $p_{\max}$.

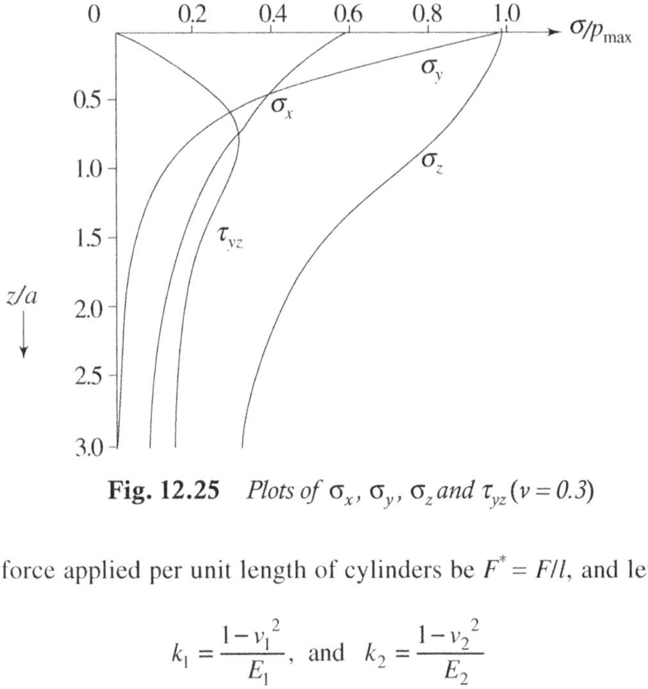

Fig. 12.25 *Plots of σ_x, σ_y, σ_z and τ_{yz} ($\nu = 0.3$)*

Let the force applied per unit length of cylinders be $F^* = F/l$, and let

$$k_1 = \frac{1-\nu_1^2}{E_1},\ \text{and}\ k_2 = \frac{1-\nu_2^2}{E_2} \tag{12.31}$$

Then

$$b = \sqrt{\left[\frac{4F^*(k_1 + k_2)r_1 r_2}{r_1 + r_2}\right]} \tag{12.32}$$

where r_1 and r_2 are the radii of the cylinders. If both cylinders have the same elastic constants and $v = 0.3$, then

$$b = 1.52\sqrt{\frac{F^* r_1 r_2}{E(r_1 + r_2)}} \tag{12.33}$$

In the case of two equal radii $r_1 = r_2 = r$,

$$b = 1.08\sqrt{\frac{F^* r}{E}} \tag{12.34}$$

For the case of contact of cylinder with a plane surface,

$$b = 1.52\sqrt{\frac{F^* r}{E}} \tag{12.35}$$

Substituting for b from Eq. (12.32) into Eq. (12.26), one gets

$$p_{\max} = \sqrt{\frac{F^*(r_1 + r_2)}{\pi^2 (k_1 + k_2) r_1 r_2}} \tag{12.36}$$

If the materials of both cylinders are the same and $v = 0.3$

$$p_{\max} = 0.418\sqrt{\frac{F^* E(r_1 + r_2)}{r_1 r_2}} \tag{12.37}$$

In case of contact of a cylinder with a plane surface,

$$p_{\max} = 0.418\sqrt{\frac{F^* E}{r}} \tag{12.38}$$

Based on the plot of τ_{yz}, the maximum shearing stress occurs at a depth $z = 0.78b$, and its magnitude is $0.301\ p_{\max}$.

Instead of maximum shear stress theory for failure of materials, sometimes the octahedral shear stress theory is used. From Eq. (1.44a).

$$\tau_{oct} = \frac{1}{3}\left[(\sigma_1-\sigma_2)^2+(\sigma_2-\sigma_3)^2+(\sigma_3-\sigma_1)^2\right]^{1/2}$$

where σ_1, σ_2 and σ_3 are principal stresses.

<u>*(i) Two Spheres in Contact*</u> $\sigma_x = \sigma_y$, and σ_z are the principal stresses at the centre of the area of contact. Since these are compressive stresses, arranging them algebraically,

$$\sigma_1 = \sigma_2 = \sigma_x = \sigma_y, \text{ and } \sigma_3 = \sigma_z$$

$$\therefore \quad \tau_{oct} = \frac{1}{3}\left[(\sigma_x-\sigma_z)^2+(\sigma_z-\sigma_x)^2\right]^{1/2}$$

$$= \frac{1}{3}\left[2\sigma_x^2+2\sigma_z^2-4\sigma_x\sigma_z\right]^{1/2}$$

$$= \frac{\sqrt{2}}{3}(\sigma_x-\sigma_z) \tag{12.39}$$

At $z = 0$, substituting from equations (12.19) and (12.20),

$$\tau_{oct} = \frac{\sqrt{2}}{3}\left[-\frac{1+2v}{2}p_{max}+p_{max}\right]$$

$$= \frac{\sqrt{2}}{6}(1-2v)p_{max} \tag{12.40}$$

With $\quad v = 0.3,$

$$\tau_{oct} = 0.094\,p_{max}$$

<u>*(ii) Two Cylinders in Contact*</u> At $z = 0$; from equations (12.27), (12.28), and (12.29),

$$\sigma_x = -2vp_{max};\ \sigma_y = -p_{max};\ \sigma_z = -p_{max}$$

Arranging algebraically,

$$\sigma_1 = \sigma_x,\ \sigma_2 = \sigma_3 = \sigma_z$$

$$\therefore \quad \tau_{oct} = \frac{1}{3}\left[(\sigma_x-\sigma_z)^2+(\sigma_z-\sigma_x)^2\right]^{1/2}$$

$$= \frac{\sqrt{2}}{3}(\sigma_x-\sigma_z) = \frac{\sqrt{2}}{3}(1-2v)p_{max} \tag{12.41}$$

Example 12.1 *Two carbon steel balls, each 25 mm in diameter are pressed together by a force F = 18N. At the centre of the area of contact, determine the values of the principal stresses, the maximum shear stress, and the octahedral shear stress.*

For carbon steel, E = 207 GPa, and v = 0.292.

Solution From Eq. (12.17)

$$a = \sqrt[3]{\frac{3F}{8}\left[\frac{2(1-\nu^2)/E}{2/d}\right]}$$

$$a = \sqrt[3]{\frac{3F}{8}\left[\frac{(1-\nu^2)d}{E}\right]}$$

Substituting the given values,

$$a = \sqrt[3]{\left[\frac{3\times 18\times 0.915\times 25\times 10^{-3}}{8\times 207\times 10^{9}}\right]}$$

$$= 10^{-4}\sqrt[3]{0.7459} = 9.07\times 10^{-5}\,\text{m} = 0.091\ \text{mm}$$

From Eq. (12.18),

$$p_{\max} = \frac{3F}{2\pi a^2}$$

$$= \frac{3\times 18\times 10^{10}}{2\times\pi\times 9.07^2} = 1045\,\text{MPa}$$

From Eq. (12.24),

$$\sigma_x = \sigma_y = -\frac{1+2\nu}{2}p_{\max}$$

$$= -\,0.792\,p_{\max} = -\,828\ \text{MPa}$$

$$\sigma_z = -\,p_{\max} = -\,1045\ \text{MPa}$$

Arranging algebraically,

$$\sigma_1 = \sigma_2 = -828\ \text{MPa},\ \sigma_3 = -\,1045\ \text{MPa}$$

Maximum shear stress is

$$\tau_{\max} = \frac{\sigma_1 - \sigma_3}{2}$$

$$= \frac{1}{2}(-828 + 1045) = 108.5\ \text{MPa}$$

Octahedral shearing stress is

$$\tau_{\text{oct}} = \frac{1}{3}\left[(\sigma_1-\sigma_2)^2 + (\sigma_2-\sigma_3)^2 + (\sigma_3-\sigma_1)^2\right]^{1/2}$$

$$= \frac{1}{3}\left[(\sigma_2-\sigma_3)^2 + (\sigma_3-\sigma_2)^2\right]^{1/2}$$

$$= \frac{\sqrt{2}}{3}(\sigma_2-\sigma_3)$$

$$= \frac{\sqrt{2}}{3}(-828 + 1045) = 102.3 \text{ MPa}$$

Also, from Eq.(12.y)

$$\tau_{oct} = \frac{\sqrt{2}}{6}(1 - 2v)P_{max}$$

$$= \frac{\sqrt{2}}{6}(1 - 0.584) \times 104.5 \times 10^7$$

$$= 102.5 \text{ MPa.}$$

Example 12.2 *In Example 12.1, one of the steel balls is replaced by a flat carbon plate. For $F = 18N$, determine the principal stresses, the maximum shearing stress, and the octahedral shearing stress, at the centre of the contact area.*

Solution From Eq. (12.17), with $d_2 = \infty$,

$$a = \sqrt[3]{\frac{3F}{8}\left[\frac{2(1-v)^2 d_1}{E}\right]}$$

Substituting the given values from Example 12.1,

$$a = \sqrt[3]{\left[\frac{3 \times 18 \times 2 \times 0.915 \times 25 \times 10^{-3}}{8 \times 207 \times 10^9}\right]}$$

$$= 10^4 \sqrt[3]{1.492} = 11.43 \times 10^{-5} \text{ m} = 0.1143 \text{ mm}$$

From Eq. (12.18),

$$p_{max} = \frac{3F}{2\pi a^2}$$

$$= \frac{3 \times 18 \times 10^{10}}{2 \times \pi \times 11.43^2} = 658 \times 10^6 \text{ Pa}.$$

From Eq. (12.24),

$$\sigma_x = \sigma_y = -\frac{1 + 2v}{2} p_{max}$$

$$= -0.792 p_{max} = -521 \times 10^6 \text{ Pa}$$

$$\sigma_z = -p_{max} = -658 \times 10^6 \text{ Pa}.$$

Arranging algebraically,

$$\sigma_1 = \sigma_2 = -521 \times 10^6 \text{ Pa}, \ \sigma_3 = -658 \times 10^6 \text{ Pa}.$$

Maximum shear stress is

$$\tau_{max} = \frac{\sigma_1 - \sigma_3}{2}$$

$$= \frac{1}{2}(-521 + 658)10^6 = 68.5 \text{ MPa}.$$

From Eq. (12.y), the octahedral shear stress is

$$\tau_{\text{oct}} = \frac{\sqrt{2}}{6}(1 - 2\nu)p_{\max}$$

$$= \frac{\sqrt{2}}{6} \times 0.416 \times 658 \times 10^6$$

$$= 64.5 \text{ MPa.}$$

Example 12.3 *In Example 12.2, determine the maximum shear stress and the maximum octahedral shear stress. At what distance from the contact surface do they occur?*

Solution The maximum shear stress and octahedral shear stress occur approximately at half the radius of the contact area, i.e., at $z = \frac{1}{2}a = 5.7 \times 10^{-5}$ m. At this point, from equations (12.19) and (12.20)

$$\sigma_x = \sigma_y = -p_{\max}\left\{\left[1 - \frac{1}{2}\tan^{-1}(2)\right](1.292) - \frac{1}{2 \times \frac{5}{4}}\right\}$$

$$= -p_{\max}\left\{\left[1 - \frac{1}{2} \times 1.107\right](1.292) - 0.4\right\}$$

$$= -0.177p_{\max} = -116 \text{ MPa}$$

and

$$\sigma_z = -p_{\max}\left[\frac{1}{1 + \frac{1}{4}}\right]$$

$$= -0.8p_{\max} = -526 \text{ MPa}.$$

Hence, the maximum shear stress is

$$\tau_{\max} = \frac{1}{2}(\sigma_x - \sigma_z)$$

$$= \frac{1}{2}(-116 + 526) = 205 \text{ MPa}.$$

The Octahedral shear stress is

$$\tau_{\text{oct}} = \frac{\sqrt{2}}{3}(\sigma_x - \sigma_z)$$

$$= \frac{\sqrt{2}}{3}(-116 + 526) = 193 \text{ MPa}.$$

II FRACTURE MECHANICS

12.7 BRITTLE FRACTURE

It is generally known that materials always show a strength that is much smaller than what might be expected from the analysis of molecular forces. For example, for a glass specimen, the theoretical strength in tension based on the analysis of molecular forces is about 11 GPa. But tensile tests conducted on glass rods reveal a strength of only 180 MPa. This discrepancy between theory and experiments was attributed to the fact that glass in its natural state contains a large number microscopic crack-producing regions of high stress concentration. Consequently, the theoretical strength would be much higher than the experimental results. Figure 12.26 shows a glass plate with a narrow crack of length $2a$. Let a uniform tension σ be applied at the two ends of the plate.Considering the crack as a microscopic alliptical hole. It was shown theoretically, based on the strain energy principle, that the stress σ required to extend the crack spontaneously is inversely proportional to the square root of the length of the crack. Experimental investigations made on glass sheets in which cracks of known length were made with a glass cutter's diamond showed a very satisfactory agreement.

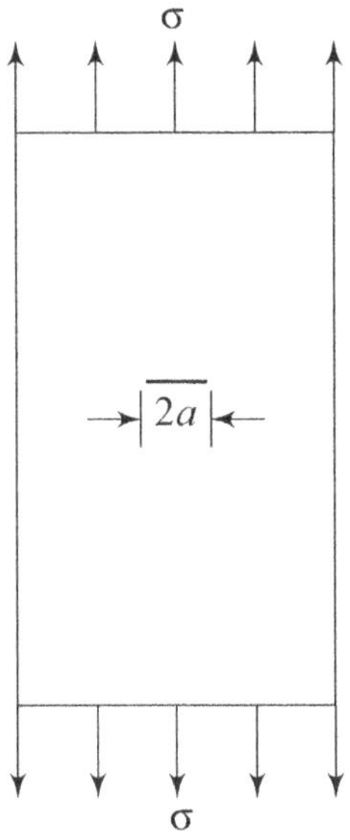

Fig. 12.26 *Glass plate with a crack of length 2a*

Previous discussions on stress concentrations revealed that very few problems involving regular geometrical irregularities could be solved theoretically to determine stress concentration factors. Most of the factors used in design calculations are based on the results of experimental investigations. The specimens needed for experimental investigations have to be prepared very carefully since they involve factors like root radius, notch depth, fillet radius, etc. In order to use these factors in practice, the designer has to know precisely the geometrical parameter present in his structural or machine member, which may not be easy. When there exists a crack, or a flaw, or an inclusion, the elastic stress concentration factor approaches infinity as the root radius approaches zero; and renders the stress concentration factor useless. Further, in the case of ductile materials, zones of high stresses make the material yield with the stresses getting redistributed. Hence, a new approach is required while dealing with cracks in structural or machine members.

In this context, a designer is interested in two factors associated with the problem of a crack in a specimen.

(a) The state of stress field in the close vicinity of the crack tip

(b) If a crack already exists, the energy required to produce a spontaneous crack extension thus creating a new fracture surface. This knowledge will help in calculating the average stress necessary to initiate a crack.

12.8 STRESS INTENSITY FACTOR

Consider a plate of uniform thickness having a centrally located crack. The plate is subjected to a uniform tensile stress σ applied at the ends. The stress field in the vicinity of the crack tip has been obtained theoretically. These are expressed by the following equations, and are with reference to Fig. 12.27. The crack of length $2a$ is a through crack in the plate of thickness t.

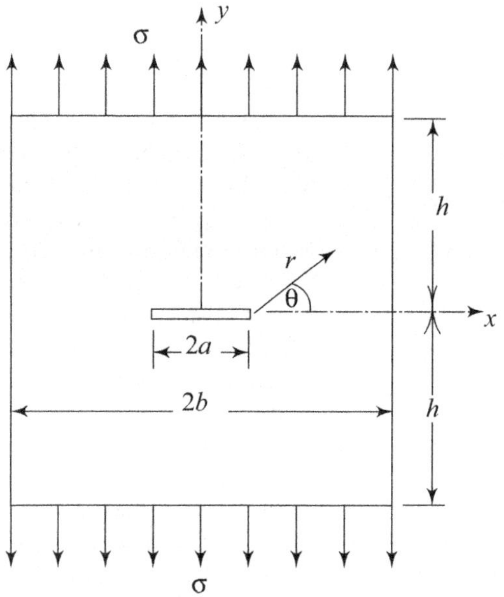

Fig. 12.27 *Plate with a through crack of length 2a*

$$\sigma_x = \frac{K}{\sqrt{2\pi r}}\left[\cos\frac{\theta}{2}\left(1 - \sin\frac{\theta}{2}\sin\frac{3\theta}{2}\right)\right] \tag{12.42a}$$

$$\sigma_y = \frac{K}{\sqrt{2\pi r}}\left[\cos\frac{\theta}{2}\left(1 + \sin\frac{\theta}{2}\sin\frac{3\theta}{2}\right)\right] \tag{12.42b}$$

$$\tau_{xy} = \frac{K}{\sqrt{2\pi r}}\left[\sin\frac{\theta}{2}\cos\frac{\theta}{2}\cos\frac{3\theta}{2}\right] \tag{12.42c}$$

Equations (12.42a, b, and c) show that the elastic normal and elastic shear stresses in the vicinity of the crack tip depend on the radial distance r from the tip, the orientation θ of the point of interest, and the factor K. This means that the state of stress at a given point in the vicinity depends completely on the factor K called the *stress intensity factor.*

However, this factor K depends on the nature of loading, the configuration of the stressed body (i.e., the location of crack in the plate, ratio of crack length to the width of the plate, etc.), and the mode of crack opening. The fracture modes will be discussed separately.

For a central crack of length $2a$, in an infinite plate subjected to a uniform tensile stress σ as shown in Fig. 12.27, the stress intensity factor K is given by

$$K_0 = \sigma\sqrt{\pi a} \tag{12.43}$$

where K is in (N/mm^2) $\sqrt{mm}$ or MPa $\sqrt{m}$. Values of K have been determined for a variety of situations employing both theory of elasticity approach and numerical techniques. As mentioned earlier, the value of K depends on the type of loading, and the geometry of the specimen. For example, If h/b = I, and a/b = 0.5, the magnitude of K_0 gets modified and becomes

$$K_1 = 1.32\sigma\sqrt{\pi a}$$

In order to take care of this dependence of K_0 on the type of loading and the geometry, Eq. (12.43) is modified as

$$K_1 = \alpha\sigma\sqrt{\pi a} \tag{12.44}$$

Figures 12.28(a) and (b) show graphically the values of $K_I/K_0 = \alpha$, where K_0 is taken as the base unit, for several values of h/b and a/b. The subscript I in K indicates that it is mode I fracture, and the meaning of this will be discussed subsequently.

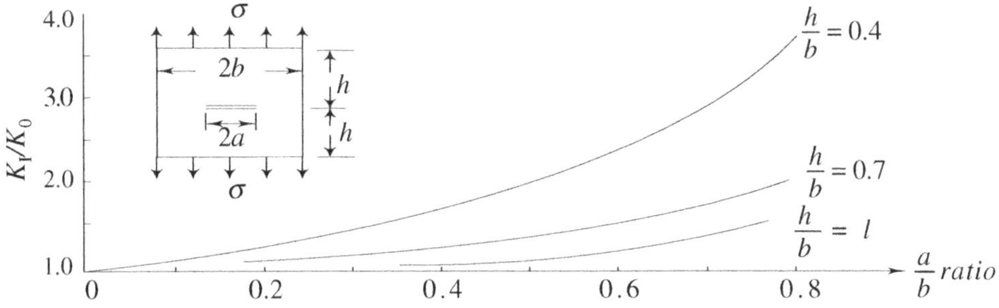

(a)

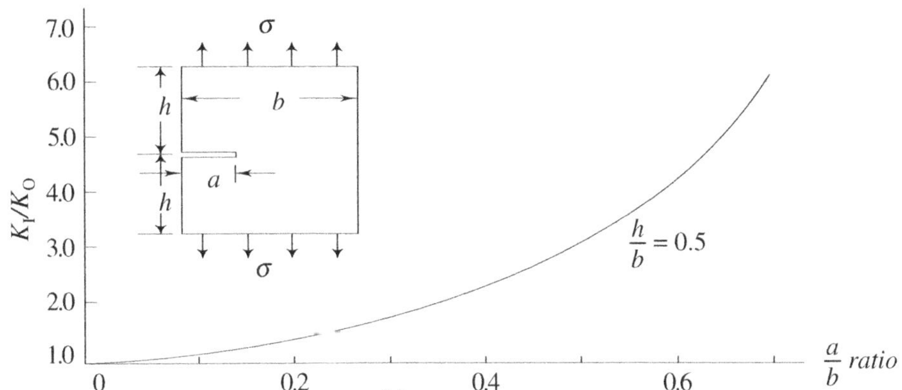

(b)

Fig. 12.28 *Values of K_I/K_0 as a function of a/b for different $\frac{h}{b}$ values. (a) Central crack of length 2a, (b) Edge crack of length a.*

12.9 FRACTURE TOUGHNESS

The previous discussion dwelt on the stresses induced in a specimen with a central crack subjected to external loading. Closely associated with this aspect is the inherent behavioural property of the material of the specimen. This aspect deals with the strength of the material. This is characterized by the *critical stress intensity factor*, also called *fracture toughness*. This is designated by the symbol K_c.

Through carefully controlled testing of the specimen of a given material, for a known applied stress, the critical crack length a_c which suddenly propagates is noted. This critical crack length gives the critical value of K_c by the equation.

$$K_{\mathrm{Ic}} = \alpha\sigma\sqrt{\pi a_c} \tag{12.45}$$

K_{Ic} is a basic material parameter called *fracture toughness*. These tests are usually conducted on single edge-notch specimens subjected to mode I, i.e., the opening mode (discussed in section 12.11), and under plane strain conditions.

If K_{Ic} is known, then it is possible to compute from Eq. (12.45), the maximum allowable stress to prevent brittle fracture for a given flaw size. For a given flaw size, the allowable stress is directly proportional to K_{Ic}, and for a given operating stress the maximum allowable crack size is proportional to the square of K_{Ic}. Therefore, increasing the value of K_{Ic} has a much larger influence on allowable crack size than on allowable stress. Although the fracture toughness K_{Ic} is a basic material property in the same sense as yield strength, it varies as a function of strain rate and temperature. This dependence on the strain rate and temperature decreases as the temperature decreases. Figure 12.29 illustrates graphically the relationship between crack length a and maximum allowable stress σ to prevent sudden extension of crack for two materials, one with high K_{Ic} and the other with lower K_{Ic}.

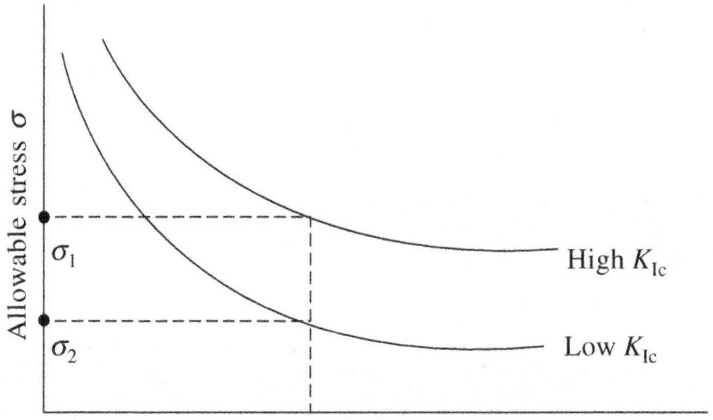

Fig. 12.29 *Crack length and allowable stress for high and low K_{Ic} materials*

As the figure illustrates, for a given crack length a_1, the maximum allowable stress σ_1 is higher for a material with high K_{Ic} than the allowable stress σ_2 for a material with low K_{Ic}.

Example 12.4 *An off-shore drilling platform has a steel sheet 35-mm thick, 12-m wide, and 20-m long. The steel sheet is subjected to a tensile stress in the direction of its length. The operating temperature is below its ductile-to-brittle transition temperature. Tests have revealed that under the conditions, the material has a fracture toughness factor K_{Ic} = 28.5 MPa $\sqrt{m}$. The sheet has a 60 mm long central transverse crack. Calculate the tensile stress for catastrophic failure. If the yield strength for the material is 240 MPa, how does the failure stress compare with it?*

Solution Making reference to Fig. 12.27, $2a = 60$ mm, $2b = 12$ m, and $2h = 20$ m,

Hence, the ratio of crack length to width of the plate is $\frac{a}{b} = \frac{30 \times 10^{-3}}{6} = 0.005.$

Further, $\frac{h}{b} = \frac{10}{6} = 1.67$.

Since $\frac{a}{b}$ is very small, the crack may be considered to be present in a very long plate, and centally located. For this case, Eq. (12.44) can be used with $\alpha = 1$. This gives for σ, the value

$$\sigma = K_I / \sqrt{\pi a}$$

Since fracture occurs when $K_I = K_{Ic}$, one gets

$$\sigma = \frac{K_{Ic}}{\sqrt{\pi a}} = \frac{28.5 \times 10^6}{\sqrt{\left[\pi \times 30 \times 10^{-3}\right]}}$$

$$= 92.8 \times 10^6 \text{ Pa} = 92.8 \text{ MPa}$$

This is the stress value at which catastrophic failure will occur. The ratio of this stress value to yield strength is

$$\frac{\sigma_{yp}}{\sigma} = \frac{240}{92.8} = 2.59, \text{ or } \frac{\sigma}{\sigma_{yp}} = 0.386$$

Thus, catastrophic failure will occur at $0.386\,\sigma_{yp}$.

Example 12.5 *A 10 m wide plate used in a heavy-machine-shop construction operation had a catastrophic failure during assembly when the sheet was subject to a stress of 90 MPa. The ambient temperature was cold. The critical stress intensity factor for the material was 20 MPa $\sqrt{m}$. It was suspected that an existing crack, presumably in the middle went undetected. Determine the maximum length of the crack that could have escaped the crack-detector's attention.*

Solution Assuming that the plate was long and the length of the crack was small compared to the width, Eq. (12.44) can be used with $\alpha = 1$. Hence,

$$\sigma = \frac{K_{Ic}}{\sqrt{\pi a}}$$

with $\sigma = 90$ MPa and $K_{Ic} = 20$ MPa $\sqrt{m}$,

$$90 \times 10^6 = \frac{20 \times 10^6}{\sqrt{\pi a}}$$

or $$\sqrt{\pi a} = \frac{20}{90} = 0.222$$

and $$a = 0.071\text{m or } 71 \text{ mm.}$$

∴ Length of the centrally located crack = $2a$ = 142 mm.

12.10 FRACTURE CONDITIONS

Section 12.7 mentioned brittle fracture without explaining what we mean by brittle fracture. It is generally known that there are materials like copper, mild-steel etc., which clearly have a defined yield points stress, maximum stress, and ultimate stress. These qualifications or distinctiveness are based on the stress–strain curves obtained usually during a tensile test. Figure 12.30 is a typical curve obtained from standard tensile test of a ductile material. Point P in this figure is called the *proportional limit.* This is the point at which the curve begins to deviate from a straight line. Point E is called the elastic limit. At this point of stressing, if the load is gradually removed, the specimen will regain its original length without any permanent set, Hooke's law, which states that stress is proportional to strain, applies only up to the proportional limit P. Many materials reach a point at which the strain begins to increase very rapidly without a corresponding increase in stress. This point is called the *yield point.* Not all materials have an obvious yield point. For this reason, *yield strength* σ_{yp} is often defined by an offset method. This is shown in Fig. 12.30, In this method, the yield strength corresponds to a definite amount of permanent set. This is usually 0.2 or 0.5 per cent of the original gauge length, although 0.01, 0.1 and 0.5 per cent are used. The other points U and F correspond to ultimate strength and the fracture or breaking stress.

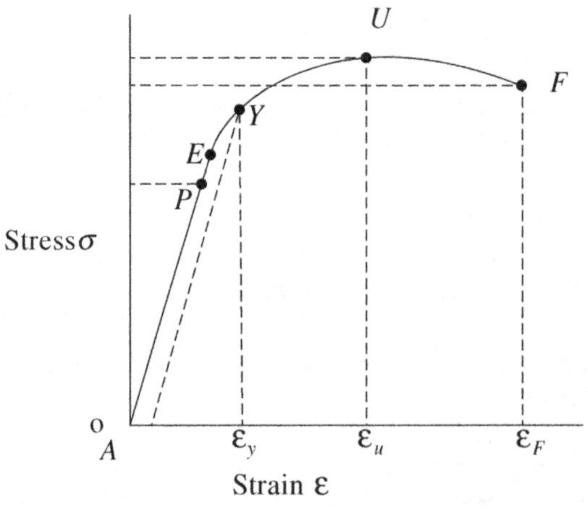

Fig. 12.30 *Tensile test diagram for a ductile material*

Materials which exhibit definite yield zones are called ductile materials. Before fracture occurs, they exhibit strong yield characteristics. Materials which do not have yield points are called brittle materials. These materials fail catastrophically after reaching a finite stress state. Glass, cast iron, are examples. In the case of ductile materials, the yield strength is temperature dependent. There exists a temperature for a given material, wherein below that temperature the material suddenly exhibits a brittle nature without any yield characteristic. This temperature is called the *transition temperature,* or *ductile-to-brittle temperature.* Tables of transition temperatures for various materials are not available, possibly because of great variations in their values even for a single material. Cold temperature is definitely an influencing factor for brittle fracture. So, operations below room temperature is an indicator of possible brittle fracture.

The term *relatively brittle* is used in test procedures. This term means fracture without yielding occurring throughout the fractured cross section. The fracture mechanic concept is correct only for linear elastic materials i.e., conditions in which no yielding occurs. But Equations. (12.42 a, b, and c) show that as r approaches zero near the crack tip, the stresses become very high and yielding occurs. However, if the yield zone is very small compared to the crack width (generally of the order of 0.1), the elastic solutions i.e., Liner Elastic Fracture Mechanics (LEFM) solutions obtained for stress intensity factors can be used.

It was also stated that the values of stress intensity factors are valid under plane strain conditions. This means that the thickness of the specimen is critical. Thin specimens do not exhibit flat fractured surfaces. They reveal ductile-brittle mode (mixed mode) of failures, and fracture stress is a function of the thickness of the specimen. As the thickness increases, the value of fracture stress becomes constant. Figure 12.31 exhibits this phenomenon. The minimum thickness to obtain plane strain conditions and valid $K_{\rm Ic}$ measurement is

$$t = 2.5\left(\frac{K_{\rm Ic}}{\sigma_{\rm yp}}\right)^2 \tag{12.46}$$

where $\sigma_{\rm yp}$ is 0.2 per cent offset yield strength (see Sec.12.14).

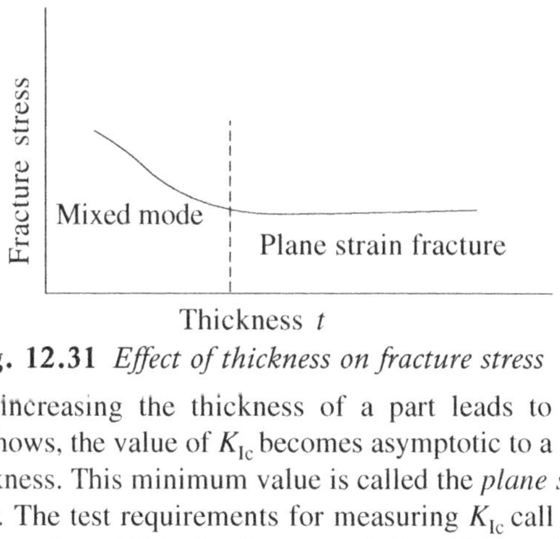

Fig. 12.31 *Effect of thickness on fracture stress*

In general, increasing the thickness of a part leads to a decrease in $K_{\rm Ic}$. As Fig. 12.31 shows, the value of $K_{\rm Ic}$ becomes asymptotic to a minimum value with increasing thickness. This minimum value is called the *plane strain critical stress-intensity factor.* The test requirements for measuring $K_{\rm Ic}$ call for plane strain values; and therefore the published values invariably refer to plane strain values.

When a crack is visible and its length can be measured, this data can be used along with its location in the member. When a crack is not visible, the designer has to assume for $2a$ the longest length that goes undetected by any of the crack detection techniques. For its locations, the designer has to assume the worst conceivable locations, since more than one location for the crack may be critical.

12.11 FRACTURE MODES

In our discussion so far, attention has been focussed on *opening mode* or the *first mode*. This was the reason for putting the subscript Ic to the critical stress intensity factor K. Generally, three ways of separating a plate are considered in facture mechanics. These are shown in Figures 12.32(a), (b) and (c).

Figure 12.32(a) is commonly called the opening mode and is designated by I, the first mode. It has an edge crack and the forces attempt to extend the crack. Figures 12.32(b) and (c) are called the *shearing modes*. In Fig. 12.32(b) the displacements stay within the plane of the plate, and are designated as mode II. In Fig. 12.32(c), the displacements are out of plane, and are called mode III. Mode III is called the *tearing mode*. In our discussion here, the attention has been mainly on mode I, because considerable amount of analysis and experimental investigations have been done on this mode.

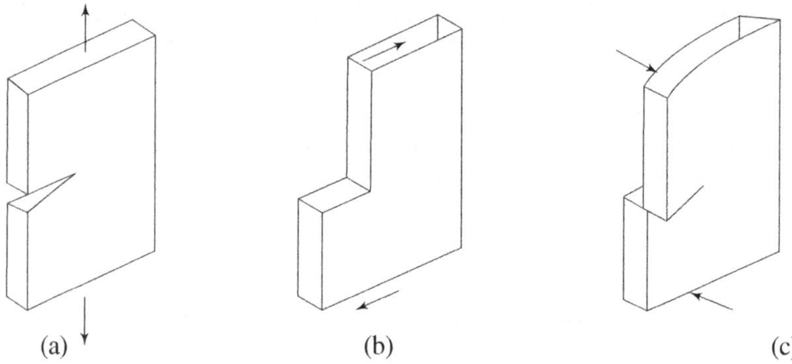

Fig. 12.32 *Fracture modes (a) Opening mode; (b) Shearing mode ; (c) Tearing/Shearing mode*

Example 12.6 *A plate of 1.5 m width and 3 m length is required for construction operations. The expected load in the longitudinal direction is 4 MN. Experimental methods to detect through thickness edge cracks are valid only for cracks longer than 2.7 mm. Two steel plates m and n are being considered for this purpose . Steel-m has yield strength of 850 MPa, and steel-n has yield strength of 1500 MPa. The corresponding critical stress intensity factors for the two materials are: for m,* $K_{Ic} = 100$ *MPa* $\sqrt{m}$, *and for n,* $K_{Ic} = 60$ *MPa* $\sqrt{m}$ *. A factor of safety of 1.5 is to be used. Minimum weight is important. Which of the two materials should be selected? Inspection did not reveal any apparent cracks in the two sheets.*

Solution (a) We shall first determine the thickness of each sheet based on the yield strengths of the materials.

Steel-m : $$\frac{\sigma_{yp}}{1.5} \times 1.5 \times t = F$$

or $$t = \frac{1.5F}{1.5 \times \sigma_{yp}}$$

$$= \frac{1.5 \times 4}{1.5 \times 850} = 4.7 \times 10^{-3} \text{ m or } 4.7 \text{ mm}$$

Steel-n : $$t = \frac{1.5 \times 4}{1.5 \times 1500} = 2.67 \times 10^{-3} \text{ m or } 2.67 \text{ mm}$$

(b) We shall next determine the thickness based on the critical stress that each sheet can bear without crack growth. Since inspection did not reveal any apparent cracks, we shall assume a crack in each sheet whose maximum length goes undetected; i.e., 2.7 mm.

Based on Fig. 12.27, for both materials,

$$\frac{h}{b} = \frac{3/2}{1.5} = 1; \ \frac{a}{b} = \frac{2.7}{1.5 \times 10^3} = 1.8 \times 10^{-3} = 0.0018$$

For these values, the curve in Fig. 12.28 gives a value for K_1/K_o as 1.1.

Steel-m : With K_{Ic}=100 MPa$\sqrt{\text{m}}$, and Eq. (12.44),

$$100 = 1.1\, \sigma \sqrt{\pi a}$$

or $$\sigma = \frac{100}{1.1\sqrt{\pi a}}$$

$$= \frac{100}{1.1\sqrt{\pi \times 2.7 \times 10^{-3}}}$$

$$= 987 \text{ MPa}$$

This crictical stress for crack extension is greater than the yield stress for the material. Hence the thickness based on σ_{yp} prevails, which is t = 4.7mm.

Steel-n: With $K_{Ic} = 60$ MPa$\sqrt{\text{m}}$, and Eq. (12.44),

$$60 = 1.1\, \sigma \sqrt{\pi a}$$

or $$\sigma = \frac{60}{1.1\sqrt{\pi a}}$$

$$= \frac{60}{1.1\sqrt{\pi \times 2.7 \times 10^{-3}}}$$

$$= 592 \text{ MPa}$$

With a factor of safety = 1.5, the allowable critical stress for steel-n is $\frac{592}{1.5} = 395$ MPa. This value is lower than $\sigma_y/1.5$ = 1000 MPa. To carry a load of 4 MN, the thickness required is therefore

$$t = \frac{4}{1.5 \times 395} = 6.75 \times 10^{-3} \text{ m} = 6.76\text{mm}$$

Hence, the *n*-steel with a lower a K_{Ic} and a higher yield strength requires a thickness of 6.76 mm, whereas, the *m*-steel with a higher K_{Ic} and lower yield strength requires a thickness of only 4.7 mm. So, the *m*-steel is recommended for the task.

Example 12.7 *In Example 12.4 dealing with off-shore platform, the thickness of the steel sheet was 35mm and the value of K_{Ic} was given as 28.5 MPa $\sqrt{m}$. What should be the 0.2 per cent yield strength to ensure plane-strain condition according to Eq. (12.46)?*

Solution $t = 35mm$, $K_{Ic} = 28.5$ MPa $\sqrt{m}$. Substituting,

$$35 \times 10^{-3} = 2.5\left(\frac{28.5 \times 10^6}{\sigma_0}\right)^2$$

or $$0.014\ \sigma_{yp} = 28.5 \times 10^6$$

or $$\sigma_{yp} = 241 \text{ MPa}$$

This agrees well with the value given in the example.

Example 12.8 *A rotating disk with a bore radius c and an outer radius b has a small radial crack of length a at the bore. Determine the critical speed for the disk based on (i) the yield stress σ_{yp}; and (ii) the critical stress intensity factor K_{Ic}.*

Solution From Eq. (8.68), for a disk rotating with an angular velocity of ω rad/sec, the circumferential stress σ_θ at a radial distance *r* from the centre is

$$\sigma_\theta = \frac{3+\nu}{8}\rho\omega^2\left[b^2 + c^2 + \frac{b^2c^2}{r^2} - \frac{1+3\nu}{3+\nu}r^2\right]$$

where ρ is the mass density of the material. This stress reaches its maximum value at the inner radius *c*, Eq. (8.70), and is equal to

$$\sigma_{\max} = \frac{3+\nu}{4}\rho\omega^2 b^2\left[1 + \frac{1-\nu}{3+\nu}\left(\frac{c}{b}\right)^2\right] \quad \text{(a)}$$

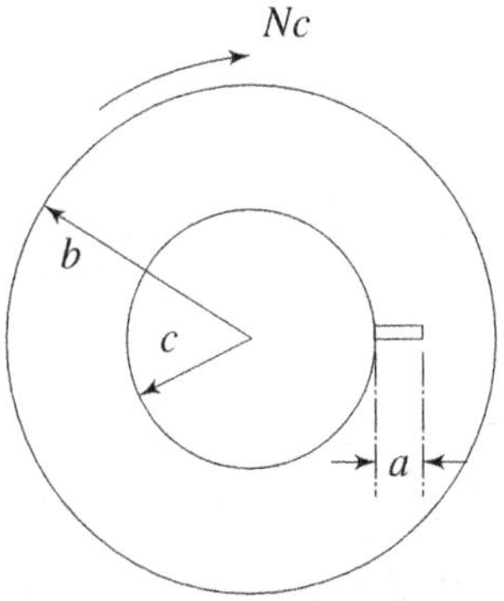

Fig. 12.33 *Rotating disk with a radial crack at the bore*

(i) With $\sigma_{\max} = \sigma_{yp}$, Eq. (a) gives for ω^2 the value

$$\omega_1 = \frac{2\sqrt{\sigma_{yp}}}{\sqrt{(3+\nu)\rho\left[b^2 + \frac{1-\nu}{3+\nu}c^2\right]}}$$

or

$$\omega_1 = \frac{2\sqrt{\sigma_{yp}}}{\sqrt{\rho\left[(3+\nu)b^2 + (1-\nu)c^2\right]}}$$

(ii) From Eq. (12.44)

$$K_{Ic} = \alpha\sigma\sqrt{\pi a}$$

or
$$\sigma = \frac{K_{Ic}}{\alpha\sqrt{\pi a}} \quad \text{(c)}$$

Substituting this into Eq. (a),

$$\frac{K_{Ic}}{\alpha\sqrt{\pi a}} = \frac{3+v}{4}\rho\omega_2^{\,2}b^2\left[1+\frac{1-v}{3+v}\left(\frac{c}{b}\right)^2\right]$$

$$= \frac{1}{4}\rho\omega_2^{\,2}\left[(3+v)b^2 + (1-v)c^2\right]$$

or
$$\omega_2 = \frac{2\sqrt{K_{Ic}}}{(\pi a)^{\frac{1}{4}}\sqrt{\alpha\rho\left[(3+v)b^2+(1-v)c^2\right]}} \quad \text{(c)}$$

Putting $\omega = \frac{2\pi N}{60}$, where N is the rpm, one can get the corresponding critical speeds N_1 and N_2.

Example 12.9 *In Example 12.8 what is the ratio of the critical speed N_1 based on the yield stress to the critical speed N_2 based on the critical stress intensity factor?*

$$N_1 = \frac{30}{\pi}\omega_1, \text{ and } N_2 = \frac{30}{\pi}\omega_2 \quad \textit{giving}\ \frac{N_1}{N_2} = \frac{\omega_1}{\omega_2}$$

Solution From Equations (b) and (c),

$$\frac{N_1}{N_2} = (\pi a)^{\frac{1}{4}}\sqrt{\frac{\alpha\sigma_y}{K_{Ic}}}$$

If $\alpha = 1.12$, $\sigma_y = 1515$ MPa, $a = 2.54$ mm, and $K_{Ic} = 50$ MPa$\sqrt{m}$,

$$\frac{N_1}{N_2} = \left(\pi\times 2.54\times 10^{-3}\right)^{\frac{1}{4}}\sqrt{\frac{1.12\times 1515}{50}}$$

$$= 0.3\times 5.83 \quad \text{(b)}$$

$$= 1.75$$

This example shows that if N_1 is the rpm decided by the yield strength criterion, there is the danger of catastrophic failure when the speed reaches $0.57N_1$.

Example 12.10 *A cylinder subjected to internal pressure p has an inner radius c and an outer radius b. The cylinder has a small radial crack of length a at the bore. The inner radius is fixed and the outer radius is to be determined.*

(i) The value of the outer radius b is to be determined according to the maximum shear stress theory ignoring the crack. A factor of safety n is involved.
(ii) The design is to be based on the critical stress intensity factor K_{Ic}.

Solution (i) To apply the maximum shear stress theory, a point at the inner radius is considered. At this point, based on Example 8.1, and equations (8.13) and (8.14),

$$\sigma_r = -p;\ \sigma_\theta = p\frac{b^2+c^2}{b^2-c^2};\ \sigma_z = p\frac{c^2}{b^2-c^2}$$

The maximum and minimum pressures are $\sigma_1 = \sigma_\theta$ and $\sigma_3 = \sigma_r$.

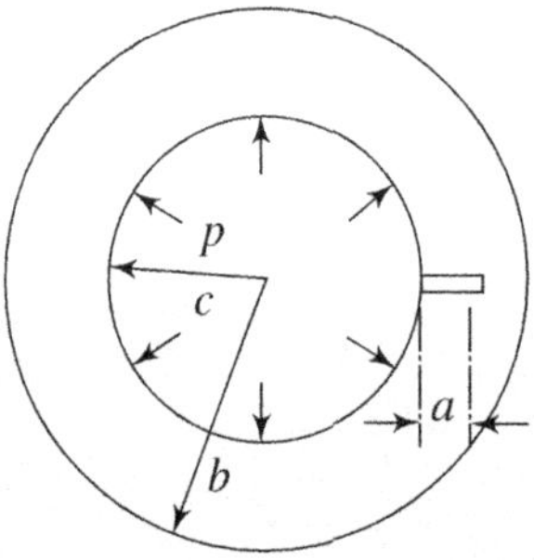

Fig. 12.34 *Cylinder with an internal crack under pressure*

Hence, $$\tau_{max} = \frac{1}{2}(\sigma_1 - \sigma_3) = p\frac{b^2}{b^2-c^2}$$

Equating this to $\frac{1}{2n}\sigma_{yp}$,

$$p\frac{b^2}{b^2-c^2} = \frac{1}{2n}\sigma_{yp}$$

or $$b^2 = \frac{\sigma_{yp}}{\sigma_{yp}-2np}c^2$$

(ii) To prevent crack extension, it is critical to consider σ_θ.
From Eq. (12.44), for an edge crack of length a,

$$K_{Ic} = \alpha\sigma\sqrt{\pi a}$$

or $$\sigma = \frac{K_{Ic}}{\sigma\sqrt{\pi a}}$$

Equating this to $\frac{1}{n}\sigma_\theta$,

$$\frac{1}{n}p\frac{b^2+c^2}{b^2-c^2} = \frac{K_{Ic}}{\alpha\sqrt{\pi a}}$$

or, $$b^2 = \frac{nK_{Ic}+p\alpha\sqrt{\pi a}}{nK_{Ic}-p\alpha\sqrt{\pi a}}c^2$$

12.12 PLANE STRESS AND PLANE STRAIN

While discussing stress concentration, it is helpful to consider load-path or load-flow lines in a body such as a wide plate, with and without geometrical discontinuities. These are similar to stream lines in fluid flow. In a pipe of uniform cross-section, the steady flow of a fluid can be represented by streamlines which are all parallel to the flow direction. If some sort of obstruction to the flow exists then the streamlines get crowded near the obstruction and the velocity of flow near the obstruction will no longer be uniform. Similarly, in the case of a body of uniform section with no discontinuities, the load lines will be uniform and all parallel, when the body is loaded longitudinally, Fig. 12.35(a). If there is a a geometrical discontinuity, such as a notch, the load lines get crowded near the notch tip and the stresses near that region will no longer be uniform, Fig. 12.35(b).

The load lines also indicate the direction of the load or the stress. In Fig. 12.35(a), the load lines are all straight indicating the uniaxial state of stress. However, in Fig. 12.35(b), the load lines bend near the notch and the tangents to the lines give the directions of the resultant stresses. As seen in Fig. 12.35(c), The tangent at A to one of the lines has two components, one in x-direction and another in y-direction. This means that though the member is subjected to uniaxial

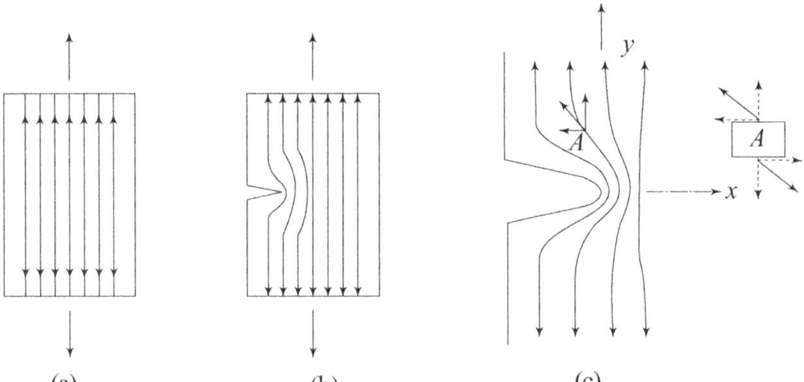

Fig. 12.35 *(a) Load lines in a uniform bar; (b) Bar with a notch; (c) Load lines indicating bi-axial state of stress*

loading, at point A, the state of stress is bi-axial. Figure12.36 shows qualitatively the bi-axial nature of the stress distribution near the notch section of a uniaxially loaded member. At the root radius of the notch $\sigma_x = 0$, since the surface of the notch is stress-free. However, as x increase, σ_x increase, reaches a maximum and at a far distance from the notch tip becomes zero. At the notch tip, σ_y is maximum and becomes uniform at the far end of x.

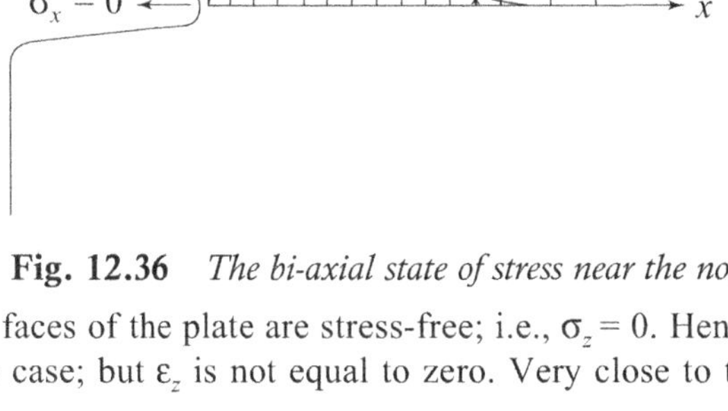

Fig. 12.36 *The bi-axial state of stress near the notch*

The two faces of the plate are stress-free; i.e., $\sigma_z = 0$. Hence, the situation is a plane stress case; but ε_z is not equal to zero. Very close to the notch tip,

$$\varepsilon_z = -\nu\frac{\sigma_x}{E} - \nu\frac{\sigma_y}{E} = -\frac{\nu}{E}(\sigma_x + \sigma_y) \tag{12.46}$$

and contraction in the plate thickness occurs. However, this is a localized effect in a wide plate, as shown in Fig. 12.37.

In the case of a thick specimen, the material near the crack tip is heavily constrained in the thickness direction (i.e., in the z direction) to contract. A small cylindrical material surrounding the crack tip will therefore experience σ_z in the z direction, as shown in Fig. 12.38. Since the faces of the plate are stress-free, σ_z will be zero at these faces. A sufficiently thick plate with a crack will therefore be in a state plane strain. The stress σ_z in the thickness direction that is required to completely prevent ε_z will be

$$\frac{\sigma_z}{E} = \frac{\nu}{E}\left(\sigma_x + \sigma_y\right)$$

i.e,
$$\sigma_z = \nu\left(\sigma_x + \sigma_y\right) \qquad (12.47)$$

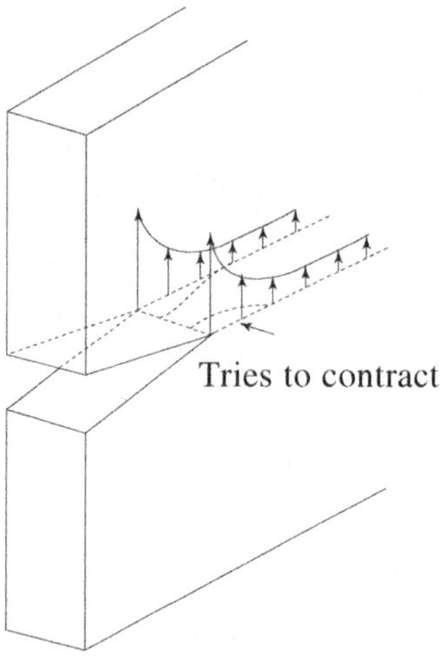

Fig. 12.37 *Contraction near the notch tip*

As mentioned, the two faces of the plate are stress-free and the value of σ_z is zero at the two faces. But, it builds up rapidly inside. Consequently, a small dimple appears near the crack tip in the two faces. Inside the plate near the crack tip, there will be a triaxial state of stress taking into account σ_x, σ_y from Fig. 12.36 and σ_z as per Eq. (12.47).

In the case of a thin plate, there is not enough material surrounding the notch tip to constrain or prevent contraction in the z-direction. So, ε_z is not zero, but σ_z is zero. Hence, this is a case of plane stress; Fig 12.39.

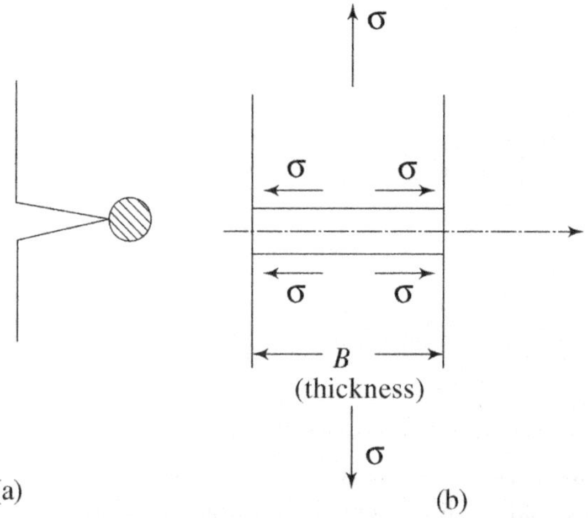

Fig. 12.38 *(a) Cylindrical material surrounding the crack tip; (b) Stresses preventing contraction*

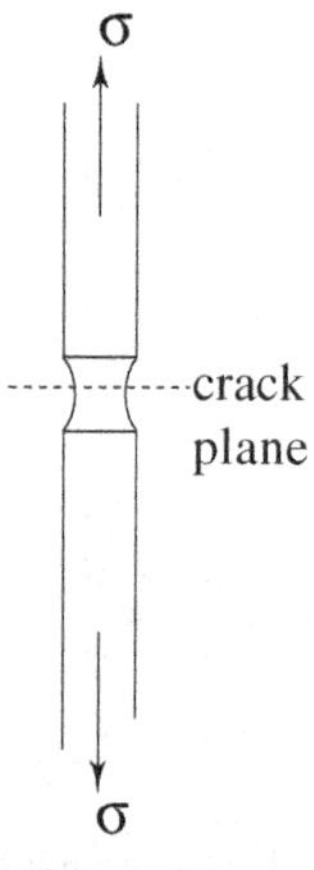

Fig. 12.39 *Thin plate; free contraction, plane stress*

12.13 PLASTIC COLLAPSE AT A NOTCH

The presence of a high state of stress near the notch tip suggests the occurrence of plastic yielding near the tip. In Chapter 4, several theories of yielding were discussed. Among these, the maximum shear stress theory and the octahedral shear stress theory are applicable to a large number of materials. According to the maximum shear stress theory yielding will occur at a point when

$$\frac{1}{2}(\sigma_1 - \sigma_3) = \frac{1}{2}\sigma_{yp} \tag{12.48}$$

where σ_1 and σ_3 are respectively the maximum and the minimum principal stresses at the point, and σ_{yp} the yield point stress for the material. According to octahedral shearing stress theory (also called the distortion energy theory), yielding will occur at a point when

$$\tau_{oct} = \frac{1}{3}\left[(\sigma_1 - \sigma_2)^2 + (\sigma_2 - \sigma_3)^2 + (\sigma_3 - \sigma_1)^2\right]^{\frac{1}{2}} = \frac{\sqrt{2}}{3}\sigma_{yp} \tag{12.49}$$

where σ_1, σ_2, and σ_3 are the principal stresses arranged algebraically, and σ_{yp}, the yield point stress for the material.

In the case of a thick plate, near the crack tip, $\sigma_x = \sigma_y, \tau_{xy} = 0$ according to equations (12.42), and

$$\sigma_z = v\,(\sigma_x + \sigma_y) \text{ according to Eq. (12.47)}$$

These are also the principal stresses, since $\tau = 0$ in the plane of symmetry, i.e., along the x-axis. Thus, near the crack tip

$$\sigma_1 = \sigma_2 = \sigma_y,\ \sigma_3 = \sigma_z = v\,(\sigma_x + \sigma_y) = 2v\sigma_y \tag{12.50}$$

Assuming $v = 0.33$, yielding occurs according to the maximum shear stress theory when

$$\frac{1}{2}(\sigma_y^{xy} - 2v\sigma_y) = \frac{1}{2}\sigma_{yp}$$

i.e., $$0.34\sigma_y = \sigma_{yp}, \quad \text{or } \sigma_y = 3\sigma_{yp} \tag{12.51}$$

This means that in the case of plane strain (thick plate), yielding occurs when $\sigma_y = 3\sigma_{yp}$. At low loads, the local stress is less than $3\sigma_{yp}$ and hence the material remains elastic. As load increases, σ_y becomes equal $3\sigma_{yp}$ and yielding occurs. According to the octahedral shearing theory yielding occurs when (with $v = 0.33$).

$$\frac{1}{3}\left[(\sigma_y - 2v\sigma_y)^2 + (2v\sigma_y - \sigma_y)^2\right]^{1/2} = \frac{\sqrt{2}}{3}\sigma_{yp}$$

i.e., $$\sqrt{2}(0.34\sigma_y) = \sqrt{2}\sigma_{yp}, \text{ or } \sigma_y = 3\sigma_{yp}.$$

This is the same as the maximum shearing stress theory. Figure 12.40 (a) represents the situation.

In the case of a thin plate, $\sigma_z = \sigma_3 = 0$ and therefore it is a plane stress case. The maximum shear is $\frac{1}{2}(\sigma_1 - \sigma_3) = \frac{1}{2}\sigma_y$. Plastic yielding occurs when $\sigma_y = \sigma_{yp}$. As loading increases, the plastic zone keeps spreading until the entire remaining section yields unless fracture occurs earlier. Figure 12.40(b) depicts the situation.

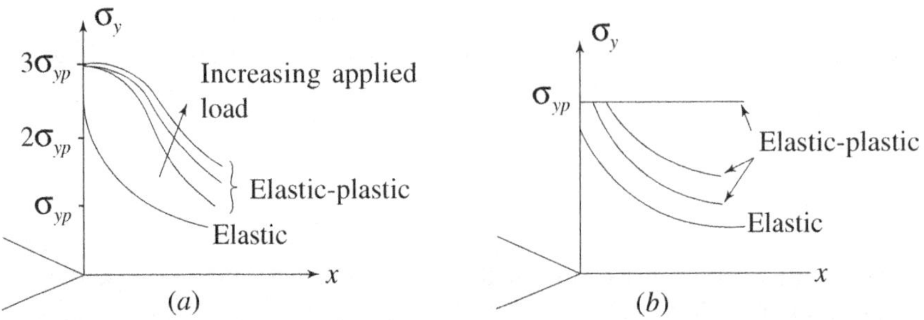

Fig. 12.40 (*a*) *Yielding at notch tip under plane strain* (*b*) *Yielding at notch tip under plane stress*

It is important to observe that in plane strain cases the notch tip stresses are much higher than σ_{yp} when yielding occurs. In plane stress cases they are limited to σ_{yp}. Thus, plane strain condition is more severe and can more easily lead to fracture and cracks.

From the foregoing discussions it becomes clear that if the stress distributions as shown in figures 12.40(a) and (b) can be reached before fracture, then *plastic collapse* can occur. Consider the case where there is no strain hardening. The maximum stress that the cross section across the notch can carry will be limited to the yield point stress. As the load increases, the yield area across the section keeps enlarging until the entire area cannot have stress greater than σ_{yp}. Such a situation is called *plastic collapse*. Thus, in plane stress where the stress in the entire cross section is equal to yield strength at the time of collapse, the maximum load carrying capacity (for a plate with single edge notch) is

$$P_{\max} = B(W - a)\sigma_{yp} \tag{12.52}$$

where B is the thickness of the plate, W is the width, and a is the crack length. This failure load is called the *collapse load* or the *limit load.* The nominal stress in the section *where there is no crack*, under the limit load is

$$\sigma_{\text{nom}} = \frac{P_{\max}}{WB}$$

$$= \frac{(W - a)}{W}\sigma_{yp} \tag{12.53}$$

As can be seen, the nominal stress keeps decreasing linearly with increasing crack length.

In plane strain case, or in general non-plane stress case, the stress distribution after the onset of yielding is not uniform, Fig. 12.40(a). The stress peak is local and the average stress across the section cannot become much higher than in the case of plane stress, and Eq. (12.52) becomes applicable.

In the case of a work-hardening material, when tearing or plastic collapse commences at the notch tip, the stresses in most of the ligament are still close to σ_{yp}, because the strain gradient is very steep. Thus, the *average stress* can be higher than σ_{yp}, but will be less than the ultimate stress σ_{ult}. This average stress across the ligament is called the *collapse strength* σ_{col}. In general, for a work-hardening material Eq. (12.53) changes into

$$\sigma_{\text{nom}} = \frac{(W - a)}{W}\sigma_{\text{col}} \tag{12.54}$$

It should be noted that the foregoing discussion is strictly for uniform applied loading.

Example 12.11 *Calculate the theoretical stress concentration factor of an elliptical notch with semi-major axis of 5cm perpendicular to the applied load, and semi-minor axis of 1cm. What is the strain concentration factor in the elastic case? What is the stress concentration factor after the notched section has fully yielded in plane stress assuming no work hardening?*

Solution The theoretical stress concentration factor K_t' for an elliptical notch is

$$K_t' = \frac{\sigma^*}{\sigma_{\text{nom}}} = \left(1 + 2\frac{a}{b}\right)$$

where a and b are respectively the semi-major and semi-minor axes of the ellipse. Here, $a = 5$ cm and $b = 1$ cm. Hence,

$$K_t' = 1 + 2\left(\frac{5}{1}\right) = 11\cdot$$

When the member is still in an elstic state, stress is proportional to strain, and hence

$$\varepsilon^* = \frac{\sigma^*}{E}, \text{ and } \varepsilon_{\text{nom}} = \frac{\sigma_{\text{nom}}}{E}$$

$$\therefore \qquad (K_t)'_{\varepsilon} = (K_t)'_{\sigma} = 11$$

When the notched section has fully yielded with no work hardening, the entire section across the notch is experiencing uniform stress and there is no stress concentration. Hence, $K_t = 1$.

Example 12.12 *For Exercise 12.11, calculate the nominal stress in the full section at the time of collapse if the yield stress is 350 MPa. The width of plate is 30 cm, and thickness is 1.25 cm. Calculate the collapse load.*

Solution

$$\sigma_{\text{nom}} = \frac{(W - 2a)}{W}\sigma_{yp}$$

$$= \frac{(30 - 10)}{30} \times 350 = 233.3 \text{ MPa}$$

$$\text{Collapse load} = Wt\sigma_{\text{nom}}$$

$$= (30\times10^{-2})(1.25\times10^{-2})(233.3\times10^{6})$$

$$= 874875\ N \simeq 875 \text{ kN}$$

Example 12.13 *As mentioned earlier in this chapter, if a crack appears in practice one sometimes drills a stop hole at the crack tip as a temporary repair. Suppose a crack has started at the edge of a strip, and its length is a. The crack tip radius is almost zero. A hole of diameter d is drilled with its centere coinciding with the crack tip. Assume that the crack with the stop hole is an ellipse. Calculate the theoretical stress concentration factor before and after drilling the stop hole. If the crack is 2.5 cm long, determine the diameter of the hole to be drilled to give a theoretical stress concentration factor of 5.*

Solution For an elliptical hole in an infinite plate, the theoretical stress concentration factor is given by

$$K'_t = 1 + 2\frac{a}{b}$$

where a and b are the semi-major and semi-minor axes of the elliptical hole. This can be recast in terms of the radius of curvature ρ of the ellipse at the end of the major axis. The radius of curvature is given by

$$\rho = \frac{b^2}{a}$$

Using this and substituting for b

$$K'_t = 1 + 2a\frac{1}{\sqrt{a\rho}}$$

or

$$K'_t = 1 + 2\sqrt{\frac{a}{\rho}}$$

In the case of a circle, $\rho = R$ the radius of the circle. If d is the diameter of the hole drilled

$$K'_t = 1 + 2\sqrt{\frac{2a}{d}}$$

For the present example, before the drilling of the stop hole,

$$K'_t = 1 + 2\frac{a}{0} = \infty$$

After drilling the hole, if the stress concentration factor is 5, then

$$5 = 1 + 2\sqrt{\frac{2a}{d}}$$

or

$$\frac{2a}{d} = 4$$

i.e.,

$$d = \frac{a}{2}$$

Hence, to bring down the stress concentration factor from ∞ to a finite value of 5, the diameter of the stop hole should be $\frac{a}{2}$.

12.14 EXPERIMENTAL DETERMINATION OF K_{Ic}

The American Society for Testing and Materials (ASTM) has set standard test methods to determine the values of plane strain fracture toughness of metallic materials. Among the several standard specimens recommended, one of them, namely the three-point bending specimen is shown in Fig. 12.41. To ensure that cracking occurs within a certain envelope and to reduce scatter, starter notches are generally used. The specimens are then fatigue pre-cracked prior to testing to simulate an ideal plane crack with essentially zero tip radius to agree with the assumptions of LEFM. To ensure plane strain conditions, the specimen dimensions must be large enough. The standard recommendations according to ASTM are:

$$a \geq 2.5\left(\frac{K_{Ic}}{\sigma_{yp}}\right)^2$$

$$B \geq 2.5\left(\frac{K_{Ic}}{\sigma_{yp}}\right)^2 \qquad (12.55)$$

$$W \geq 5.0\left(\frac{K_{Ic}}{\sigma_{yp}}\right)^2$$

Since the value of K_{Ic} is not known prior to testing, some estimate based on other experiments is used, or use of large thickness specimens is recommended. At the end tests, the value of K_{Ic} obtained is used to validate the dimensions of the specimen according to Eq. (12.55).

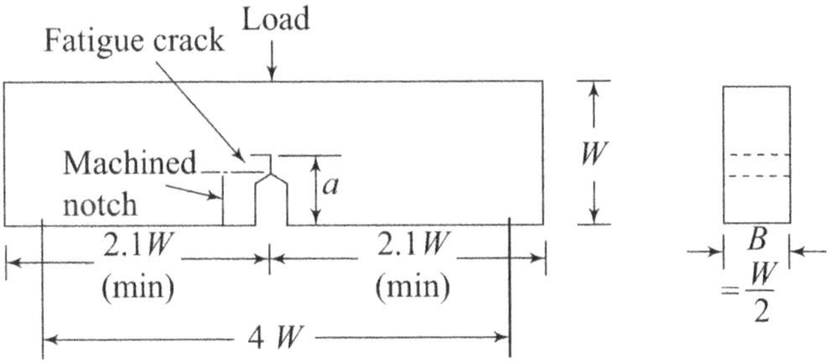

Fig. 12.41 *ASTM specimen for three-point bend test*

Table 12.1 gives the representative values of plane strain fracture toughness for selected engineering alloys.

Table 12.1

Material	*Modulus (MPa)*	*Yield stress σ_{yp} (MPa)*	*Toughess K_{Ic} (MPa$\sqrt{m}$)*
Steels			
Medium carbon	2.1×10^5	2.6×10^2	54
High strength alloys		14.6×10^2	98
Maraging steel		18.0×10^2	76

(*continued*)

Aluminium alloys			
2024 T3	70×10^4	3.45×10^2	44
2024 T8		4.2×10^2	27
7075 T6		5.4×10^2	30
Titanium alloys			
Ti-6A-4V	1.0×10^5	10.6×10^2	73
(high strength)		11.0×10^2	38

12.15 STRAIN–ENERGY RELEASE RATE

It is obvious that a body with a crack or a void is less stiff than a similar body without a void. Under uniaxial loading, *stiffness M* of a given member is defined as the force or load necessary to cause a unit deflection (i.e. displacement) under the load or in the direction of loading. Consider a body with *a* crack of length *a* and subjected to a load *P* as shown in Fig. 12.42(a). Let the body be of unit thickness.

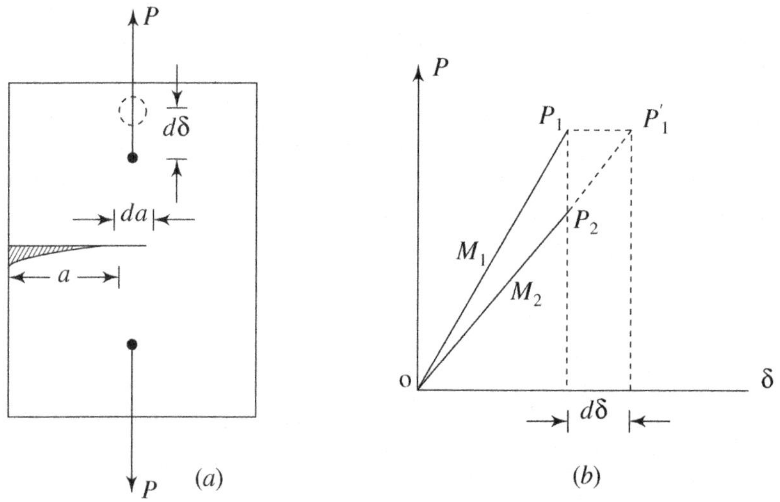

Fig. 12.42 *Single edge-crack extention*

As the load P on the body is gradually increased, displacement of the point of application occurs, and for a linearly elastic body, the load-displacement line OP_1 will be as shown in Fig. 12.42(b). The elastic strain energy U stored is equal to the work done by the load P, i.e.,

$$U = \frac{1}{2} P\delta \tag{12.56}$$

where δ is the displacement of the point of application. In terms of stiffness, since

$$\delta = \frac{P}{M} \tag{12.57}$$

the energy is

$$U = \frac{1}{2}\frac{P^2}{M} \tag{12.58}$$

where M is the stiffness of the body with a crack of length a. Let the crack length a be increased by an amount δa. As a result of this, the stiffness gets reduced from M_1 to M_2. There are now two cases to consider, Fig 12.43(a) and (b).

(i) The loading grips are held fixed (i.e., after the initial displacement δ under the load P_1) and the crack is extended.

(ii) The load P_1 is held constant and the crack is extended. Due to this, the stiffness gets reduced and the load moves down.

(i) In the case of fixed grip, with the additional cut δa, the load P_1 gets reduced to P_2 corresponding to the reduced stiffness M_2, but the original displacement δ_1 remains unchanged, i.e.,

$$\delta_1 = \frac{P_1}{M_1} = \delta_2 = \frac{P_2}{M_2} \tag{12.59}$$

Further, due to additional crack length δa, the strain energy gets reduced such that from Eq. (12.58)

$$\left(\frac{\partial U}{\partial a}\right)_\delta = \frac{1}{2}\ \frac{2P}{M}\left(\frac{\partial P}{\partial a}\right) + P^2\frac{\partial\left(\frac{1}{M}\right)}{\partial a} \tag{12.60}$$

where the subscript δ indicate that it is fixed-grips. The quantity $\frac{\partial U}{\partial a}$ is called the strain–energy release rate.

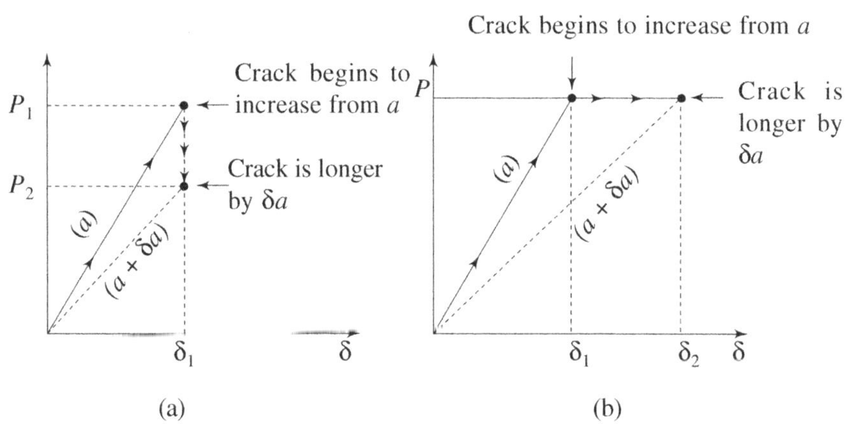

Fig. 12.43 *Load-extension plots for crack extension (a) Fixed-grip (fixed displacement); (b) Constant load*

Differentiating Eq. (12.57)

$$\frac{\partial \delta}{\partial a} = 0 = \frac{1}{M}\frac{\partial P}{\partial a} + P\frac{\partial\left(\frac{1}{M}\right)}{\partial a}$$

or
$$\frac{\partial P}{\partial a} = -PM\frac{\partial\left(\frac{1}{M}\right)}{\partial a}$$

Substituting in Eq. (12.60)

$$\left(\frac{\partial U}{\partial a}\right)_\delta = \frac{1}{2}\left[-\frac{2P}{M}PM\frac{\partial\left(\frac{1}{M}\right)}{\partial a} + P^2\frac{\partial\left(\frac{1}{M}\right)}{\partial a}\right]$$

$$= -\frac{1}{2}P^2\frac{\partial\left(\frac{1}{M}\right)}{\partial a} \tag{12.61}$$

The expression given by Eq. (12.61) is the *strain–energy release rate* under fixed-grip condition.

(ii) In the case of constant load P, from Fig.12.37(b), the change in strain–energy due to the extension of the crack is

$$\mathrm{d}U = U_2 - U_1 = \frac{1}{2}P(\delta_2 - \delta_1)$$

Since
$$\delta = \frac{P}{M},\quad \left(\frac{\partial \delta}{\partial a}\right)_P = P\frac{\partial\left(\frac{1}{M}\right)}{\partial a};$$

and
$$\delta_2 = \delta_1 + \frac{\partial \delta_1}{\partial a}\mathrm{d}a$$

$\therefore$
$$\mathrm{d}U = \frac{1}{2}P\left[\left(\delta_1 + \frac{\partial \delta_1}{\partial a}\mathrm{d}a\right) - \delta_1\right]$$

$$= \frac{1}{2}P\frac{\partial \delta_1}{\partial a}\mathrm{d}a$$

$$= \frac{1}{2}P\left[P\frac{\partial\left(\frac{1}{M}\right)}{\partial a}\right]\mathrm{d}a$$

i.e.,
$$\left(\frac{\partial U}{\partial a}\right)_P = \frac{1}{2}P^2\frac{\partial\left(\frac{1}{M}\right)}{\partial a} \tag{12.62}$$

This is identical to Eq.(12.61) excepting for the sign. Hence, the strain–energy release rate is independent of the type of load application (e.g. fixed-grip, constant load, combinations of load change and displacements, etc.). From figures 12.43(a) and (b), at instability, i.e., at the instant that the crack length is about to get extended, the *critical strain–energy release rate* G_c is

$$\frac{dU}{da} = G_c = \frac{1}{2}P^2 \frac{\partial\left(\frac{1}{M}\right)}{\partial a} = \frac{1}{2}P^2 \frac{\partial C}{\partial a} \tag{12.63}$$

The factor $C = \frac{1}{M}$ is called the *compliance* of the cracked plate, which depends on the crack size. Compliance C is the deflection per unit load on the specimen. Once the compliance versus crack length relationship is established for a given specimen configuration, G_c can be obtained by noting the load at fracture. It is necessary that the plastic deformation at the load tip is kept to a minimum.

The compliance coefficients are generally expressed in the dimensionless form EBC, where E is Young's modulus, B is the specimen thickness, and C, the compliance at a given slit length a. A set of compliance measurements is made on a specimen, and the slit is extended by small increments between each pair of consecutive measurements. The slit length for each measurement must be measured accurately. The procedure is repeated until the slit length is greater than the longest crack to be used in the test specimen.

12.16 MEANING OF ENERGY CRITERION

The change in strain energy due to extension of crack can be interpreted as the energy necessary to create a fracture over δa. Consequently, one can write

$$U^* = U^*_1(\text{body with no crack}) + U^*_2(\text{due to crack})$$

Consider a large plate of length L, width W and thickness B with a small central crack of length $2a$. If the loading is uniform tension, then the elastic strain energy of the uncracked body is

$$U^*_1 \text{ (body with no crack)} = \frac{1}{2}\frac{\sigma}{E} LBW \tag{12.64}$$

The stress due to a crack depends upon the crack tip stresses. These crack tip stresses are proportional to the applied stress σ. Thus the strain energy due to crack will be proportional to $\frac{\sigma^2}{E}$. The energy will also be proportional to the thickness B (thicker the plate greater will be the energy). Hence, U_2^* will be proportional to $\frac{\sigma^2}{E}B$ Further, the energy due to crack dependes on the crack size a also. U_1^* and U_2^* should have the same dimensions of energy. Hence, the crack size a should appear as a^2 in U_2^* ; i.e.,

$$U_2^* = C\frac{\sigma^2}{E}Ba^2 \tag{12.65}$$

where C is a dimensionless constant of proportionality. A detailed analysis shows that $C = \pi$. The total strain energy of a plate of *unit thickness*, having a centre crack of length $2a$ is therefore

$$U = U_1 + U_2 = \frac{1}{2}\frac{\sigma^2}{E}LW + \frac{\pi\sigma^2}{E}a^2 \tag{12.66}$$

$$\therefore \quad \frac{dU}{da} = \frac{2\pi\sigma^2 a}{E} \tag{12.67}$$

This is for a crack with two tips. Since all considerations are for one crack tip,

$$\frac{dU}{da} = \frac{\pi\sigma^2 a}{E} \tag{12.68}$$

per crack tip, per unit thickness. The fracture energy per unit crack extension is called *fracture resistance* and is denoted by R, while the energy release rate is denoted by G_c. Thus, we have from equations (12.63) and (12.68)

$$G_c = R \text{ and } R = \frac{\pi\sigma^2 a}{E} \qquad 12.69)$$

Equation (12.69) shows that fracture occurs when $(\pi\sigma^2 a)$ reaches a certain value, namely, ER. From Eq. (12.43), the factor $\pi\sigma^2 a$ is equal to the square of the stress intensity factor K_1. Hence, Eq. (12.69) tells that fracture occurs when $K_1{}^2$ reaches a certain value, i.e. ER. In other words,

$$\text{fracture if : } K_{Ic} = \sqrt{ER} = \text{toughness} \tag{12.70}$$

$$\text{fracture resistance } R = \frac{K_{Ic}{}^2}{E} \tag{12.71}$$

Example 12.14 *A 75-cm wide steel plate has a central crack of length 2a = 10 cm. The plate is 5 mm thick. The plate is pulled to fracture and the fracture load is 800* kN. *Determine the stress intensity factor assuming* $\frac{a}{W}$ *as small. Also, determine the value of fracture resistance R.* E *for the material is* 207 *GPa.*

Solution Since $\frac{a}{W}$ is small, $\alpha = 1$ in Eq. (12.44). Thus,

$$K_1 = \sigma\sqrt{\pi a}$$

The nominal stress σ of the uncracked specimen at the time of fracture is

$$\sigma = \frac{800,000}{(75\times10^{-2})(5\times10^{-3})}$$

$$= 2133\times10^5\, N/m^2 = 213.3 \text{ MPa}$$

$$K_{Ic} = 213.3\times10^6\sqrt{\left[\pi\times5\times10^{-2}\right]}$$

$= 84.5$ MPa $\sqrt{\text{m}}$. This is *fracture toughness.*

From Eq. (12.61), the fracture resistance is

$$R = \frac{\pi\sigma^2 a}{E}$$

$$= \frac{\pi \times (213.3)^2 \times 10^{12} \times 5 \times 10^{-2}}{207 \times 10^9}$$

$$= 34.5 \times 10^3 \text{ N/m}$$

Also, $$R = \frac{K_{Ic}^{\ 2}}{E}$$

$$= \frac{(84.5)^2 \times 10^{12}}{207 \times 10^9} = 34.5 \times 10^3 \text{ N/m}$$

The *residual strength* is the fracture stress σ_{fr}; i.e., the nominal stress at which failure takes place (or the remaining strength due to the presence of crack).

Thus, $$K_{\text{Ic}} = \alpha \sigma_{\text{fr}} \sqrt{\pi a}$$

∴ $$\sigma_{\text{fr}} = \frac{K_{\text{Ic}}}{\alpha\sqrt{\pi a}}$$

$$= \frac{\text{Toughness}}{\alpha\sqrt{\pi a}}$$

$$= \frac{84.5}{1 \times \sqrt{\pi \times 5 \times 10^{-2}}} = 213 \text{ MPa}$$

Example 12.15 *Using the result of the previous example, calculate the residual strength of a plate with an edge crack of length a = 5 cm. The width of the plate W = 12.5 cm. Check for collapse. Use α =2.1, and σ_{yp} = 480 MPa.*

Solution The residual strength σ_{fr} is

$$\sigma_{fr} = \frac{K_{Ic}}{\alpha\sqrt{\pi a}}$$

$$= \frac{84.5}{2.1 \times \sqrt{\pi \times 5 \times 10^{-2}}} = 101.5 \text{ MPa}$$

The nominal stress at the time of collapse, from Eq. (12.53) is

$$\sigma_{\text{nom}} = \frac{(W-a)}{W}\sigma_{\text{yp}}$$

$$= \frac{(12.5-5)}{15.2} \times 480 = 288 \text{ MPa}$$

Since 288 >101.5, plastic collapse does not occur.

12.17 DESIGN CONSIDERATION

The conditions for fracture in a component depend on the interaction of material properties, such as the toughness, with the design stress and crack size. For a large plate with a central crack, stress intensity factor K_c is given by

$$K_c = \sigma\sqrt{\pi a} \tag{a}$$

where σ is the design stress and *a* is the flaw size. In the process of using this equation in the design process, the selection of the material generally depends on the environmental conditions in which the designed product will be functioning. For example, the conditions may be such as to require a corrosion resistant material. Once a selection like this is made, the value of the critical stress intensity factor K_c is essentially fixed. In addition, if the situation allows for the presence of a relatively large crack—one that can be readibly detected and repaired—the design stress is fixed and must be less than $K_c/\sqrt{\pi a}$. For instance, assume that for the wing skin of a military aircraft, a certain aluminium alloy is selected because of its high strength and light weight. As a consequence of this, the value of K_c is fixed. Added to this, if the design stress σ is set at a high level to increase the payload capacity of the aircraft, then the allowable flaw size is given by $K^2c/(\pi\sigma^2)$. If this flaw size goes undetected due to the limitations of the inspection process, a catastrophic fracture may occur. This flaw may get covered up by a rivet head, and the crack may get extended from the rivet hole and cause failure. The significance of Eq. (a) lies in the fact that it is essential to decide what is most important in the design of a component. Is it the material selection because of the environment, availability, etc., or the high level of design stress because of weight, size, and cost consideration, or the flaw size that must be tolerated for safe functioning of the component? Once any two combinations to the three variables (fracture toughess, design stress, and the flaw size) is identified, the value of the third variable is fixed.

12.18 ELASTO-PLASTIC FRACTURE MECHANICS (EPFM)

Consider a body *B* having *linear or non-linear* elastic properties containing a crack or a void. Let the body have a volume *V*, loaded by surface traction $\overline{F}$ on the boundary S_F, and prescribed displacements $\overline{D}$ on the boundary S_D, Fig. 12.44 (a). Under the action of external forces and prescribed displacements, the body will undergo deformation and store strain energy. The energy stored is equal to the work done by the internal stresses during the deformation process. In the case of a *linearly elastic body*, the elastic energy *per unit volume* at any point of the body is given by Eq. (11.8); i.e.,

$$W = \frac{1}{2}\left(\overset{*}{\sigma}_x \overset{*}{\varepsilon}_x + \overset{*}{\sigma}_y \overset{*}{\varepsilon}_y + \overset{*}{\sigma}_z \overset{*}{\varepsilon}_z + \overset{*}{\tau}_{xy} \overset{*}{\gamma}_{xy} + \overset{*}{\tau}_{yz} \overset{*}{\gamma}_{yz} + \overset{*}{\tau}_{zx} \overset{*}{\gamma}_{zx}\right) \tag{12.72}$$

where $\overset{*}{\sigma}_x, \overset{*}{\sigma}_y$ - - - $\overset{*}{\tau}_{zx}, \overset{*}{\varepsilon}_x, \overset{*}{\varepsilon}_y$ - - - $\overset{*}{\gamma}_{zx}$ are the final or terminal values reached at the end of gradual loading. In the case of a non-linearly elastic body, let the stress–strain curve be as shown in Fig. 12.44 (b). Consider an elementary rectangular volume of the body with sides Δx, Δy and Δz. The stresses acting on the rectangular faces are shown in the figure. Due to σ_x acting on the area $\Delta y \Delta z$, the energy stored is equal to the work done by it and is equal to

$$\text{Lt.} \int (\sigma_x \Delta y \Delta z) \Delta \varepsilon_x \Delta x = \text{Lt.} \int (\sigma_x \Delta \varepsilon_x) \Delta x \Delta y \Delta z = \int (\sigma_x d\varepsilon_x) dx\, dy\, dz$$

where $\Delta \varepsilon_x \Delta x$ is the elementary extension in the x direction (refer sec. 4.2).

Similarly, the work done by other forces $\sigma_y \Delta x \Delta z$, $\sigma_z \Delta x \Delta y$, etc, can be written. Assuming that deformations are small and that superposition principle is applicable, the elastic strain energy stored in the elementary volume is

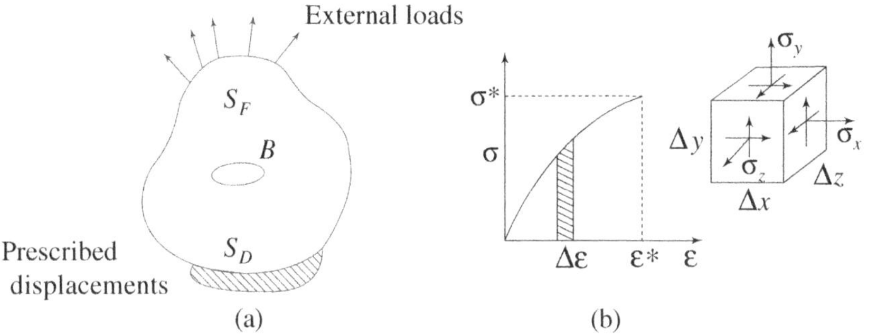

Fig. 12.44 *(a) Elastic body with cavity; (b) Non-linear elastic curve*

$$\int_0^{\sigma^*, \varepsilon^*} \left(\sigma_x d\varepsilon_x + \sigma_y d\varepsilon_y + \sigma_z d\varepsilon_z + \tau_{xy}\gamma_{xy} + \tau_{yz}\gamma_{yz} + \tau_{zx}\gamma_{zx}\right) dxdydz \qquad (12.73)$$

The quantity in the parenthesis under the integral sign is the *strain energy per unit volume*, also called *strain energy density*, at the point considered. The limits of the integration are from zero to final values at the end of loading; i.e., σ_x^*, σ_y^* - - - τ^*_{zx}. In the case of a linearly elastic solid, the strain energy density given by Eq. (12.73) reduces to that given by Eq. (12.72). The strain energy density at any point is denoted by W, where

$$W = \int (\Sigma \sigma_x d\varepsilon_x) \qquad (12.74)$$

The summation sign under the integral stands for the expanded version given in Eq. (12.73). The total strain energy stored in the body is therefore,

$$U = \int_V W dV = \int_V \left[\int \Sigma \sigma_x d\varepsilon_x \right] dV \qquad (12.75)$$

Now consider the body B with the cavity. Let ΔB be a small elementary volume *adjacent to the cavity,* Fig. 12.45(a).

Let the elementary volume ΔB of the body be isolated from B and let free-body diagrams of the newly created void, and that of ΔB be drawn as in Fig. 12.45(b). Only a part of the cavity, and the elementary volume are shown enlarged in the figure. The elementary part ΔB will be having surface fractions T^*, and the surface of the newly created cavity ΔC (i.e., the space that was occupied by ΔB) will be having equal and opposite surface fraction $\boldsymbol{T}^*$. This is similar to action and reaction discussed in reference to Fig.1.2.

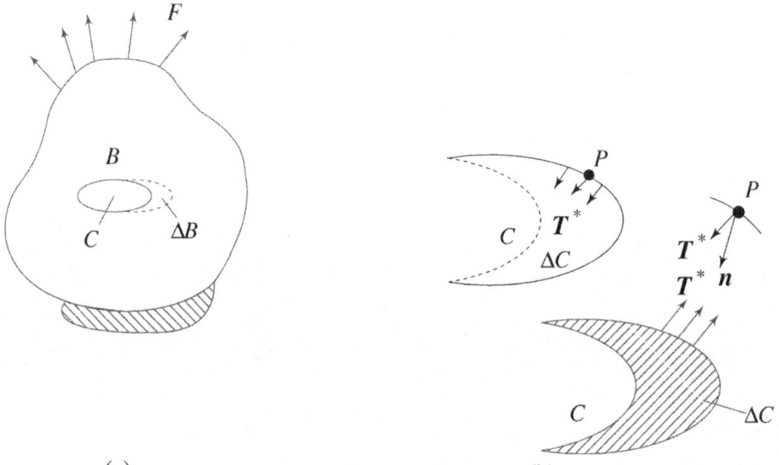

(a) (b)

Fig. 12.45 *(a) Body with cavity ; (b) Newly created cavity and small volume removed*

The total elastic strain energy of the original body is now equal to the strain energy of the body with the newly created cavity, plus the strain energy stored in the elementary volume ΔB that is removed; i.e.,

$$U = \int_V W dV = \int_{V-\Delta B} W dV + \int_{\Delta B} W dV \tag{12.76}$$

Consider a point P and an elementary area δs surrounding it on the surface of the newly created cavity. Let $\boldsymbol{n}$ be the normal to this area and $\boldsymbol{T}^*$ the traction Fig. 12.45 (b). The traction vector $\boldsymbol{T}^*$ will have components T_x^*, T_y^*, and T_z^* in x, y, and z directions. During the loading process, the point P will have undergone displacements u_x^*, u_y^*, and u_z^* in x, y, and z directions. The traction vector $\boldsymbol{T}^*$ and the displacement vector $\boldsymbol{u}$ are the final or the terminal values at the end of loading the body, B. In the case of a lineraly elastic body, $\boldsymbol{T}$ and $\boldsymbol{U}$ are proportional to each other. In the case of a non-linearly elastic body, they are not proportional .

Let n_x, n_y, and n_z be the direction cosines of the normal $\boldsymbol{n}$; and let $\sigma_x^*, \sigma_y^*, \sigma_z^*, \tau_{xy}^*, \tau_{yz}^*$, and τ_{zx}^*, the rectangular stress components at P. Then, from equations (1.9),

$$T_x^* = n_x\sigma_x^* + n_y\tau_{yx}^* + n_z\tau_{zx}^*$$

$$T_y^* = n_x\tau_{xy}^* + n_y\sigma_y^* + n_z\tau_{zy}^* \tag{12.77}$$

$$T_z^* = n_x \tau_{zx}^* + n_y \tau_{zy}^* + n_z \sigma_z^*$$

During the loading process, the work done by the traction $\boldsymbol{T}$ acting on the area δs is equal to [similar to Eq. (12.74)],

$$\Delta W^* = \int_0^{T^*, u^*} \left(T_x du_x + T_y du_y + T_z du_z\right) ds \tag{12.78}$$

This expression is valid for both linear and non-linear elastic bodies. The total work done by traction forces acting on the entire surface area of the new cavity during the loading process is

$$W^* = \int_{\Delta S} \Delta w \, ds = \int_{\Delta S} \left[\int_0^{T^*, u^*} \left(T_x du_x + T_y du_y + T_z du_z\right) \right] ds \tag{12.79}$$

Now we try to make the newly created cavity traction free so that it becomes a virtual extension of the original cavity. This is easily achieved by applying equal and opposite traction forces T^* at the surface of the new cavity. During this process, the forces T^* applied will do work *on the body* and this is equal to Eq. (12.79). It is important to recollect what has been done so far.

We started with a body B (linearly or non-linearly elastic), having a cavity C and loaded by surface traction $\boldsymbol{F}$ on S_F and prescribed displacements $\boldsymbol{D}$ on $S_{D;}$ Fig. 12.44(a). During the deformation process, the elastic body stored strain energy U given by Eq. (12.75). Next, an elementary volume of body ΔB adjacent to the cavity was identified and this was isolated from the parent body. Free-body diagrams of the body with the old cavity C and the newly created cavity ΔC, and the elementary volume ΔB were drawn. The elementary body ΔB was acted upon by surface traction $\boldsymbol{T}^*$, and the surface of the elementary cavity ΔC had surface traction equal and opposite to $\boldsymbol{T}^*$. The elastic strain energy of the original body B was decomposed into two parts: (a) that of the body B with the newly created cavity ΔC; and (b) that of the isolated body ΔB. Finally, in order to make the surface of ΔC, traction free, we apply gradually, equal and opposite forces $\boldsymbol{T}^*$, so that we have now a body with an extended cavity $C + \Delta C$. During the process of applying $\boldsymbol{T}$ to the surface of ΔC, work is done on the body and the energy stored due to this is given by Eq. (12.79).

The strain energy stored now in the body B–ΔB, i.e., in the body with extended cavity is

$$U' = \int_V W dV - \int_{\Delta B} W dV + \int_{\Delta S} \Delta W^* ds \tag{12.80}$$

Hence, the decrease in energy in the process of creating a void or a cavity is

$$-\Delta U = \int_{\Delta B} W dV - \int_{\Delta S} \Delta W^* ds \tag{12.81}$$

12.19 PLANE BODY

Let the body B considered be a plane body. Equation (12.81) can then be written as

$$-\Delta U = \int_{\Delta A} W dA - \int_{\Delta S} \Delta W^* ds \tag{12.82}$$

where ΔA is the area of the material removed in forming a void and ΔS represents the newly created traction-free boundary surface. Now in the limit, let the cavity or the void considered become a crack of length a. For an infinitesimal crack extension, the rate of change of energy with crack growth can be expressed as

$$-\frac{\partial U}{\partial a}=\iint_A \frac{\partial W}{\partial a}\,dxdy-\int_\Gamma \frac{\partial W^*}{\partial a}\,ds \tag{12.83}$$

12.20 GREEN'S THOREM

Let Γ be the closed boundary of a domain A, and let $P(x,y)$ and $Q(x,y)$ be two functions that are continuous together with their partial derivatives $\frac{\partial Q}{\partial x}$ and $\frac{\partial P}{\partial y}$ in the domain A and the boundary Γ, Fig. 12.46. Then,

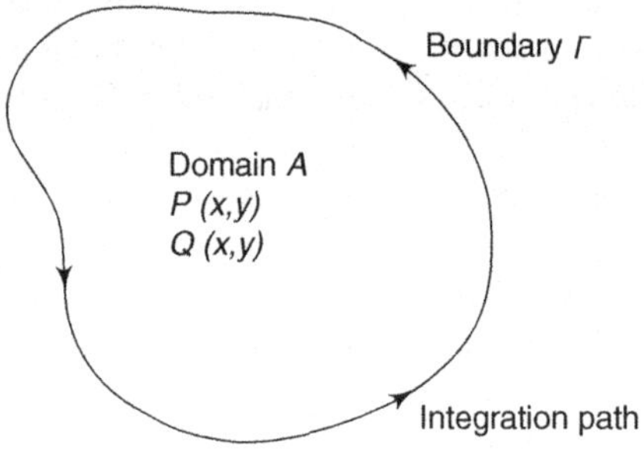

Fig.12.46 *Boundary Γ enclosing the domain A*

$$\iint_A \left(\frac{\partial Q}{\partial x}-\frac{\partial p}{\partial y}\right)dxdy=\int_\Gamma (Pdx+Qdy) \tag{12.84}$$

In this case, the direction in which the contour Γ is traversed is chosen so that the domain A remains to the left, Fig. 12.46.

12.21 THE J-INTEGRAL

Let the co-ordinate system be as shown in Fig. 12.47 such that the origin is at the crack-tip. Then, $da=dx$, and Eq. (12.83) can be written as

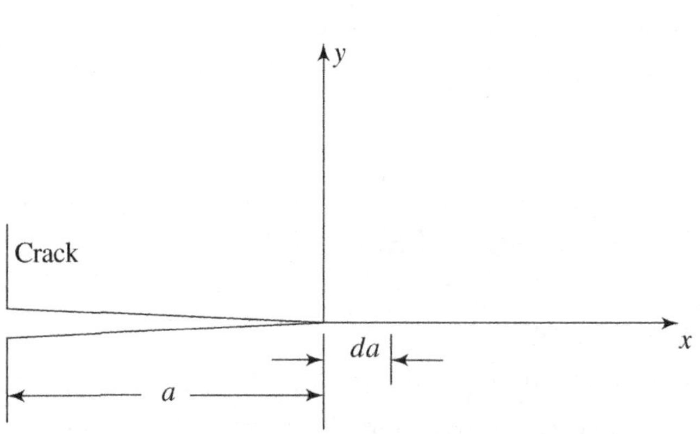

Fig. 12.47 *Body with extended crack*

$$-\frac{\partial U}{\partial a}=\iint_A \frac{\partial W}{\partial x}\,dxdy-\int_\Gamma \frac{\partial W^*}{\partial x}\,ds \tag{12.85}$$

Using Green's theorm, the area integral in Eq. (12.85) can be converted to line integral giving

$$-\frac{\partial U}{\partial a}=\int_\Gamma W dy-\int_\Gamma \frac{\partial w}{\partial x}\,ds \tag{12.86}$$

The quantity $\left(-\frac{\partial U}{\partial a}\right)$ is called the J-integral, i.e.,

$$J=-\frac{\partial U}{\partial a}=\int_\Gamma W dy-\int_\Gamma \frac{\partial w}{\partial x}\,ds \quad \text{(unit: Nm}^{-1}\text{)} \tag{12.87}$$

J is thus the drop in potential energy per unit virtual extension of crack.

An important consequence of Eq. (12.87) is its applicability to plastic behaviour under certain restrictions. The main restriction is that the body must be subjected to monotonically increasing loading and must not experience any unloding. J is thus a measure of the input work to the system and not the amount of work recoverable on unloading.

12.22 PATH INDEPENDENCE OF THE *J*-INTEGRAL

Consider Eq. (12.85) with reference to the closed path Γ shown in Fig. 12.48 (a). The first term on the right-hand side of the expression, i.e.,

$$\iint_A \frac{\partial W}{\partial x}\,dxdy$$

becomes for a plane body from equations (12.73) and (12.74)

$$\frac{\partial W}{\partial x}=\sigma_x\frac{\partial \varepsilon_x}{\partial x}+\tau_{xy}\frac{\partial \gamma_{xy}}{\partial x}+\sigma_y\frac{\partial \varepsilon_y}{\sigma_x} \tag{12.88}$$

Since,

$$\varepsilon_x=\frac{\partial u_x}{\partial x},\ \varepsilon_y=\frac{\partial u_y}{\partial y},\ \gamma_{xy}=\frac{\partial u_x}{\partial y}+\frac{\partial u_y}{\partial x},$$

Equation (12.88) becomes

$$\frac{\partial W}{\partial x}=\sigma_x\frac{\partial}{\partial x}\left(\frac{\partial u_x}{\partial x}\right)+\tau_{xy}\frac{\partial}{\partial x}\left(\frac{\partial u_x}{\partial y}+\frac{\partial u_y}{\partial x}\right)+\sigma_y\frac{\partial}{\partial x}\left(\frac{\partial u_y}{\partial y}\right)$$

$$\therefore \iint_A \frac{\partial W}{\partial x}\,dxdy=\iint_A\left[\sigma_x\frac{\partial}{\partial x}\left(\frac{\partial u_x}{\partial x}\right)+\tau_{xy}\frac{\partial}{\partial x}\left(\frac{\partial u_x}{\partial y}+\frac{\partial u_y}{\partial x}\right)+\sigma_y\frac{\partial}{\partial x}\left(\frac{\partial u_y}{\partial y}\right)\right]dxdy$$

$$=\iint_A\left[\frac{\partial}{\partial x}\left(\sigma_x\frac{\partial u_x}{\partial x}+\tau_{xy}\frac{\partial u_y}{\partial x}\right)+\frac{\partial}{\partial y}\left(\tau_{xy}\frac{\partial u_x}{\partial x}+\sigma_y\frac{\partial u_y}{\partial x}\right)\right]dxdy \tag{12.89}$$

Now consider the second term on the right-hand side of Eq (12.85),

i.e., $\int_\Gamma \frac{\partial W^*}{\partial x} ds$.

From Eq. (12.79) for a plane body,

$$\int_\Gamma \frac{\partial W^*}{\partial x} ds = \int_\Gamma \left[T_x \frac{\partial u_x}{\partial x} + T_y \frac{\partial u_y}{\partial x} \right] ds$$

From Eq. (12.77)

$$\int_\Gamma \frac{\partial W^*}{\partial x} ds = \int_\Gamma \left[\frac{\partial u_x}{\partial x} \left(n_x \sigma_x + n_y \tau_{yx} \right) + \frac{\partial u_y}{\partial x} \left(n_x \tau_{xy} + n_y \sigma_y \right) \right] ds \qquad (12.90)$$

From Fig. 12.48(b) with unit normal,

$$n_x ds = ds \cos\theta = dy,\ n_y ds = ds \sin\theta = -dx.$$

Substituting these in Eq. (12.90)

$$\int_\Gamma \frac{\partial W^*}{\partial x} ds = \int_\Gamma \left(\sigma_x \frac{\partial u_x}{\partial x} + \tau_{xy} \frac{\partial u_y}{\partial x} \right) dy - \int_\Gamma \left(\tau_{xy} \frac{\partial u_x}{\partial x} + \sigma_y \frac{\partial u_y}{\partial x} \right) dx$$

Using Green's thorem, the above expression can be written as

$$\int_\Gamma \frac{\partial W^*}{\partial s} ds = \iint_A \frac{\partial}{\partial x} \left(\sigma_x \frac{\partial u_x}{\partial x} + \tau_{xy} \frac{\partial u_y}{\partial x} \right) + \frac{\partial}{\partial y} \left(\tau_{xy} \frac{\partial u_x}{\partial x} + \sigma_y \frac{\partial u_y}{\partial x} \right) dxdy$$

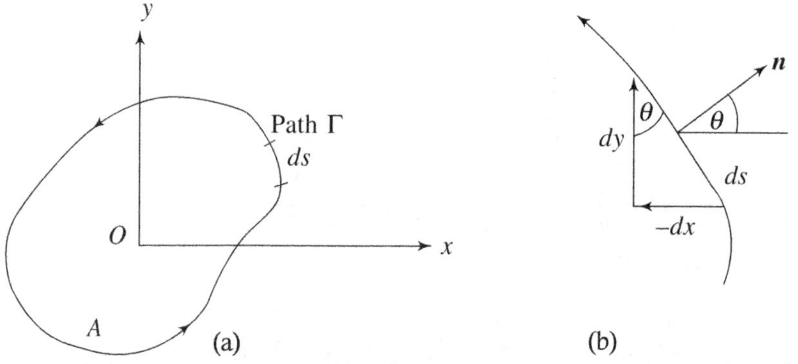

Fig. 12.48 *(a) Closed contour Γ surrounding the domain A; (b) segment ds of the path with normal **n***

This expression is identical to Eq. (12.89). Hence, *for a closed contour Γ as in Fig. 12.48(a), the J-integral is zero,* i.e.,

$$J = \int_\Gamma W dy - \int_\Gamma \frac{\partial W^*}{\partial x} ds = 0 \qquad (12.91)$$

Consider Fig. 12.49(a) which shows a body with a crack. A *closed contour ABCDEFA*, which includes the two flanks of the crack, *CD* and *AF* is shown. This contour consists of two paths Γ_1 and Γ_2, and the two flanks.

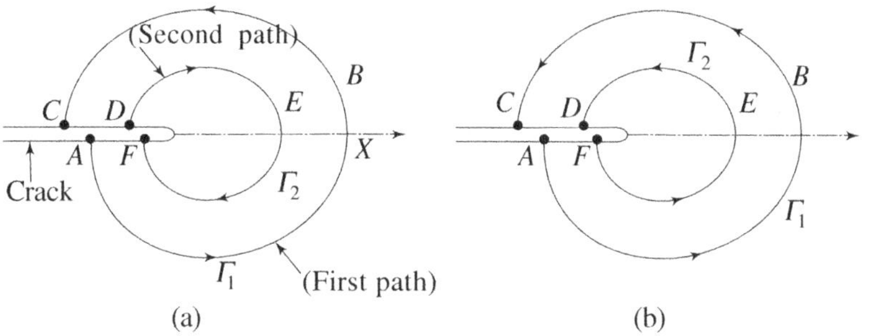

Fig. 12.49 *Closed contours for a body with a crack (a) Two paths in different directions; (b) Two paths in the same direction*

The two paths Γ_1 and Γ_2 of the contour are in opposite direction to each other. Since the *J*-integral for a closed contour is zero, we have

$$J = J_{ABC} + J_{CD} + J_{DEF} + J_{FA} = 0$$

Along the flanks *CD* and *FA* of the crack, $y = dy = 0$ and the traction force $\boldsymbol{T}$ is also zero. Hence, for the flanks, the J-integral according to Eqs. (12.87) and (12.78), is zero. Accordingly,

$$J_{ABC} = -J_{DEF}$$

The second path *DEF* is opposite in direction to the first path *ABC*. If the path *DEF* is in the same direction (i.e., the domain or the area being to the left of the traversing direction), Fig. 12.49 (b),

$$J_{\Gamma_1} = J_{\Gamma_2} \tag{12.92}$$

This implies that the J-integral is path independent when applied around a crack tip from one crack surface to the other.

12.23 J-INTEGRAL AS A FRACTURE CRITERION

The path independency of the *J*-integral can be used as a fracture criterion in the same manner that the stress intensity factor is used. From Eq. (12.87), *J* is the drop in potential energy per unit virtual extension of crack; i.e.,

$$J = -\frac{1}{B}\left(\frac{\partial U}{\partial a}\right) \tag{12.93}$$

where *B* is the specimen thickness. The procedure indicated by Eq. (12.93) is as follows.

First, load displacement diagrams are obtained for a number of pre-cracked specimens. Let the crack-lengths be a_1, a_2, a_3, etc. Figure 12.50(a) depicts these. The energy per unit thickness U_1 delivered to the specimen at a given level of displacement δ is obtained as the area under the load-displacement curve. U_1 is then plotted as a function of crack-length for several constant values of displacement δ_1, δ_2, δ_3, etc., Fig. 12.50 (b).

The negative slopes of U_1–a curves are plotted against displacement for any desired crack length between the shortest and the longest used in testing, Fig. (12.50(c).

The slopes represent $-\dfrac{\partial U_1}{\partial a}$ at a given value of displacement which is obviously J. A knowledge of the displacement δ on the onset of crack extension enables the determination of J_C from the $J - \delta$ calibration curve for each initial crack length. Alternatively, if J_C is an appropriate criterion of crack extension, then δ'_1, δ'_2 and δ'_3 are the displacements on the onset of crack extensions for the respective crack length δ_1, δ_2, and δ_3.

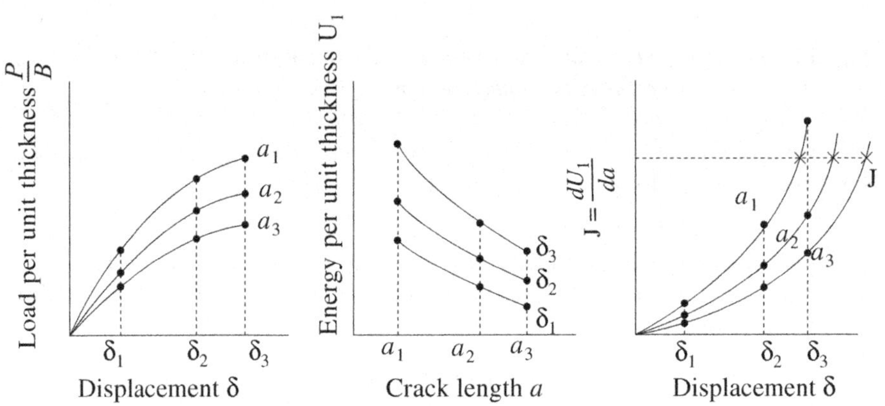

Fig. 12.50 *(a) Load-vs-displacement for different slit lengths a_1, a_2, a_3, (b) Energy-vs-slitlength; (c) J-vs-displacement*

12.24 ASTM-STANDARD TEST FOR J_{Ic}

The American society for Testing and Materials has standardized a test method to determine J_{1C} as a measure of toughness. The objective is to determine the value of J at the initiation of crack growth. It is not intended to characterize crack growth beyond the initiation stage. The recommended specimens are the notched bend and compact tension. Figure 12.51 shows the sketch of the notched bend specimen.

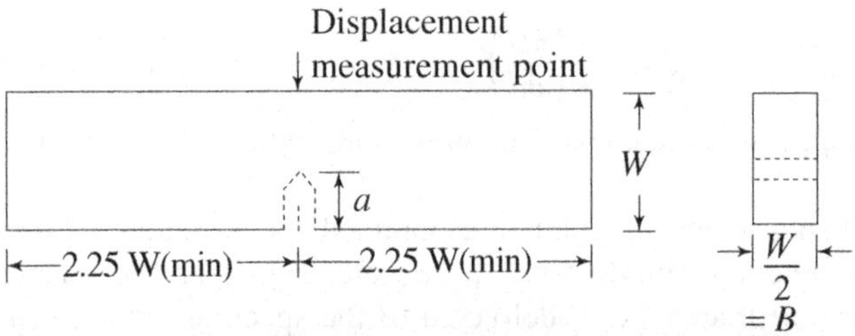

Fig. 12.51 *Notched three-point bend specimen*

The specimen has a deep initial crack $\left(\frac{a}{W} \geq 0.5\right)$. In order to obtain a valid J_{1c} value, the crack length a, the initial uncracked ligament dimension $(W - a) = b$, and the width B must satisfy the condition

$$b, B > 25\left(\frac{J_{Ic}}{\sigma_f}\right) \tag{a}$$

in order to ensure that the crack tip stress/strain field is characterized by path-independent integrals. Evaluations of the J-integral are made from load versus load–displacement curves using the area under the load displacement curve. For the three-point bend specimen, the J-integral is given by

$$J = \frac{A}{Bb} \cdot f\left(\frac{a_0}{W}\right)$$

where A is the area under the load versus load-point displacement diagram, B is the specimen thickness, b is the initial uncracked ligament $(W - a_o)$, W is the width of the specimen, and a_0 is the original crack size. For the three point bend specimen,

$$f\left(\frac{a_0}{W}\right) = 2$$

The initial values of J_{1c} obtained from the measurements of the area under the load versus displacement curve is validated by the condition (a).

12.25 RELATIONSHIPS OF K_c, G_c, AND J

It is obvious that the changes involved in the process of extension of a crack in a loaded body is intimately connected with the stress field existing in the neighborhood of the crack tip. This means that the critical stress intensity factor K_{1c}, the critical strain energy release rate G_c, and the J-integral are related. The relationships are as follows:

$$J = G_c = \frac{K_{Ic}^{\ 2}}{E}\left(1 - v^2\right) \quad \text{for plane strain} \tag{12.94 a}$$

$$J = G_c = \frac{K_{Ic}^{\ 2}}{E} \quad \text{for plane stress} \tag{12.94 b}$$

Problems

12.1 A 20-mm long cast iron rod of 25-mm diameter is pressed on to a thick copper plate with a force of 20 N. Determine the width of the contact area, the maximum pressure at the centre of the contact area, and the octahedral stress at the centre of the contact area. The elastic constants for the materials are: cast iron–E = 41.4 GPa, v = 0.211; copper–E = 44.7 GPa, v = 0.326.

[Ans: 5.24×10^{-2} mm; $p_{max} = 24.3$ MPa; $\tau_{oct} = 4$ MPa.]

12.2 For two spheres in contact under pressure, show that the maximum shear stress on the z-axis occurs very nearly at half the distance of the radius of the contact area and its value is $0.31 p_{max}$.

12.3 For two cylinders pressed together, show that the maximum shear occurs at a depth of $z = 0.78b$ and its magnitude is $0.301\ p_{max}$, where b is the half-width of the contact area. Consider $\tau_{yz} = \frac{1}{2}(\sigma_y - \sigma_z)$.

12.4 An aluminium plate of 1.5-m width and 3-m length is required to support a force of 2 MN in the 3-m direction. Inspection procedures can detect a through-thickness edge cracks longer than 2.7 mm. Al-2024 and Al–7178 are the materials under consideration. Al-2024 has a value of 26 MPa $\sqrt{m}$ for K_{Ic}, and a yield stress $S_y = 455$ MPa. For Al-7178, $K_{Ic} = 33$ MPa $\sqrt{m}$, and $S_y = 490$ MPa. Weight is a major consideration. Using a factor of safety of 1.5, select the proper sheet and its thickness.

[Ans: Use Al-7178, t = 6.1 mm]

12.5 A steel sheet that is 16 m long and 8 m wide is found to have a central transverse crack of 40-mm length. The material of the sheet has a fracture toughness factor $K_{Ic} = 25$ MPa $\sqrt{m}$. Determine the maximum longitudinal stress the sheet can withstand without the danger of catastrophic failure.

[Ans: 3.15MPa]

12.6 A cylinder with an internal radius of 5 cm and external radius of 6.5 cm has a radial crack of 2-mm length on the outer periphery. The material has a yield strength of 490MPa. The two ends of the cylinder are closed. Determine the maximum internal pressure that can be applied without yielding or fracture occurring. Consider points at the inner and outer boundaries. A factor of safety 2 is used.

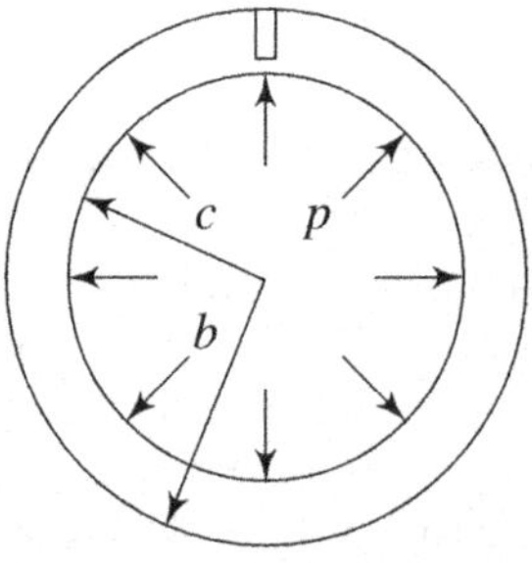

Fig. 12.35 *Tube with an external crack under internal pressure*

$$\left[\begin{array}{l}\text{Ans} : p = 50 \text{ MPa or } p = 1.943\,K_{Ic} \\ \text{if } K_{Ic} < 25.7 \text{ MPa}\sqrt{\text{m}}\end{array}\right]$$

12.7 Calculate the theoretical stress concentration factor for an elliptical notch with a major axis equal to 8 cm, and a minor axis equal to 0.7 cm. Loading is perpendicular to the major axis. What is the radius of curvature at the ends of the major axis? Assume that there is no yielding.

$$\left[\text{Ans} : K_t' = 23.86,\ \rho = 0.031 \text{ cm}\right]$$

12.8 For Problem 12.7, calcute the nominal stress in the full section at the time of collapse if the yield strength is 525 MPa. What is the fracture load? Width $W = 25$ cm, and thickness $t = 0.15$cm.

$$\left[\text{Ans} : P_{max} = 133.9 \text{ kN},\ \sigma_{nom} = 355 \text{ MPa}\right]$$

12.9 Calculate the fracture toughness of a material for which a plate test with central crack gives the following information: Width $W = 50$ cm, thickness $B = 1.9$ cm, crack length $2a = 5$ cm, failure load $P = 1335$ kN. The yield strength is $\sigma_{yp} = 480$MPa. Is this plane Strain? Check for collapse.

$$\left[\begin{array}{l}\text{Ans} : \text{Toughness} = 39.2 \text{ MPa}\sqrt{\text{m}}; \\ \text{Yes; No collapse.}\end{array}\right]$$

12.10 Given a toughness of $K = 77$ MPa $\sqrt{\text{m}}$, and an yield strength of $\sigma_{yp} = 520$ MPa, determine the residual strength of a centre cracked plate of 45 cm width and crack length $2a = 7.5$ cm. Check for collapse. $\alpha = 1.01$.

$$\left[\begin{array}{l}\text{Ans} : \sigma_{fr} = 222 \text{ MPa}; \\ \text{No collapse.}\end{array}\right]$$

APPENDIX

The strain compatibility condition for the two-dimensional case is, from Eq.(2.56 a)

$$\frac{\partial^2 \varepsilon_{xx}}{\partial y^2} + \frac{\partial \varepsilon_{yy}}{\partial x^2} = \frac{\partial^2 \gamma_{xy}}{\partial x \partial y} \tag{a}$$

This can be converted to stress compatibility equation for the plane stress case using the stress–strain relations:

$$\varepsilon_{xx} = \frac{1}{E}(\sigma_x - v\sigma_y), \quad \varepsilon_{yy} = \frac{1}{E}(\sigma_y - v\sigma_x) \tag{b}$$

$$v_{xy} = \frac{1}{G}\tau_{xy} = \frac{2(1+v)}{E}\tau_{xy} \tag{c}$$

Substituting in Eq. (a),

$$\frac{\partial^2}{\partial y^2}(\sigma_x - v\sigma_y) + \frac{\partial^2}{\partial x^2}(\sigma_y - v\sigma_x) = \frac{\partial^2}{\partial x\,\partial y}\left[2(1+v)\tau_{xy}\right] \tag{d}$$

The equations of equilibrium in the absence of body forces from equations (1.65) are

$$\frac{\partial \sigma_x}{\partial x} + \frac{\partial \tau_{xy}}{\partial y} = 0 \tag{e}$$

$$\frac{\partial \sigma_y}{\partial y} + \frac{\partial \tau_{xy}}{\partial x} = 0 \tag{f}$$

Differentiating Eq. (e) with respect to x, and Eq. (f) with respect to y and adding, one gets

$$2\frac{\partial^2 \tau_{xy}}{\partial x \partial y} = -\frac{\partial^2 \sigma_x}{\partial x^2} - \frac{\partial^2 \sigma_y}{\partial y^2}$$

Substituting the above in Eq. (d),

$$\frac{\partial^2 \sigma_x}{\partial y^2} + \frac{\partial^2 \sigma_y}{\partial x^2} - v\left(\frac{\partial^2 \sigma_y}{\partial y^2} + \frac{\partial^2 \sigma_x}{\partial x^2}\right) = -(1+v)\left(\frac{\partial^2 \sigma_x}{\partial x^2} + \frac{\partial^2 \sigma_y}{\partial y^2}\right)$$

i.e.,
$$\left(\frac{\partial^2}{\partial x^2}+\frac{\partial^2}{\partial y^2}\right)(\sigma_x+\sigma_y)=0 \tag{g}$$

Equation (g) is the stress equation of compatibility. The usual method of solving an elasticity problem is by introducing a function ϕ of x, and y, that satisfies the equations of equilibrium (e) and (f), the compatibility condition (g), and the appropriate boundary conditions. Let a function ϕ (x, y) be choosen such that

$$\sigma_x=\frac{\partial^2\phi}{\partial y^2};\ \sigma_y=\frac{\partial^2\phi}{\partial x^2};\ \tau_{xy}=-\frac{\partial^2\phi}{\partial x\partial y} \tag{h}$$

As can be checked, this function satisfies the equations of equilibrium. In order that it may satisfy the compatibility condition (g), we should have

$$\left(\frac{\partial^2}{\partial x^2}+\frac{\partial^2}{\partial y^2}\right)\left(\frac{\partial^2\phi}{\partial x^2}+\frac{\partial^2\phi}{\partial Y^2}\right)=0 \tag{j}$$

i.e.,
$$\frac{\partial^4\phi}{\partial x^4}+2\frac{\partial^4\phi}{\partial y^4}+\frac{\partial^4\phi}{\partial y^4}=0 \tag{k}$$

A function ϕ (x, y) which satisfies Eq. (k) satisfies the equations of equilibrium and the compatibility condition. If it satisfies in addition, boundary conditions of a given problem, then such a function is the proper function for that problem. We shall transform Eq. (k) into polar coordinates to solve axi-symmetric problems. Let the stress function in polar coordinates be $\phi(r, \theta)$, and let

$$\sigma_r=\frac{1}{r}\frac{\partial\phi}{\partial r}+\frac{1}{r^2}\frac{\partial^2\phi}{\partial\theta^2}$$

$$\sigma_\theta=\frac{\partial^2\phi}{\partial r^2} \tag{m}$$

$$\tau_{r\theta}=\frac{1}{r^2}\frac{\partial\phi}{\partial\theta}-\frac{1}{r}\frac{\partial^2\phi}{\partial r\partial\theta}=-\frac{\partial}{\partial r}\left(\frac{1}{r}\frac{\partial\phi}{\partial\theta}\right)$$

The function ϕ (r, θ) so defined by Eq. (m) satisfies the equations of equilibrium, Eq. (1.70), in polar coordinates.

Equation (g) is the stress equation of compatibility expressed in Cartesian co-ordinates. It can easily be converted into polar co-ordinates. We have,

$$r^2=x^2+y^2 \text{ and } \theta=\arctan\frac{y}{x}$$

from which

$$\frac{\partial r}{\partial x}=\frac{x}{r}=\cos\theta,\ \frac{\partial r}{\partial y}=\frac{y}{r}=\sin\theta$$

$$\frac{\partial \theta}{\partial x} = -\frac{y}{r^2} = -\frac{\sin\theta}{r}, \quad \frac{\partial \theta}{\partial y} = \frac{x}{r^2} = \frac{\cos\theta}{r}$$

Let the stress function in polar coordinates be $\phi(r, \theta)$. For this function

$$\frac{\partial \phi}{\partial x} = \frac{\partial \phi}{\partial r}\frac{\partial r}{\partial x} + \frac{\partial \phi}{\partial \theta}\cdot\frac{\partial \theta}{\partial x} = \frac{\partial \phi}{\partial r}\cos\theta - \frac{1}{r}\frac{\partial \phi}{\partial \theta}\sin\theta$$

Symbolically,

$$\frac{\partial}{\partial x} = \left(\frac{\partial}{\partial r}\cos\theta - \frac{1}{r}\sin\theta\frac{\partial}{\partial \theta}\right)$$

$$\therefore \quad \frac{\partial^2 \phi}{\partial x^2} = \frac{\partial}{\partial x}\left[\frac{\partial \phi}{\partial x}\right] = \left(\frac{\partial}{\partial r}\cos\theta - \frac{1}{r}\sin\theta\frac{\partial}{\partial \theta}\right)\left(\frac{\partial \phi}{\partial r}\cos\theta - \frac{1}{r}\sin\theta\frac{\partial}{\partial \theta}\right)$$

$$= \frac{\partial^2 \phi}{\partial r^2}\cos^2\theta - 2\frac{\partial^2 \phi}{\partial r\partial \theta} + \frac{\partial \phi}{\partial r}\frac{\sin^2\theta}{r} + 2\frac{\partial \phi}{\partial \theta}\frac{\sin\theta\cos\theta}{r^2} +$$

$$\frac{\partial^2 \phi}{\partial \theta^2}\frac{\sin^2\theta}{r} \qquad \text{(n)}$$

In the same manner one gets

$$\frac{\partial^2 \phi}{\partial y^2} = \frac{\partial^2 \phi}{\partial r^2}\sin^2\theta + 2\frac{\partial^2 \phi}{\partial r\partial \theta}\frac{\sin\theta\cos\theta}{r} + \frac{\partial \phi}{\partial r}\frac{\cos^2\theta}{r}$$

$$-2\frac{\partial \phi}{\partial \theta}\frac{\sin\theta\cos\theta}{r^2} + \frac{\partial^2 \phi}{\partial \theta^2}\frac{\cos^2\theta}{r^2} \qquad \text{(p)}$$

Adding together equations (n) and (p)

$$\frac{\partial^2 \phi}{\partial x^2} + \frac{\partial^2 \phi}{\partial y^2} = \frac{\partial^2 \phi}{\partial r^2} + \frac{1}{r}\frac{\partial \phi}{\partial r} + \frac{1}{r^2}\frac{\partial^2 \phi}{\partial \theta^2}$$

Using this, Eq. (j) can be written as

$$\left(\frac{\partial^2}{\partial x^2} + \frac{\partial^2}{\partial y^2}\right)\left(\frac{\partial^2 \phi}{\partial x^2} + \frac{\partial^2 \phi}{\partial y^2}\right) = \left(\frac{\partial^2}{\partial r^2} + \frac{1}{r}\frac{\partial}{\partial r} + \frac{1}{r^2}\frac{\partial^2}{\partial \theta^2}\right)\left(\frac{\partial^2 \phi}{\partial r^2} + \frac{1}{r}\frac{\partial \phi}{\partial r} + \frac{1}{r^2}\frac{\partial^2 \phi}{\partial \theta^2}\right) = 0 \qquad \text{(q)}$$

Any function $\phi(r, \theta)$ satisfying this equation will satisfy equilibrium equations and compatibility condition. If the function in addition, satisfies the boundary conditions for a given problem, then it is the proper stress function for that problem.

WIDE PLATE WITH A SMALL CIRCULAR HOLE

Consider a wide plate with a small circular hole of radius a, subjected to a uniform tensile stress σ, Fig. A-1.

If a large circle of radius b is drawn concentric with the hole, then the stress distribution around the circumference of the circle is the one caused just by σ, without being affected by the hole, since the hole is very small and the boundary

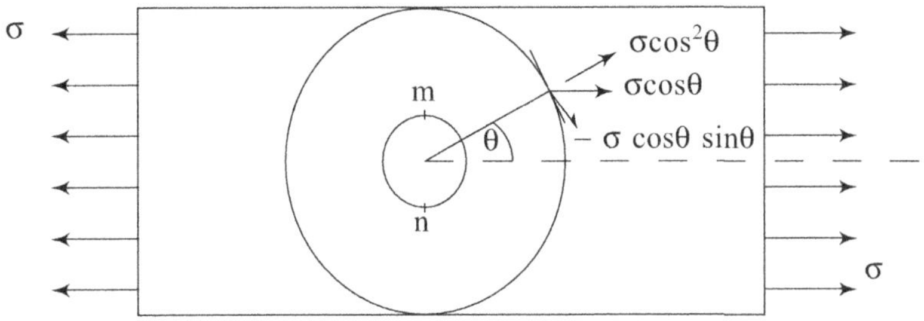

Fig. A-1. *A wide plate with a small hole subjected to a tensile stress σ.*

of the circle is far removed from the hole. The stress distribution around the big circle can be determined from statics as was done in Section 12.2(a). At an angle θ, on a small area $bd\theta$, the stress is $\sigma \cos\theta$. This can be resolved into two components; one radial : $\sigma \cos^2\theta$, and the other tangential:– $\sigma \cos\theta\sin\theta$. This large circular thick plate with a small hole can be isolated and analysed as equivalent to the original problem.

The ring is now subjected to the following stresses:

$$\text{radial:} \quad \sigma \cos^2\theta = \frac{1}{2}\sigma(1+\cos 2\theta)$$

$$\text{tangential:} \; -\sigma \cos\theta \sin\theta = -\frac{1}{2}\, \sigma \sin 2\theta \tag{r}$$

The case of uniform radial stress $\frac{1}{2}\sigma$ on the ring can be solved using equations (8.16) and (8.17). The remaining parts consisting of the varying radial stress $\frac{1}{2}\sigma \cos 2\theta$, and the tangential stress $\frac{1}{2}\sigma \sin 2\theta$ can be analysed through the stress function method.

Let the stress function be of the form $\phi = f(r) \cos 2\theta$. This has to satisfy the compatibility condition given by Eq. (p). Substituting

$$\left(\frac{\partial^2}{\partial r^2} + \frac{1}{r}\frac{\partial}{\partial r} + \frac{1}{r^2}\frac{\partial^2}{\partial \theta^2}\right)$$

$$\times\left\{\frac{\partial^2}{\partial r^2}[f(r)\cos 2\theta] + \frac{1}{r}\frac{\partial}{\partial r}[f(r)\cos 2\theta] + \frac{1}{r^2}\frac{\partial^2}{\partial \theta^2}[f(r)\cos 2\theta]\right\} = 0$$

i.e,
$$\left(\frac{\partial^2}{\partial r^2} + \frac{1}{r}\frac{\partial}{\partial r} + \frac{1}{r^2}\frac{\partial^2}{\partial \theta^2}\right)$$

$$\times\left[\frac{\partial^2 f(r)}{\partial r^2}\cos 2\theta + \frac{1}{r^2}\frac{\partial f(r)}{\partial r}\cos 2\theta - \frac{4 f(r)}{r^2}\cos 2\theta\right] = 0$$

Cancelling cos 2θ, and observing that the differential equation involves only $f(r)$, the partial differential equation becomes an ordinary differential equation, which is

$$\left(\frac{d^2}{dr^2}+\frac{1}{r}\frac{d}{dr}-\frac{4}{r^2}\right)\left(\frac{d^2 f}{dr}+\frac{1}{r}\frac{df}{dr}-\frac{4f}{r^2}\right)=0$$

The general solution is

$$f(r)=Ar^2+Br^4+C\frac{1}{r^2}+D$$

The stress function is therefore

$$\phi=f(r)\cos 2\theta=\left(Ar^2+Br^4+C\frac{1}{r^2}+D\right)\cos 2\theta$$

The corresponding stresses from Eq. (m) are

$$\sigma_r = \frac{1}{r}\frac{\partial\phi}{\partial r}+\frac{1}{r^2}\frac{\partial^2\phi}{\partial\theta^2} = -\left(2A+\frac{6C}{r^4}+\frac{4D}{r^2}\right)\cos 2\theta$$

$$\sigma_\theta = \frac{\partial^2\phi}{\partial r^2} = \left(2A+12Br^2+\frac{6C}{r^4}\right)\cos 2\theta \qquad \text{(s)}$$

$$\tau_{r\theta} = -\frac{\partial}{\partial r}\left(\frac{1}{r}\frac{\partial\phi}{\partial\theta}\right) = \left(2A+6Br^2-\frac{6C}{r^4}-\frac{2D}{r^2}\right)\sin 2\theta$$

The constants of integration are now determined from the conditions: (i) that the edge of the inner hole is free from external forces, and (ii) the outer boundary is subjected to stresses given Eq. (r).

These conditions give the following equations.

$$2A+\frac{6C}{b^4}+\frac{4D}{b^2}=-\frac{1}{2}\sigma$$

$$2A+\frac{6C}{a^4}+\frac{4D}{a^2}=0$$

$$2A+6Bb^2-\frac{6C}{b^4}-\frac{2D}{b^2}=-\frac{1}{2}\sigma$$

$$2A+6Ba^2-\frac{6C}{a^4}-\frac{2D}{a^2}=0$$

Solving these and putting $\frac{a}{b}$ 0 because of a very wide plate, one obtains

$$A=-\frac{\sigma}{4},\ B=0,\ C=-\frac{a^4}{4}\sigma,\ D=\frac{a^2}{2}\sigma.$$

Substituting these in Eq. (s) we get the stresses in the large disc (equivalently in the plate) due to the varying radial stresses $\frac{1}{2}\sigma\cos 2\phi$ and the tangential stresses $\frac{1}{2}\sigma\sin 2\phi$. Remembering that in addition to these, the disc is subjected to the uniform radial stress $\frac{1}{2}\sigma$ on the outer boundary, whose solution is given by equations (8.16) and (8.17), the final solutions are

$$\sigma_r = \frac{\sigma}{2}\left(1 - \frac{a^2}{r^2}\right) + \frac{\sigma}{2}\left(1 + \frac{3a^4}{r^4} - \frac{4a^2}{r^2}\right)\cos 2\theta$$

$$\sigma_\theta = \frac{\sigma}{2}\left(1 + \frac{a^2}{r^2}\right) - \frac{\sigma}{2}\left(1 + \frac{3a^4}{r^4}\right)\cos 2\theta \tag{t}$$

$$\tau_{r\theta} = -\frac{\sigma}{2}\left(1 - \frac{3a^4}{r^4} + \frac{2a^2}{r^2}\right)\sin 2\theta$$

When r is very large, the stresses approach the values given by Eq. (r). At the edge of the small hole

$$\sigma_r = \tau_{r\theta} = 0, \quad \text{and} \quad \sigma_\theta = \sigma(1 - 2\cos 2\theta) \tag{u}$$

It can be seen that σ_θ is greatest when $\theta = \frac{\pi}{2}$ or $\theta = \frac{3\pi}{2}$, i.e., at the ends m and n of the diameter perpendicular to the direction of σ. At these points, $\sigma_\theta = 3\sigma$. When $\theta = 0$ or $\theta = \pi$, $\sigma_\theta = -\sigma$.

INDEX

Made in United States
Orlando, FL
04 November 2026

90200704R00292